DEEP-SKY WONDERS

Published by Firefly Books Ltd. 2011

First printing

Publisher Cataloging-in-Publication Data (U.S.)

French, Sue.
Deep-sky wonders : a tour of the universe with *Sky & Telescope*'s Sue French./ Sue French.
[ ] p. : col. photos., charts ; cm.
Includes index.
Summary: Wonders of the deep sky and detailed telescope instructions for observations.

ISBN-13: 978-1-55407-793-9
ISBN-10: 1-55407-793-1

1.Astronomy – Observers' manuals. 2. Telescopes – Observers' manuals. 3. Sky and telescope. I. Title.
522 dc22 QB64.F746 2011

Library and Archives Canada Cataloguing in Publication

French, Sue (Sue C.)
Deep-sky wonders : A tour of the universe with *Sky & Telescope*'s Sue French / Sue French. –

Includes index..
ISBN-13: 978-1-55407-793-9
ISBN-10: 1-55407-793-1

1. Astronomy--Observers' manuals.
2. Telescopes--Amateurs' manuals. I. Title.

QB64.F74 2011 522 C2011-902335-0

Published in the United States by
Firefly Books (U.S.) Inc.
P.O. Box 1338, Ellicott Station
Buffalo, New York 14205

Published in Canada by
Firefly Books Ltd.
66 Leek Crescent
Richmond Hill, Ontario L4B 1H1

Cover and interior design:
Janice McLean / Bookmakers Press Inc.

Printed in China

The Publisher gratefully acknowledges the financial support for our publishing program by the Government of Canada through the Canada Book Fund as administered by the Department of Canadian Heritage.

Front Cover: NOAO / AURA / NSF / Travis Rector / Brenda Wolpa

Back cover, clockwise from bottom left: POSS-II / Caltech / Palomar; Robert Gendler; Brian Lula; Johannes Schedler

Spine: Robert Gendler

DEEP-SKY WONDERS

A Tour of the Universe
With *Sky & Telescope*'s **Sue French**

FIREFLY BOOKS

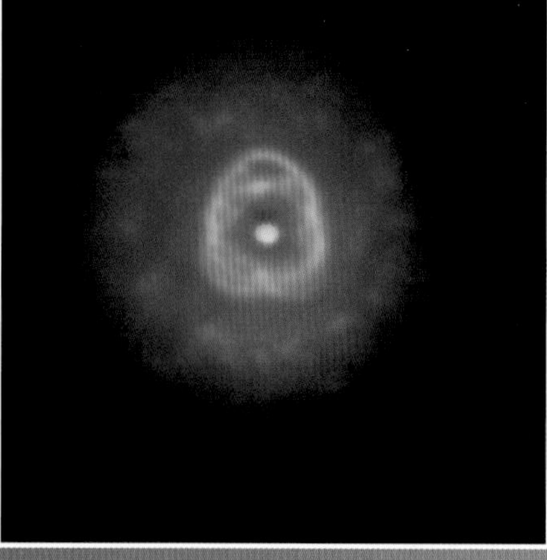

SUMMER

July

August

September

AUTUMN

October

November

December

Foreword

The universe is a very big place. Point a telescope anywhere in the night sky, and you will see an overabundance of stars. At first glance, they all look alike, except that some are brighter than others. In fact, just like people in a crowd, each star is a unique individual, with a history and characteristics all its own. But a typical backyard scope shows literally millions of stars, so it's impossible to get to know each one personally — you have to pick and choose.

Scattered thinly among the stars are the nuggets that deep-sky stargazers treasure most: star clusters, nebulae, and galaxies. You might see only one such nonstellar object for every thousand stars. Still, that adds up to an overwhelming number.

Faced with this embarrassment of riches, it's hard to steer a course between two opposing hazards. Many stargazers learn how to find a few showpiece objects and end up visiting those over and over. That's fine as long as you find new things to appreciate on each visit — otherwise, you will soon get bored, no matter how spectacular the objects.

Other stargazers explore the sky afresh on each outing, scanning around at random until they find something interesting. That's fine too, but only if you take some time to appreciate each new find. Note its position, find out its name, and consider why it's different from (or similar to) other objects of its kind. If you just take a quick look and move on, each new discovery starts to look the same.

I do my fair share of random exploration as well as revisiting old standbys, but my favorite way to explore the sky is with a tour guide. And ever since I started editing Sue French's columns for *Sky & Telescope* magazine, I've never lacked for things to see and do. I'm responsible for the charts and illustrations in her articles, and I figure that the only way to find out for sure whether these are helpful to our readers is to follow her guided tours myself.

Messier 41 is the splashiest star cluster in the constellation Canis Major, the Big Dog. Its brightest star is a slowly dying giant that has used up the hydrogen fuel in its core. Several such aged golden gems ornament this large and beautiful treasure chest of sparkling suns. Photo: POSS-II / Caltech / Palomar

Sue can fairly be described as a fanatical observer; she rarely misses an opportunity to use her telescope on a clear, moonless night. She's equally obsessive about writing, often spending weeks of research on the science, history, and folklore behind the objects she describes. She hobnobs with the world's greatest amateur observers, and many professionals, at national star parties and on the Internet. Yet as a frequent volunteer at public viewing sessions and as a former planetarium educator, she's keenly aware of the beginner's perspective.

Few other writers can match Sue's knowledge of the sky, and nobody else selects celestial targets with such originality and flair. Most of her tours include fresh insights about well-known showpieces as well as obscure but fascinating objects from catalogs about which few people have ever heard. In every tour, she tries to include some objects that will be easy and gratifying for beginners and at least one that will challenge the most skilled observers — and everything in between.

A few years ago, *Sky & Telescope* collected Sue's first 60 columns in a book called *Celestial Sampler*. It's a wonderful book but somewhat limited, because most of those tours included only observations with small telescopes (no bigger than 4 inches of aperture).

This new compilation takes that to a higher level. It offers 23 of her best small-scope columns and 77 new tours that include observations with instruments ranging from handheld binoculars to a 15-inch reflector. It truly does have something for everyone. No matter what your equipment or skill level, this book is your guide to many long, happy, and fruitful nights of observing. It's a book you can start with, grow with, and continue to learn from for the rest of your life.

— Tony Flanders, Associate Editor
Sky & Telescope magazine

Winter

Winter Wonders

The renowned Double Cluster isn't the only deep-sky attraction high overhead in January.

January nights often turn bitter cold for those of us in midnorthern latitudes, but they offer wonders well worth the frigid fingers and numbed nose. The crisp, dry air often tempts us with exceptional transparency, luring us out into the star-pierced night. So bundle up and join me for a winter sky tour that will sweep us across a swath of sky from northern Perseus into the neighboring constellation Cassiopeia.

We'll begin with **M76**, a small planetary nebula that goes by such nicknames as the Little Dumbbell, Barbell, Cork, and Butterfly. To locate M76, start at 4th-magnitude, blue-white Phi (ϕ) Persei. Placing Phi at the southern edge of a low-power field, you should see a distinctly yellow-orange, 7th-magnitude star to the north. M76 is just 12' west-northwest of this star.

Through my 105mm refractor at 127x, M76 most closely suggests a cork to me. It is fairly bright and bar-shaped, running northeast to southwest with a slight pinch in the middle. The nebula looks patchy, and the southwest lobe appears to be the brighter one. M76 lies across the hypotenuse of a right triangle formed by three very faint field stars. If you are battling light pollution, an oxygen III or narrowband filter can help.

Under the dark skies of the northern Adirondack Mountains in New York, I can just glimpse what seems to be a dull zone separating the two lobes of the nebula. Its double nature earned M76 two separate designations in Johan L. E. Dreyer's *New General Catalogue of Nebulae and Clusters of Stars* (London, 1888). The southwest section is NGC 650, while the northeast section is NGC 651. I can sometimes descry vague hints of the diaphanous wings that loop outward from the long sides of the nebula. In long-exposure images, these wings do make M76 look very much like a butterfly.

The fabulous **Double Cluster** is our next stop, a naked-eye wonder known since antiquity. Use the January all-sky star map on page 302 to find it. Look for the two stars labeled γ and δ in W-shaped Cassiopeia. Imagine a line from the first through the second, continue farther for twice that distance, and you'll be led right to the Double Cluster. It's a hazy patch embedded in the dim band of the Milky Way.

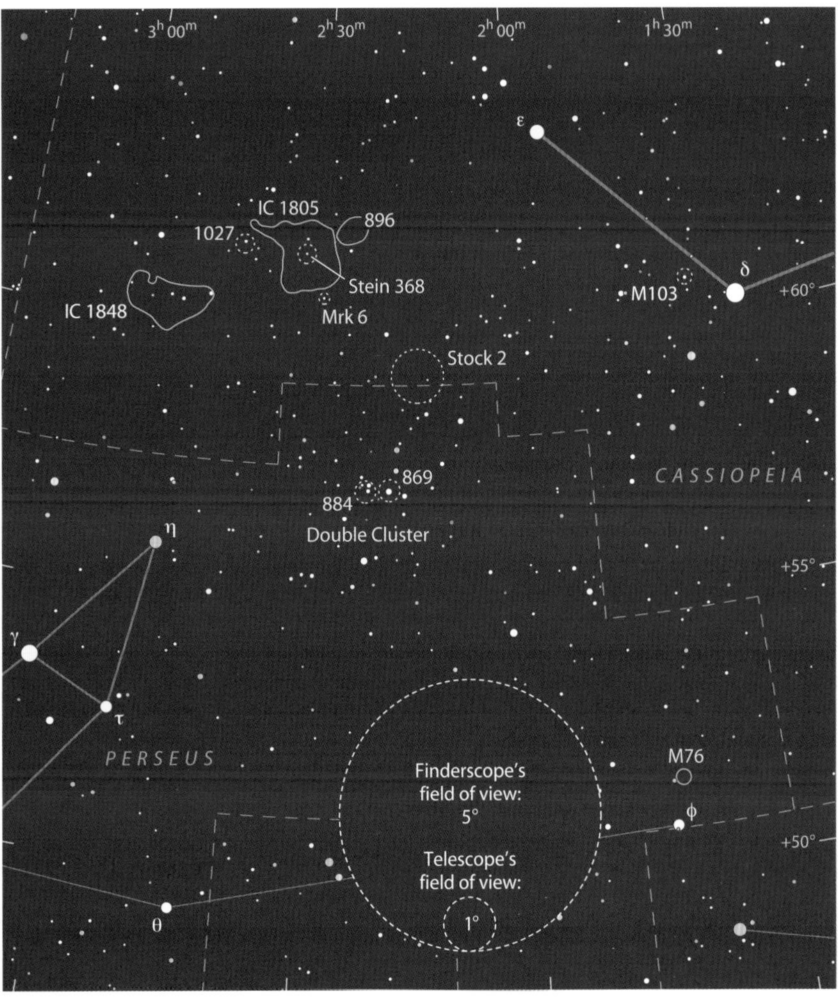

Adapted from *Sky Atlas 2000.0* data

Clusters and Nebulae Where Perseus and Cassiopeia Meet

Object	Type	Magnitude	Size/Sep.	Dist. (l-y)	Right Ascension	Declination	*MSA*	*U2*
M76	Planetary nebula	10.1	1.7'	4,000	1h 42.3m	+51° 35'	63	29R
NGC 884	Open cluster	6.1	30'	7,000	2h 22.3m	+57° 08'	62	29L
NGC 869	Open cluster	5.3	30'	7,000	2h 19.1m	+57° 08'	62	29L
Stock 2	Open cluster	4.4	60'	1,000	2h 15.6m	+59° 32'	46	29L
IC 1805	Open cluster	6.5	20'	6,000	2h 32.7m	+61° 27'	46	29L
IC 1805	Emission nebula	—	96' × 80'	6,000	2h 32.8m	+60° 30'	46	29L
Stein 368	Double star	8.0, 10.1	10"	6,000	2h 32.7m	+61° 27'	46	29L
NGC 896	Emission nebula	7.5	20'	6,000	2h 24.8m	+62° 01'	46	29L
NGC 1027	Open cluster	6.7	20'	3,000	2h 42.6m	+61° 36'	46	29L
Mrk 6	Open cluster	7.1	6'	2,000	2h 29.7m	+60° 41'	46	29L

M76's size refers to the length of this nebula's distinctive, bright bar. The columns headed *MSA* and *U2* give the chart numbers of objects in the *Millennium Star Atlas* and *Uranometria 2000.0*, 2nd edition, respectively.

Although the pair can be crammed into a 1° field of view, it is better appreciated with a short-focal-length telescope giving fields 1.5° wide or more. Having ample dark sky around the clusters helps to offset their sparkling beauty. These clusters form a true pair, for they are known to have approximately the same age and distance from us.

The eastern member of the Double Cluster is **NGC 884**. With my 105mm scope at 68x, I see about 80 bright to very faint stars within this group. Two small knots of bright stars lie close together, southwest of center. NGC 884 is flecked with a handful of orange stars, two of which seem orphaned between the clusters. The orange stars are variables, and through a small telescope, it may be easier to perceive their colors when they are near maximum light.

The western cluster is designated **NGC 869**. This one appears a shade smaller than its companion and more highly concentrated. It has two fairly bright stars, and I count about 60 fainter ones. The central bright star has a distinctive bowl of stars to its southeast. With his 102mm refractor, California amateur Ron Bhanukitsiri sees this as a glowing eye and eyebrow.

After the Double Cluster, let's take stock — **Stock 2**, to be precise. A curving chain of 6th- to 8th-magnitude stars winds 2° northward from NGC 869 to Stock 2. Together, the Double Cluster and Stock 2 are a compelling sight through 50mm binoculars. Be sure to use a low-power, wide-field eyepiece for telescopic views, since Stock 2 is about 1° across.

My little refractor at 47x shows several dozen stars in a loosely scattered group. The brightest stars make a stick-figure man with his head to the west and spread legs to the east. His arms are upraised and curved as though showing off his muscles. For this reason, Massachusetts amateur John Davis has dubbed Stock 2 the Muscle Man Cluster, and the name seems very apt.

Next, we'll move 3° northeast of Stock 2 to the nebulous cluster **IC 1805**. At 68x, it's a coarse group of 40 stars about 13' across. A circlet of stars containing the lucida of the group (an old term for brightest member) sits at the heart of the cluster with arms of stars of 8th magnitude and fainter radiating from it. The central star is the double **Stein 368** and shows a faint companion 10" to the east.

Dropping the magnification to 17x gives my scope a 3.6° field of view, and I

The Double Cluster is one of the night sky's finest spectacles. Photo: Robert Gendler

First published January 2003

The central rectangle of the planetary nebula M76 is what most observers see at a casual glance, and this odd shape has inspired such names as the Cork and the Little Dumbbell. William C. McLaughlin, using a 12.5-inch f/9 reflector in Oregon, combined 21 CCD images to create this view.

can see extensive nebulosity in and around the cluster. Using an oxygen III filter makes it much more apparent. The brightest areas include the cluster itself and a wide patch running east and then curving north of the cluster. Starting at the eastern side of the wide patch, a fainter loop curves south of the cluster and then up its west side. Both loops are very patchy and together span about 1½°. A small, bright, detached portion, **NGC 896**, lies 1° northwest of the cluster. Under dark skies, this large complex has been seen in instruments as small as 30mm binoculars.

Just 1.2° east of IC 1805 is another cluster, **NGC 1027**. At 87x, I see one 7th-magnitude star loosely surrounded by approximately 40 faint to extremely faint stars within 17'. The cluster's brighter members appear to spiral out from the central star for 1½ turns.

An interesting knot of stars lies a little under 1° south-southwest of IC 1805. At 87x, **Markarian 6** (Mrk 6) is not an obvious cluster, but it has a noteworthy shape. Look for four stars of magnitude 8.5 to 9.7 in a slightly curved line running nearly north-south. Five dimmer stars join the one at the southern end to make an arrowhead, while the three to the north form the arrow's tail.

I've seen so many enchanting sights in this area of sky that I wish I had more space here to share them. But part of the beauty of observing is in knowing that there's always more to see.

Northern Nights

Where Camelopardalis and Cassiopeia meet, the sky is star-sparse but rich in other sights.

It often seems that we eagerly await seasonal sights appearing in the east while taking for granted those that stay with us all night and all year long. Such is the fate of Camelopardalis, the Giraffe, which is circumpolar for midnorthern latitudes.

Camelopardalis is a "modern" constellation whose origin dates back just to the early 17th century. Its creation is generally attributed to Flemish cartographer Petrus Plancius, who used it to fill an oddly shaped region of faint stars that had not been assigned to any constellation. Gazing northward, we see no clear outline of the Giraffe — only his spots scattered across the sky.

In this January sky tour, we'll look at the area around southwestern Camelopardalis, where the hindquarters of the Giraffe meet the queenly toes of neighboring Cassiopeia. Small-scope enthusiasts will find it rich in deep-sky wonders large and small, effortless and challenging.

Let's start with **Stock 23**, an open cluster that straddles the official boundary between these two constellations. Stock 23 is sometimes called Pazmino's Cluster, after New York amateur John Pazmino, who chanced upon the group with a friend's 4.3-inch refractor while sweeping for the Double Cluster in Perseus. His serendipitous sighting was brought to the attention of amateur observers in *Sky & Telescope*'s March 1978 Deep-Sky Wonders column.

Stock 23 is visible through binoculars as a little group of four stars in an irregular trapezoid resembling Draco's head. With my 105mm refractor at 87x, I count 27 stars of magnitude 7½ and fainter, spread across 13'. Most are arrayed in an oval outline running northwest to southeast, with two extensions springing from the southwest side. The unusual shape begs for games of dot-to-dot. To me it looks like the face of a lop-eared rabbit with ears at half-mast. The base of one ear is a nearly matched close double star, Σ362.

Stock 23 is involved in nebulosity that I have never been able to detect. While scanning the area, however, I've noticed a rather bright nebula 2° west-northwest. **IC 1848** appears about 1½° long and ¾° wide and is best seen with a low-power, wide-angle eyepiece. It is visible through my little refractor at 17x and 28x even without a filter. A narrowband light-pollution filter improves the view a bit, while both oxygen III and hydrogen-beta filters do better still. Your results may vary depending on the size of your

Far-North Sights for a January Evening

Object	Type	Magnitude	Size	Dist. (l-y)	Right Ascension	Declination	*MSA*	*U2*
Stock 23	Open cluster	5.6	14'	—	3^h 16.3^m	+60° 02'	45	28R
IC 1848	Emission nebula	7.0	100' × 50'	6,500	2^h 53.5^m	+60° 24'	45	28R
IC 1848	Open cluster	6.5	18'	6,500	2^h 51.2^m	+60° 24'	45	28R
Cr 34	Open cluster	6.8	24'	6,500	2^h 59.4^m	+60° 34'	45	28R
Tr 3	Open cluster	7.0	23'	—	3^h 12.0^m	+63° 11'	45/32	17L
Kemble 1	Asterism	4.0	150'	—	3^h 57.4^m	+63° 04'	31/43	16R
NGC 1502	Open cluster	5.7	7'	2,700	4^h 07.8^m	+62° 20'	43	16R/28L
NGC 1501	Planetary nebula	11.5	52"	4,200	4^h 07.0^m	+60° 55'	43	28L
IC 342	Spiral galaxy	8.3	21'	11 million	3^h 46.8^m	+68° 06'	31	16R

Angular sizes are from catalogs or photographs; most objects appear somewhat smaller when a telescope is used visually. Approximate distances are given in light-years based on recent research. The columns headed *MSA* and *U2* give the chart numbers of objects in the *Millennium Star Atlas* and *Uranometria 2000.0*, 2nd edition, respectively.

Having roughly the same angular size as Jupiter, NGC 1501 looks like a pale, oval glow in telescopes. The 14th-magnitude central star seems bright in this CCD close-up, taken with a 20-inch RC Optical Systems telescope, but this star is a challenge visually. North is to upper left. Photo: Adam Block / NOAO / AURA / NSF

scope and the darkness of your sky. The nebula is patchy and brightest in wide bands along the north, east, and west sides.

The nebula enshrouds two open clusters, neither particularly obvious. The western one shares the designation **IC 1848** with the nebula and surrounds the widely spaced bright double star Σ306 AG. At 68x, I see many very faint stars scattered across 18' with the densest areas lying south and east. To the east, **Collinder 34** (Cr 34) is larger and centered on a pair of stars (magnitudes 8 and 9) with radial strings of stars branching outward from it.

The cluster **Trumpler 3** (Tr 3), found 3.2° north of Stock 23, is a little more obvious. At 87x, I am able to see a glimmering splash of 35 stars, 9th magnitude and fainter, in a loose swarm with indefinite borders. Three of the brightest stars form a north-south line west of center; a fourth lies in the eastern edge of the cluster.

If you scan 5° east from Trumpler 3, you'll come to one of the most noteworthy asterisms in the sky. Canadian Lucian J. Kemble stumbled across this group by pure accident while scanning the sky with 7x35 binoculars. He called it a "beautiful cascade of faint stars tumbling from the northwest down to the open cluster NGC 1502." Kemble sent his description and a drawing to Walter Scott Houston, who published them in his December 1980 Deep-Sky Wonders column in *Sky & Telescope*. In later columns, Houston referred to this charming alignment of stars as **Kemble's Cascade**, and the name stuck.

Kemble's Cascade is a 2½° line of 7th- to 9th-magnitude

A sparkling star cluster within a much larger nebula, IC 1848 glows with the characteristic red hue of ionized hydrogen. Since human eyes are largely color-blind with faint, diffuse objects, most people see the nebula as a soft, grayish glow. This view is 2⅓° across. Photo: Sean Walker

Seen nearly face-on, the sprawling spiral arms of the galaxy IC 342 span an area of sky almost as large as the Moon. Fine details of their structure stand out well in this image by Robert Gendler of Avon, Connecticut. He used a 12.5-inch telescope and an SBIG ST-10E camera.

stars. One 5th-magnitude star punctuates the middle of the chain. In my little refractor, a wide-angle eyepiece yielding 17x and a 3.6° field gives a beautiful view, revealing about 20 stars. Near the southeast end of the Cascade, **NGC 1502** is a pretty cluster with many faint stars crowded around a pair of yellow-white, 7th-magnitude suns (Σ485). At 68x, I count 25 stars gathered into a squat triangle.

Dropping 1.4° south of NGC 1502, we find the little planetary nebula **NGC 1501**. It appears small, round, and fairly bright through my 105mm scope at 68x. This planetary takes on a little more character with each increase in aperture. Features to look for include a somewhat dark center, brighter northeast and southwest edges, a faint central star, and a slightly oval shape.

Our final stop is **IC 342**, or Caldwell 5, one of the closest galaxies beyond our own Local Group. Starting at the northwest end of Kemble's Cascade, move 1.8° north to a 4th-magnitude orange star with a 7th-magnitude yellow star 20' east. North 1.6° and a little west of the orange star, you'll find a similarly spaced pair consisting of a 6th-magnitude white star and a 7th-magnitude deep yellow star. The parallelogram formed by the four should be easily visible in a finder. IC 342 is centered 54' north of the white star. Placing this star at the southern edge of a low-power field will put the galaxy near the northern edge.

For a good view of this large, low-surface-brightness galaxy, be sure to center it in your field of view. In my 105mm scope at 28x, this pretty galaxy is a vaporous phantom spangled with faint stars. It appears oval, its long dimension running north and south with a length of 12'. From a dark-sky site with his 105mm refractor, noted observer Stephen James O'Meara has been able to trace out IC 342's three main spiral arms.

IC 342 is a member of the Maffei 1 Group of galaxies. It is relatively nearby at 11 million light-years, rivals our own galaxy in luminosity, and is one of the brightest galaxies in the northern sky. Its diminished glory as seen through telescopes is largely due to our vantage point. IC 342 lies only 10.6° above the galactic plane of the Milky Way, where it is highly obscured by intervening clouds of gas and dust.

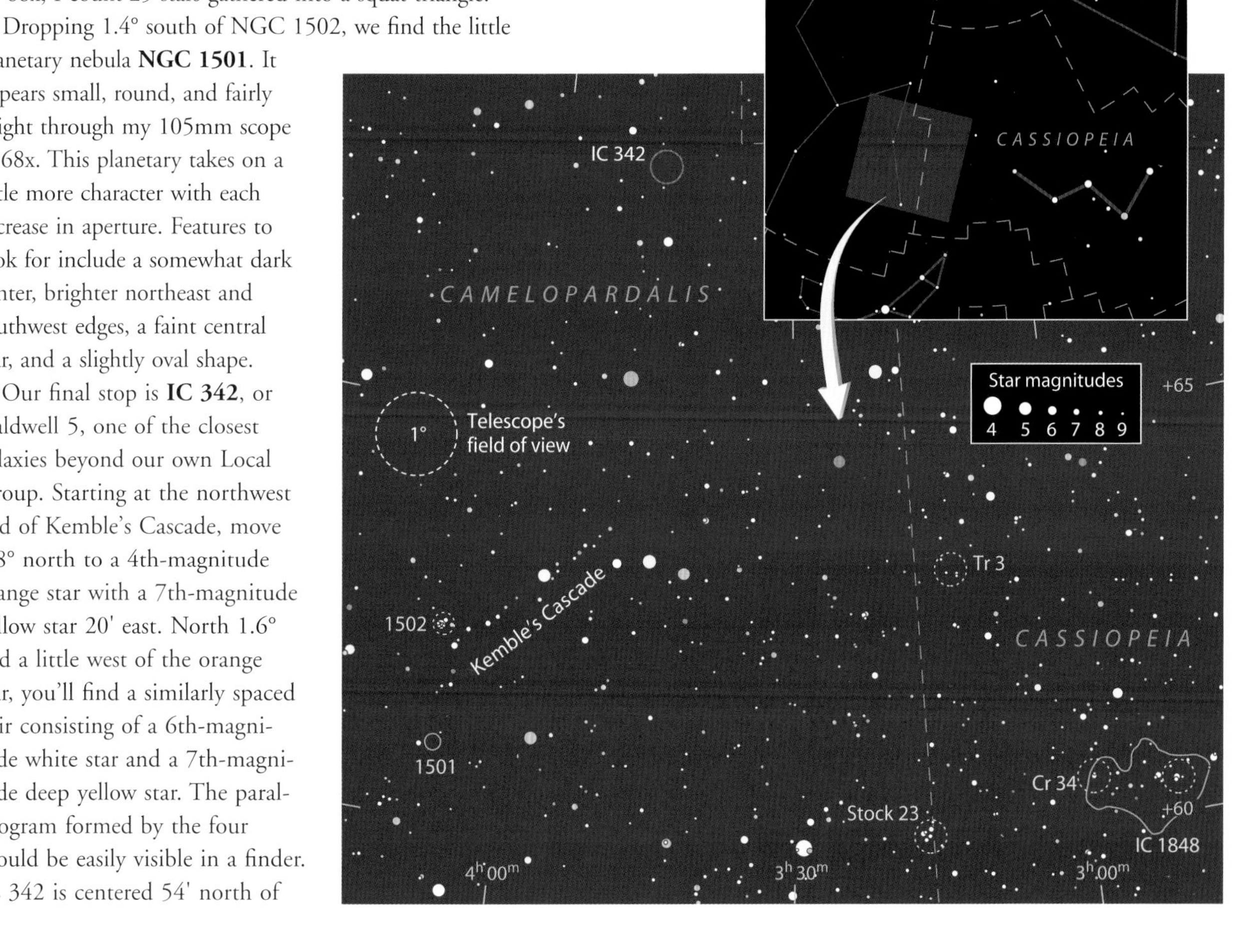

First published January 2004

Cruising Down the River

The winding star path of Eridanus, the River, leads you past some interesting deep-sky sights.

The source of the celestial river Eridanus lies near Rigel, the brightest beacon of the constellation Orion. The mighty stream actually begins at Cursa, or Beta (β) Eridani, and continues in a meandering course that makes this the sixth-largest constellation in the sky. But we'll confine our tour to the northern reaches of Eridanus and start near the River's source.

"Double, double toil and trouble" is a phrase from William Shakespeare's *Macbeth*, and it may well describe our hunt for the Witch Head Nebula, **IC 2118**. The Witch Head is a visual paradox, an object that can be seen in binoculars yet remains difficult through any size instrument.

To locate the Witch Head with a telescope, place Cursa at the eastern edge of your lowest-power eyepiece's field of view and scan southward. Small rich-field telescopes give a decided advantage when you're searching for this large

Along the River's Banks

Object	Type	Magnitude	Size/Sep.	Right Ascension	Declination	*MSA*	*U2*
IC 2118	Reflection nebula	—	180' × 60'	$5^h 04.8^m$	−7° 13'	279	137L
o^2 Eridani	Triple star	4.4, 9.5, 11.2	83", 9"	$4^h 15.3^m$	−7° 39'	282	137R
NGC 1535	Planetary nebula	9.4	48" × 42"	$4^h 14.3^m$	−12° 44'	306	137R
NGC 1247	"Flat" galaxy	12.5	3.4' × 0.5'	$3^h 12.2^m$	−10° 29'	309	138R
NGC 1300	Barred spiral galaxy	10.4	6.2' × 4.1'	$3^h 19.7^m$	−19° 25'	332/333	156R
NGC 1297	Lenticular galaxy	11.8	2.2' × 1.9'	$3^h 19.2^m$	−19° 06'	332/333	156R
h3565	Double star	5.9, 8.2	8"	$3^h 18.7^m$	−18° 34'	332/333	156R
NGC 1407	Elliptical galaxy	9.7	4.6' × 4.3'	$3^h 40.2^m$	−18° 35'	332	156R
NGC 1400	Lenticular galaxy	11.0	2.3' × 2.0'	$3^h 39.5^m$	−18° 41'	332	156R

Angular sizes are from catalogs or photographs; most objects appear somewhat smaller when a telescope is used visually. Approximate distances are given in light-years based on recent research. The columns headed *MSA* and *U2* give the chart numbers of objects in the *Millennium Star Atlas* and *Uranometria 2000.0*, 2nd edition, respectively.

The celestial river Eridanus flows southward with many a twist and turn. Even northern observers can easily see its first great bend, shown here, on January evenings.

Robert Gendler captured IC 2118, the intricate Witch Head Nebula, in this 10-frame mosaic. The bright star near the right edge is 3rd-magnitude Lambda (λ) Eridani. This 3°-wide field has south up.

nebula. I tackled the Witch Head one year while at the annual Winter Star Party in the Florida Keys. Although the sky is not exceptionally dark there, the southerly observing site places the nebula much higher in the sky than it is from my home in upstate New York. At 17x, with a true field 3.6° across, my 105mm refractor shows a faint glow about 2° long running north-south and curving a little eastward at its northern end. The nebula is better defined along its eastern side, which is irregular with a prominent bulge near the center. Observers with large scopes (and necessarily narrower fields) must observe the Witch Head a piece at a time, and they still find it a challenge.

The Witch Head is an unusually blue nebula that shines primarily by the reflected glow of nearby Rigel. It is not especially bright in the wavelengths usually emitted by gaseous nebulae. Despite this, many observers report a favorable response to various nebula filters, so it seems the nebula is guilty of yet another duplicity. Give IC 2118 a try, and see whether you are charmed or cursed by the Witch.

Now let's float downstream to the remarkable triple star **Omicron² (o²) Eridani**. The primary is a 4th-magnitude golden star, but its faint companions are what make this such a noteworthy system. The 10th-magnitude secondary, found a generous 83" to the east-southeast, is one of the few white dwarfs visible in a small telescope. This minute star is only 1½ times the diameter of Earth, yet it weighs in at half the mass of our Sun. A penny minted from its material would have the heft of 40,000 American pennies!

Omicron's third component makes this system even more extraordinary: an 11th-magnitude red-dwarf star lying 9" north-northwest of the white dwarf. A magnification of about 70x is enough to split the pair when the atmosphere is steady. The red dwarf is around one-fifth as massive as the Sun and one-quarter as big across. These dwarf stars are visible in backyard scopes thanks only to their proximity to Earth. At a distance of 16½ light-years, they (with Omicron² itself) are among our nearest neighbors.

From here, we'll follow one of the river bends and drop 5° southward to the planetary nebula **NGC 1535**. The 5th-magnitude star 39 Eridani, very similar in hue to Omicron², lies halfway along the sweep. Look for NGC 1535 in the eastern side of a 1° triangle of 8th-magnitude stars. English amateur William Lassell observed this planetary on January 7, 1853, and described it as "the most interesting and extraordinary object of the kind I have ever seen. A bright well-defined star, perhaps 11th magnitude, right in the centre of a circular nebula, whose edge was its brightest part; and this nebula again placed upon a larger and fainter [nebula], concentric and equally symmetrical."

Lassell made this observation at 565x with his home-built 24-inch equatorial reflector. Through a small telescope at low power, the object looks like a bluish star. At 127x, my little refractor shows a small, slightly oval disk. Occasionally, a brighter point sparkles in the center. A 6-inch scope begins to show the double nature praised by Lassell. The central star is clearly visible through my 10-inch reflector at 170x. A bright bluish ring with a slightly darkened center is surrounded by a thin fainter halo. Some observers perceive shades of green in the annulus. New Mexico sky hound Greg Crinklaw has nicknamed this pretty gem Cleopatra's Eye.

Moving westward, we come to the nearly edge-on spiral galaxy **NGC 1247**. "Flat" galaxies are among my favorite observing targets. NGC 1247 is one of the entries in the *Revised Flat Galaxy Catalogue*, a compilation by Igor D. Karachentsev and his colleagues of 4,236 galaxies that appear at least seven times longer than they are wide.

NGC 1247 lies 2° south-southwest of Zeta (ζ) Eridani. The galaxy rests one-third of the way from a star to its west-northwest to a wide pair east-southeast, all 10th magnitude. Suspecting that my quarry might be too faint for the small refractor, I pursued it with my 10-inch scope. I swept the galaxy up at 44x and then examined it more closely at 170x. This spindle-shaped galaxy runs east-northeast to west-southwest for about 3'. It grows slightly brighter toward the middle and has a very faint star superposed north of center. The sky was a bit hazy when I looked at this galaxy, so I suspect it can

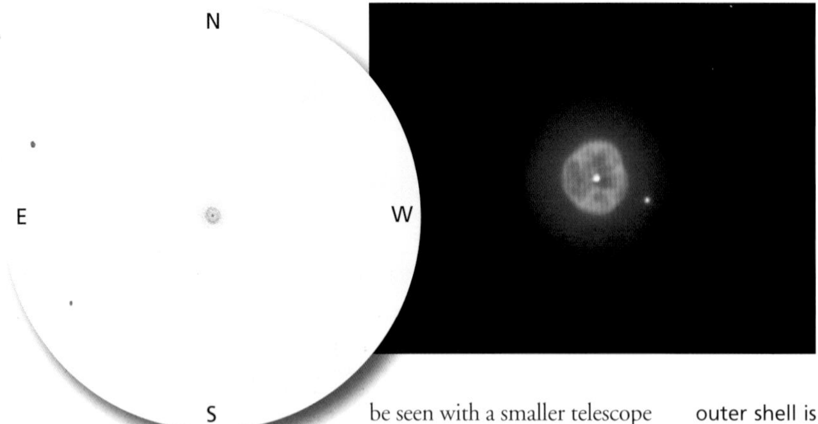

Far left: Using a 6-inch reflector at Jyväskylä, Finland, Jere Kahanpää made this sketch of NGC 1535. He called the central star easy at 266x, noting that the eastern edge of the inner ring appeared slightly brighter than the opposite edge. The field shown here is 11' wide, with north up. Left: Floating in deep space like some giant jellyfish, the planetary nebula NGC 1535 shows fine details in this close-up that no astronomer ever sees when looking into a telescope's eyepiece. The outer shell is just 50" across. Adam Block used a 20-inch reflector and CCD camera for both images on this page. Photo: Adam Block / NOAO / AURA / NSF

be seen with a smaller telescope on a good night. I'd be interested to know whether it can be captured with less aperture.

Just past NGC 1247, the River loops south and then east following a line of nine stars all designated Tau (τ), as though Johann Bayer was trying to conserve Greek letters when he worked his way through this lengthy constellation. The gorgeous barred spiral **NGC 1300** sits 2⅓° due north of reddish orange Tau^4. This nearly face-on galaxy can be seen in scopes as small as 4 inches in aperture, but it probably takes at least a 10-inch to show the bar.

I spotted NGC 1300 in my 10-inch reflector at 70x and then studied it at 118x and 170x. The galaxy shows a moderately bright core in a faint bar that runs east-southeast to west-northwest. The dimmer halo is oval with its long axis having the same alignment as the bar. Some patchiness in the halo hints at the galaxy's spiral structure. At 118x, I see **NGC 1297** in the same field of view just 20' to the north-northwest. This small, round, faint galaxy grows brighter toward the center and has a 14th-magnitude star at its northern edge. Using a wide-angle eyepiece to reduce the magnification to 70x puts the pretty double star **h3565** in the northern part of the field. Its close-set 6th- and 8th-magnitude components look white and gold to me.

Let's continue farther east to **NGC 1407**, located 3½° north-northeast of Tau^5 and 1½° southeast of 20 Eridani. NGC 1407 is the largest and brightest elliptical galaxy in the Eridanus A group, and it has been seen in scopes as small as 55mm in aperture. Although it appears brighter in a larger scope, the view is essentially the same through most small to medium amateur telescopes: a small, round galaxy with a bright core and stellar nucleus. **NGC 1400**, 12' to the southwest, is a smaller and slightly dimmer version of its neighbor. Although NGC 1400 has a much lower radial velocity than NGC 1407 and the other members of the system, it is believed to be a true member of the group. The reason for its peculiar velocity is as yet unknown. Eridanus A is about 65 million light-years away and is a subgroup of the Eridanus Galaxy Cluster. It is thought that these subgroups will eventually merge into one massive cluster.

If you have a star chart that plots some of the fainter galaxies clustered around NGC 1407, perhaps you'd like to give them a try. These are some of the easiest, in order of decreasing surface brightness: NGC 1452, 1394, 1391, 1440, 1393, and 1383.

Seen almost face-on from our vantage point, the beautiful galaxy NGC 1300 has a fine example of the elongated hub that gives barred spirals their name. This view has north toward the upper right and spans about 7' from left to right. A much wider field would have included the galaxy NGC 1297 as well. Photo: Nicole Bies / Esidro Hernandez / Adam Block / NOAO / AURA / NSF

First published January 2005

Heavenly Beauties

Star-studded January skies offer deep-sky treats for every size telescope.

High in the evening sky at this time of year, the magnificent double star **Almach** marks the eastern end of the glittering sweep of stars that dominates the constellation Andromeda. In 1804, German-born British astronomer William Herschel called Almach one of the most beautiful objects in the heavens and said that the "striking difference in the colour of the two stars, suggests the idea of a sun and its planet, to which the contrast of their unequal size contributes not a little." Herschel saw the 2nd-magnitude primary as reddish white and described the 5th-magnitude companion as "fine light sky-blue, inclining to green."

Also known as Gamma (γ) Andromedae, Almach has always been a favorite of double-star enthusiasts, though not everyone perceives the same colors. Through my 105mm refractor at 87x, I see the pair as gold and blue, but the companion appears white through my 10-inch reflector. Almach's primary is a K3 yellow-orange giant that would nearly fill the orbit of Mercury if it replaced our Sun.

The remarkable edge-on galaxy **NGC 891** sits just 3.4° east of Almach. My 105mm scope at 87x shows a faint 6'-long streak tipped north-northeast in a starry field. Due north of the galaxy's center, a 12th-magnitude star decorates its western side. The view of NGC 891 improves greatly with increased aperture. Showing a very mottled face and a brighter core, the galaxy is quite striking in my 10-inch reflector at 118x. It measures about 1.6' x 11' and sports an 11th-magnitude star within its southern tip. At 171x, the core is a 2½'-long flattened oval that gives NGC 891 a slight central bulge. A dusky lane, most evident across the core, splits the galaxy lengthwise, and a 13th-magnitude star is pinned to the eastern edge of the galaxy's northern end.

The flat disks of spiral galaxies appear more or less round when seen face-on but elongated when tipped to our line of sight. NGC 891 is oriented exactly edge-on, a fortuitous circumstance that gives us a pencil-thin profile to enjoy. The obscuring lane that bisects the galaxy is composed of tiny dust grains no bigger than particles of smoke. Our own galaxy would look very much like NGC 891 if we could view it from a similar vantage point.

Now let's turn to the open cluster **NGC 752** (Caldwell 28), found 4.6° south-southwest of Almach. In 14x70 binoculars, NGC 752 is a very pretty cluster of about 60 moderately bright to faint stars spread over nearly a degree. The wide double star **56 Andromedae** (56 And) sits off its south-southwestern flank and is plotted on the chart on page 20. My little refractor at 47x reveals about 90 stars, many arranged in crisscrossing chains. The group's brightest star is golden, and the 6th-magnitude components of 56 Andromedae shine gold and orange. Switching to my 10-inch scope at 44x, I see NGC 752 as bright and gorgeous with stragglers nearly filling the 87' field. Prominent curves of stars dominate the scene, and there are several wide pairs. Many stars now show hints of color, notably three in a little triangle

January Gems

Object	Type	Magnitude	Size/Sep.	Right Ascension	Declination	*MSA*	*U2*
Almach	Double star	2.3, 5.0	9.7"	2^h 03.9^m	+42° 20'	101	44L
NGC 891	Galaxy	9.9	11.7' × 1.6'	2^h 22.6^m	+42° 21'	101	44L
NGC 752	Open cluster	5.7	75'	1^h 57.6^m	+37° 50'	123	62L
56 And	Double star	5.8, 6.1	201"	1^h 56.2^m	+37° 15'	123	62L
M33	Galaxy	5.7	71' × 42'	1^h 33.9^m	+30° 40'	146	62L
NGC 604	Diffuse nebula	10.5	1.0' × 0.7'	1^h 34.5^m	+30° 47'	146	62L
Collinder 21	Asterism	8.2	7.0'	1^h 50.2^m	+27° 05'	145	80L
Σ172	Double star	10.2, 10.4	17.7"	1^h 50.0^m	+27° 06'	145	80L
β1313	Double star	8.6, 9.4	0.6"	1^h 50.2^m	+27° 02'	145	80L
NGC 672	Galaxy	10.9	6.0' × 2.4'	1^h 47.9^m	+27° 26'	146	80L
IC 1727	Galaxy	11.5	5.7' × 2.4'	1^h 47.5^m	+27° 20'	146	80L

Angular sizes or separations are from recent catalogs. The visual impression of an object's size is often smaller than the cataloged value and varies according to the aperture and magnification of the viewing instrument. The columns headed *MSA* and *U2* give the chart numbers of objects in the *Millennium Star Atlas* and *Uranometria 2000.0*, 2nd edition, respectively.

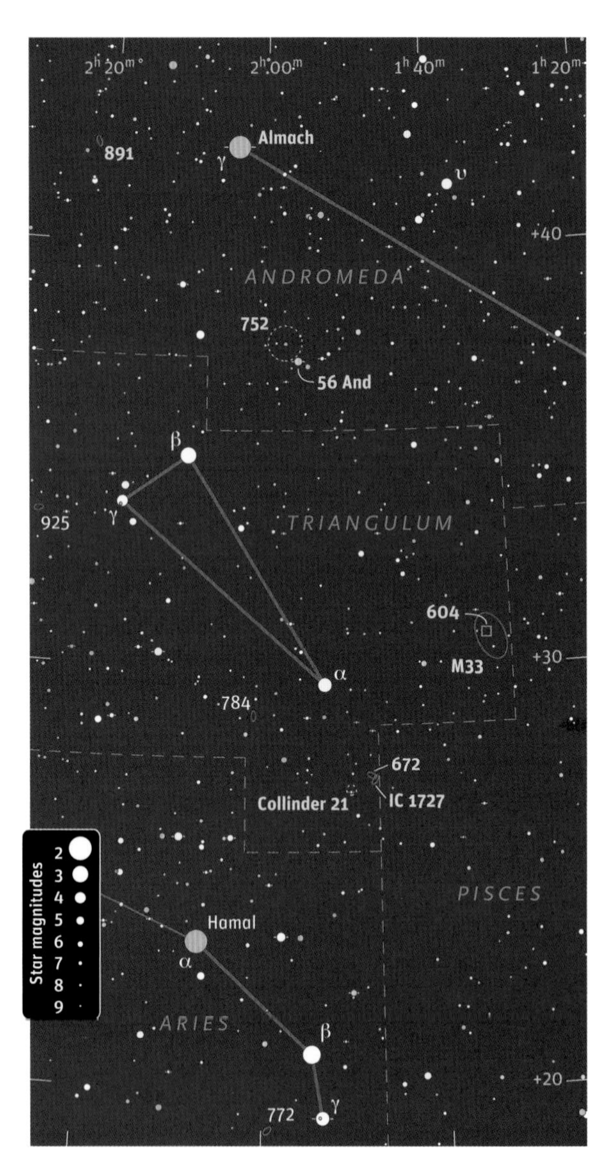

south of center that are dressed in shades of gold. The colorful nature of this cluster occurs because it's more than a billion years old, so many of its stars have evolved into red giants.

South of NGC 752, we find the three-starred asterism that defines the constellation Triangulum, with Alpha (α) Trianguli at its pointy tip. The spiral galaxy **M33**, sometimes called the Pinwheel Galaxy, is 4.3° west-northwest of Alpha. It lies 2.6 million light-years away and is part of our Local Group of galaxies. By way of comparison, NGC 891 is 12 times more distant. Despite its relative nearness, the Pinwheel can be a difficult galaxy to spot. Its light is spread over a large area, thus giving it a low surface brightness. Nonetheless, M33 has been spotted with the unaided eye under very dark skies, and it's one of the few galaxies that will reveal spiral structure in modest-size backyard scopes.

Occasionally, people at a star party tell me they're unable to find M33. When I show them the galaxy, they usually say that they weren't expecting to see something that looked so large and faint. With a wide-angle eyepiece giving 68x, my 105mm scope has a 1.15° field of view, and the galaxy's extremely dim outer halo stretches north-northeast to south-southwest across three-quarters of the field. Its width is a bit more than half its length. A large, weakly brighter inner halo runs north-south and surrounds a small east-west core.

My 10-inch scope at 70x unveils the subtle spiral structure of the Pinwheel. One arm emerges from the western side of the core and wraps around to the north. A balancing arm begins east of the core and wraps south. Several superposed stars are scattered across the galaxy's face, and some very faint ones sparkle in its core. The inner halo and core are quite mottled, and the 2' core harbors a faint stellar nucleus.

M33 houses several starclouds and nebulae bright enough to carry their own NGC or IC designations. The most conspicuous one is the giant star-forming region **NGC 604**, which is located 12' northeast of M33's nucleus and right beside an 11th-magnitude star. Through my small refractor at low power, the pair looks almost like a double star whose secondary won't quite focus. But NGC 604 becomes a distinct little smudge at 87x. My 10-inch scope at 118x shows a 1' northwest-southeast oval with a broadly brighter center. NGC 604 is an exceptionally large nebula, spanning about 1,500 light-years — 50 times bigger than the Orion Nebula, M42.

A trio of interesting targets lies 2½° south-southwest of Alpha Trianguli. The most obvious one is **Collinder 21**, visible in my 105mm scope at 68x as a baker's dozen of stars strung along the outline of a 6' circle whose northeast quadrant is severely dented. On the circle's northwestern rim, the double star Σ172 gleams with a nicely matched pair of 10th-magnitude suns. Observers with 10-inch or larger scopes might like to try splitting

The attribute that makes the spiral galaxy M33 in Triangulum one of the finest galaxies accessible to Northern Hemisphere astrophotographers also makes it the bane of visual observers: size. Because M33 appears larger than the Moon, its light is spread rather thin and its visibility is more critically dependent on sky transparency than are smaller objects that have a higher surface brightness. The field is ¾° wide. Photo: Sean Walker

First published January 2006

Aligned edge-on to our line of sight, the spiral galaxy NGC 891 in Andromeda presents a slender profile bisected by a distinct dust lane. It is visible even in a small backyard telescope, but an 8-inch aperture is typically needed to catch sight of the dust lane. The field is ¼° wide. Photo: Adam Block / NOAO / AURA / NSF

the group's brightest star, a yellow-white gem on the southern rim. **β1313** consists of two nearly equal components a mere 0.6" apart.

Brian Skiff of Lowell Observatory conducted a photometric investigation of Collinder 21 with the observatory's 21-inch reflector. His study suggested that most of the stars lie at different distances and do not form a true cluster. And Skiff later confirmed this by analyzing the data from proper-motion surveys, which show that most of the stars are moving in different directions. He also found that the reddish 10th-magnitude star on the circlet's northern rim is a variable star well in the background of the asterism's other stars.

Just 37' northwest of Collinder 21, we come to the galaxy **NGC 672**. The brightest star of Collinder 21 and another of similar magnitude due north make a squat isosceles triangle with NGC 672. My small refractor at 87x shows a faint 2' galaxy stretched east-northeast to west-southwest. In my 10-inch reflector at 118x, NGC 672 is framed by a triangle of 13th-magnitude stars, two of which straddle its western end. The galaxy is about 5' long and one-third as wide, and it grows gently brighter toward an irregular, blotchy, elongated core. NGC 672 shares the field with the much lower-surface-brightness galaxy **IC 1727**, 8' southwest, which appears 2' long and half as wide. NGC 672 lies at a distance of about 26 million light-years and forms a gravitationally interacting pair with IC 1727.

Chasing the Starry Cetus

We can sail December's celestial seas to plumb the depths of Cetus.

I have boarded the Argo-Navis, and joined the chase against the starry Cetus far beyond the utmost stretch of Hydrus and the Flying Fish.

— Herman Melville, *Moby-Dick*, 1851

As the whale is an enormous creature, so, too, is his star figure vast. Cetus is the fourth-largest constellation, outspread only by Hydra, Virgo, and Ursa Major. As the stars pursue their westward course each night, Cetus seems to swim backwards across the sky. Farther east and higher than Cetus's tail, the head lingers longest, and the V of the Hyades conveniently points to its circlet of stars.

We'll begin this star-hop with the double star **Nu (ν) Ceti**, which lies about halfway between Gamma (γ) and Xi² (ξ²) in the circlet. In my 105mm refractor, the lovely yellow primary holds a faint companion to its eastern side. I've been able to split the pair at 47x when the atmosphere is steady, but I've needed nearly twice as much magnification when the seeing is poor.

In his classic *Bedford Catalogue*, William H. Smyth reports that the companion star to Nu Ceti "can only be seen by glimpses, on ardent gazing," with his 5.9-inch refractor in 1833. In the *Monthly Notices of the Royal Astronomical Society*, William Noble states that he found this same star comparatively conspicuous in his 4.2-inch glass in 1873, while Thomas William Webb found it easy with a 5.5-inch in 1861. Noble wrote, "The object in question must be a well-marked variable." I have been unable to find any source that substantiates this variability. It's easy to believe that the disparity in accounts was simply due to differences in observer, instrument, and sky, but it may be interesting to

More of the Whale, but Not the Tail

Object	Type	Magnitude	Size/Sep.	Right Ascension	Declination	*MSA*	*U2*
ν Ceti	Double star	5.0, 9.1	7.9"	2^h 35.9^m	+5° 36'	239	99L
Cosmic Question Mark	Asterism	3.9	2.1° × 0.7°	2^h 36.3^m	+6° 42'	239	99L
γ Ceti	Multiple star	3.6, 6.2, 10.2	2.3", 14'	2^h 43.3^m	+3° 14'	238	119L
M77	Spiral galaxy	8.9	7.1' × 6.0'	2^h 42.7^m	–0° 01'	262	119L
NGC 1055	Spiral galaxy	10.6	7.6' × 2.6'	2^h 41.7^m	+0° 27'	262	119L
NGC 1087	Spiral galaxy	10.9	3.7' × 2.2'	2^h 46.4^m	–0° 30'	262	119L
NGC 1090	Spiral galaxy	11.8	4.0' × 1.7'	2^h 46.6^m	–0° 15'	262	119L
NGC 1094	Spiral galaxy	12.5	1.5' × 1.0'	2^h 47.5^m	–0° 17'	262	119L
84 Ceti	Double star	5.8, 9.7	3.6"	2^h 41.2^m	–0° 42'	262	119L
ο Ceti	Variable star	3½–9	—	2^h 19.3^m	–2° 59'	264	119R

Angular sizes are from recent catalogs. The visual impression of an object's size is often smaller than the cataloged value and varies according to the aperture and magnification of the viewing instrument. The columns headed *MSA* and *U2* give the appropriate chart numbers in the *Millennium Star Atlas and Uranometria 2000.0*, 2nd edition, respectively. All the objects this month are in the area of sky covered by Chart 4 in *Sky & Telescope's Pocket Sky Atlas*.

keep an eye on this double nonetheless.

Nu Ceti sits at the bottom of a delightful asterism that California amateur Dana Patchick dubs the **Cosmic Question Mark**. Nu is the question mark's point, while five 6th- and 7th-magnitude stars to the north complete the inquiring figure. This 2°-tall bit of celestial punctuation stands out nicely in my 8x50 finder and is fine quarry for small binoculars.

Gamma (γ) Ceti is a more challenging double star than Nu Ceti. My little refractor at 127x barely separates the white primary from its pale yellow attendant to the west-northwest. A 10th-magnitude red-dwarf star 14' to the northwest shares the same proper motion (apparent motion on the celestial sphere) as this double and may be a physical member of the system. If so, the red dwarf is at least 21,000 astronomical units from its companions and would take a minimum of 1½ million years to orbit them. This nearby system is 82 light-years away.

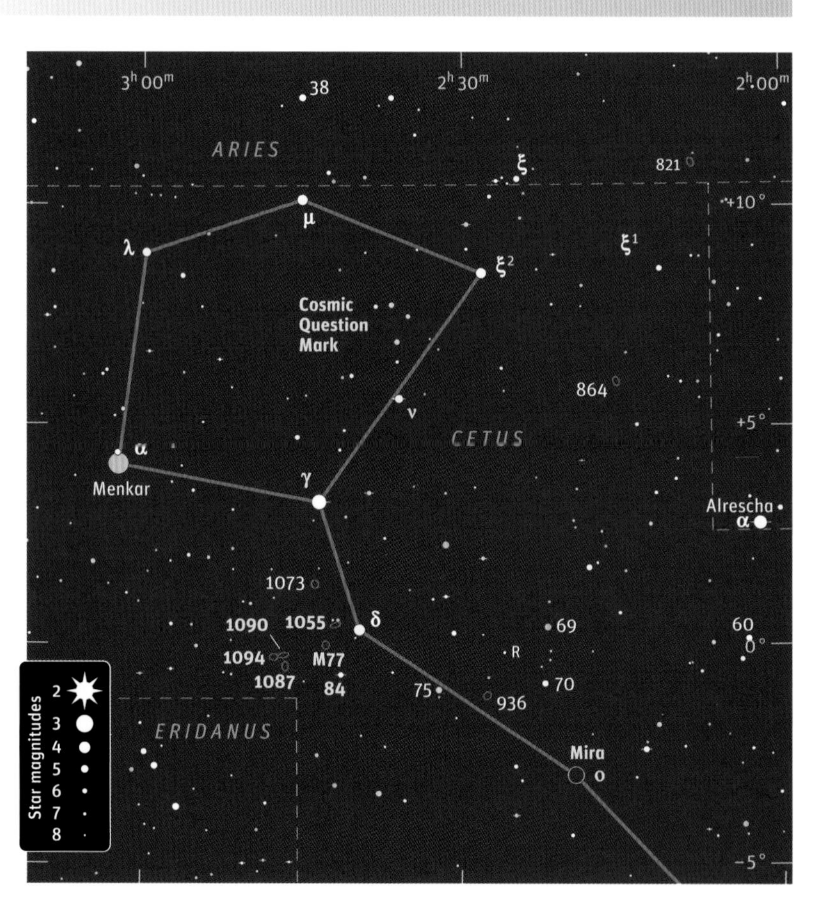

The only Messier object in Cetus is the spiral galaxy **M77**, located 52' east-southeast of Delta (δ) Ceti. My 105mm scope at 127x shows a 2½' halo that's slightly oval and tipped north-northeast. The halo surrounds a small bright core and a nearly stellar nucleus. A star similar in brightness to the nucleus lies just off the galaxy's east-southeastern edge. My 10-inch reflector at 213x reveals subtle spiral arms adorning the halo. The eastern arm wraps north around the galaxy, while the western arm wraps south.

Now look for an east-west pair of stars, magnitudes 7 and 8, about halfway along and north of a line connecting M77 and Delta. The spiral galaxy **NGC 1055** lies 6' south of this wide pair. In my small refractor at 87x, I see a 4' x 1' spindle canted east-southeast. A faint star sits at the north-northwestern edge of its weakly brighter core.

NGC 1055 and M77 are thought to form an interacting pair about 50 million light-years away. The group of galaxies they belong to also includes **NGC 1087**. Look ½° east of

M77 for a widely spaced, north-south pair of stars, magnitudes 8.2 and 8.8. Placing the brighter star in the northwestern part of a low-power field will bring NGC 1087 into view, where you'll see it 6' south of a 10th-magnitude star. My 105mm scope at 87x displays a moderately faint, 2' x 1.2' north-south oval of uniform brightness. **NGC 1090**, ¼° to the north, shares the field but is extremely faint.

My 10-inch reflector at 115x brightens the galaxies and reveals a third, **NGC 1094**, in the same field of view. Although NGC 1094 has a higher surface brightness than NGC 1090, it's more difficult to spot because of its diminutive size. Look for it 4½' south of a 9.7-magnitude star. Boosting the magnification to 213x makes it easier to notice, but it still shows no detail. NGC 1090 is fairly elongated and stretches east-southeast to west-northwest for 2'. NGC 1087 appears 2½' long and half as wide. It grows gently brighter toward the center and looks highly textured.

While in this region, you can visit the pretty double star **84 Ceti** resting ¾° south-southwest of M77, where it's the area's brightest star. My 10-inch scope at 166x shows a yellowish primary with a considerably fainter orange companion to the northwest. This is another nearby system, a mere 71 light-years distant.

Our final target will be **Omicron** (ο) **Ceti**, which was the first known periodic variable star. The discovery of its changeable nature is generally credited to the Dutch cleric and amateur astronomer David Fabricius, who caught it at maximum in 1596 but assumed that it was a nova. It was first lettered Omicron on Johann Bayer's 1603 *Uranometria* and given its common name, Mira (The Wonderful), by Johannes Hevelius in 1642.

The periodicity of Mira was first determined in 1638 by Johann Fokkens Holwarda of the Netherlands. The star varies from average values of about magnitude 3½ to 9 and back in 11 months, but individual highs and lows may differ by more than a magnitude. To find the current magnitude of Mira, use the Light Curve Generator on the website of the American Association of Variable Star Observers (AAVSO). Enter "Mira" in the star name box, and hit enter on your keyboard. The resulting graph will tell you not only how bright Mira is now but also whether it's fading or brightening.

Mira is an unstable red-giant star whose outer layers undergo slow and semiregular expansion and contraction. Other stars that exhibit similar properties are now known as Mira-type variables. At maximum size, Mira is a whale of a star — more than 330 times bigger across than our Sun.

We are going a-whaling,
and there is plenty of that yet to come.

— Herman Melville

Although Cetus is the fourth-largest constellation, covering more than 1,200 square degrees, it contains only one Messier object — the spiral galaxy M77. In 1913, spectra of M77 obtained by V. M. Slipher at Lowell Observatory revealed the galaxy to be receding from us at more than 1,000 kilometers per second, setting the stage for Edwin Hubble's revelation during the 1920s that the universe is expanding. The field here is 11' wide with north up. Photo: Robert Gendler

First published January 2007

Perseus's Flying Mantle

Riding high overhead on January evenings, the Milky Way in Perseus is a rich hunting ground for deep-sky enthusiasts.

Perseus hosts a magnificent array of deep-sky wonders, mostly because his star-figure overlays the plane of our galaxy. As Garrett P. Serviss wrote in his 19th-century classic *Round the Year with the Stars*, Perseus "wraps the glory of the Milky Way around him like a flying mantle." The constellation's brightest star,

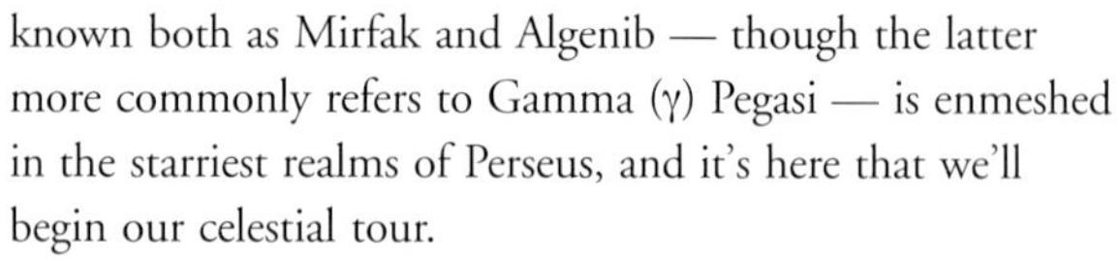

known both as Mirfak and Algenib — though the latter more commonly refers to Gamma (γ) Pegasi — is enmeshed in the starriest realms of Perseus, and it's here that we'll begin our celestial tour.

Mirfak (α Persei) is the namesake star of the Alpha Persei Cluster, also known as **Melotte 20** (Mel 20). Gazing up with unaided eyes from my semirural home, I see several faint stars around and spreading southeast of Mirfak. A touch of haze hints at unresolved stars. The cluster is huge and makes a superb binocular target. With my 12x36 image-stabilized binoculars, I count 20 bright and 70 faint stars covering 4° x $2\frac{1}{2}$°. Many are arranged in a striking S shape that snakes from yellow-white Mirfak to a loop containing orange Sigma (σ) Persei. Sparser chains enhance the view, including one that trends northwest from Mirfak to a Sun-yellow star with a wide companion.

Melotte 20 is a true cluster about 600 light-years away and 50 million years young. It's nestled within a more loosely knit and much larger stellar family born at the same time. Although the association's stars are moving together through space, they form an unbound group lacking the gravitational coherence to hold itself together.

Smaller and more compact, the open cluster **M34** makes an isosceles triangle with Kappa (κ) Persei and Beta (β) Persei, or Algol, a remarkable eclipsing binary star. The cluster's misty glow is easy to sweep up in a small finder, while a large one plucks out several stars. My 105mm refractor at a magnification of 127x shows a $\frac{1}{2}$°-square group of 75 mixed bright and faint stars. A rectangular core, aligned with one of the square's diagonals, contains the majority of the bright members, many arranged in pairs. The cluster's deep yellow lucida sits in the middle of the square's southern side, while its second-brightest star is a glowing orange ember on

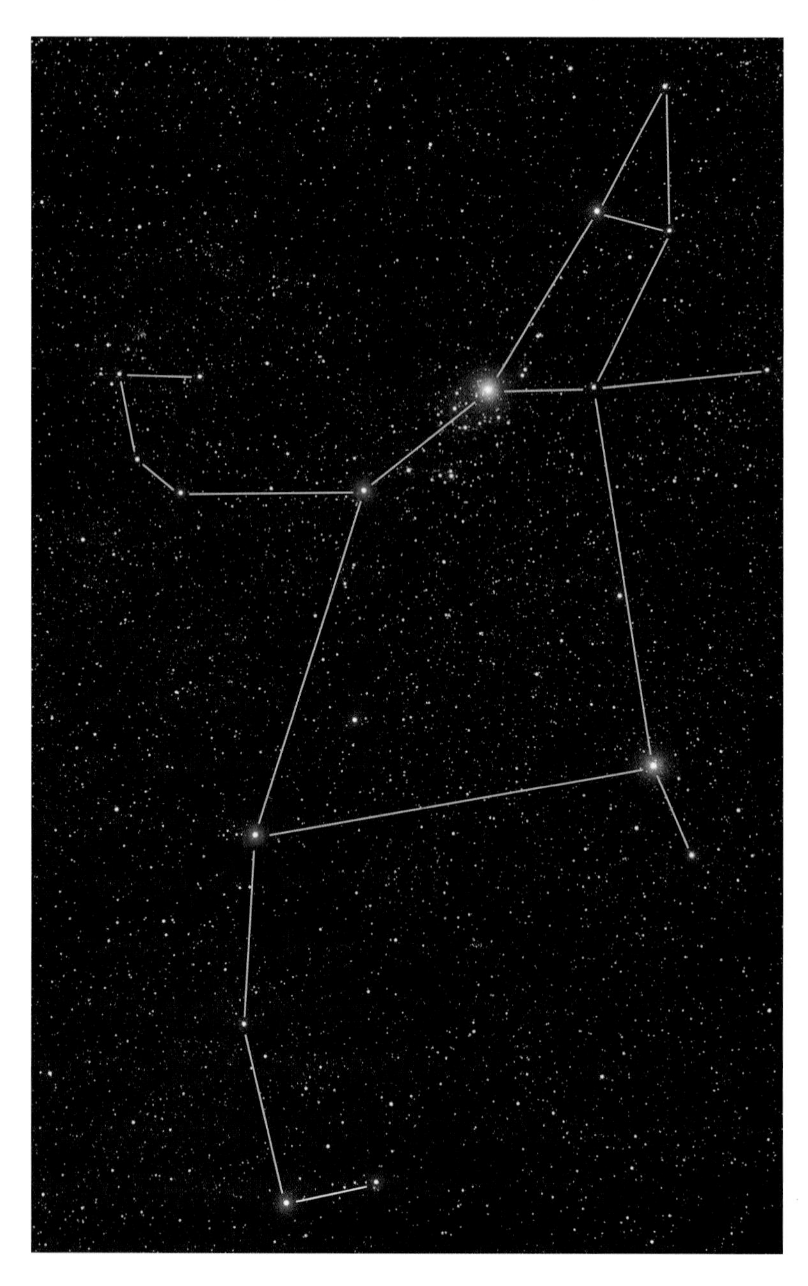

Straddling the plane of our galaxy, the constellation Perseus is home to a multitude of star clusters. Chief among them for naked-eye and binocular observers is Melotte 20, a sprawling, 5°-diameter collection of stars surrounding Mirfak (Alpha Persei). Photo: Akira Fujii

The brightest galaxy within the borders of Perseus is NGC 1023. While you can spot it with a modest-aperture telescope, it's better suited for viewing with 8-inch and larger instruments. A companion galaxy, seen as a brightening on the eastern (left) end of NGC 1023, has rarely been noted by backyard observers. The field is ½° wide; north is up. Photo: POSS-II / Caltech / Palomar

the opposite side. M34 is five times older than Melotte 20, and its stars look fainter largely because we survey them across a gulf of 1,600 light-years.

Sweeping ½° east of M34's brightest star will bring you to a 9' curve of five stars, magnitudes 9 to 11. The brightest and westernmost stars form a 2' east-west pair, and the little-known planetary nebula **Abell 4** lies only 1.6' north-northwest of the latter. With my 10-inch reflector, this elusive nebula is invisible at low power. At 166x, with an oxygen III or a narrowband filter, Abell 4 is intermittently seen with averted vision as a featureless disk. It's easier at 213x. The oxygen III filter makes the view a bit too dark for my taste, but it works well with a bigger scope.

Images of Abell 4 reveal a 16th-magnitude galaxy perched a scant 48" to its west-northwest. This edge-on spiral is often incorrectly identified as CGCG 539-91, which really refers to Abell 4 (despite being a designation from a galaxy catalog). The galaxy is properly known by the mind-numbing designation 2MASX J02452000+4233270. Although virtually impossible to remember, that moniker lets us know that this galaxy is listed in the Two Micron All Sky Survey extended-source catalog and gives its precise sky coordinates. I haven't heard of anyone observing this galaxy in anything smaller than a 24-inch scope.

Now drop 2½° south from M34 to 12 Persei, which is visible to the unaided eye in a moderately dark sky. Just northeast of this yellow sun, the double star **Σ292** is a treat in my little refractor at 17x. Its 7.6-magnitude, blue-white primary is accompanied by an 8.2-magnitude, white companion 23" toward 12 Persei.

Seeing beyond

Stars, dust, and gas concentrated within the plane of our Milky Way Galaxy generally block the view of the universe beyond. But when looking in the direction of Perseus, we're viewing toward regions where this obscuring material is sometimes thin enough to let distant galaxies peek through. Such is the case with NGC 1023, pictured above.

Prizes in Perseus

Object	Type	Magnitude	Size/Sep.	Right Ascension	Declination	*MSA*	*U2*
Mel 20	Open cluster	2.3	5°	3^h 24.3^m	+49° 52'	78	43L
M34	Open cluster	5.2	35'	2^h 42.1^m	+42° 45'	100	43R
Abell 4	Planetary nebula	14.4	22"	2^h 45.4^m	+42° 33'	100	43R
Σ292	Double star	7.6, 8.2	23"	2^h 42.5^m	+40° 16'	100	43R
NGC 1023	Galaxy	9.4	7.4' × 2.5'	2^h 40.4^m	+39° 04'	100	61L
NGC 1245	Open cluster	8.4	10'	3^h 14.7^m	+47° 14'	78	43L
NGC 1193	Open cluster	12.6	3'	3^h 05.9^m	+44° 23'	99	43R

Angular sizes and separations are from recent catalogs. Visually, an object's size is often smaller than the cataloged value and varies according to the aperture and magnification of the viewing instrument. Right ascension and declination are for equinox 2000.0. The columns headed *MSA* and *U2* give the appropriate chart numbers in the *Millennium Star Atlas* and *Uranometria 2000.0*, 2nd edition, respectively. All the objects this month are in the area of sky covered by Chart 13 in *Sky & Telescope's Pocket Sky Atlas*.

Despite its rich assortment of deep-sky objects, including the famous Double Cluster, only two objects in Perseus — the open cluster M34 and the planetary nebula M76 — are logged in the famous catalog of 18th-century French comet hunter Charles Messier. Easily visible in binoculars, M34 is shown here in a field ¾° wide with north up. Photo: Bernhard Hubl

After admiring Σ292, place it in the eastern side of a low-power field and hop 1¼° south to **NGC 1023**, the brightest galaxy in Perseus. Its east-west elongated blur is tricky to spot in my 105mm scope at 17x, chiefly due to a pair of distracting stars off its southern flank. At 87x, however, NGC 1023 presents a bright, elongated core that intensifies toward a starlike nucleus. The galaxy is 1.3' wide, and very faint arms stretch it to a length of about 3.6'. Through my 10-inch scope at 220x, the galaxy covers 7.5' x 1.7' with a 12th-magnitude star off each tip. The brightening contours of the core grow rounder near the center, and a 13.9-magnitude star sits just west of the core.

In photographs, a dwarf galaxy seems to dangle from the eastern end of NGC 1023. Despite knowing its exact location, I've been unable to distinguish NGC 1023A with my 10-inch reflector. The best I can say is that NGC 1023 seems to be a little wider at that point than I'd expect it to be without a companion. The area of subtly enhanced width lies near a 13.7-magnitude star south of NGC 1023's eastern end.

Next, we'll visit a third open cluster, even smaller and more compact than M34. **NGC 1245** is located a little shy of halfway from Kappa to Mirfak and slightly west of an imaginary line connecting them. Through my little refractor at 17x, the cluster is a granular glow with an 8th-magnitude star on its south-southeastern edge. Each increase in magnification teases out more stars until, at 122x, I count 25 sparkling points in a misty net 8' across. My 10-inch scope at 213x resolves much of the dappled haze into a glittering group of 70 stars. NGC 1245 is a shining swarm of several hundred suns 9,800 light-years away and three times older than M34.

The much fainter cluster **NGC 1193** powders the sky 47' southwest of Kappa. Its delicate haze sports a 12th-magnitude star on its western edge when seen through my 105mm refractor at 87x. A wide pair of moderately bright stars rests 4.5' west-northwest, the dimmer one orange. Although NGC 1193 is a very rich cluster, its stars appear greatly dimmed by its vast distance of 17,000 light-years. Only a dozen stars can be pried out of the haze with a 10-inch scope at high power in a dark and steady sky. NGC 1193 is also an exceptionally ancient open cluster, generally thought to be 8 billion years old.

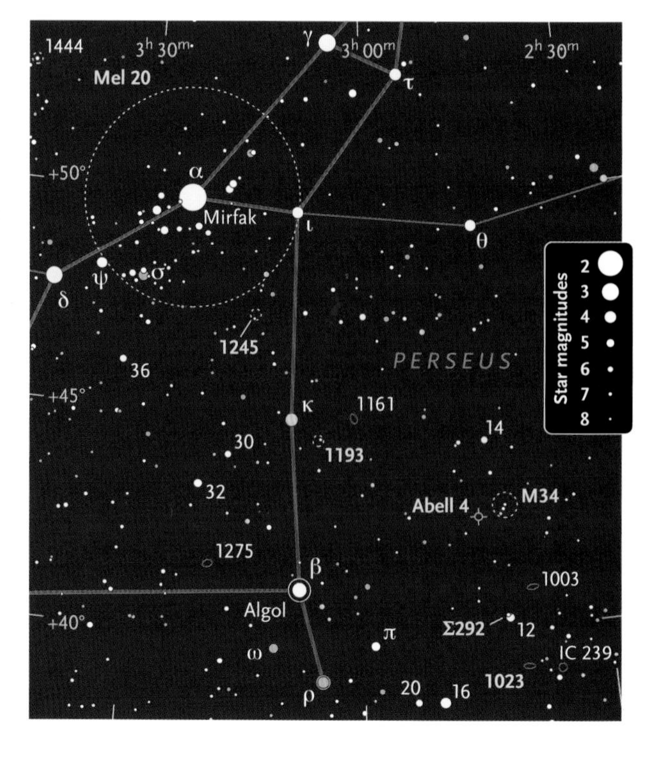

First published January 2008

The Segment of Perseus

The Perseus Milky Way contains an abundance of fascinating star clusters and nebulae.

While various sources agree that the Segment of Perseus is an eye-catching asterism, they often differ on which stars it includes. Following its trail back in time, the earliest sources I uncovered create the pattern from just three stars — Gamma (γ), Alpha (α), and Delta (δ) Persei — forming a segment of a very large circle concave toward Ursa Major.

The French astronomer Joseph-Jérôme Lefrançais de Lalande is the earliest author I found who actually pinned a name on the asterism. He called it the Sash of Perseus in his 1764 book *Astronomie*, but the French appellation morphed into the Arc of Perseus in the mid-1800s. Meanwhile, the Segment of Perseus surfaced in the December 1830 issue of *The Imperial Magazine*, though the citation makes it clear that the term was already well known.

Let's start our tour of deep-sky wonders 1° east of Gamma, the Segment's northernmost star, where we'll find **NGC 1220**. This open cluster is a small, fairly faint, misty spot in my 105mm refractor at a magnification of 47x. Its soft glow hovers 8' north of a 21'-long isosceles triangle that points south-southeast and is fashioned by three 7th- and 8th-magnitude stars. At 87x, the cluster is slightly granular. Two faint points of light are intermittently visible, one of them at the southern edge of the 1½' haze.

Through my 10-inch reflector at 43x, the triangle star closest to NGC 1220 looks orange, and a pair of faint stars lies halfway between it and the cluster. At 115x, the haze appears grainy and elongated north-south. Six stars are visible, including one off the western side. Boosting the magnification to 213x reveals 10 stars in the core group, with one in the north being a close double. Slightly detached, a few additional stars cradle the group's western flank.

Only 60 million years old, NGC 1220 is a youthful cluster. It lies about 5,900 light-years away in the Perseus Arm of the

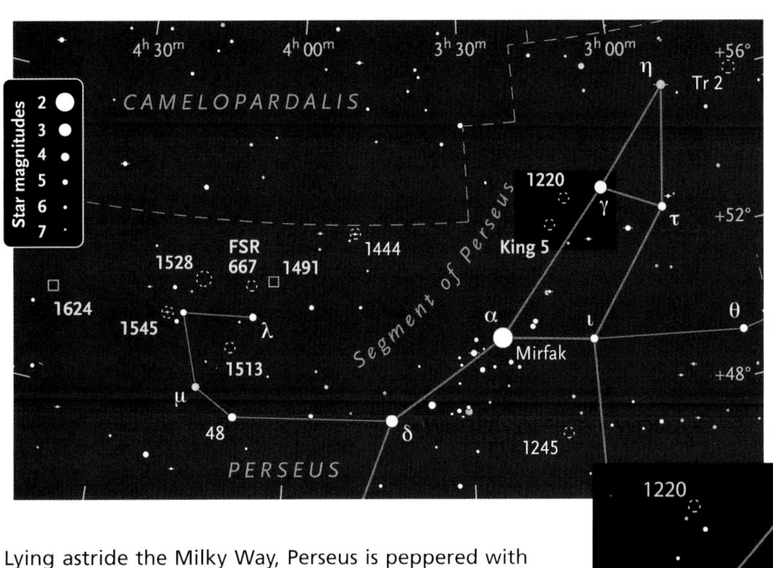

Lying astride the Milky Way, Perseus is peppered with nebulae and star clusters. The close-up at right shows stars to magnitude 9.0, as does the detailed chart of the eastern region on page 28.

Swarms of Stars and Clouds of Light

Object	Type	Magnitude	Size	Right Ascension	Declination
NGC 1220	Open cluster	11.8	2'	$3^h\ 11.7^m$	+53° 21'
King 5	Open cluster	—	6'	$3^h\ 14.7^m$	+52° 42'
NGC 1513	Open cluster	8.4	12'	$4^h\ 09.9^m$	+49° 31'
NGC 1491	Emission nebula	8.5	21'	$4^h\ 03.6^m$	+51° 18'
FSR 667	Open cluster	8.9	7'	$4^h\ 07.2^m$	+51° 10'
NGC 1528	Open cluster	6.4	21'	$4^h\ 15.3^m$	+51° 13'
NGC 1545	Open cluster	6.2	18'	$4^h\ 21.0^m$	+50° 15'
NGC 1624	Nebula & cluster	11.8	5'	$4^h\ 40.6^m$	+50° 28'

Angular sizes and separations are from recent catalogs. Visually, an object's size is often smaller than the cataloged value and varies according to the aperture and magnification of the viewing instrument.

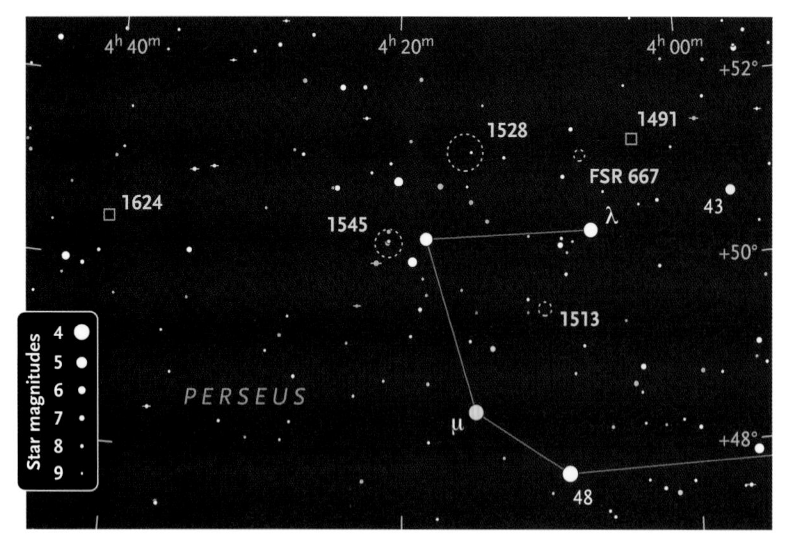

Milky Way Galaxy. That's the next spiral arm outward from the Orion (or Local) Arm, which shelters our Sun on its inner edge.

Nearby **King 5**, on the other hand, is an aged group about 1 billion years old. Look for it 47' southeast of NGC 1220 and 9' west-southwest of an 8.7-magnitude star. This is a ghostly cluster in my little refractor. At 87x, it is barely visible with averted vision as a 3½' gauzy glow with an 11th-magnitude star guarding its northeastern edge. Through my 10-inch scope at 43x, this star forms an equilateral triangle with dimmer stars 4' south-southeast and southwest, the former just beyond the mist and the latter within it. At 115x, I see 15 stars over fleecy haze. The most prominent patch takes refuge in the triangle and holds a relatively bright star in its southern edge. At 213x, King 5 is rich in faint stars. I count 25 to 30 in a 4' x 6' group elongated east-west, but the western region is somewhat underpopulated.

Lightweight clusters hold themselves together for only a few hundred million years. King 5 has survived longer because it contains a respectable 6,000 stars, whose mutual gravity hinders star loss. But King 5 does show signs of age in its distribution of stars. Over the cluster's lifetime, its heftiest surviving members have settled into the core, while low-mass stars have migrated outward. King 5 also resides in the Perseus Arm, about 6,200 light-years distant.

Now swoop down to Delta Persei, the southernmost star in the Segment. The 4th-magnitude star 48 Persei lies 4.3° due east and makes a right triangle with two stars of similar brightness, Mu (μ) and Lambda (λ). The open cluster **NGC 1513** lies a little more than halfway from Mu to Lambda.

In my 105mm refractor at 17x, NGC 1513 is an easily visible, 8' spot of fog with a 9.6-magnitude star at its north-northeastern edge. Four faint and a few very faint stars are resolved at 47x, and 87x teases out 16 stars. A curve of brighter stars wraps around the eastern half of the cluster, about 8' from its edge.

My 10-inch reflector at 115x displays 40 stars, mostly in a shape that looks like a duck to me. The 9.6-magnitude star and a companion north-northwest mark the tip of the duck's tail. A faint star at the cluster's southern edge is the tip of his beak. His body fills the western half of NGC 1513, and an 11th-magnitude star off the group's western side is "beneath" his belly. The duck's head contains most of the bright stars and occupies the cluster's southeast quadrant. Can you see my quacking friend?

NGC 1513 is 4,300 light-years away, on the opposite side of the Orion Arm from our solar system.

A line from Mu through Lambda extended for half again that distance takes you to the brightest patch in the emission nebula **NGC 1491**. In my little refractor at 47x, this is a fairly bright nebula about 4' long, elongated north-south, involving an 11th-magnitude star offset to the east. A narrowband nebula filter improves the view and brings out fainter nebulosity eastward. Using the filter with my 10-inch scope, I see broad swaths of nebulosity that extend north, east, and southeast from the bright area and stretch the nebula to ¼°. It grows still larger through an oxygen III filter and fades gradually into the background.

In 2007, the results of a systematic survey for infrared

Above: NGC 1624 is a classic example of a young star cluster still surrounded by the cloud of glowing gas from which it was born. Photo: Sean Walker / Sheldon Faworski. Right: Mario Weigand's lovely photograph shows the contrast between the delicate stardust of NGC 1528 (upper right) and the brighter, more coarsely arranged stars of NGC 1545 (lower left).

clusters by Froebrich, Scholz, and Raftery were published in the *Monthly Notices of the Royal Astronomical Society*. The survey includes the probable cluster **FSR 667**, which is an interesting visual target. This same group was pointed out to me several years ago by California amateur Dana Patchick, who happened upon it in 1980 with his 8-inch reflector. Patchick calls it the Squiggle Cluster and says, "It is a nice curving chain of about a dozen stars, with the brightest being in the middle at magnitude 10.6 and the rest ranging on down to about magnitude 12.7."

Look for the Squiggle 49' north and a shade east of Lambda. It's a little fuzzy spot in my 105mm scope at 17x, while at 127x, I see 12 stars mainly in a 7'-long sinusoidal curve running north-south. The brightest star shows a close companion. The Squiggle is a catchy little group well worth a visit.

Sweeping 1.3° east from the Squiggle takes us to **NGC 1528**. Through my 105mm scope at 47x, this very pretty, irregular collection of 45 stars spans ⅓°. Two 9th-magnitude stars off the group's northeastern side become glittering eyes for some celestial creature with a star-spangled carapace. My 10-inch scope shows off 75 stars and calls attention to a few reddish gems among them. Our jeweled critter gains a four-starred proboscis between and beyond the eyes.

Dropping 1.3° southeast of NGC 1528 takes us to the open cluster **NGC 1545**, dominated by the colorful triple star South 445 at its center. The widely spaced components form a skinny isosceles triangle pointing west-southwest. Through my little refractor at 68x, the 7th-magnitude primary is orange, the 8th-magnitude secondary north-northwest is yellow, and the 9th-magnitude companion at the triangle's pointy end seems bluish. About 30 stars, mostly faint, run outward from this triple in several branching arms. At the cluster's northern edge, the double star Σ519 sports an orange 8th-magnitude primary with a 9th-magnitude attendant to its north.

The distance to FSR 667 is undetermined, but the last three NGC objects we visited all lie within the Orion Arm of our galaxy, between us and NGC 1513.

Our final target is **NGC 1624**, an emission nebula and open cluster lying 3.1° east of NGC 1545. Through my 105mm scope at 28x, it's an obvious little fuzzlet centered on one faint star. A magnification of 127x unveils five faint stars caught in a filmy net about 4' across. A sixth star is perched on the nebula's west-northwestern rim.

NGC 1624 is by far the most distant object in our sky tour. It's thought to be about 20,000 light-years away, beyond even the Perseus Arm, in the Outer Arm of our galaxy!

The big picture

Eyepieces with wide apparent fields of view (anywhere from 65° to 100°) are especially helpful for viewing large star clusters. They let you use enough magnification to resolve the individual stars while still showing enough of the surrounding sky to make the clusters stand out well as entities in their own right.

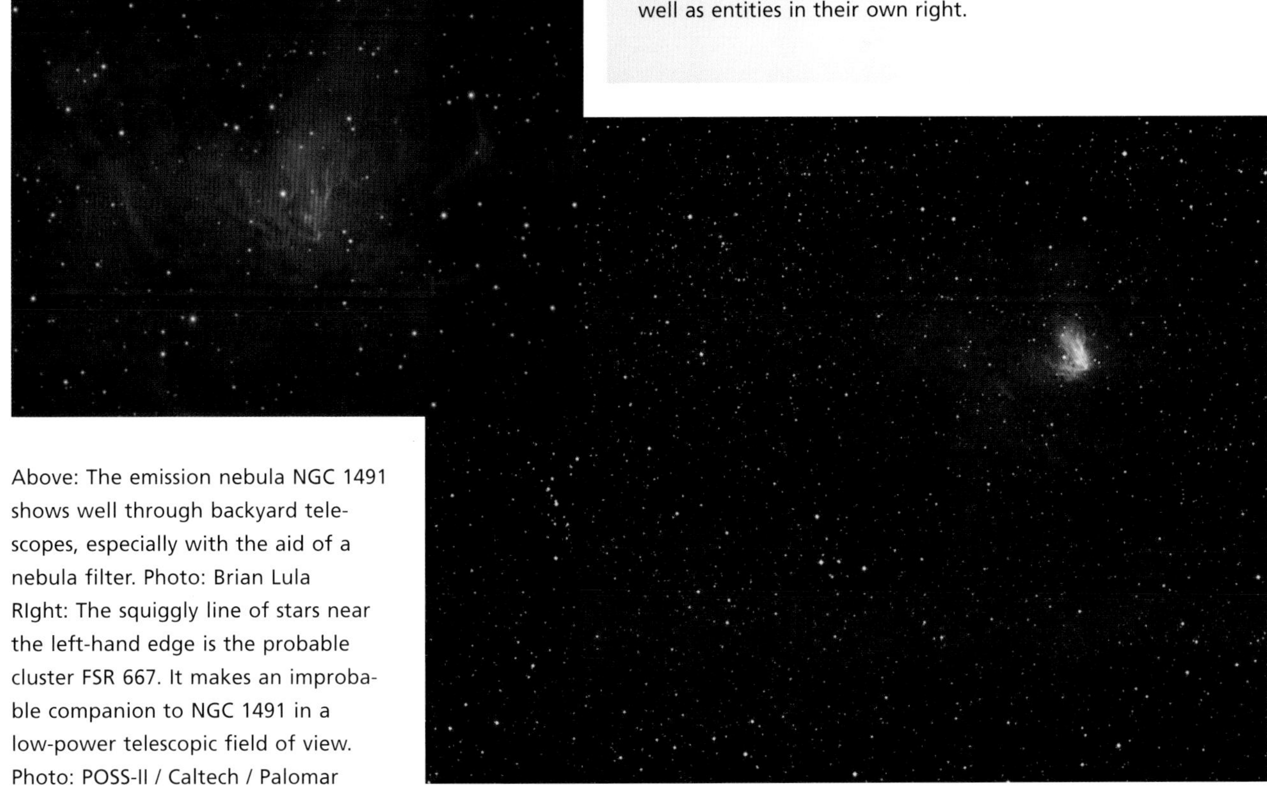

Above: The emission nebula NGC 1491 shows well through backyard telescopes, especially with the aid of a nebula filter. Photo: Brian Lula
RIght: The squiggly line of stars near the left-hand edge is the probable cluster FSR 667. It makes an improbable companion to NGC 1491 in a low-power telescopic field of view. Photo: POSS-II / Caltech / Palomar

First published January 2009

Auriga, the Charioteer

Some of the sky's finest nebulae
and star clusters adorn this constellation.

Thou hast loosened the necks of thine horses, and goaded
their flanks with affright,
To the race of a course that we know not on ways
that are hid from our sight.
As a wind through the darkness the wheels of their chariot
are whirled,
And the light of its passage is night on the face of the world.
— Algernon Charles Swinburne, *Erechtheus*, 1876

According to one myth, Erechtheus (or Erichthonius) was the mortal son of the Greek god Hephaestus. Erechtheus created the first four-horse chariot (*quadriga*) to ride beneath the heavens, which impressed the gods and earned him a place among the stars as the constellation Auriga.

We'll start our tour of Auriga's starry realm with **Melotte 31** (Mel 31). This elongated group of 35 stars spans 2¼° with golden 16 Aurigae at its center. In a suburban sky, Melotte 31 looks like a hazy glow to the unaided eye, but a rural sky may allow you to resolve a few of its stars.

The group is easily visible through a finderscope, binoculars, or a small telescope at low power. The bright stars strung from 16 to 19 Aurigae are particularly eye-catching. *Sky & Telescope* senior editor Alan MacRobert has long referred to this cute asterism as the Leaping Minnow, while California amateur Robert Douglas calls it Auriga's Frying Pan. Despite the striking countenance of Melotte 31, its stars seem to be largely unrelated.

Melotte 31 is named for the British astronomer Philibert Jacques Melotte, who included it as one of the 245 objects listed in his 1915 *A Catalogue of Star Clusters shown on the Franklin-Adams Chart Plates*. Melotte is also well known for his discovery of Pasiphaë, one of Jupiter's moons.

Through my 105mm refractor at 28x, the Minnow shares the field of view with the emission nebula **IC 410** and its embedded open cluster **NGC 1893**. This coarse gathering of suns is framed by a triangle of 9th-magnitude stars and ensnared in the eastern reaches of a gauzy mist. The cluster appears about 12' across while the nebula overspreads at least 19'. Boosting the magnification to 76x, I count forty 9th- to 13th-magnitude stars. The nebula is patchy and irregular, with a dimmer bay in its eastern side and a dark blotch just west of the cluster's center.

NGC 1893 reveals 60 stars, and it doubles in size when seen through my 10-inch reflector at 70x. Many of the bright stars follow a pattern that reminds me of a pair of crossed candy canes, and the brightest star in the northeastern part shines with a golden hue. IC 410 stretches westward to a 9th-magnitude star near the cluster's edge and faintly beyond toward the nice double star **Espin 332**. The pair consists of an 8.9-magnitude primary with a 9.5-magnitude secondary to its southwest.

Suspecting a small brighter spot in the nebula about one-third of the way from the golden star to the 10th-magnitude star at the center of the

Deep photographs show that IC 405, IC 410, and IC 417 (right, center, and far left) are actually the brightest parts of a single nebula. The bright stars of Melotte 31 shine between IC 405 and IC 410. Photo: Robert Gendler

cluster, I zoomed in on the area with higher powers. They gave a much better view, and I could easily see a patch of enhanced brightness. A faint star rests inside, and a dimmer one nuzzles the southern edge. This bright region marks the head of Simeis 130, one of IC 410's cometary nebulae. Nicknamed Tadpoles, they are sites of denser gas and dust being eroded by stellar winds and radiation from the cluster. I didn't notice the head of the other Tadpole, Simeis 129, located 4' northwest, nor did I see the Tadpoles' tails trailing away from the cluster.

About ¾° northwest of the Minnow, we find the eruptive variable AE Aurigae, a blue-white star that fluctuates irregularly between magnitude 5.4 and 6.1. This runaway star was ejected from the Orion star-forming complex approximately 2.5 million years ago. It's thought that a close encounter between two binary systems led to some star swapping that resulted in the eccentric binary Iota (ι) Orionis and two high-speed escapees, AE Aurigae and Mu (μ) Columbae.

AE Aurigae now serves as the chief source of illumination for the emission/reflection nebula **IC 405**, which it chanced upon only in astronomically recent times. The German astronomer Max Wolf noted in 1903 that the nebular material surrounding AE Aurigae "looks like a burning body from which several enormous curved flames seem to break out like gigantic prominences." He thought this "flaming star" worth study, and its nebula thus became known as the Flaming Star Nebula.

With my 105mm refractor at 17x, nebulous haze is fairly obvious near AE Aurigae and the 7.7-magnitude, pale yellow star 8' to its northwest. If you have a hydrogen-beta filter, IC 405 is one of the relatively rare objects you can add to its trophy case — but a narrowband filter can also be of help. In my mirror-imaged view, I faintly see a 1½° J of nebulosity, especially when I scan east-west across it. The J dangles upside down in the sky, but only the bright region in its hook forms the Flaming Star Nebula.

Now let's move eastward to the 5th-magnitude star Phi (φ)

Below right: The nebula IC 410 and its embedded star cluster NGC 1893 are gorgeous in David Jurasevich's hydrogen-alpha image. The author saw the southern Tadpole (Simeis 130) through her 10-inch reflector, but not the fainter northern Tadpole (Simeis 129).

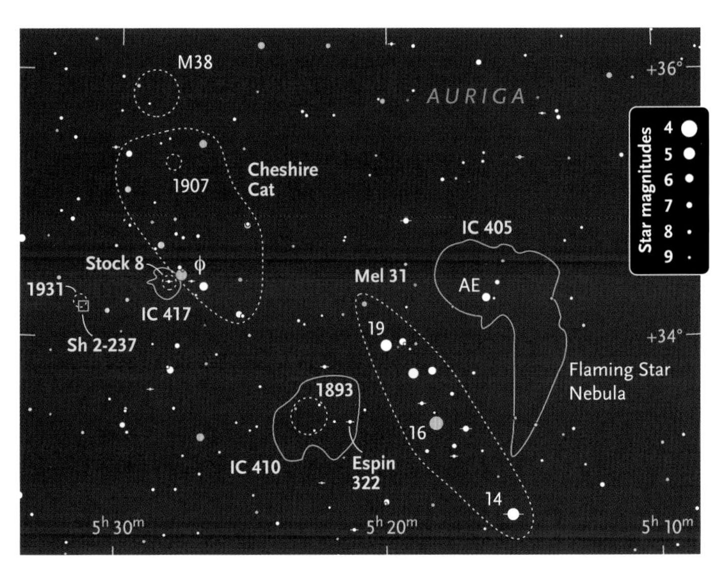

Chariot of Stars, Clouds of Fire

Object	Type	Magnitude	Size/Sep.	Right Ascension	Declination
Mel 31	Asterism	—	135'	$5^h 18.2^m$	+33° 22'
NGC 1893 / IC 410	Cluster / nebula	7.0	40' × 30'	$5^h 22.6^m$	+33° 22'
Espin 332	Double star	8.9, 9.5	14.8"	$5^h 21.4^m$	+33° 23'
IC 405	Bright nebula	—	30' × 20'	$5^h 16.6^m$	+34° 25'
Cheshire Cat	Asterism	3.9	90'	$5^h 27.3^m$	+34° 52'
Stock 8 / IC 417	Cluster / nebula	—	15'	$5^h 28.1^m$	+34° 25'
NGC 1931 / Sh 2-237	Cluster / nebula	—	7'	$5^h 31.4^m$	+34° 15'

Angular sizes and separations are from recent catalogs. Visually, an object's size is often smaller than the cataloged value and varies according to the aperture and magnification of the viewing instrument.

Aurigae. Phi gleams in the smile of a 1.5° asterism that New York amateur Ben Cacace dubbed the **Cheshire Cat**. There are six stars in the wide grin, tipped north-northeast, and two eye stars to their west. Phi and the northern eye both glow yellow-orange in my 105mm scope at 17x. Since the vanishing cat's dimmest star is magnitude 6.9, the asterism is an easy target for most binoculars.

The opulent open cluster Messier 38 decorates the northern corner of the Cheshire Cat's mouth. Off the lip of the Cheshire Cat, right next to Phi Aurigae, the open cluster **Stock 8** is wrapped in the nebulous cloak of **IC 417**. Through my 105mm refractor at 47x, they appear as a hazy patch with several faint stars, the brightest one shining at 9th magnitude near the center. This star becomes a double (Σ707) at 76x, with the 11th-magnitude companion 18" southeast of the primary. The sparse cluster shows 11 stars and is elongated north-south about 6½', while the nebula is a little longer and extends farther east of the cluster than west.

Sharpless 2-237 is so bright that it almost hides NGC 1931, its embedded star cluster. Photo: Al and Andy Ferayorni / Adam Block / NOAO / AURA / NSF

As with IC 405 and 410, IC 417 (right of center) and Sharpless 2-237 are the two brightest parts of a single huge nebula. Photo: POSS-II / Caltech / Palomar

Stock 8 looks much richer when viewed through my 10-inch reflector at 118x. I see 35 to 40 moderately bright stars loosely strewn across 11' of sky. IC 417 engulfs the more crowded regions of the cluster and covers about 8'.

Viewed through my little refractor at 47x, IC 417 shares the field with the smaller but more obvious nebula **Sharpless 2-237** (Sh 2-237). The 11th-magnitude star nestled in its heart is almost overpowered by the glow of the nebula, but it shows up much better when I increase the power to 87x. It sits at the northwest corner of a 3½' box that it forms with three dimmer stars. The nebula is very bright close to its star and fades sharply outward to a diameter of perhaps 3½'.

NGC 1931, the cluster associated with Sharpless 2-237, begins to emerge in my 10-inch scope at high power. The bright star is shown to be a quadruple with the three brightest members arranged in a tiny triangle and the fourth component to their northeast. Several additional stars straggle south through west-southwest of the group.

The three clusters highlighted here are among the youngest in the sky. Their eldest members are a mere 4 million years old, and starbirth is still ongoing.

First published January 2010

Touring Orion's Sword

En garde! The smallest of telescopes can survey Orion's sword from end to end.

The wonderfully bright constellation Orion, the Hunter, strides high across the southern sky in the depths of winter. The February all-sky star map on page 303 shows Orion above the "Facing South" label and standing athwart the celestial equator. Betelgeuse and Bellatrix mark Orion's shoulders, while Rigel and Saiph mark his legs or feet. In the middle is Orion's belt, the eye-catching diagonal row of three stars.

From rural or suburban locations, you can see what seems to be a line of three or four faint stars dangling below Orion's belt. This is Orion's sword. It is encrusted with a fascinating complex of nebulae and star clusters waiting to be explored with a small telescope.

M42. The central star of Orion's sword may seem unusual even to the unaided eye. If your night is very dark and your vision sharp, a close look reveals that this star appears slightly fuzzy. A glance through binoculars or a good finderscope confirms the suspicion. A look through a telescope reveals the famous Great Orion Nebula, a vast stellar nursery of glowing gas and dark dust lit by the young stars within. At first you may see only a dim, fan-shaped glow enveloping a few stars. But examine, study, and be patient. Focus your attention on different aspects, return as a friend to visit it on other nights, and you will learn to see the many faces of this intricate spectacle.

First try a low-power, wide-field view. You'll see that the brightest portion of the nebula surrounds the brightest star within it. Fainter arcs of nebulosity extend northwest and south-southeast to enclose an even fainter region glowing between them. Through a 105mm telescope well away from light pollution, I can see these curving arms meet to form a closed loop about ⅔° wide that sideswipes bright Iota (ι) Orionis and its associated stars to the south. This is definitely a sight worth investing some observing practice to achieve!

Now try a medium magnification of about 40x to 80x. The brightest area of the nebula takes on a mottled appearance, full of swirling detail that can appear more three-dimensional than in any photograph. Many people can see the dis-

Orion's sword from top to bottom. This field is 2° tall, with north up and east left. The bright heart of M42, the Great Orion Nebula, is overexposed; for details within it, see the higher-power view on page 34. Photo: Robert Gendler

Hunting the Hunter

Object	Type	Magnitude	Distance (l-y)	Right Ascension	Declination
M42	Diffuse nebula	3.0	1,300	5^h 35.0^m	–5° 25'
Trapezium	Multiple star	5.1, 6.4, 6.6, 7.5	1,300	5^h 35.3^m	–5° 23'
M43	Diffuse nebula	9.0	1,300	5^h 35.5^m	–5° 17'
NGC 1977/5/3	Cluster + nebula	~4	1,600	5^h 35.3^m	–4° 49'
NGC 1981	Open cluster	4.2	1,300	5^h 35.2^m	–4° 26'
ι Orionis	Triple star	2.9, 7.0, 9.7	1,800	5^h 35.4^m	–5° 55'
Σ747	Double star	4.7, 5.5	1,800	5^h 35.0^m	–6° 00'
Σ745	Optical double	8.4, 8.7	—	5^h 34.8^m	–6° 00'

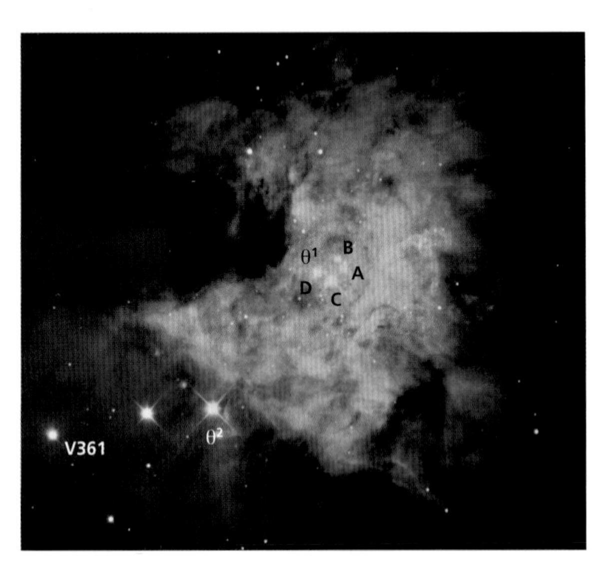

This close-up of the inner Orion Nebula, left, roughly matches the greenish hue seen by many visual observers. The stars of the Trapezium appear nearly lost in the nebula, but they are its brightest feature visually. Use the diagram below to hunt for the fifth and sixth Trapezium stars, E and F, on a night of good seeing. Photo: Lick Observatory

tinct greenish hue of this region, which is quite unlike the pink or red that nebulae usually display in photographs. Our eyes are less sensitive than color film to a nebula's strong, deep red hydrogen emission but respond well to the green emission from a nebula's doubly ionized oxygen.

Notice the dark protrusion of interstellar dust jutting into the bright part of the nebula from the northeast. This dark nebula is known as the Fish's Mouth. At powers of 125x and up, the mottling of the bright area becomes more obvious. It has been likened to a mackerel sky.

The Trapezium. At medium and high powers, the brightest star stands clearly exposed for the multiple system it is. This is Theta[1] (θ^1) Orionis, the famous Trapezium. Even a 60mm scope will reveal its four brightest components arranged in the shape of a tiny trapezoid (though in a telescope so small, one star appears very faint). The four are designated A, B, C, and D, from west to east. C is the brightest; B is the faintest.

With a larger scope at high power, you may be able to see two more members of the Trapezium with some difficulty. The atmospheric seeing must be very steady so that the stars focus down to sharp points. With a 90mm telescope, you may be able to spot star E; it's more or less

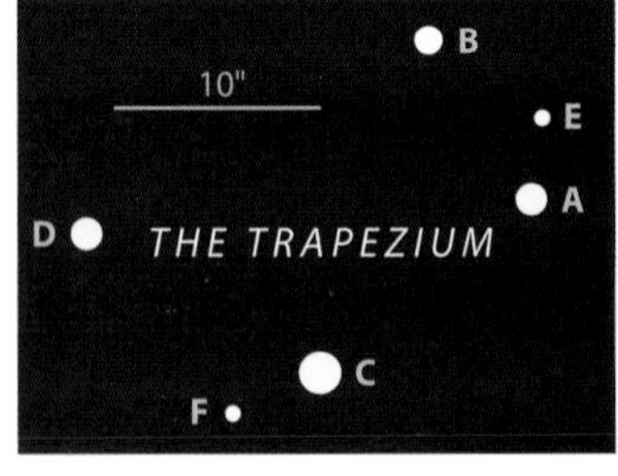

between A and B, as indicated in the diagram at left. Star F is more difficult and may require at least a 4-inch or larger scope, since it is easily lost in the glare of C. On a night of mediocre seeing, no telescope will reveal it at all.

M43. We're not finished with the Orion Nebula yet. A 7th-magnitude star lies just north of the Fish's Mouth. In a dark sky, a 60mm scope will show a small, roundish glow surrounding it, slightly off center. Through my 105mm refractor, the nebula, M43, takes on the "comma" shape familiar from photographs. The tail of the comma, to the northeast, appears fainter than its large dot.

NGC 1977 (with NGC 1975 and NGC 1973). Move just ½° north, and you come to a large, sparse bunch of stars enveloped in very faint nebulosity. My 105mm shows two 5th-magnitude stars lined up nearly east-west and 15 fainter ones scattered around. The two brightest display a slight color contrast. The western star, 42 Orionis, shows the blue-white color common to many of the massive young stars in Orion. The eastern one, 45 Orionis, is yellow-white.

In a dark sky, I can see that the whole group is embedded in a dim, elongated glow showing some internal structure. The brightest part of this reflection nebula curves through the two brightest stars and goes well past them on both sides. A dark patch lies north of these, with some fainter nebulosity extending beyond it.

First published February 2000

NGC 1981. Just north of NGC 1977, we come to another loose, sparse star cluster that's an easy find through binoculars or a small telescope. About a dozen 6th- to 10th-magnitude stars seem to trace the path of a bouncing ball about ⅓° wide from east to west. The three stars forming the eastern bounce are the brightest and are aligned almost north-south, forming the dim north end of Orion's sword. Through a 105mm telescope, about a dozen fainter stars can be counted in the area.

Iota (ι) Orionis. The bottom end of the sword is marked by its brightest star, 3rd-magnitude Iota Orionis, ½° south of the Orion Nebula. Iota is an attractive triple star for small telescopes that appears white, blue, and orange-red to many observers.

The brighter of Iota's two companions is magnitude 7.7 and lies 11" to the southeast; it can be seen at 50x. Both it and the primary are hot, blue-white stars, but when they are seen close together like this, the dimmer star usually appears much bluer by contrast. Iota's second companion is 11th magnitude; you'll probably need a 105mm or larger scope to see it 50" to the east-southeast. It is often perceived as some shade of red.

Struve 747. Just southwest of Iota is the wide double star Struve 747 (Σ747), a pair of blue-white gems 36" apart shining at magnitudes 4.8 and 5.7. Steadily held binoculars will resolve them.

Just west of this pair, by less than half their distance from Iota, is a dimmer but eye-catching double, **Σ745**. It consists of 8th- and 9th-magnitude stars 28" apart. This is probably not a true binary. One star is roughly 1,800 light-years distant, while the other appears to be much closer.

This tour of Orion's sword has spanned less than 2° of sky. If your telescope can yield a magnification as low as 25x in a good eyepiece, you can encompass the entire area in one field of view. It's one of the most magnificent showpieces of the night. Northern observers may shy away from the icy darks of January and February, but the winter sky beckons with some of the most beautiful vistas in the heavens. It is well worth bundling up to spend an evening among them.

Hodierna's Auriga

Three clusters in Auriga have a connection to 17th-century Sicily and an astronomer who was hunting for nebulae.

From midnorthern latitudes, Auriga, the Charioteer, can be found straight overhead on February evenings. Its distinctive pentagon of bright stars, with brilliant Capella — also known as Alpha (α) Aurigae — at one corner, makes this an easy constellation to spot. Auriga is extremely rich in deep-sky treasures, many easily visible through small telescopes.

Notice that one of the stars in the pentagon seems to be shared with the neighboring constellation Taurus, the Bull. Long ago, it was not uncommon for constellations to share a star, but when the International Astronomical Union published the constellation boundaries in 1930, this star was placed within Taurus and is now officially Beta (β) Tauri.

Here, we focus on **M36**, **M37**, and **M38** — three open clusters discovered by the Sicilian Giovanni Hodierna (1597-1660), who was also known as Gioanbatista Odierna. A tract containing his observations was published in 1654 and rediscovered only recently. Hodierna classified these star clusters as *nebulosae* — patches of the sky that appear as small clouds to the naked eye but show themselves to be made of stars close together when seen through a telescope.

We'll start with M38, buried deep in the heart of the pentagon. It is located 7.2° due north of Beta Tauri and

Following in the Footsteps of Galileo

Object	Type	Magnitude	Distance (l-y)	Right Ascension	Declination
M38	Open cluster	6.4	4,300	5^h 28.7^m	+35° 50'
NGC 1907	Open cluster	8.2	4,500	5^h 28.0^m	+35° 19'
M36	Open cluster	6.0	4,000	5^h 36.0^m	+34° 08'
Σ737	Double star	9.1, 9.2	4,000	5^h 36.4^m	+34° 08'
Sei 350	Double star	10.3, 10.3	4,000	5^h 36.2^m	+34° 07'
B226	Dark nebula	N/A	N/A	5^h 37.0^m	+33° 45'
M37	Open cluster	5.6	4,400	5^h 52.5^m	+34° 33'

The open clusters M38 (top) and NGC 1907 (bottom) lie buried in the heart of Auriga. M38 has more than 300 stars, some arranged in chains radiating from the center. NGC 1907 is a compact cluster of more than 100 stars about ½° south-southwest of its companion. Photo: Sean Walker and Sheldon Faworski

makes a nearly equilateral triangle with that star and Iota (ι) Aurigae. M38 is visible through a finder as a small hazy spot. With my 105mm refractor at 68x, I see about 60 stars of 8th magnitude and fainter in an area around 20' across, but the boundaries are ill defined. The cluster has a distinctive shape, with a lone 9.7-magnitude star centered in a 5' hole nearly devoid of stars. Many of the brighter stars radiate from this central dark zone in four arms, making a cross with the longer bar running west-southwest to east-northeast. M38's brightest member, an 8.4-magnitude yellowish star lying at the east-northeast edge of the cluster, marks the foot of the cross. Increasing the magnification to 87x brings out some fainter stars, swelling the star count to about 80.

At 87x, my scope has a 53' true field, which lets me shoehorn M38 into the same view with neighboring **NGC 1907**. This little star cluster lies south-southwest of M38 and is about 6' across. Through my refractor, I see a rich sprinkling of very faint stars over a hazy background, with one slightly brighter star east of center. A matched pair of 10th-magnitude stars adorns the south-southeast edge.

M36 lies a mere 2.3° southeast of M38. It makes a very squat isosceles triangle with Beta Tauri and Theta (θ) Aurigae and is visible through a finder. My refractor at 87x shows 50 moderately bright to faint stars in a loose group about 15' across. Multiple curving arms composed of the brighter stars radiate from the center. You can see a pair of 9th-magnitude stars (**Σ737**) east-southeast of center and a wider pair of 10th-magnitude stars (**Sei 350**) south of center. The stars of M36 look as if they were scooped from a spot just south of the cluster where we find a void of about the same size. This is the dark nebula **B226**, which starts about 10' from the cluster's southern edge.

Our final open cluster, M37, is 3.7° east-southeast of M36 and is the brightest of the bunch. Look for it approximately as far outside Auriga's pentagon as M36 is inside it. M37 is the stunner of Hodierna's three Auriga clusters. My refractor at 87x reveals an extremely rich flurry of faint stars gathered in clumps scattered around a central, orangish 9.1-magnitude star. Dark lanes and patches abound, threaded among the teeming swarms of stars. This beautiful confusion of stars is

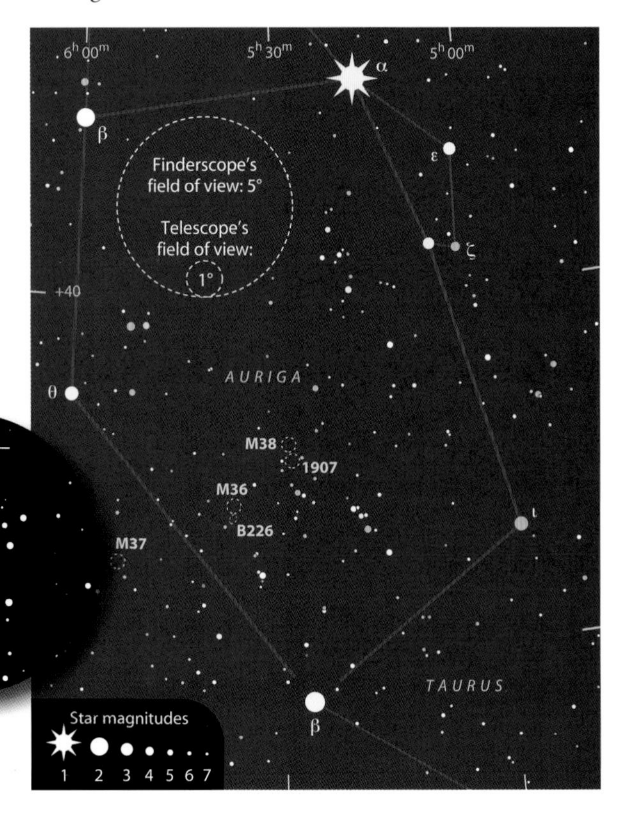

On this chart (and in all photographs), north is up and east is left. The dashed circles show fields of view for a typical finderscope (5°) or a small telescope with a low-power eyepiece (1°). To find north through your eyepiece, nudge your telescope toward Polaris; new sky enters the view from the north edge. (If you're using a right-angle star diagonal, it probably gives a mirror image. Take it out to see an image matching the map.) Detail: Small-scope users looking for an observing challenge can turn their gaze toward the center of M36, where the two close double stars Σ737 and Sei 350 reside.

First published February 2001

Top: Like its companions in Auriga, the open cluster M36 looks somewhat like a hazy star against the backdrop of the Milky Way. Increased magnification opens up its crowded core. Bottom: The most beautiful and richly populated cluster in Auriga is M37. Look for an orange star at its heart, surrounded by clumps of stars and dark lanes. Photos: Preston Scott Justis

one of my favorite deep-sky sights.

Giovanni Hodierna was one of the first astronomers in Sicily to grasp the importance of Galileo's new ideas. His instruments were simple Galilean refractors, the only known example having a magnification of 20x and a limiting magnitude of about 8. He systematically swept the sky, planning to publish a sky atlas with 100 maps on which he would include his nebulous objects. Hodierna never completed the work, and only a few samples of his observations remain, which nevertheless appear to include 46 objects. Some are mere asterisms, and some are not described well enough to identify, but of the "nebulae" he observed, at least 10 seem to be original discoveries. This is quite extraordinary when we consider that during the same youthful era of the telescope, the rest of the astronomical community discovered just one such object, the Orion Nebula.

Hodierna believed that "all the admirable objects that can be seen in the sky" could be resolved into stars if it were not for the limitations of his telescope. He speculated that the apparent brightness of these stars could depend not only on their intrinsic brightness but also on their distances. He entertained the idea that the distribution of stars might seem disordered to our view because they were ordered about some other spot in the universe. As a religious man in an era when putting the Sun at the center of the universe was still dangerous, Hodierna was quite daring to discuss, even hypothetically, a center much farther away.

Due to Hodierna's relative isolation and a general lack of interest in stellar astronomy at the time, his discoveries and ideas have remained nearly unknown. Let's remember them as we gaze at the clusters of Auriga and appreciate the fact that he would have welcomed the humblest of our telescopes as marvelous instruments.

The Giant's Shield

This string of stars in Orion invites a closer look.

Now near the Twins, behold Orion rise;
His arms extended measure half the skies:
His stride no less. Onward with steady pace
He treads the boundless realms of starry space.

— Marcus Manilius, *Astronomica*

Unlike the original Latin text written in the 1st century AD, which merely has Orion's arms stretched across a great expanse of sky, this charming 17th-century translation by Thomas Creech takes poetic license with the Hunter's fabled size. Nevertheless, we can't help being impressed by the majesty of this imposing constellation.

Our eyes are irresistibly drawn to the bright stars of Orion's torso, yet his outline includes many stars of lesser light. Let's turn, then, to his shield, which is also depicted as a lion skin draped over the Hunter's outstretched arm. On our February all-sky star map on page 303, you'll see it as a curve of 3rd- to 5th-magnitude stars lying along the meridian. Often bypassed in favor of the bright treasures of his sword, Orion's shield offers its own rewards.

We'll begin our dalliance with the deep sky at the open star cluster **NGC 1662**. It shares an 8x50 finder field with 4.6-magnitude Pi^1 (π^1) Orionis and looks like a little nebula holding a few faint stars. At 47x, my 105mm refractor uncovers nine fairly bright stars surrounded by 12 fainter ones within 20'. My 10-inch Newtonian reflector at 70x shows 30 stars, with most of the brightest outlining a heart-

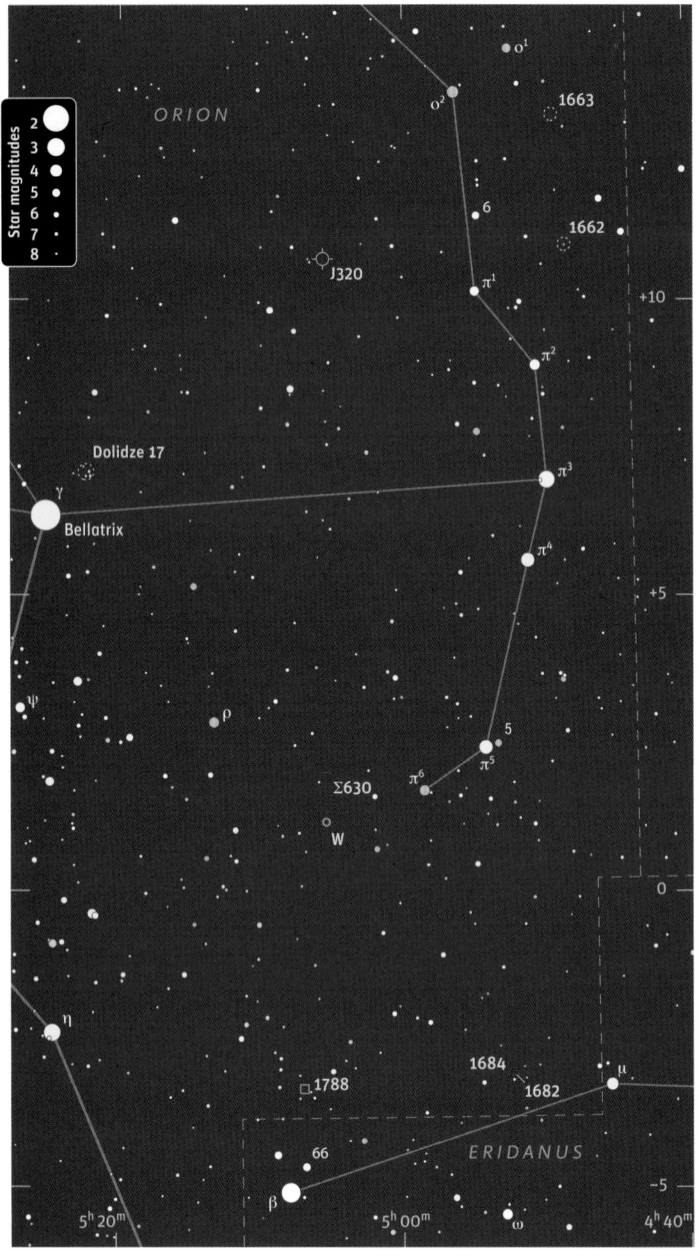

shaped figure whose point lies outside the cluster to the west-southwest. The two lobes of the heart are joined by a loop of four stars that are all part of the multiple star h684. The primary is markedly yellow, and the south-central component shows a fainter, fifth companion 10" to its northwest. The brightest star in the northern lobe of the heart is also notably yellow.

California amateur Russell Sipe sees a quite different shape within NGC 1662. He envisions a Klingon battle cruiser (from the television series "Star Trek"), with the loop forming the forward-jutting body of the spacecraft and the brightest stars of the lobes outlining its bent wings (see illustration, bottom left).

From here, move 2.2° north to **NGC 1663**. It's not a conspicuous cluster in my small refractor. A magnification of 87x shows three 10th- and 11th-magnitude stars curving along its southwestern border. Several very faint stars trace a sinusoidal curve weaving across the northern part of the cluster. The rest of the group is relatively barren and shows only a few stars of 12th magnitude and fainter. In my 10-inch reflector at 44x, a W-shaped asterism reminiscent of Cassiopeia lies southwest of the cluster in the 1.5° field of view. At 118x, I count only 18 stars in this sparse group.

A recent study indicates that NGC 1663 may be an open-cluster remnant. Such remnants are the residue of old clusters that have lost most of their members. If it's a real cluster and not just a chance alignment of stars, NGC 1663 would be about 2 billion years old and 2,000 light-years distant. The three bright stars are likely foreground objects.

Our next target is the planetary nebula **Jonckheere 320** (J320), located 2.7° east-northeast of Pi^1 at the pointy end of a long, skinny isosceles triangle formed with 6 Orionis and Pi^1. A finder will easily encompass these as well as a

"Star Trek" fan Russell Sipe sees the sparse cluster NGC 1662 as the running lights of a Klingon battle cruiser. The image at right shows the cluster in a ⅓°-wide field with north up. Photo, far right: POSS-II / Caltech / Palomar. Right: *Sky & Telescope* illustration; Chart, above: Adapted from *Sky Atlas 2000.0* data

pair of 8th-magnitude stars 15' east of the nebula. Training a low-power eyepiece on the 8th-magnitude pair will show you a slightly dimmer star 12' west of the southern one. J320 makes a squat trapezoid with these three stars.

This planetary is easily visible in my 105mm refractor even at 28x, but it looks like a star. It begins to appear just nonstellar at 87x and shows a very small disk at 127x. The nebula responds very well to an oxygen III (greenish) filter, with which it looks about as bright as the 8.8-magnitude star 4.6' to its south-southeast. In my 10-inch at low power, it is the easternmost and brightest of three "stars" in a 3' arc. The view at 170x reveals a little blue-gray oval running east-southeast to west-northwest. Observers with large telescopes should look for very faint extensions aligned approximately north-south. J320 is a fairly large planetary, almost a light-year across. The nebula's petite apparent size is due to its great distance, about 20,000 light-years.

Now let's make a brief stop near Orion's shoulder, where we'll find the cute asterism **Dolidze 17**. Look 1° northwest of Bellatrix, or Gamma (γ) Orionis, for a group of five stars of about 8th magnitude in a shape that looks a bit like a staple. Just east of the group, another star of similar brightness completes a pattern that reminds me of a piece of Halloween candy corn 16' long. This sweet treat is both large and bright enough to show up in the smallest of telescopes.

Another treat for small scopes is the colorful stretch of sky that reaches from 5 Orionis to the southeast. With a 3.6° field at 17x, my little refractor can encompass the entire region in one view. Starting at this reddish orange star and sweeping southeast, I see bluish white Pi^5 (π^5) golden orange Pi^6 (π^6), the double star **Σ630**, and reddish **W Orionis**. At this low magnification, Σ630 shows the 7th-magnitude blue-white primary holding an 8th-magnitude white secondary close to its northeastern side. W Orionis is a carbon star. Its atmospheric carbon-bearing molecules act like a filter and make this one of the reddest stars in the sky. The star's brightness varies over a cycle of about seven months superposed on a grander period of nearly seven years. The short cycle currently has W Orionis fluctuating between about magnitude 5½ and 7½.

A few degrees south of W Orionis, we come to the bright reflection nebula **NGC 1788**. The simplest way to find it is to start at 3rd-magnitude Beta (β) Eridani. In a finderscope, you can see 5th-magnitude 66 Eridani 30'

Like oncoming headlights on a foggy night, the stars of the cluster NGC 1788 shine out from the nebulosity that engulfs them. This is a 12-minute exposure by George Normandin using the 20-inch reflector and SBIG ST-9E CCD camera of Kopernik Observatory in Vestal, New York. North is up, and the field is 16' square.

Decorations of Orion's Shield

Object	Type	Magnitude	Size/Sep.	Right Ascension	Declination	*MSA*	*U2*
NGC 1662	Open cluster	6.4	20'	4ʰ 48.5ᵐ	+10° 56'	208	97L
NGC 1663	Open cluster	9.5	9'	4ʰ 49.4ᵐ	+13° 08'	208	97L
Jonckheere 320	Planetary nebula	11.9	26" x 14"	5ʰ 05.6ᵐ	+10° 42'	207	97L
Dolidze 17	Asterism	6.2	12'	5ʰ 22.4ᵐ	+7° 07'	230	97L
Σ630	Double star	6.5, 7.7	14"	5ʰ 02.0ᵐ	+1° 37'	255	117L
W Orionis	Carbon star	5½–7½	—	5ʰ 05.4ᵐ	+1° 11'	255	117L
NGC 1788	Reflection nebula	—	5.5' x 3.0'	5ʰ 06.9ᵐ	–3° 20'	255	117L
NGC 1684	Galaxy	11.7	2.4' x 1.6'	4ʰ 52.5ᵐ	–3° 06'	256	117L
NGC 1682	Galaxy	12.6	0.8' x 0.8'	4ʰ 52.3ᵐ	–3° 06'	256	117L

Angular sizes or separations are from recent catalogs. The columns headed *MSA* and *U2* give the chart numbers of objects in the *Millennium Star Atlas* and *Uranometria 2000.0*, 2nd edition, respectively.

First published February 2005

The galaxies NGC 1684 (left of center) and NGC 1682 (3' to its right) show little structure in this plate from the Second Palomar Observatory Sky Survey. A few arcminutes north of NGC 1682, the even fainter galaxy NGC 1683 can also be seen in this image. Photo: POSS-II / Caltech / Palomar

north-northwest. From there, the nebula is 1.3° due north. Beta and NGC 1788 share the field at 22x in my small refractor, and the nebula is easily visible surrounding and spreading southeast of a 10th-magnitude star. Examined at 87x, the brightest part of the nebula is its southeastern section, which grows brighter toward a faint star. With my 10-inch scope at 200x, this pretty nebula appears about 4' long (northwest to southeast) and 2' wide. Another faint star is embedded in the north-central portion of the nebula.

NGC 1788 is known to contain a number of low-mass stars that have not yet settled onto the main sequence as hydrogen-burning stars. Some may even be young brown dwarfs too small to sustain nuclear fusion.

Our final stop is the galaxy pair **NGC 1684** and **NGC 1682**, found 3.6° west of NGC 1788. If you'd like to start at a closer and brighter landmark, scan 1.8° eastward from 4th-magnitude Mu (μ) Eridani. NGC 1684 is the brightest galaxy in the general vicinity of Orion's shield. My 105mm scope can pick it out even before it rises above the light dome of a nearby city. It looks like a faint oval glow, longer in the east-west direction. A power of 87x shows that the galaxy grows brighter toward the center and is about 1½' long. NGC 1682 is faintly visible as a tiny spot west of its larger neighbor, and an 8th-magnitude star lies 4' south of the pair.

Through my 10-inch reflector at 170x, the brighter galaxy shows a large oval core and a small, bright nucleus. Its small, round companion grows brighter toward the middle and harbors a stellar nucleus. The galaxies are separated by about 3' center to center. When you gaze at these distant swarms of stars, you are seeing light that began its journey 200 million years ago — when dinosaurs ruled Earth.

Icy Blue Diamonds

The Pleiades offer something for everyone.

And now the stately-moving Pleiades,
In that soft infinite darkness overhead
Hang jewel-wise upon a silver thread.
— Marjorie Lowry Christie Pickthall,
"Stars," 1925

The **Pleiades** have commanded attention from cultures across the globe, and little wonder. Each time these sparkling beauties make their first foray into the evening sky, my heart is stirred in warm welcome despite the promise of cold weather that they bring. During February evenings, they ride high in the south and invite us to admire their shimmering splendor.

Although the Pleiades are also known as the Seven Sisters, the number of stars visible to the unaided eye varies from person to person and according to sky conditions. I have seen 11 Pleiads from my semirural home, while former *Sky & Telescope* columnist Walter Scott Houston managed to spot 18 in exceptionally dark Arizona skies more than 70 years ago.

Collectively regarded as the Seven Sisters of Greek mythology, their names were not applied to specific stars until Renaissance times. Oddly, the sisters' names were not given to the seven brightest stars, which form a tiny dipper asterism, but, rather, to two faint and five relatively bright stars in the dipper's bowl. The one called Sterope actually consists of two stars a bit too close together to resolve without optical aid. The stars of the dipper's handle are Atlas and Pleione, the Pleiads' father and mother. Pleione huddles close to her brighter husband and is easily lost in his glare when the sky is poor or the observer is hampered by nearsightedness.

The Pleiades are stunning through binoculars, which greatly increase the number of visible stars. In the incomparable *Burnham's Celestial Handbook*, Robert Burnham writes, "In a dark sky the 8 or 9 bright members glitter like an array of icy blue diamonds on black velvet; the frosty impression is increased by the nebulous haze which swirls about the stars and reflects their gleaming radiance like pale moonlight on a field of snow crystals."

This nebulous haze is most apparent encircling Merope and spreading southward from it, where it's designated NGC 1435 and commonly called the **Merope Nebula**. A smaller patch known as the **Maia Nebula** (NGC 1432) surrounds its namesake star. Take care that you don't fall victim to false nebulae, since the fog of your breath on your eyepieces in chill winter air will cause nebulae to blossom around all the stars. A quick check of Atlas and Pleione will help keep you honest, since the small amount of real haze near this pair is unlikely to be seen visually. Surprisingly, the nebulous cloud and cluster are merely undergoing a chance encounter as they travel in different directions through space.

The Pleiades harbor one of my favorite and most challenging targets, Barnard's Merope Nebula, or **IC 349**, which should not be confused with the large cloud of well-known nebulosity called the Merope Nebula. I first became interested in Barnard's Merope Nebula in 1999 when I read that no amateur sightings of it were known. IC 349 is a small, brighter patch within the Merope Nebula discovered in 1890 by Edward Emerson Barnard. It's only 30" in diameter and lies 36" south-southeast of Merope. Since Barnard was able to visually spot the nebula with the 12-inch refractor at Lick Observatory, near San Jose, California, I thought I might stand a chance with my 14.5-inch reflector.

I made an occulting bar for my 10.4mm eyepiece (212x) to block the glare of Merope. The bar was simply a thin strip of aluminum foil placed across the diameter of the field-stop ring and affixed with rubber cement. I rotated the tube of my telescope so that IC 349 would fall between two of the four diffraction spikes caused by the spider holding the scope's secondary mirror. I left the scope's drive off and turned the eyepiece until the occulting bar was almost east-west in the field and angled such that Merope would gradually slide deeper behind the bar as the star drifted across the field of view. On my first night, I thought I could see IC 349. On the second night, I could frequently spot it, and

Perhaps the most celebrated deep-sky object, the Pleiades can be easily seen with the unaided eye even in less than ideal skies. The surrounding nebulosity, however, is more finicky — its visibility is highly dependent on sky conditions. This view shows a field 1.7° wide with north up. Photo: Robert Gendler

February Fascinations

Object	Type	Magnitude	Size/Sep.	Right Ascension	Declination	*MSA*	*U2*
Pleiades	Open cluster	1.5	2°	3h 47.5m	+24° 06'	163	78R
Merope Nebula	Reflection nebula	—	30'	3h 46.1m	+23° 47'	163	78R
Maia Nebula	Reflection nebula	—	26'	3h 45.8m	+24° 22'	163	78R
IC 349	Reflection nebula	—	30"	3h 46.3m	+23° 56'	163	78R
ΟΣΣ 38	Double star	6.8, 6.9	134"	3h 44.6m	+27° 54'	140	78R
Dolidze 14	Asterism	5.0	12'	4h 06.8m	+27° 34'	139	78L
PGC 1811119	Galaxy	~16	28" × 16"	4h 06.65m	+27° 32.4'	139	78L
NGC 1514	Planetary nebula	10.9	2.0' × 2.3'	4h 09.3m	+30° 47'	139	60L

Angular sizes or separations are from recent catalogs. The visual impression of an object's size is often smaller than the cataloged value and varies according to the aperture and magnification of the viewing instrument. The columns headed *MSA* and *U2* give the chart numbers of objects in the *Millennium Star Atlas* and *Uranometria 2000.0*, 2nd edition, respectively.

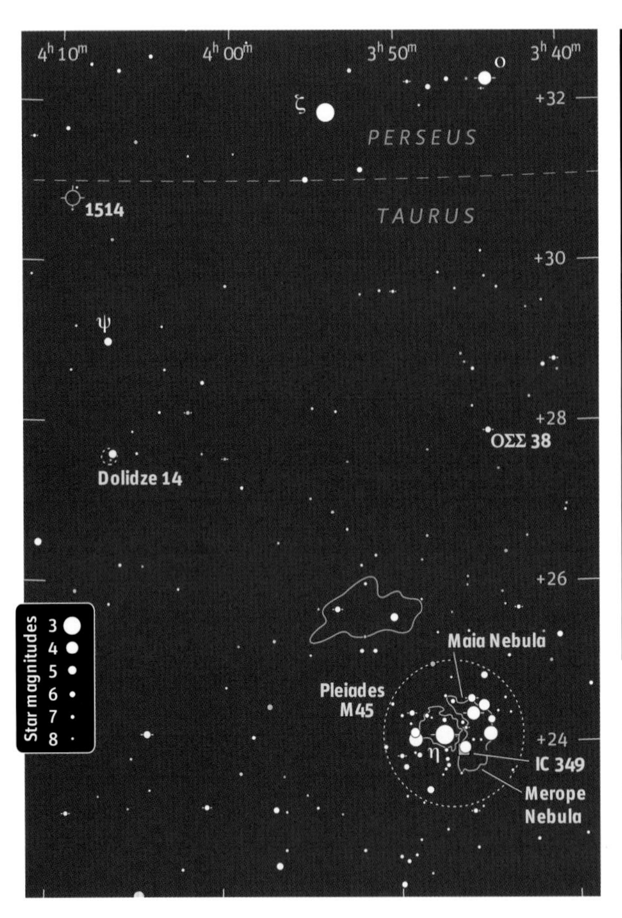

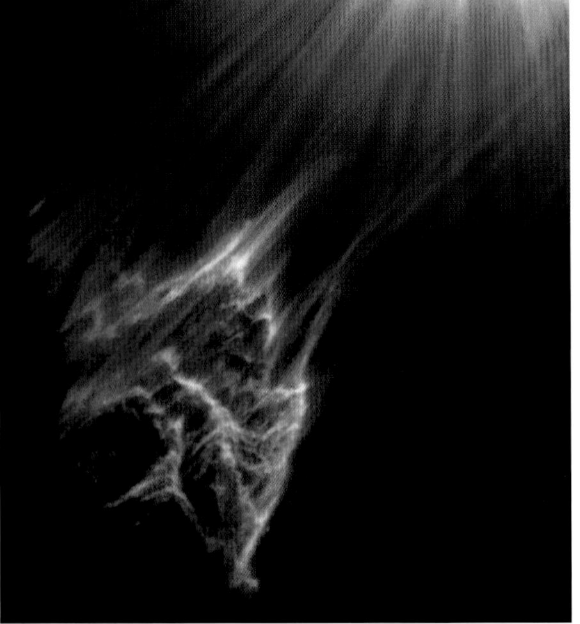

Above: Discovered in 1890, IC 349 is a tiny patch within the Merope Nebula located only 36 arcseconds south-southeast of Merope (just out of the frame at upper right). It is often lost in the star's overexposed image in photographs of the Pleiades such as the one on page 41. Photo: NASA / Hubble Heritage Team / STScI / AURA

on the third, I could see a small patch brighter than the surrounding nebulosity on every pass. Fearing that I might be seeing the effects of stray light, I examined the area between the opposite diffraction spikes (those north of Merope). I saw nothing there during several transits of Merope as it gradually moved out from behind the occulting bar.

After these observations, I sent out a special request to other experienced observers. Positive sightings were reported by Jay Reynolds Freeman of California with a 14-inch scope; Jay LeBlanc of Arizona with a 17.5-inch; and the late Michael Kerr of Australia with a 25-inch.

Other interesting targets share this region of Taurus with the Pleiades. The double star ΟΣΣ **38** lies just 3.4° north of Taygeta and is the brightest star in the area. Its widely spaced 7th-magnitude suns can be split even in a finderscope. Through my 105mm refractor at 17x, the northeastern component looks yellow and the southwestern one appears white.

Putting ΟΣΣ 38 in the northern part of a low-power field and sweeping 4.9° eastward takes you to the unusual star

First published February 2006

group **Dolidze 14**. It's dominated by 41 Tauri, a blue-white 5th-magnitude star that makes Dolidze 14 easy to recognize. My little refractor at 17x also shows an east-west line of three fairly bright stars — yellow, golden, and yellow-white. At 87x, a dozen faint stars join the group for an overall size of 12'. Arizona astronomer Brian Skiff has determined that Dolidze 14 is an asterism of mostly unrelated stars.

Deep photographs show an extremely faint background galaxy within the apparent borders of Dolidze 14. **PGC 1811119** sits halfway between and a little north of the two westernmost stars in the east-west line. Can you spot this galaxy visually?

From Dolidze 14, move 1.4° north to 5th-magnitude Psi (ψ) Tauri, then place Psi at the western edge of a low-power field and scan another 1.8° north. Here you'll find a nearly north-south pair of 8th-magnitude stars with the planetary nebula **NGC 1514** in between. With my 105mm scope at 17x, the planetary is distinguishable as a tiny, round, faint halo engulfing a 9th-magnitude star. At 87x, the nebula appears brighter around the perimeter and about 2' in diameter. NGC 1514 is a pretty sight in my 10-inch reflector at 139x. The nebula appears slightly oval, and its rim is brighter in two large arcs, while the interior of the nebula seems slightly mottled. It's easy to see why some observers have nicknamed this delicate planetary the Crystal Ball Nebula.

Hazy Gleams and Starry Streams

Perseus holds a wealth of deep-sky delights for February nights.

Regions of lucid matter taking forms,
Brushes of fire, hazy gleams,
Clusters and beds of worlds, and bee-like swarms
Of suns, and starry streams.
— Alfred, Lord Tennyson, 1833

Southern Perseus is an area lush with the hazy gleams of diffuse nebulae. The most renowned is **NGC 1499**, called the California Nebula because it resembles that state's shape. Edward Emerson Barnard discovered NGC 1499 visually with the 6-inch Cooke refractor at Tennessee's Vanderbilt Observatory in 1885. Barnard saw only the brightest part of the nebula, but photographs later revealed its larger extent. Covering a substantial swath of sky, the nebula is best viewed in wide-field instruments. Former *Sky & Telescope* columnist Walter Scott Houston glimpsed it from Connecticut in the early 1980s with 6x30 binoculars. He and Arizona astronomer Brian Skiff were among the first observers to see the California Nebula by simply looking through a nebula filter with the unaided eye.

The California Nebula is quite easy to locate, as it sprawls just north of the 4th-magnitude star Xi (ξ) Persei, or Menkib. My 105mm refractor at 17x includes both the nebula and Xi Persei in its 3.6° field. While a narrowband filter improves the view, NGC 1499 responds much better to a hydrogen-beta filter, which renders it an easy target for my little scope. The nebula stretches east-southeast to west-northwest for 2½° and is one-third as wide. It has distinct northern and southern borders and is much dimmer through the center.

Don't let these accounts lull you into thinking that the California Nebula is a snap to observe. Unless your sky is very dark, the nebula is likely to remain completely invisible without a filter. If your field of view isn't wide enough to encompass the whole thing and some sky around it, you may need to scan north-south across the nebula to detect its edges. NGC 1499 has a very low surface brightness. Even with a filter, a wide field, and a reasonably dark sky, this can be a difficult object for novice observers. But don't give up! Visit the nebula from time to time, and slowly scan the field. Remember that a faint object in a moving field of view is often easier to spot than it is if the field is stationary.

A lovely nebulous complex is found 4.7° east of the California Nebula's eastern tip. Sometimes called the Northern Trifid, for its resemblance to the Trifid Nebula in Sagittarius, **NGC 1579** is easily spotted in my 10-inch reflector at 44x. At 118x, it spans about 6' and appears quite irregular in shape and intensity. The nebula has a bright heart and several stars rooted in its fringes. NGC 1579 glows within the surrounding dark nebula LDN 1482. The main source of its illumination is a young, high-luminosity star embedded in a small region of ionized hydrogen, both of which are heavily obscured by a dusty shell. Infrared studies have detected about 35 fainter stars within the nebula. The group is thought to be about 2,000 light-years away and less than 1 million years old.

Moving 4.8° west of the western tip of the California Nebula, we come to the intriguing star cluster **NGC 1342**. Its streaming chains of stars seem to evoke many engaging

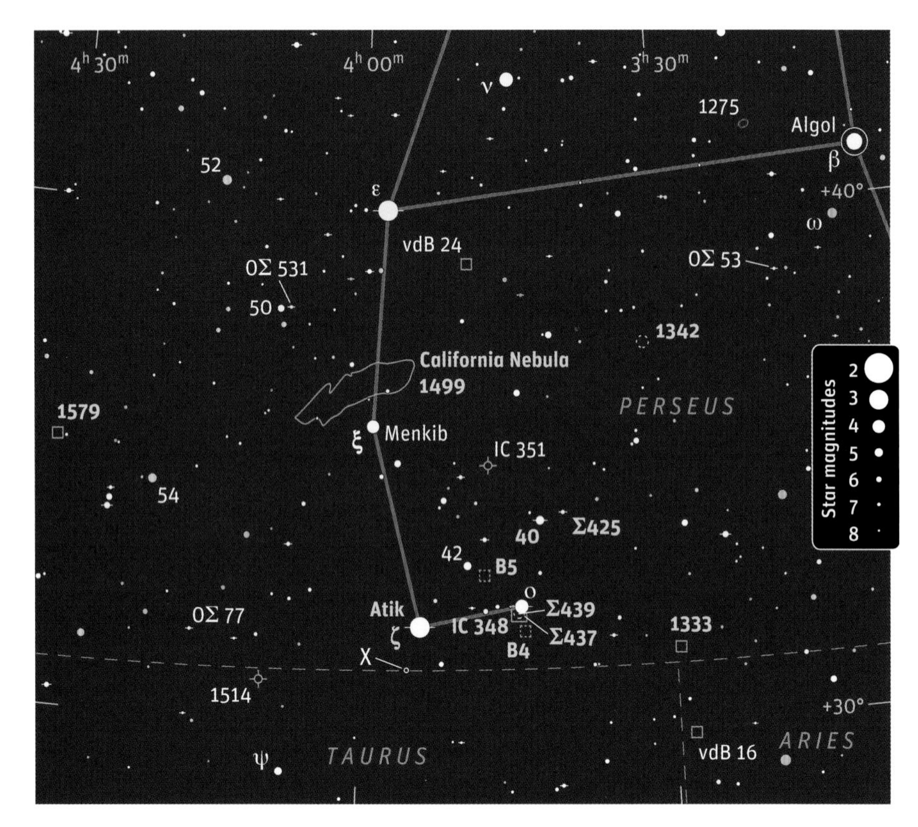

images. Canadian amateur Steve Irvine sees a miniature constellation Gemini through his 4.9-inch Maksutov telescope. In my 105mm refractor at 47x, I see 30 stars arranged in a figure that reminds me of Kokopelli, the humpbacked flute player of Anasazi lore. His head lies to the west and his body to the east, and his flute juts northward. My 10-inch reflector draws out additional stars that turn Kokopelli's hair and body into the flowing veil and folded skirt of a kneeling nun. The northern stars now outline hands upraised in prayer. What figure does your imagination create?

Now jump down to the pretty quintuple star **Zeta (ζ) Persei**, also called Atik. Spaciously separated from the blue-white primary, a nearly matched pair of 10th-magnitude stars lies a little west of south. A much closer 9th-magnitude star sits south-southwest of the primary, and an 11th-magnitude companion is a bit north of west at an intermediate distance. I observed Zeta at about 70x with both my 105mm and 10-inch scopes on the same night, but I couldn't see the dimmest star with the 105mm. Only the closest companion and the brighter member of the 10th-magnitude pair are physically bound to the primary. The others merely lie along the same line of sight.

Zeta Persei is the brightest member of Perseus OB 2, which at 1,000 light-years is one of the nearest stellar associations dominated by high-luminosity stars of spectral types O and B. Zeta is a blue supergiant that has exhausted the supply of hydrogen fuel in its core and is evolving away from the main sequence of stable stars. The brightest main-sequence star of Perseus OB 2 is **40 Persei**, which also boasts a companion star. The bright, blue-white primary and its 10th-magnitude companion appear widely separated at 47x. While in the area, sweep ½° west-northwest of 40 Persei to **Σ425**. Its yellowish 7½-mag-

A Portfolio of Deep-Sky Wonders in Southern Perseus

Object	Type	Magnitude	Size/Sep.	Right Ascension	Declination	*MSA*	*U2*
NGC 1499	**Emission nebula**	**6.0**	**160' × 40'**	**4h 00.5m**	**+36° 33'**	**117**	**60**
NGC 1579	**Nebulous complex**	**—**	**12' × 8'**	**4h 30.1m**	**+35° 17'**	**116**	**60L**
NGC 1342	**Open cluster**	**6.7**	**17'**	**3h 31.6m**	**+37° 23'**	**119**	**60R**
ζ Persei	**Multiple star**	**2.9, 9.2, 10.0, 10.4, 11.2**	**13", 120", 98", 33"**	**3h 54.1m**	**+31° 53'**	**140**	**60R**
40 Persei	**Double star**	**5.0, 10.0**	**26"**	**3h 42.4m**	**+33° 58'**	**118**	**60R**
Σ425	**Double star**	**7.5, 7.6**	**2.0"**	**3h 40.1m**	**+34° 07'**	**118**	**60R**
B4	**Dark nebula**	**—**	**46' × 29'**	**3h 44.0m**	**+31° 48'**	**140**	**60R**
B5	**Dark nebula**	**—**	**1.5°**	**3h 47.9m**	**+32° 54'**	**140**	**60R**
IC 348	**Cluster & nebula**	**7.3**	**8'**	**3h 44.6m**	**+32° 10'**	**140**	**60R**
Σ439	**Triple star**	**9.3, 9.5, 10.3**	**0.5", 24"**	**3h 44.6m**	**+32° 10'**	**140**	**60R**
Σ437	**Double star**	**9.8, 10.0**	**11"**	**3h 44.1m**	**+32° 07'**	**140**	**60R**
NGC 1333	**Reflection nebula**	**5.7**	**6' × 3'**	**3h 29.3m**	**+31° 24'**	**141**	**60R**

Angular sizes and separations are from recent catalogs. The visual impression of an object's size is often smaller than the cataloged value and varies according to the aperture and magnification of the viewing instrument. The columns headed *MSA* and *U2* give the appropriate chart numbers in the *Millennium Star Atlas* and *Uranometria 2000.0*, 2nd edition, respectively. All the objects this month are in the area of sky covered by Chart 13 in *Sky & Telescope's Pocket Sky Atlas.*

nitude components are much closer than the tightest pair of the Zeta Persei system, but I can separate them at 87x with my little refractor.

The three primary stars mentioned above are all found on Plate 3 (Perseus and Taurus region) of Barnard's 1927 *Photographic Atlas of Selected Regions of the Milky Way* (www.library.gatech.edu/barnard). But Barnard's focus for the 6-hour 41-minute exposure was an amazing "region of obscure and obscuring nebulosity" near Omicron (ο) Persei. The portion of the plate within the borders of Perseus contains the first five entries in Barnard's catalog of dark nebulae, B1 through B5. With my 105mm scope at 28x, I find **B4** and **B5** to be the most obvious members of the crew. B5 is centered 1° northeast of Omicron. Its 45' x 15' oval leans northeast against the 5th-magnitude star 42 Persei. B4 covers more than a degree of sky just south of Omicron and has a large indentation in its western side.

By boosting the magnification to 153x, I can see the little gas-choked star cluster **IC 348** on the northern edge of B4. It shows 10 stars mostly gathered into a 5' oval ring. If I keep Omicron out of the field, I can see faint nebulosity enveloping the brightest star and the two closest to it, with a third star on the nebula's western edge. The brightest star and the one to its northeast form the triple **Σ439** AB-C. Only 0.5" apart, the A and B components are too snug for my little scope to split. **Σ437**, a matched set of 10th-magnitude stars, sits a short distance west-southwest of IC 348 in the same field of view. My 10-inch scope adds a wisp of nebulosity around the cluster's southernmost star. Gauzy mist enfolds both patches and connects their eastern sides.

About 1½° west of B4, you'll come to a 37' isosceles triangle of 7th- and 8th-magnitude stars. It points west-southwest to another reflection nebula, a little over one triangle length farther away. This is **NGC 1333**, a hazy oval with a 10½-magnitude star embedded in its northeastern end. It covers about 5' x 2½' with its southwestern end reaching out to a 12th-magnitude star. The nebula is patchy in my 10-inch scope at 115x and appears brightest about one-quarter of the way from the brighter star to the dimmer one. Many of the nebulae in this area belong to the Perseus OB 2 molecular cloud. Despite the name, its relationship to the Perseus OB 2 association is unknown because of its highly uncertain distance.

Known to some as the Northern Trifid, after its resemblance to the familiar Trifid Nebula in Sagittarius, NGC 1579 in Perseus is an easy target for modest backyard telescopes. Located in a relatively star-poor region, it is often overlooked. The field here is 17' wide with north up. Photo: R. Jay GaBany

First published February 2007

Starry Nights

Located along the edge of the Milky Way, Canis Major is home to a variety of starry delights for deep-sky observers.

I heard the trailing garments of the Night
Sweep through her marble halls!
I saw her sable skirts all fringed with light
From the celestial walls!

— Henry Wadsworth Longfellow,
"Hymn to the Night"

These words were penned in 1839 by American poet Henry Wadsworth Longfellow as he tried to capture his impression of the scene through his chamber window. Although he was viewing summer's Night, her winter skirts are fringed with even brighter lights. Chief among them is Alpha (α) Canis Majoris, or **Sirius**, the most dazzling star in our night sky.

Sirius holds a white-dwarf companion on the doorstep of its brilliant glare. Since Sirius has long been called the Dog Star, its little escort, a white-dwarf star discovered in 1862, has become known as the Pup. Their separation has been increasing since 1993 and stood at about 8" in 2008, with the Pup following its big brother across the sky.

The problem is that Sirius outshines the white dwarf by some 10,000 times. I've disentangled the light of the

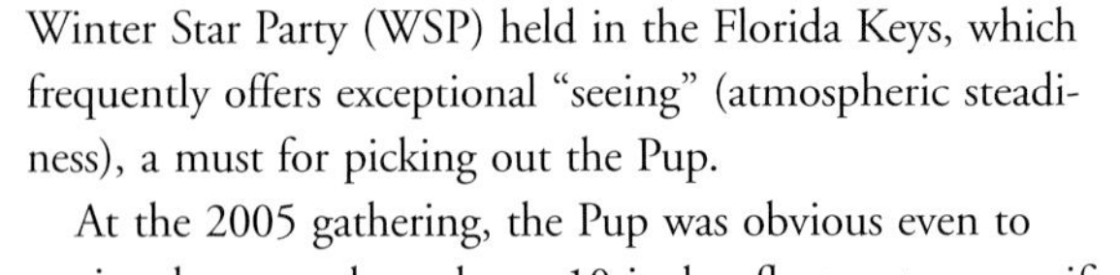

8.5-magnitude Pup from its –1.5-magnitude primary at the Winter Star Party (WSP) held in the Florida Keys, which frequently offers exceptional "seeing" (atmospheric steadiness), a must for picking out the Pup.

At the 2005 gathering, the Pup was obvious even to novice skygazers through my 10-inch reflector at a magnification of 320x. Although the pair was wider in 2006, the Pup wasn't quite as conspicuous because softer seeing slightly blurred the star images. The Pup will reach its greatest apparent distance from Sirius in 2022 at 11.3". When will you first pry it from the overpowering glow of its neighbor? You don't need a 10-inch scope to try. Observers have reported success with half that aperture.

Canis Major hosts many other deep-sky wonders worthy of notice. If you'd like to try splitting a less demanding double star, consider 5th-magnitude **Mu (μ) Canis Majoris** (μ CMa). It lies a little more than halfway from Gamma (γ) to Theta (θ) Canis Majoris and a little west of an imaginary line connecting them. My 105mm refractor at 127x reveals a beautiful golden primary accompanied by a blue-white, 7th-magnitude attendant 3.2" to its north-northwest. Although this pair is considerably tighter than Sirius and the Pup, its components are less difficult to split because their magnitude difference isn't nearly as great.

Dropping 1.6° due south from Sirius brings us to a deep yellow, 7th-magnitude star with the spiral galaxy **NGC 2283** 11' to its east-northeast. My little refractor at 87x discloses a small smudge with three stars involved. Even in my 10-inch scope, the galaxy simply displays a uniform glow, but it adds one more superposed star. Visually, NGC 2283 could easily be mistaken for a nebula. In fact, it's plotted as one in my old Skalnaté Pleso *Atlas of the Heavens*, and it's included as Cederblad 86 in Sven Cederblad's 1946 catalog of galactic nebulae. The galaxy's spiral structure and several additional foreground stars, not visible through my scope, should make this a lovely target for skilled astrophotographers.

Plunging 2.5° farther south brings us to Canis Major's only Messier object, the open cluster **M41**, visible as a hazy patch to the unaided eye under moderately dark skies. M41 was first cataloged by Giovanni Hodierna (aka Gioanbatista

Home to the brilliant beacon Sirius, the night's brightest star, Canis Major is an easy constellation to find on February evenings. The author describes a variety of telescopic sights located less than the width of an outstretched hand from Sirius. Photo: Akira Fujii

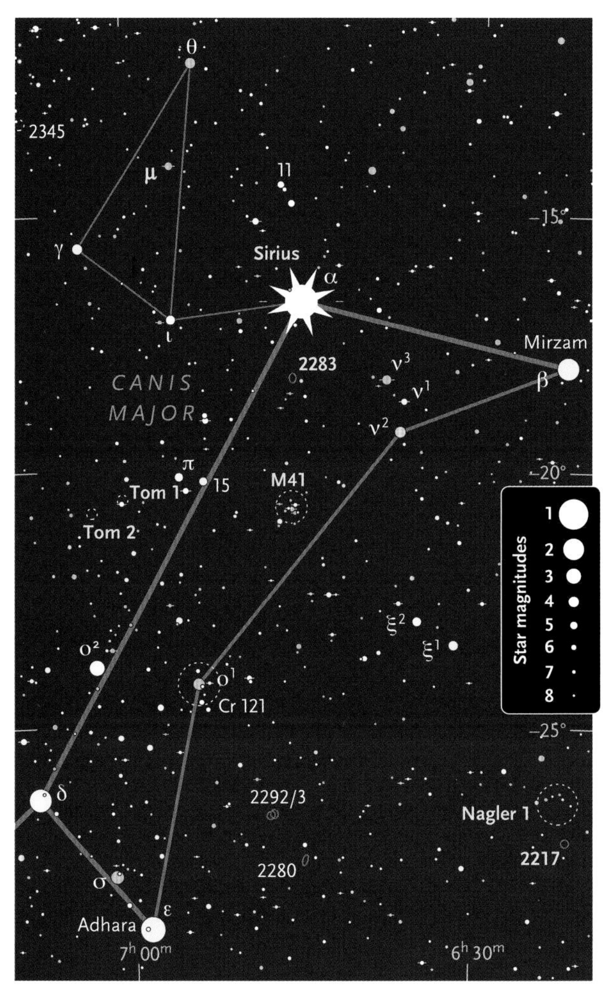

Visible as a hazy glow to the unaided eye under a dark sky, the open star cluster Messier 41 is an ideal target for all sizes of binoculars and telescopes. It's the sole Messier object in Canis Major, despite the constellation's location along the cluster-rich Milky Way. The field is 1.4° wide with north up. Photo: POSS-II / Caltech / Palomar

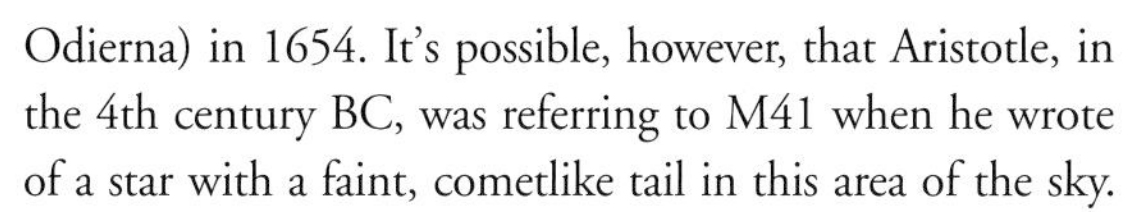
Odierna) in 1654. It's possible, however, that Aristotle, in the 4th century BC, was referring to M41 when he wrote of a star with a faint, cometlike tail in this area of the sky.

My 15x45 image-stabilized binoculars show a group of 30 stars, 7th magnitude and fainter, gathered into a 40' swath of sky. The 6th-magnitude star 12 Canis Majoris sits off the group's southeastern edge. Through my 105mm scope at 47x, M41 boasts 80 mixed bright and faint stars, many strung in jewel-bright chains radiating from the center. I can see several yellow and orange gems sprinkled throughout the cluster.

Now sweep 3.4° eastward to a gold and white pair of 7th-magnitude stars 2½' apart. These point northwest to **Tombaugh 1** (Tom 1), one of the open clusters Clyde Tombaugh discovered while searching for trans-Neptunian planets. I viewed Tombaugh 1 during the 2006 WSP

Starry Delights for February Nights

Object	Type	Magnitude	Size/Sep.	Right Ascension	Declination	*MSA*	*U2*
Sirius	**Double star**	**–1.5, 8.5**	**8.4"**	**6h 45.2m**	**–16° 43'**	**322**	**135R**
μ CMa	**Double star**	**5.3, 7.1**	**3.2"**	**6h 56.1m**	**–14° 03'**	**298**	**135R**
NGC 2283	**Galaxy**	**11.5**	**3.4' × 2.7'**	**6h 45.9m**	**–18° 13'**	**322**	**154L**
M41	**Open cluster**	**4.5**	**39'**	**6h 46.1m**	**–20° 46'**	**322**	**154L**
Tom 1	**Open cluster**	**9.3**	**6'**	**7h 00.5m**	**–20° 34'**	**(322)**	**154L**
Tom 2	**Open cluster**	**10.4**	**3'**	**7h 03.1m**	**–20° 49'**	**(321)**	**154L**
Nagler 1	**Asterism**	**—**	**16' × 48'**	**6h 22.4m**	**–26° 28'**	**347**	**154R**
NGC 2217	**Galaxy**	**10.7**	**4.5' × 4.2'**	**6h 21.7m**	**–27° 14'**	**347**	**154R**

Angular sizes and separations are from recent catalogs. Visually, an object's size is often smaller than the cataloged value and varies according to the aperture and magnification of the viewing instrument. The columns headed *MSA* and *U2* give the chart numbers in the *Millennium Star Atlas* and *Uranometria 2000.0*, 2nd edition, respectively. Chart numbers in parentheses indicate that the object is not plotted. All the objects this month are in the area of sky covered by Chart 27 in *Sky & Telescope's Pocket Sky Atlas*.

While the central core on the spiral galaxy NGC 2217 is visible in small telescopes, the author has been unable to glimpse the galaxy's outer ring, which is prominent in this long-exposure image. The field is ⅓° wide with north up. Photo: POSS-II / Caltech / Palomar

through Betsy Whitlock's 105mm refractor. It was easily visible as a small hazy spot at 17x, while at 28x, a lone star appeared in the center. At 47x, the cluster revealed several diamond-dust stars over a 5' patch of mottled haze bracketed by a pair of 11th-magnitude suns. A dozen stars glimmered at 87x, and only a touch of unresolved haze lingered. At the same magnification with a nearby 3.6-inch refractor belonging to Joe Bergeron of New York, 9 or 10 stars showed faintly. My 10-inch reflector at 118x uncovered approximately 35 stars in a vaguely triangular group.

Just 40' east-southeast, **Tombaugh 2** (Tom 2) was visible in the same low-power field through Whitlock's scope. This cluster is more elusive than its neighbor, so use the 8½-magnitude star 4' to its north to help pinpoint its location. At 87x, the cluster is easier to see but merely shows a 2' misty spot with an 11th-magnitude star off its southeastern edge. In my 10-inch scope at 118x, the haze is prettily dusted with extremely faint chips of light.

Tombaugh 2 may be associated with the Canis Major Dwarf Galaxy, which is being cannibalized by our Milky Way. The Canis Major Dwarf is our closest companion galaxy, about 25,000 light-years from the Sun and 42,000 light-years from our galactic center.

More power, more stars

With a given telescope, you'll see fainter stars when you increase the magnification. For scopes 8 inches or less in aperture, the improvement is dramatic from the lowest magnifications up to about 150x, above which the effect is more subtle. Larger apertures show noticeable improvements in the visibility of faint stars up to about 250x. As with many aspects of visual astronomy, of course, your mileage may vary.

A few years ago, optical designer Al Nagler told me of an interesting asterism that he came across in this part of the sky. You can find it by placing Zeta (ζ) Canis Majoris at the southern edge of a binocular field, which will put **Nagler 1** in the north.

My 15x45 binoculars display Nagler 1's nice 16' x 48' chevron of stars, pointing northward and scattered in magnitude from 7 to 10. The first star down the eastern side from the chevron's point is double. Through a telescope, the three brightest stars are clothed in shades of yellow-orange and red-orange. With a little imagination and a wide enough field of view, you can double the chevron's dimensions by extending its arms through additional stars, though the western side is considerably more prominent.

Dropping 1° south of Nagler 1, you'll see a 35'-long right triangle composed of two yellow-white, 8th-magnitude stars and an exceptionally red 9th-magnitude star. The galaxy **NGC 2217** rests 16' north of the star at the pointy end; it occupies one corner of a nearly rectangular trapezoid formed with the triangle's stars. The galaxy appears small, round, and faint through my 105mm at 87x. It shows some brightening toward the center and harbors a nearly stellar nucleus.

My 10-inch reflector at 171x reveals a 1' oval tipped east-southeast enfolding a round core. A pair of faint stars sits off the western side. These stars are embedded in a faint 4' x 3' ring that leans north-northeast and is completely invisible in my scope. If you manage to spot the diaphanous outer annulus of this distant city of stars, well done!

I felt her presence, by its spell of might,
Stoop o'er me from above;
The calm, majestic presence of the Night,
As of the one I love.

— Henry Wadsworth Longfellow, "Hymn to the Night"

First published February 2008

Copeland's Lost Trigon

How well can you replicate 50-year-old observations with modern telescopes?

The March 1957 issue of *Sky & Telescope* featured an article on Monoceros by Leland S. Copeland. Copeland is well known for the many nicknames he bestowed on deep-sky wonders. Some have become quite popular, and some have passed into obscurity. Copeland's Trigon, perhaps due to confusion over its position, seems to have been lost altogether.

Let's retrace Copeland's tour of Monoceros, trying to see what he saw in these objects and saving for last the Trigon, which surely deserves a revival. All the observations that I'll share along the way were made with my 105mm refractor.

Copeland starts his journey with the appealing triple star **Beta (β) Monocerotis** (β Mon). At a magnification of 47x, I see just two of its components, with the brighter star harboring a slightly fainter companion to the southeast. At 87x, the companion is revealed as a close pair, the dimmer star east-southeast of the brighter one. All three components are blue-white, 5th-magnitude suns.

From Beta, Copeland launches the observer 2.2° north to the open cluster **NGC 2232**, which he calls the Double Wedge. At 47x, the group shows 25 stars mainly fanning southward from 5th-magnitude 10 Monocerotis. Five stars frame 10 Monocerotis: one north, one south, one east-northeast, and a pair west-southwest. To the northwest, a blunt wedge of six stars points toward the fan. The wedge includes the star 9 Monocerotis at its northern end and has about 25 faint stars sprinkled across it. Many sources don't include this second wedge as part of NGC 2232, but in their book *Star Clusters*, Brent Archinal and Steven Hynes argue for a size and position encompassing both.

NGC 2301, Copeland's Golden Worm, is next on the circuit. At 47x, this cluster displays 15 moderately faint to faint stars in a wavy, north-south line about ⅓° long (see the photograph at the top of page 51). A shorter and less conspicuous bar runs east from its center. The two strings are joined at the group's lucida (brightest star), which is surrounded by a knot of faint stars over dappled haze. This haze is resolved into a dense packet of twenty 10th- to 12th-magnitude suns at 87x, and the total star count reaches 40.

Copeland describes the cluster as "a curving group," so perhaps the Worm is the bent line stretching south and east that bears several golden-hued stars.

Now we move to **M50**, nicknamed the Coil by Copeland. Even at 17x, this is an easily visible collection of a dozen faint to very faint stars tangled in an unresolved mist covering 11' x 9'. The cluster is surrounded by a detached, somewhat pentagonal halo of stars about 23' across. M50 is a splendid group at 87x. Five stars in the shape of a staple rest at the heart of the group. They're surrounded by a void, which is in turn enveloped by an ovoid of 23 moderately bright to faint stars, including some nice pairs at its southern end. Appearing less detached at this magnification, the halo presents a large loop of stars northwest and three arcs evenly spaced around the rest of the cluster that unwrap counterclockwise in my mirror-reversed view. But in all these curves,

The Rosette Nebula surrounds the star cluster NGC 2244, described on page 50. You can't see the nebula's colors through a telescope, but its swirls of light and its wealth of dark filaments make the Rosette a magnificent sight when viewed under dark skies through a nebula filter. Photo: *Sky & Telescope* / Sean Walker

Can you see Copeland's Coil among the stars of M50? Photo: Robert Gendler

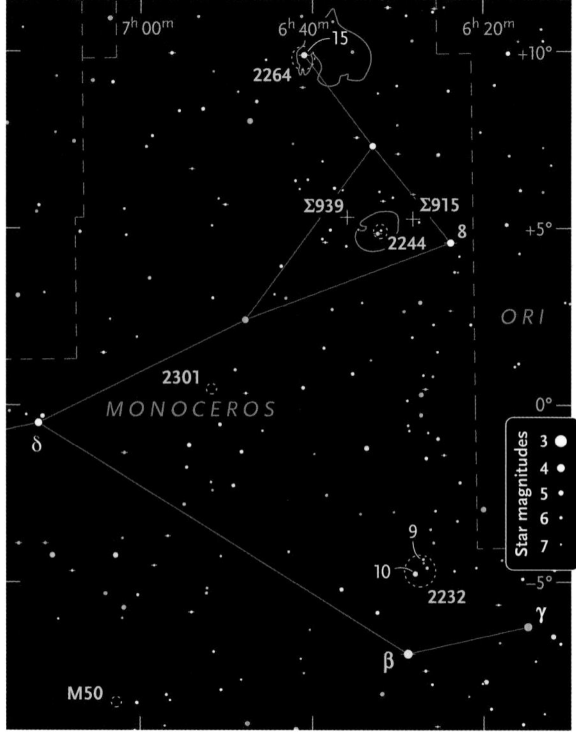

where is Copeland's Coil? M50's bright stars are twisted into an S shape (mirrored for me) that would make one full coil of a spring. Do you see it?

Next, we'll visit **NGC 2264**, widely known by the name Copeland bestowed: the Christmas Tree Cluster. It's no wonder this name stuck, for the resemblance is amazing. At 47x, about 54 lights adorn its fir-tree profile, which spans 26' from base to tip. Icy blue 15 Monocerotis (a variable star also known as S Monocerotis) marks the bottom of the tree's trunk and holds close a much dimmer companion north-northeast. Although the faint stars south of the tree's tip aren't considered part of the cluster, I can imagine them fashioning a large five-pointed star crowning the tree. Since the tree hangs tip-south in the sky, it can sometimes be seen upright when viewed through a telescope that inverts the view, as shown in the photograph on page 51, lower right.

After admiring the Christmas Tree, Copeland sweeps us over to Epsilon (ε) Monocerotis and NGC 2244. Epsilon, more commonly known as **8 Monocerotis** (8 Mon), is an easily split double star. At 47x, I see the white primary accompanied by a markedly dimmer, yellow companion to the north-northeast. The open cluster **NGC 2244**, which Copeland calls the Harp, is 2.1° east of this star. I'm not sure how Copeland envisioned the Harp, but here's my best guess. At 47x, I count 25 mixed bright and faint stars in the central region. Most of the bright ones inhabit a 20' bar that tilts north-northwest and might represent the shorter, heavier side of an upside-down concert harp. Many more stars are scattered around this, but it's difficult to tell where the cluster ends and the rich star field begins. The most conspicuous bunch stretches south-southwest from the bar's northern end and may define the harp's longer and more fragile side. What do you think?

Retracing Copeland's Tour of Monoceros

Object	Type	Magnitude	Size/Sep.	Right Ascension	Declination
β Mon	Triple star	4.6, 5.0, 5.3	AB 7.1", BC 3.0"	6h 28.8m	–7° 02'
NGC 2232	Open cluster	4.2	53'	6h 27.3m	–4° 46'
NGC 2301	Open cluster	6.0	15'	6h 51.8m	+0° 28'
M50	Open cluster	5.9	15'	7h 02.8m	–8° 23'
NGC 2264	Open cluster	4.1	40'	6h 41.0m	+9° 54'
8 Mon	Double star	4.4, 6.6	12.1"	6h 23.8m	+4° 36'
NGC 2244	Open cluster	4.8	30'	6h 32.3m	+4° 51'
Σ915	Triple star	7.6, 8.5, 11.2	6.0", 39.2"	6h 28.2m	+5° 16'
Σ939	Triple star	8.4, 9.2, 9.4	30.6", 39.5"	6h 35.9m	+5° 19'

Angular sizes and separations are from recent catalogs. Visually, an object's size is often smaller than the cataloged value and varies according to the aperture and magnification of the viewing instrument.

Which of NGC 2301's yellow stars do you think is responsible for the cluster's nickname, the Golden Worm? Photo: Bernhard Hubl

NGC 2244 sits at the heart of the Rosette Nebula, which Copeland says "cannot be seen through ordinary telescopes." But he didn't have access to the nebula filters that we enjoy today. At 17x, with an oxygen III filter, the Rosette is beautiful and complex. It spans about 1° and wreathes a central void that encompasses much of the bright bar of cluster stars. The nebula shows considerable variation in width and brightness and is ribboned with dusky lanes.

Finally, we come to Copeland's Trigon, which he described as a "very small delta of stars" and "a miniature marvel of three relatively bright stars, close together." He writes that it's a little west and north of NGC 2244, and it's so plotted on his accompanying chart. Here we find the triple star **Σ915**, but this cannot be the Trigon. The two bright components make a nice double at 47x, but the third star is faint and so far southeast of the pair that they make a very skinny triangle.

Later in the text, however, Copeland equates the Trigon with the triple star **Σ939**, which is a little east and north of NGC 2244. Visiting this triple, I can just distinguish its cute little triangle at 17x, but I find it more appealing at 47x. The 8th-magnitude primary shows 9th-magnitude companions east-southeast and northeast in a lovely delta shape. Surely this is Copeland's Trigon, worthy of both attention and a name of its own. And as Copeland asserts, some of us will take special possession of this delightful denizen of the deep sky.

Finding Beta

The faint stars of Monoceros can be hard to recognize. One easy way to start is by following the line of Orion's belt toward Sirius. About halfway, and slightly to the north, are two similarly bright stars. The star nearer Orion is Gamma (γ) Monocerotis and the one closer to Sirius is Beta (β), the starting point of this star-hop.

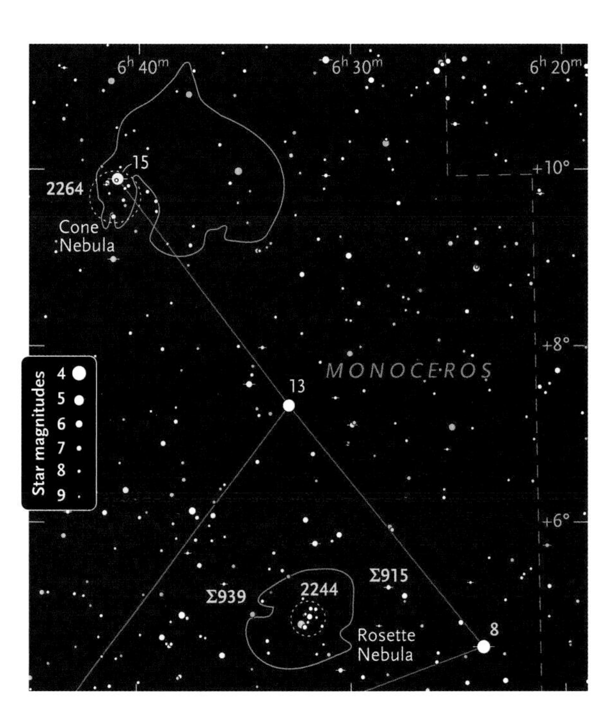

NGC 2264 is shown here with south up, as it typically appears through a reflecting telescope, to emphasize the resemblance to a Christmas tree. The nebulosity surrounding the cluster is hard to see through the eyepiece of a telescope even with the aid of a nebula filter. The dark intrusion that meets the tip of the tree is the Cone Nebula. Photo: Cord Scholz

First published February 2009

Auriga's Riches

Many wonderful but little-known clusters and nebulae adorn the celestial Charioteer.

Straddling the treasure-laden realm of the Milky Way, Auriga has a wealth of deep-sky delights. We've toured several of those nestled within the constellation's distinctive pentagon. Let's appraise a sample of the jewels and diaphanous silks outside the pentagon, starting in the far eastern reaches of Auriga.

This area of the sky is home to 11 stars that bear the Greek-letter designation Psi (ψ), but Psi^{10} Aurigae actually resides within the neighboring constellation Lynx and is more commonly known as 16 Lyncis. Two of the stars are close together and share the designation Psi^{8}. Even if the pair appears single to your unaided eye, its combined magnitude of 5.6 still makes Psi^{8} the faintest of the bunch, while Psi^{2} is brightest at magnitude 4.8. How many Psis can you spy with your eyes?

The arresting open cluster **NGC 2281** sits 50' south-southwest of Psi^{7} Aurigae. Through my 105mm refractor at 47x, NGC 2281 is a 20' group of 35 moderately bright to faint stars. Many of the bright ones outline a paisley-shaped loop concave to the north-northeast. The pattern of stars at the loop's narrow end reminds me of the constellation Delphinus or, with a few additional stars, the Christian ichthys fish symbol. The fish sports a golden nose, while through my 10-inch reflector the root of his tail shows a similar hue. A colorful pentagon of stars surrounds NGC 2281, three of the corners marked by yellow-orange suns and another by a pair whose northern member is a smoldering orange ember.

For a deeply colorful star, let's slip 3½° southwest to the semiregular variable **UU Aurigae**. It ranges between magnitude 5 and 7 with superposed periods of 233 and 439 days. UU Aurigae is a carbon star, a cool giant that has a very

The bright, bluish stars of NGC 2281 (right) indicate that it's a relatively nearby and young cluster. It's almost 17° from the plane of the Milky Way, so the background stars are quite sparse. NGC 2192 (above) is more distant and older than NGC 2281, so its stars appear fainter, closer together, and redder. Photos: POSS II / Caltech / Palomar

Above left: Faint Sh 2-217 is centered 9' east of the bright double star John Herschel 2241. Above right: Sh 2-219 is smaller but more intense than Sh 2-217. Photos: POSS-II / Caltech / Palomar

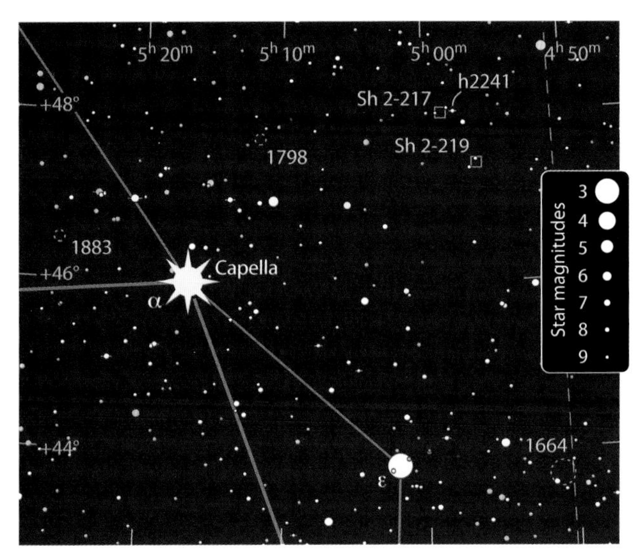

ruddy tint because most of its blue light is filtered out by carbon molecules and carbon compounds in its atmosphere.

Sweeping from Theta (θ) Aurigae to 40 Aurigae and continuing the same distance beyond brings us to a much different open cluster, **NGC 2192**. My little refractor at 28x simply shows a small fuzzy patch. At 87x, the group appears very granular and about 4' across. It's sprinkled with very faint to extremely faint stars, the brightest one in the north-northeastern edge. Even my 10-inch scope at 192x teases out only 20 stars against a patchy haze of unresolved suns.

The disparate appearance of our first two clusters is largely a function of distance. Splashy NGC 2281 is only 1,800 light-years away, while diamond-dust NGC 2192 is 4½ times more remote. Another difference between the clusters is their age. NGC 2281 is about 360 million years old, whereas NGC 2192 has been around for a whopping 2 billion years.

Our next stop is the intricate planetary nebula **IC 2149**, which lies 38' west-northwest of the reddish orange star Pi (π) Aurigae. In my 5.1-inch refractor at 63x, it's a bright, minuscule, aqua nebula with a brighter center. At 234x, the nebula is elongated, somewhat brighter near the central star, and widest at the west-southwestern end. With my 10-inch reflector at 299x, IC 2149 is brighter along the long axis, especially in the eastern half. An oxygen III filter helps improve contrast at lower magnifications.

IC 2149 was discovered by Williamina Fleming and first reported in Harvard College Observatory Circular No. 111 in 1906. While inspecting photographic plates taken with the 8-inch Draper refractor during Harvard's stellar spectrum survey, Fleming saw that this object, previously cataloged as a star, exhibited the characteristic bright lines of a gaseous nebula.

If you sweep 2.2° eastward from Omicron (ο) Aurigae, you'll come to a slightly fainter, yellow star with a star of similar brightness ½° farther east. This last star anchors the northeastern edge of the open cluster **NGC 2126**.

The cluster is a pretty dusting of minute stars through my 105mm refractor at 47x. I count 15 individuals in 6' at 127x. My 10-inch reflector at 192x reveals 23 stars, mostly gathered into a 4' triangle. A two-star tail off one side turns the triangle into a little arrow aimed north-northeast.

Two more diamond-dust clusters reside near brilliant Capella. NGC 1883 is 1.7° to its east-northeast, and NGC 1798 is 1.9° to its north-northwest.

In my 105mm refractor at 87x, **NGC 1883** is a faint, hazy patch lodged in the sharp end of a skinny triangle of stars that points south-southwest. The 2' haze appears

Delicate NGC 1664 has been likened to a kite, a clover, a flower, and a ginkgo leaf. Photo: POSS-II / Caltech / Palomar

granular, with one faint star tacking the eastern edge and a blurry blotch in the northwestern region. At 174x, the blotch gives up one very faint star. At 213x, my 10-inch scope displays a dozen pinpoint stars in a 2½' area, and two outliers to the south.

NGC 1798 is bigger and brighter than NGC 1883. Through my 105mm refractor at 122x, its misty glow shows three very faint stars and a non-stellar spot east-southeast of center, where images disclose a tiny knot of stars. My 10-inch reflector at 192x teases out 15 to 20 stars in 4½' to 5'.

Two emission nebulae are found 2½° west of NGC 1798. Both are very challenging through my 105mm scope at 76x. **Sharpless 2-217** (Sh 2-217) lies 9' east of **h2241**, an east-west pair of matched, white stars like little eyes in the dark. The feeble glow of the nebula spans roughly 4½', engulfing an 11th-magnitude star and a few fainter ones. A narrowband nebula filter helps Sh 2-217 stand out a bit better. Just ¾° southwest, **Sharpless 2-219** (Sh 2-219) covers only 1½'. A 12th-magnitude star is embedded west of center. It makes a 5' curve (concave east) with one star south and two north, magnitudes 9 to 12.

With my 10-inch scope at 118x, I noticed a small brighter patch in the southwestern part of Sh 2-217 that, in turn, seems to have a brighter spot inside. A closer look at 192x confirmed this, but I couldn't tell whether the spot was stellar. A bit of research revealed that this scrap of scrim houses a highly obscured and crowded star cluster, visible in infrared images.

Now we'll move down to Epsilon (ε) Aurigae. Normally shining at magnitude 2.9, Epsilon is eclipsed every 27.12 years by what is generally assumed to be a large, dark disk of material. The star takes half a year to fade, remains at about magnitude 3.8 for a year, then takes half a year to return to its usual brightness. The next eclipse will not begin until the year 2036.

The open cluster **NGC 1664** is found precisely 2° west of Epsilon. With its loops and chains of stars, NGC 1664 is a terrific cluster for connect-the-dot games. I've heard it described as a stingray, a kite with a long tail, a four-leaf clover, and a heart-shaped balloon on a string. Most of these images have occurred to me at one time or another, as well as an admittedly more esoteric one where I picture the stars as a ginkgo leaf with a wiggly stem. In my defense, ginkgo trees are planted along the main street of Scotia, New York, a village near my home.

In my 105mm refractor at 17x, NGC 1664 is a small group of very faint stars with a 7th-magnitude gem burnishing its southeastern side. At 87x, the 6' x 4' ginkgo leaf sits at the center of NGC 1664 with its 7'-long stem wandering toward and passing just west of the bright star. Two dozen stars are loosely scattered around our ginkgo-kite-balloon-stingray-clover, swelling the group to about ⅓°.

Clusters, Stars, and Nebulae Outside Auriga's Pentagon

Object	Type	Magnitude	Size/Sep.	Right Ascension	Declination
NGC 2281	Open cluster	5.4	25'	6^h 48.3^m	+41° 05'
UU Aurigae	Carbon star	5–7	—	6^h 36.5^m	+38° 27'
NGC 2192	Open cluster	10.9	5.0'	6^h 15.3^m	+39° 51'
IC 2149	Planetary nebula	10.6	34" × 29"	5^h 56.4^m	+46° 06'
NGC 2126	Open cluster	10.2	6.0'	6^h 02.6^m	+49° 52'
NGC 1883	Open cluster	12.0	3.0'	5^h 25.9^m	+46° 29'
NGC 1798	Open cluster	10.0	5.0'	5^h 11.7^m	+47° 42'
Sh 2-217	Emission nebula	—	7.3' × 6.3'	4^h 58.7^m	+48° 00'
h2241	Double star	9.3, 9.5	12"	4^h 57.8^m	+48° 01'
Sh 2-219	Emission nebula	—	2.0'	4^h 56.2^m	+47° 24'
NGC 1664	Open cluster	7.6	18'	4^h 51.1^m	+43° 41'

Angular sizes and separations are from recent catalogs. Visually, an object's size is often smaller than the cataloged value and varies according to the aperture and magnification of the viewing instrument.

First published February 2010

Double Your Fun

Gemini is home to a variety of paired stars and deep-sky objects.

You'll find Gemini, the Twins, high in the south on the March all-sky map on page 304. Its brightest stars, nearly at the zenith, bear the names of the mythical brothers. The star Castor marks the head of one twin, while Pollux marks the head of the other. Gemini harbors one of the finest open clusters, one of the most spectacular planetary nebulae, and one of the most fascinating multiple-star systems in the northern sky.

We'll begin with beautiful **M35**, which Massachusetts amateur Lew Gramer calls the Shoe-Buckle Cluster. It glitters on Castor's northern "shoe," between and above Eta (η) and 1 Geminorum. M35 is faintly visible to the unaided eye under dark, transparent skies and is clearly evident through a small finder as a patch of mist. I find it quite striking with 7x50 binoculars and partially resolved into moderately bright stars. In his 1888 classic, *Astronomy with an Opera-Glass*, Garrett Serviss described this cluster as "a piece of frosted silver over which a twinkling light is playing."

My 105mm refractor at 47x shows this to be an outstanding cluster with more than 100 stars in an area the size of the full Moon. There is a void near the center populated by only a few faint points of light. A pretty arc of stars starts at the north edge of M35 and curves toward the west as it reaches the void. My impression through my 6-inch reflector at 39x is a bit different. Here I note a 70-star core in a vaguely rectangular group about 20' x 25' running northwest to southeast. The bright star in the northeast side has a distinctly golden hue; it is the primary component of the double star **OΣ 134**, with a bluish companion to its south.

Recent studies give M35 an age of about 150 million years. On the cosmic or even geologic time scale, this is quite young. While gazing at M35, you might be stepping on stones older than that.

The little fuzzy spot next to M35's fringe is the open cluster **NGC 2158**. Zooming in at 127x, my little refractor shows many extremely faint points of light over a hazy, unresolved background 4' across. The brightest star lies at its southeast edge. This would be a very impressive-looking cluster if not for its distance — 12,000 light-years — more than

Gemini's best-known star cluster, M35 is the broad sprinkling of stars left of center. Its much more remote "companion" cluster, NGC 2158, often eludes observers in light-polluted skies. An arrow marks the double star OΣ 134. The field shown here is nearly 3° across, like that presented by a typical 20x spotting scope; north is up. Photo: Akira Fujii

Mining the Mythical Twins

Object	Type	Magnitude	Size/Sep.	Dist. (l-y)	Right Ascension	Declination	*MSA*	*U2*
M35	Open cluster	5.1	28'	2,600	6^h 09.0^m	+24° 21'	156	76R
OΣ 134	Double star	7.6, 9.1	31"	2,600	6^h 09.3^m	+24° 26'	156	76R
NGC 2158	Open cluster	8.6	5'	12,000	6^h 07.4^m	+24° 06'	156	76R
NGC 2266	Open cluster	9.5	6'	11,000	6^h 43.3^m	+26° 58'	154	76L
δ Gem	Double star	3.6, 8.2	5.5"	59	7^h 20.1^m	+21° 59'	152	76L
NGC 2392	Planetary nebula	9.2	47" × 43"	3,800	7^h 29.2^m	+20° 55'	152	75R
Castor	Triple star	1.9, 3.0, 8.9	4.1", 71"	52	7^h 34.6^m	+31° 53'	130	57R
YY Gem	Variable star	8.9 to 9.6	—	52	7^h 34.6^m	+31° 52'	130	57R

Approximate distances are given in light-years. The columns headed *MSA* and *U2* give the chart numbers where objects are plotted in the *Millennium Star Atlas* and *Uranometria 2000.0*, 2nd edition, respectively.

four times as far off as M35. NGC 2158 is a very old cluster with an age of around 2 billion years; the most prominent stars sprinkled across its face are red giants.

NGC 2266 is a comparable group sitting 1.8° north of Epsilon (ε) Geminorum, the star that marks Castor's knee. At 87x, I see a dozen faint to very faint stars against a background haze from which elusive pinpoints of light wink in and out of view. This pretty cluster is 6' across and somewhat triangular, with the brightest star at the southwest point. NGC 2266 contains many red-giant stars, and in color images, it is one of the most beautiful star clusters I've ever seen.

Now let's hop over to Castor's twin, Pollux, where we'll find the double star **Delta (δ) Geminorum** (δ Gem). The planet Pluto was quite close to this star when it was discovered on photographs that were taken in January 1930. At 87x with my 105mm scope, Delta is a very tight pair. The yellow-white primary has an 8th-magnitude attendant to the southwest. Although this companion is a red-dwarf star, some observers describe it as red-purple or even blue.

The intricate planetary nebula **NGC 2392**, or Caldwell 39, dwells 2.4° east-southeast of Delta. An 8th-magnitude star just north of the planetary makes the pair resemble a double star when viewed at low power. My little scope at 153x reveals a small, roundish, slightly mottled, blue-gray glow surrounding a bright central star. Some skygazers notice a blinking effect when looking at this nebula. Using averted vision makes the nebula more apparent, while staring straight at it makes it blink off.

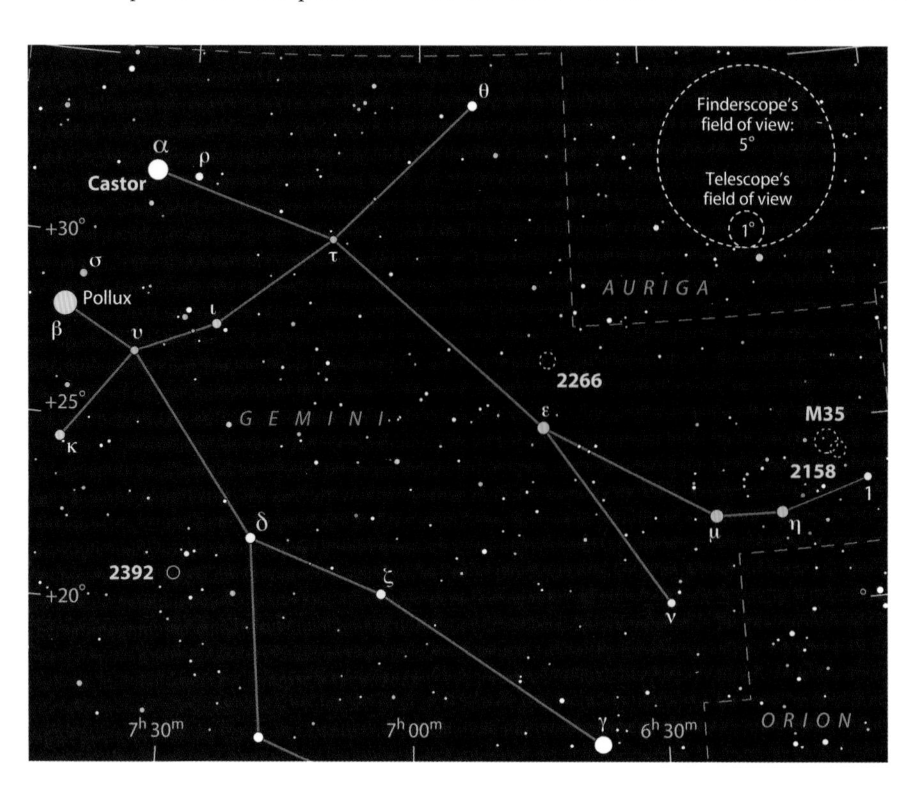

NGC 2392 is often called the Eskimo Nebula because some ground-based photos make it look like a face encircled by a furry hood. The Eskimo's fur parka may be detected as a decrease in brightness through small scopes, but you'll probably need at least an 8-inch to pick up any of the dark ring separating it from the Eskimo's face. A 6-inch scope begins to show some of the dark features in the Eskimo's face, notably a dark patch west of center that outlines part of the nose. When it comes to viewing the subtle details of the Eskimo Nebula, high power coupled with a light-pollution filter of the oxygen III or narrowband type can be a great help.

In January 2000, the newly refurbished Hubble Space Telescope turned its gaze toward the Eskimo Nebula, revealing a fascinating wealth of detail. The structure of the fur hat is thought to be caused by a slow equatorial wind

First published March 2003

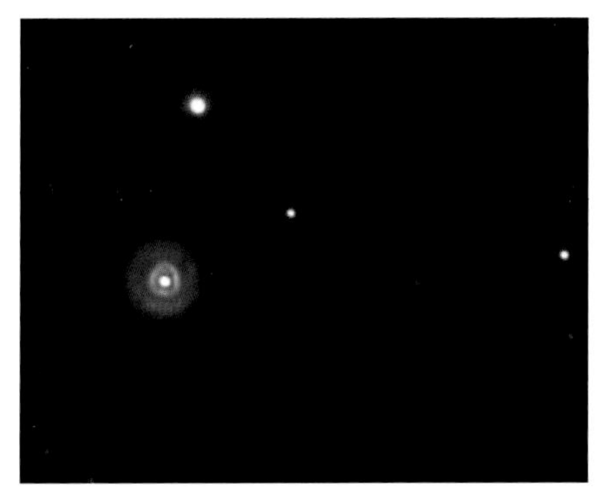

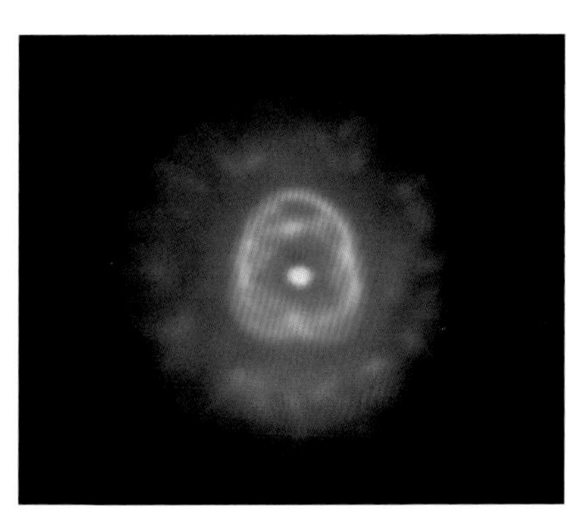

Top left and right: The Eskimo Nebula (NGC 2392) is quite tiny; the 8th-magnitude star in the wider view is just 99" north of the nebula's central star. These CCD images were made with Celestron 8-inch and Meade 16-inch telescopes. Bottom: Striking for its colors, the cluster NGC 2266 is seen here as no small scope can ever show it. Photos, top left: *Sky & Telescope* / Sean Walker; top right: Ross and Julia Meyers / Adam Block / NOAO / AURA / NSF; left: POSS-II / Caltech / Palomar

from the dying central star, while the filamentary face is molded by faster winds blowing from its poles.

Our final target is **Castor**, a system of six suns orbiting one another in an intricate ballet. Each of its three visible components hides a companion too close to be resolved through a telescope — three sets of secret twins disclosed only by their telltale spectra. Through my small refractor at 87x, 2nd-magnitude Castor A closely guards 3rd-magnitude Castor B to the east-northeast, and both appear white. Much dimmer Castor C lies a generous 71" to the south-southeast. It consists of a pair of nearly matched red-dwarf stars undergoing mutual eclipses, so we witness two during each orbital period of 19.5 hours. In each case, the stars' combined light is halved, dropping about 0.7 magnitude. As a variable star, Castor C is designated **YY Geminorum** (YY Gem).

The constellation of the Twins serves up a double share of observing pleasure with objects that are both a delight to view and a wonder to contemplate. Give them a try on your next clear night.

Perambulations in Puppis

Set your sights on the star-studded field south of Procyon and east of Sirius.

At this time of year, wandering through Puppis is one of my favorite pastimes. Its exceptionally rich star fields provide fertile hunting grounds for any deep-sky observer, with dozens of objects visible through a small scope. One of the constellation's most magnificent areas is visible to the unaided eye as a hazy patch in a dark velvet sky. It contains most of the objects we'll take a look at here.

Let's begin with **M46**, a singularly stunning star cluster. Even in 14x70 binoculars, it appears extremely rich in minute stars. But to be sure you're looking in the right spot, first center your telescope on the 4th-magnitude star Alpha (α) Monocerotis and then slew 5° due south. My 105mm refractor at 17x shows M46 as a round, densely packed swarm of faint stars over a hazy background. The brightest star is magnitude 8.7, and it's found in the western side; most of the rest are magnitude 11 or 12. Higher magnifications transform the misty backdrop into additional stars, but a small void rests at the cluster's heart.

While viewing this cluster at 87x, I can see **NGC 2438**, a planetary nebula that seems to be embedded in its northern fringes. It appears round and fairly bright, and there's an 11th-magnitude star at its southeastern edge. NGC 2438 is 1' across, a little smaller than the famous Ring Nebula in Lyra. If you have trouble spotting it, a narrowband or oxygen III light-pollution filter will dim the stars

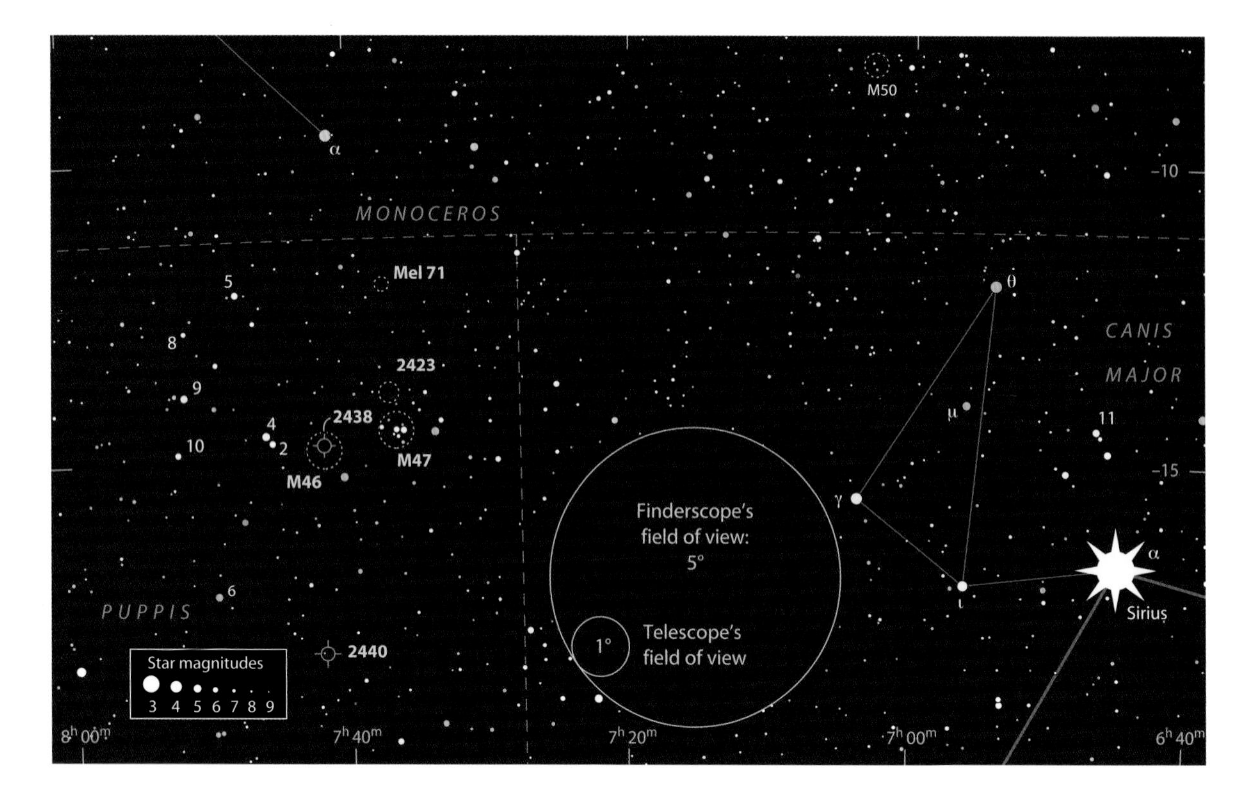

and help the nebula stand out. Despite its apparent location within M46, the planetary is probably a foreground object that happens to lie along the same line of sight.

M47 is another fascinating star cluster. When he cataloged this object in 1771, Parisian astronomer Charles Messier described it as a cluster of stars not far from M46 and containing brighter stars. But he placed it at right ascension 7h 54.8m, declination –15° 25', which marks a conspicuously blank spot in the sky. Nonetheless, Johan L. E. Dreyer assigned the designation NGC 2487 to this location when he compiled his 1888 *New General Catalogue of Nebulae and Clusters of Stars*.

In 1934, German astronomer Oswald Thomas pointed out that another object in Dreyer's compilation, NGC 2422, situated 1.3° west-northwest of M46, must be what Messier really saw. In 1959, T. F. Morris of Montreal, Quebec, offered a clever explanation for the confusion. Messier wrote that he had measured M47's location relative to the star 2 Navis (now known as 2 Puppis). The position he gave was 9^m east and 44' south of this star. If we say instead that M47 is 9^m *west* and 44' *north* of 2 Puppis, we end up fairly near the position of NGC 2422. So Messier may have applied each offset backward.

Whatever the story behind Messier's discovery, Sicilian astronomer Giovanni Hodierna (aka Gioanbatista Odierna) scooped him by more than a century. Odierna's little-known deep-sky catalog, published in 1654, both describes and maps the cluster we now call M47.

M47 is a beautiful sight in almost any instrument. Through 14x70 binoculars, 20 to 25 stars gather into a very

Sights in Northwestern Puppis

Object	Type	Magnitude	Size	Distance (l-y)	Right Ascension	Declination	*MSA*	*U2*
M46	**Open cluster**	**6.1**	**27'**	**4,500**	**7^h 41.8^m**	**–14° 49'**	**295**	**135L**
NGC 2438	**Planetary nebula**	**11.0**	**64"**	**2,900**	**7^h 41.8^m**	**–14° 44'**	**295**	**135L**
M47	**Open cluster**	**4.4**	**29'**	**1,600**	**7^h 36.6^m**	**–14° 29'**	**296**	**135L**
NGC 2423	**Open cluster**	**6.7**	**19'**	**2,500**	**7^h 37.1^m**	**–13° 52'**	**296**	**135L**
Mel 71	**Open cluster**	**7.1**	**9'**	**10,300**	**7^h 37.5^m**	**–12° 03'**	**296**	**135L**
NGC 2440	**Planetary nebula**	**9.4**	**20" × 15"**	**3,600**	**7^h 41.9^m**	**–18° 13'**	**319**	**153R**

Angular sizes are from catalogs or photographs. Visually, an object's size is often smaller than the cataloged value and varies according to the aperture and magnification of the viewing instrument. The columns headed *MSA* and *U2* give the chart numbers of objects in the *Millennium Star Atlas* and *Uranometria 2000.0*, 2nd edition, respectively.

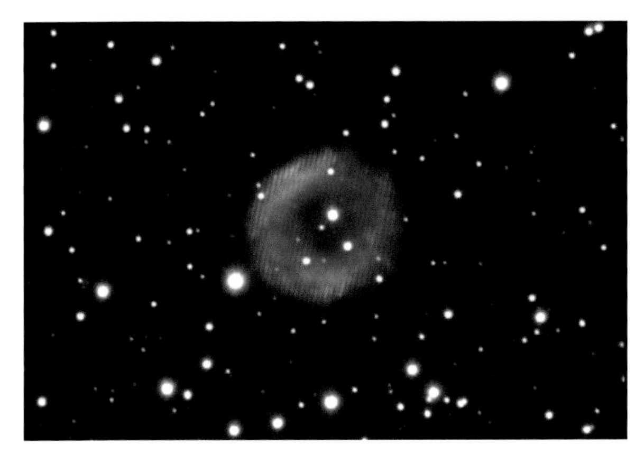

Top: Can you spot NGC 2438 as a tiny "soap bubble" in this 1°-wide photograph of the open cluster M46? Photo: George R. Viscome. Bottom: This highly magnified view of NGC 2438, the planetary in M46, brings out fine structure that no visual observer sees. The CCD image was taken from Kitt Peak in Arizona. Photo: Nicole Bies / Esidro Hernandez / Adam Block / NOAO / AURA / NSF

loose, irregular cluster ½° across. I count 48 mixed bright and faint stars in my little refractor at 17x. Most of the dominant gems sparkle with fierce, bluish white star fire, but a few orange jewels can be picked out among them.

M47 contains several multiple stars. Perhaps the prettiest for a small scope is Σ1121, near the cluster's center. It is easy to single out as the southernmost in an arc of three bright stars. Σ1121 consists of a nearly matched pair of 7th-magnitude blue-white suns nicely split at 68x.

Another cluster can be found near M47; just follow a little chain of roughly 9th-magnitude stars northward. At 87x, **NGC 2423** shows about 30 faint stars loosely scattered across 15'. The center of the group is fixed by one of its brightest members, the double star h3983 (a discovery of John Herschel). The 9.1-magnitude primary hosts a 9.7-magnitude secondary 8" west-northwest. The primary star is itself a double, but the components are too close for a small telescope to split. NGC 2423's lucida (the old-time observers' term for "brightest member") is an 8.6-magnitude star at its south-southwest border, but most of the visible stars shine at magnitude 11 or 12.

A small telescope with a short focal length can showcase all three clusters in the same field of view at 25x or less. Some wide-angle eyepieces will allow you to push the magnification as high as 45x and still fit them all in. The combination of these three disparate "star cities" makes for a captivating view.

Ranging a bit farther afield, we find the pretty little cluster **Melotte 71** (Mel 71) 1.8° north of NGC 2423. With my 105mm scope at 87x, I get the impression of a diamond crushed into fine powder with surviving chips casting glints of light. Near the southwest edge gleam the two brightest stars, of which the eastern one is a close, matched double. Dimmer specks glitter against a frosty backdrop 9' across.

By returning to M46 and dropping 3.4° due south, we can turn up another planetary nebula. Look for it just west of an 8th-magnitude orange star. At low powers, **NGC 2440** looks like a slightly fainter star, but 87x reveals a northeast-to-southwest oval that is distinctly robin's-egg blue. The planetary is bright and takes magnification well. At 153x, I can see a bright spot in the middle, but this is not the central star. With his 105mm refractor at about 400x, California amateur Ron Bhanukitsiri has been able to split this brightening into two distinct knots — a sight usually reserved for larger telescopes. Although too faint to be seen in a small telescope, the nebula's central star has one of the hottest confirmed surface temperatures. It blazes at more than 200,000°C, which is 30 times the surface temperature of our Sun.

As the lustrous Milky Way plunges through Puppis, it leaves many such wonders in its wake. Be sure to set aside the enchanting hours of a starry, moonless eve to enjoy them.

Open star clusters M47 (below center) and NGC 2423 (top) can be studied together with a low-power telescope. The field of view in this photograph is 1° tall. All images in this article have north up. Photo: George R. Viscome

First published March 2004

The Flight of the Unicorn

Monoceros's dim stars are offset by other fine sights.

Without motion I
Rise on gently curved glass wings
Into the night sky

— Carter D. Hayward

Monoceros, the Unicorn, is a relatively modern constellation. Flemish cartographer Petrus Plancius introduced it on his celestial globe of 1613, but the constellation may have earlier forebears. In *Star Names: Their Lore and Meaning* (Dover, 1963), Richard Hinckley Allen claims that French scholar Joseph Justus Scaliger found it on a Persian sphere. Astronomer Ludewig Ideler, in his 1809 work on the origin and meaning of star names, writes that he found this reference to it in a 1564 German astrology book: "The other horse under the Twins and the Crab has many stars, though not bright."

The bright star 15 Monocerotis dominates the Christmas Tree Cluster (NGC 2264), whose shape is easier to recognize visually than in photographs. About ½° to its south lies the silver-lined Cone Nebula, an object quite striking on photographs but visually a challenge. Photo: Robert Gendler

Monoceros is nearly as difficult to catch sight of as the mythical creature it represents. The Unicorn sports no star brighter than 4th magnitude, yet it's a wonderful constellation to explore. The winter Milky Way drapes Monoceros in delicate chiffon spangled with spectacular nebulae and star clusters. Let's lift ourselves on wings of glass to soar among them.

We'll start with **NGC 2264** at the Unicorn's head in northern Monoceros. This open cluster lies 3.2° south-southwest of Xi (ξ) Geminorum and includes the 4.7-magnitude star 15 Monocerotis. Dubbed the Christmas Tree Cluster by Leland S. Copeland, this striking cluster well deserves its nickname. I recall observing NGC 2264 long ago when I'd heard of the Christmas Tree but didn't know to which cluster the name referred. One look through the eyepiece and I knew this must be it!

Through 15x45 binoculars, the Christmas Tree hangs upside down from its bright trunk star, 15 Monocerotis. The triangular tree is outlined by 20 stars of 7th magnitude and fainter. The sprawling nebulosity **Sharpless 2-273** (Sh 2-273) can be seen enveloping and faintly stretching northwestward from it, with the dark nebula **Schoenberg 205/6** encroaching from the north. With my 10-inch Newtonian reflector at 44x, I can just detect the small, dark finger of the **Cone Nebula** (LDN 1613) jutting into the nebula from the south and reaching to the pointy tip of the Christmas Tree. An oxygen III filter gives the nebula a minor contrast boost by helping isolate its greenish emission, but observers have had success with other light-pollution-rejection filters as well. While you're here, spare a moment to admire 15 Monocerotis. With a spectral type of O7, it's one of the bluest stars in the sky.

Two small nebulae sit 2° west of 15 Monocerotis. In my 105mm refractor at 68x, **NGC 2245** is visible west-southwest of an 8th-magnitude star and shows a very faint star embedded in the northeast. In my 10-inch reflector at 70x, nearby **NGC 2247** shares the field. It looks larger but dimmer than its companion and has a 9th-magnitude star at its heart. Both objects are usually listed as reflection nebulae, which shine by reflected starlight. But NGC 2245 appears brighter when an oxygen III filter is used, indicating that the nebula has at least some emission component to its glow.

NGC 2261, also known as Hubble's Variable Nebula or Caldwell 46, is a remarkable object lying 1.2° south-southwest of 15 Monocerotis. Observing from England, William Herschel discovered this reflection nebula in 1783. More than a hundred years later, American astronomer Edwin Hubble, then a graduate student working at Yerkes Observatory, was given the task of investigating the nebula for possible variability. Hubble found that it changes in both apparent shape and brightness.

I can spot Hubble's Variable Nebula with my 15x45 binoculars, but it looks nearly stellar. My little refractor at 87x shows a small, fairly bright, fan-shaped nebula that looks like a broad-tailed comet. A bright stellar point can be seen in its head, and the wide tail stretches north. Through my 10-inch at 170x, the sides of the fan appear brighter than the center or top. Since noticeable changes may occur in just a few weeks, these observations are merely snapshots of how the nebula can look.

The stellar spot at the point of the fan has long been known by the variable-star designation R Monocerotis. However, the starlike point we see is actually a bright circumstellar shell of gas and dust hiding a young binary star.

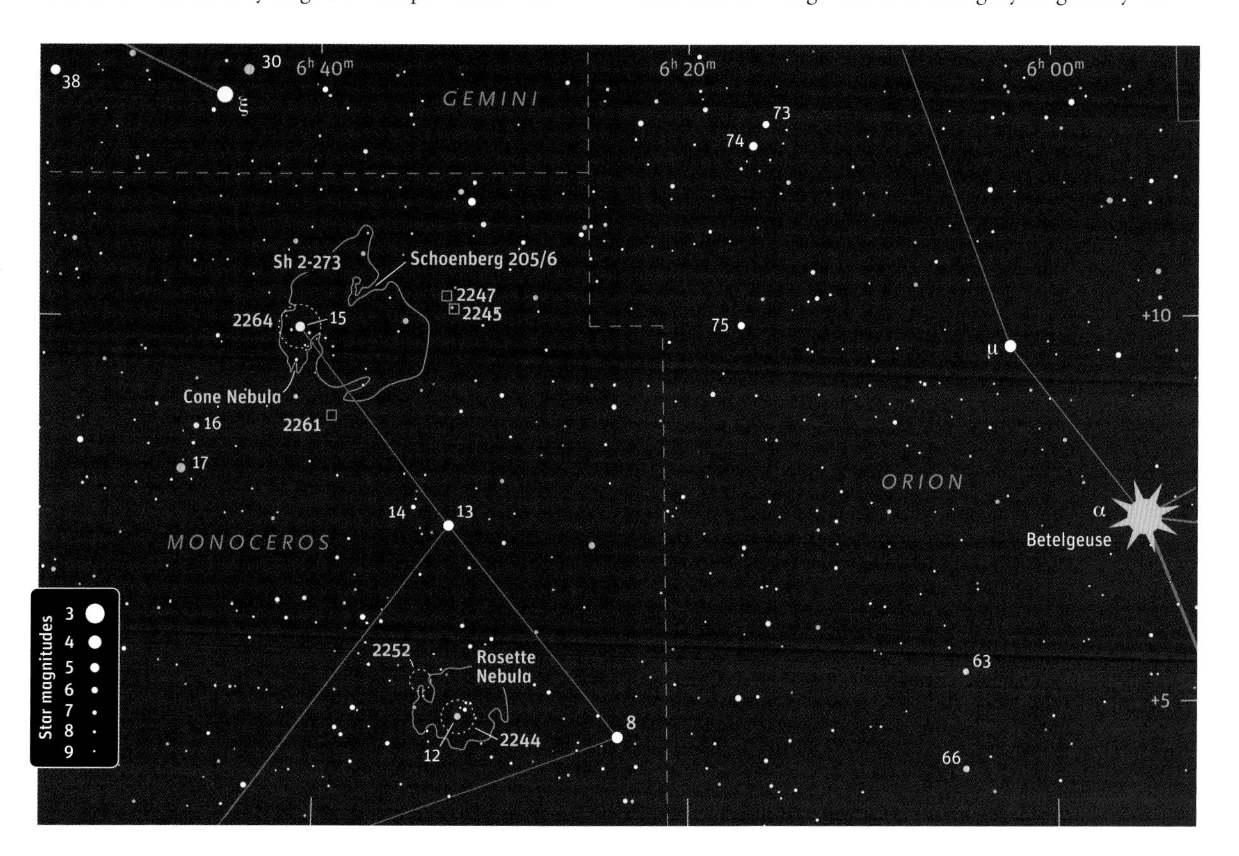

Menu for Monoceros

Object	Type	Magnitude	Size/Sep.	Right Ascension	Declination	*MSA*	*U2*
NGC 2264	Open cluster	4.1	40'	6h 41.0m	+9° 54'	202	95R
Sh 2-273	Emission nebula	—	140'	6h 37.6m	+9° 50'	202/3	95R
Schoenberg 205/6	Dark nebula	—	35' × 15'	6h 37.1m	+10° 21'	203	95R
Cone Nebula	Dark nebula	—	4.5' × 2.5'	6h 41.2m	+9° 23'	202	95R
NGC 2245	Reflection nebula	—	2'	6h 32.7m	+10° 09'	203	95R
NGC 2247	Reflection nebula	—	2'	6h 33.1m	+10° 19'	203	95R
NGC 2261	Reflection/emission nebula	10	2'	6h 39.2m	+8° 45'	202/3	95R
8 Mon	Double star	4.4, 6.6	12.5"	6h 23.8m	+4° 36'	227	95R
Rosette Nebula	Emission nebula	5?	80' × 60'	6h 31.7m	+5° 04'	227	95R
NGC 2244	Open cluster	4.4	30'	6h 32.3m	+4° 51'	227	95R
NGC 2252	Open cluster	7.7	18'	6h 34.3m	+5° 19'	227	95R

The visual magnitudes of most nebulae are poorly (if at all) known. Angular sizes or separations are from recent catalogs. The columns headed *MSA* and *U2* give the chart numbers of objects in the *Millennium Star Atlas* and *Uranometria 2000.0*, 2nd edition, respectively.

Hubble's Variable Nebula (NGC 2261) has given many sky sweepers the momentary thrill that they might have spotted a new comet. North is toward upper left. Photo: Jim Misti

Dark dust clouds in motion near the stars cast shadows on the reflection nebula and alter its appearance. On the night of January 26, 1949, Hubble's Variable Nebula had the privilege of posing for the first official photograph taken by the historic 200-inch Hale Telescope on Palomar Mountain.

Now let's drop 5.6° southwestward to the bright double star **8 Monocerotis** (8 Mon), which can be easily split through any scope at about 50x. The 4.4-magnitude white primary holds the 6.6-magnitude companion 12.5" to its north-northeast. The companion is listed as having a spectral type of F5 and should appear pale yellow. However, many observers, especially those using small telescopes, have called the fainter star blue, lilac, purple, or even mauve. I logged the star as having a bluish tint through my 6-inch reflector at 43x. If the spectral type is accurate, the perceived shades of blue must be a contrast illusion. Placing the stars farther apart with a higher magnification — or gathering more light with a larger aperture — might help dispel the illusion.

The beautiful **Rosette Nebula** (Caldwell 49) is located 2.1° east of 8 Monocerotis. Its embedded star cluster, **NGC 2244** (Caldwell 50), is bright enough for me to discern as a naked-eye hazy patch from my semirural site in upstate New York. My 15x45 binoculars show 15 stars encircled by a wide, faint, patchy annulus of nebulosity. My 10-inch reflector at 43x picks out five bright and a few fairly bright stars in two parallel, slightly curved lines. About 40 fainter stars join the scene, many spilling over into the nebulosity. With an oxygen III filter added to the eyepiece, the nebula is stunning! The Rosette spans more than 1° and displays a wealth of detail with much large-scale mottling. The annulus shows a large faint patch to the east-southeast and bright patches that hug the north, west, and southwest sides of the cluster. An irregular 20' hole darkens the center of the nebula. The cluster's bright stars are off center in this hole, so that the southernmost of them are within the nebulosity. The brightest star, yellow-orange 12 Monocerotis, is a foreground object.

Because of the Rosette Nebula's large size and patchy nature, it was discovered a piece at a time. The Rosette thus ended up with three separate designations in Johan L. E. Dreyer's 1888 *New General Catalogue of Nebulae and Clusters of Stars*: NGC 2237, NGC 2238, and NGC 2246.

The star cluster **NGC 2252** kisses the northeastern border of the Rosette. Through my little refractor at 68x, the group is loose and very irregular with 25 faint to very faint stars. The brightest are arranged in a wishbone shape, and a semicircle of stars follows the eastern part. In my 10-inch scope, the pattern of bright stars reminds me of a naked-eye view of the constellation Perseus.

When gazing at these star clusters and nebulae, we are looking toward the outer part of our Milky Way Galaxy. Those we visited near the Christmas Tree Cluster are around 3,000 light-years away in the Local Arm, where our Sun also resides. The Rosette and NGC 2252 are more distant. They lie 5,000 light-years away on the near side of an interior spur to the Perseus Arm, the next spiral arm outward.

First published March 2005

One of the sky's finest examples of a star cluster within a nebula, the Rosette has a reddish hue in color images but looks colorless to the human eye at the telescope. Note the secondary cluster, NGC 2252, on the nebula's northeastern (upper-left) fringe. From his backyard in Austin, Texas, Russell Croman recorded this 2.4°-wide field with a Tele Vue 105mm refractor.

Dog-and-Pony Show

A winter wonderland of deep-sky sights lies along the Milky Way near Sirius.

Near the end of the 19th century, small traveling circuses were known as dog-and-pony shows, named for the stars of their animal acts. At this time of the year, we have our own dog-and-pony show in the evening sky. Canis Minor, the Little Dog, seems to be riding on the back of Monoceros, the Unicorn, while Canis Major, the Big Dog, romps beside them. Since the night sky is the only place we can see such a magical pony, let's focus on the Unicorn. As befits a mythical creature, Monoceros is difficult to see — his constellation contains no star brighter than 4th magnitude.

Only one object from Charles Messier's catalog finds its home in Monoceros: the open cluster **M50**. Since it dwells in a region bereft of naked-eye stars, the stellar beacons of Canis Major are more useful as guides to M50. Following an imaginary line north-northeast from Sirius through Theta (θ) Canis Majoris and continuing for four-fifths that distance again takes you to the correct vicinity. Theta and M50 share opposite sides of a finder field, with the cluster visible as a soft glow east of a 6th-magnitude star.

Steadily supported binoculars can resolve some of M50's stars, but magnification plays a big role in perceiving them. In 7x50 binoculars, I see five stars over haze, while 15x45s show me at least 15. My 105mm refractor at 17x unveils a lovely group rich in faint stars and accented by a bright orange sun in the southern part of the cluster. I count 50 stars within 15' at 87x. Two areas devoid of stars sit south and north-northeast of the cluster's center. In my 10-inch reflector at 70x, additional stars help outline another void northwest of center. This creates a nightmarish view that reminds me of Edvard Munch's painting *The Scream*. The two northern patches are the eyes, and the elongated southern cavity is the mouth, open wide in horror.

Returning to Theta and then moving 1.8° north brings you to an asterism that astronomy author Tom Lorenzin calls the **Minus 3 Group**. Through binoculars or a scope giving an upright image, this 44' group forms the numeral 3 with a three-starred minus sign placed a bit too high in front of it. The 3 shape was independently noted by Canadian amateur Randy Pakan and is sometimes called Pakan's 3. The view through my little refractor is a mirror image, which turns Pakan's 3 into a Sigma (Σ) composed of about 20 stars. While some long-focus telescopes can't encompass the entire Minus 3 Group, Pakan's 3 is only 23' tall and a viable target for almost any scope.

A few degrees to the east, the amazing complex of interstellar gas and dust clouds popularly known as the Seagull Nebula spreads its wings across 2½° of sky. If you sweep eastward from Pakan's 3 until the asterism is just out of your finder's field, the filmy network of the Seagull Nebula will be near its center. The main swath of nebulosity is **Sharpless 2-296** (Sh 2-296), the Seagull's wings. Only a small scope with a low-power, wide-field eyepiece will embrace enough sky to take in their full span. Through my 105mm scope at 17x, I can readily see the nebula without a filter, but a narrowband nebula filter offers modest improve-

Wandering Among the Animals

Object	Type	Magnitude	Size/Sep.	Right Ascension	Declination	*MSA*	*U2*
M50	Open cluster	5.9	15'	7^h 02.8^m	–08° 23'	273	135R
Minus 3 Group	Asterism	—	44' × 23'	6^h 54.0^m	–10° 12'	298	135R
Sh 2-296	Emission nebula	—	150' × 60'	7^h 06.0^m	–10° 55'	297	135R
vdB 93	Emission/reflection nebula	—	19' × 17'	7^h 04.5^m	–10° 28'	297	135R
Ced 90	Emission/reflection nebula	—	7'	7^h 05.3^m	–12° 20'	297	135R
NGC 2343	Open cluster	6.7	6'	7^h 08.1^m	–10° 37'	297	135R
NGC 2335	Open cluster	7.2	7'	7^h 06.8^m	–10° 02'	297	135R
NGC 2353	Open cluster	7.1	18'	7^h 14.5^m	–10° 16'	297	135L
Mel 72	Open cluster	10.1	5'	7^h 38.5^m	–10° 42'	296	135L
NGC 2506	Open cluster	7.6	12'	8^h 00.0^m	–10° 46'	295	134R

Angular sizes or separations are from recent catalogs. The visual impression of an object's size is often smaller than the cataloged value and varies according to the aperture and magnification of the viewing instrument. The columns headed *MSA* and *U2* give the chart numbers of objects in the *Millennium Star Atlas* and *Uranometria 2000.0*, 2nd edition, respectively.

Popularly known as the Seagull Nebula, this vast complex of cosmic dust and gas straddling the border of Canis Major and Monoceros is a collection of objects with their own designations. It's a favorite area for astrophotographers and visual observers, especially those who enjoy hunting down the subtle glow of emission nebulosity. The field is 1¾° wide with north up. Photo: Walter Koprolin

ment. Some observers prefer the view through a hydrogen-beta filter.

Sharpless 2-296 is usually plotted on star atlases as IC 2177. Johan L. E. Dreyer included IC 2177 in his *Second Index Catalogue of Nebulae and Clusters of Stars* (1908) and noted his source as a report by Isaac Roberts in the November 1898 *Astronomische Nachrichten.* The object that Roberts discovered on a photograph, however, was actually the Seagull's head, a much smaller and somewhat brighter nebula better known today as **van den Bergh 93** (vdB 93), which surrounds the low-amplitude, 7th-magnitude variable star V750 Monocerotis. With my little refractor at 47x, vdB 93 is visible as a 7' haze mostly south and southeast of V750 Monocerotis. Elusive fainter nebulosity surrounds the star, extending the nebula to about 12'. An even smaller detached nebula, **Cederblad 90** (Ced 90), engulfs an 8th-magnitude star off the tip of the Seagull's southern wing.

Several star clusters are tangled in the Seagull's feathers. The brightest is **NGC 2343**, located 55' east and a little south of V750 Monocerotis. I see a few faint stars over mist in 14x70 binoculars, while my 105mm scope at 87x reveals 14 stars concentrated into 6'. In order of diminishing magnitude, the three brightest stars lie at the east, north, and west edges of the cluster, giving it a triangular appearance. The eastern star, Σ1028, is a double, whose yellow primary is accompanied by an 11th-magnitude companion 11' northwest.

Another notable cluster in the Seagull is **NGC 2335**, located at the bend of its northern wing. In my small refractor at 47x, it shares the field with NGC 2343. Fainter and less obvious than its companion, NGC 2335 is a gathering of 10 faint stars enmeshed in mist.

Many stars scattered throughout this area belong to the Canis Major OB1 (CMa OB1) association, a loose collection of hot young suns that provide illumination for the nebula. A physical relationship with NGC 2335 is unlikely. The cluster is much older than CMa OB1, and recent distance estimates place it beyond the complex. The situation with NGC 2343 is less certain. At about 3,400 light-years,

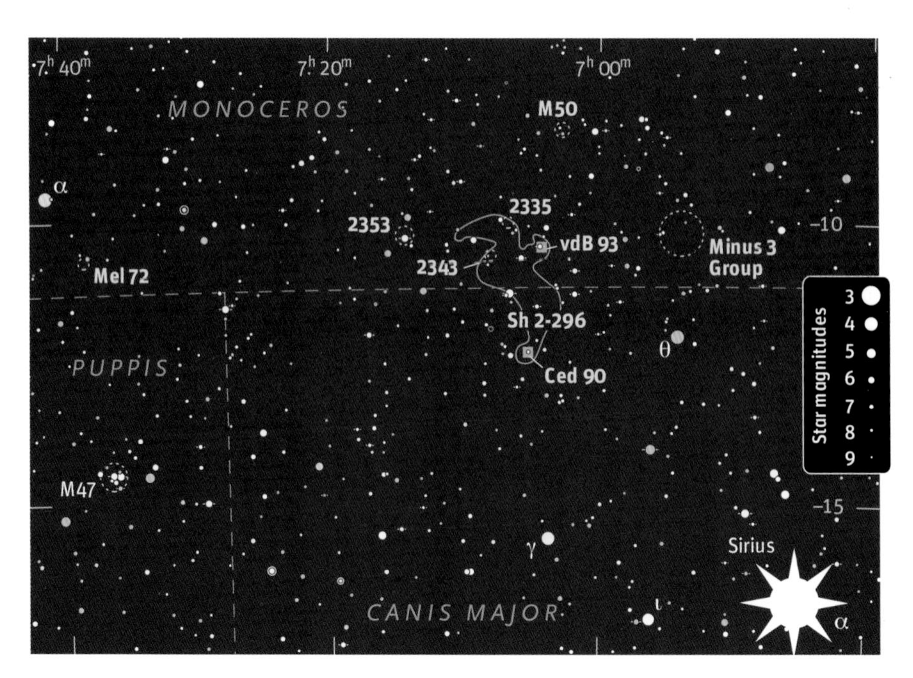

First published March 2006

The only Messier object in Monoceros is the open star cluster M50. As viewed in the author's 10-inch reflector, three regions devoid of stars give the cluster an overall appearance reminiscent of the famous Edvard Munch painting *The Scream*. The field is 40' wide with north up. Photo: Robert Gendler

this cluster lies within the possible range of distances proposed for the nebula. Age estimates for NGC 2343 vary greatly, but the cluster is probably too old to have formed from the same protostellar material as CMa OB1.

Let's leave the Seagull behind and sweep eastward through three noteworthy star clusters. If you put NGC 2343 in the southern part of a low-power field and move 1.6° eastward, you'll come to **NGC 2353**, a pretty group of many faint stars mostly huddled against its 6th-magnitude blue-white lucida. With my small refractor at 47x, I count 30 stars in 15'. An eye-catching 9th-magnitude pair, Σ1052, lies northeast of the blue-white star, and a pair of 6th-magnitude orange stars widely brackets the cluster. About 6° farther east, or 1.3° south-southwest of Alpha (α) Monocerotis, you'll find the diamond-dust cluster **Melotte 72** (Mel 72). My refractor at 87x shows minute specks sparkling over haze in a triangular patch about 5' across. An orange 7th-magnitude star rests outside the northwest edge. Continuing east for 5.3° takes us to the splendid cluster **NGC 2506**. My little scope at 87x displays 20 faint stars in two curves connected by a bar, all netted in a 7' mist of unresolved stars. Through my 10-inch reflector at 73x, the group is teeming with silvery cascades of faint stars, the more prominent ones piled into two frosty-looking knots.

Bark With the Overdog

There's a lot more to observe in Canis Major than the night sky's brightest star.

The great Overdog,
That heavenly beast
With a star in one eye,
Gives a leap in the east.

He dances upright
All the way to the west
And never once drops
On his forefeet to rest.

I'm a poor underdog,
But to-night I will bark
With the great Overdog
That romps through the dark.

— Robert Frost,
"Canis Major," 1928

Canis Major, the Big Dog, now dances upright and paws the air in our evening sky. His westering toes will soon pass their prime observing time, but the starry realm behind his head and back is well placed for pursuit yet a while. So when darkness falls, let's bark with the Dog and discover the deep-sky treasures that follow in his wake.

We'll begin with the open cluster **NGC 2345**, which lies 2.7° north-northeast of Gamma (γ) Canis Majoris. My 105mm refractor at 55x reveals a dozen moderately bright to faint stars in two diverging lines plus one outlier, all set against a misty backdrop. A 10-inch scope resolves this haze into about 30 faint stars within 12'. An 8th-magnitude foreground star lounges at the north-northeastern edge of the group. Large scopes show hints of yellow and orange among the brightest cluster members.

Now sweep 2.5° due east to the impressive emission nebula **NGC 2359**. The cloud is faint in my little refractor, but it stands out better when I add a narrowband nebula filter and is much more obvious with an oxygen III filter. In a 47x view, I see a fat arc with some mottling and two faint extensions. The brightest parts of the nebula are easily visible without a filter in my 10-inch reflector at 70x. The most prominent area is 6½' tall and shaped like the numeral 2. The bar of the 2 leans west-southwest with its curve to the north. A fainter extension strikes out to the northwest from the top of the 2. The curve of the 2 can be pictured as the dome of a Viking helmet, and the westward-reaching spikes are its horns. This depiction earned NGC 2359 the nickname Thor's Helmet. The arc of the 2 is filled with a gauzy veil of light. A subtle arm of nebu-

Running With the Big Dog

Object	Type	Magnitude	Size/Sep.	Right Ascension	Declination	*MSA*	*U2*
NGC 2345	Open cluster	7.7	12'	7h 08.3m	−13° 12'	297	135R
NGC 2359	Emission nebula	9	13' × 11'	7h 18.5m	−13° 14'	297	135L
Haffner 6	Open cluster	9.2	7'	7h 20.0m	−13° 10'	297	135L
NGC 2374	Open cluster	8.0	19'	7h 24.0m	−13° 16'	296	135L
Basel 11A	Open cluster	8.2	5'	7h 17.1m	−13° 58'	297	135L
NGC 2360	Open cluster	7.2	12'	7h 17.7m	−15° 39'	297	135L
Haffner 23	Open cluster	7.5	11'	7h 09.5m	−16° 56'	321	135R
Mink 1-13	Planetary nebula	12.6	30" × 42"	7h 21.2m	−18° 09'	(320)	153R
NGC 2362	Open cluster	3.8	7'	7h 18.7m	−24° 57'	345	153R
NGC 2367	Open cluster	7.9	5'	7h 20.1m	−21° 53'	345	153R
h3945	Optical double	5.0, 5.8	27"	7h 16.6m	−23° 19'	345	153R

Angular sizes and separations are from recent catalogs. The visual impression of an object's size is often smaller than the cataloged value and varies according to the aperture and magnification of the viewing instrument. The columns headed *MSA* and *U2* give the chart numbers in the *Millennium Star Atlas* and *Uranometria 2000.0*, 2nd edition, respectively. Chart numbers in parentheses indicate that the object is not plotted. All the objects this month are in the area of sky covered by Chart 27 in *Sky & Telescope's Pocket Sky Atlas*.

losity reaches eastward from the top of the 2, and short ones proceed east and west from the base of the 2.

A star marks each end of the 2's curve, while a third sits inside the arc. This last one is an 11½-magnitude Wolf-Rayet star, and it's the source of energy that powers NGC 2359. The fierce stellar wind flowing outward from this exceedingly hot, massive star snowplowed the gas and dust around it into the bubblelike formation at the heart of the nebula. The complex structure surrounding the bubble is formed by interactions among the star, previously ejected material, and the interstellar medium.

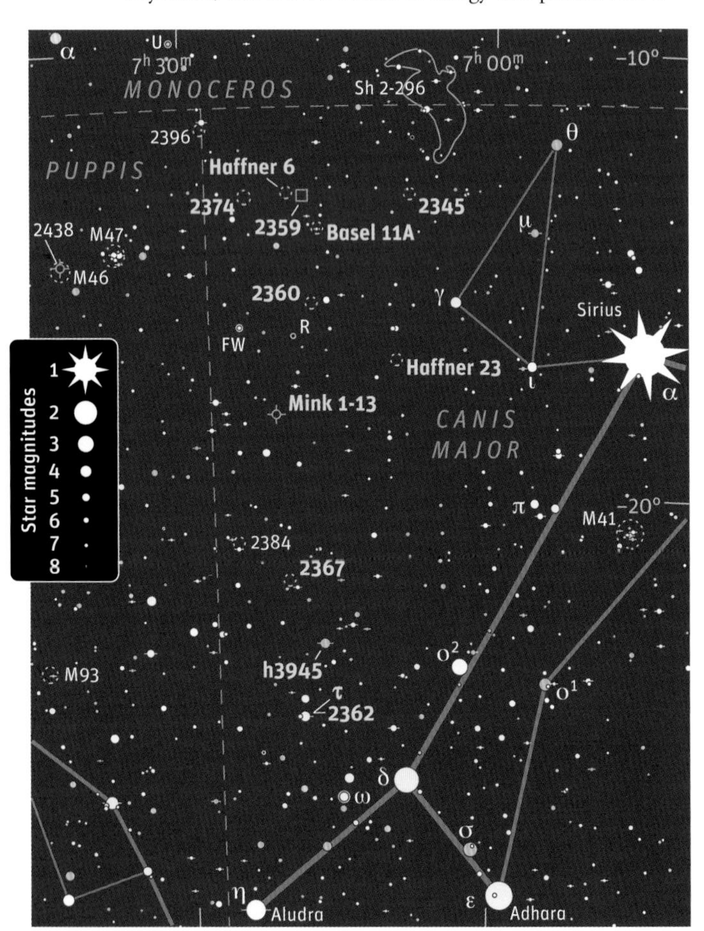

The small open cluster **Haffner 6** is centered 22' east-northeast of Thor's Helmet. Thoroughly inconspicuous in my 10-inch reflector, Haffner 6 is a 7' group rich in extremely faint stars, and it could be a nice target for a large telescope. As you might expect from its minute stars, Haffner 6 is a remote cluster; it's estimated to be 10,000 light-years away from us. It's also a very aged one, estimated to be about 670 million years old.

By moving 1° farther eastward, we come to **NGC 2374**. Observing in England, William Herschel discovered this open cluster in 1785 and described it with a shorthand notation that translates as "a cluster of pretty large scattered stars, pretty rich, about 20' long in a crooked figure." My 10-inch scope at 170x displays a 6' gathering of 25 faint stars, many zigzagging across the cluster in a W pattern. With a larger field at 64x, I see 50 moderately bright to faint stars in a broken halo detached from the central group.

Since the halo spans about 20', it must be part of the cluster that Herschel described. In my little refractor at 68x, the core shows only seven faint stars enmeshed in the dappled glow of unresolved suns.

The open cluster **Basel 11A** lies 49' south-southwest of Thor's Helmet. Observing with my 105mm refractor at 68x, I unearthed a small group containing one 8th-magnitude star with a dozen faint sparks scattered around it. A 9½-magnitude star rests at its north-northwest edge. In my 10-inch reflector, Basel 11A seems more clusterlike, with the bright luminary enclosed in a triangle of 20 faint stars.

The much prettier cluster **NGC 2360** is found 1.7° south of Basel 11A and 3.4° due east of Gamma (γ) Canis Majoris. Through 7x50 binoculars, it's a small hazy patch next to a 5th-magnitude star. My little refractor at 87x reveals 60 stars, mostly magnitude 11 and 12, in an irregular group about 12' across. Starry loops, whorls, and chains of stars ornament the cluster, and a bluish 9th-magnitude star is buttoned to its eastern edge.

Now move 2.4° west-southwest to **Haffner 23**. At 47x, my 105mm scope shows 30 faint to very faint stars loosely sprinkled across a mottled background about 13' across. The coordinates generally listed for Haffner 23 give the position of the brightest star, which shines at 9th magnitude. Visually, the cluster's center appears to be a few arcminutes northeast of this star.

For a change of pace, check out the planetary nebula **Minkowski 1-13** (Mink 1-13; PN G232.4-1.8), located 3° east-southeast of Haffner 23. You'll need a good star chart to pinpoint this tiny object, which appears almost stellar at 44x in my 10-inch scope. Increasing the magnification to 220x, I see the nebula as a small but fairly bright oval tipped north-northeast. It shines at magnitude 12.6 and sits a mere 45" north of an 11.2-magnitude star.

Swooping about 7° southward takes us to blue-white Tau (τ) Canis Majoris, cradled in the center of the beautiful cluster **NGC 2362**. My little refractor at 47x shows 25 much fainter stars tightly nestled against Tau in a richly compact group 6' across. In my 10-inch at 220x, I count 45 stars crowded into 7'. Arizona amateur deep-sky observer Wayne Johnson (aka Mr. Galaxy) describes NGC 2362 as "a swarm of bees surrounding their queen."

After an observing session at Fremont Peak, a group of California amateurs nicknamed Tau Canis Majoris the Mexican Jumping Star. They noticed that when the wind shook their telescopes, Tau seemed to move differently than the cluster stars around it. This is presumedly a visual effect due to afterimages of the bright star. Try to see this yourself by tapping the side of your scope.

At approximately 5 million years old, NGC 2362 is an exceptionally young cluster. That's only 4 percent of the age of the Pleiades in Taurus.

Powered by an extremely hot Wolf-Rayet star, the unusual emission nebula NGC 2359 in Canis Major has brighter features that resemble a Viking warrior's headgear. This has given rise to the nebula's popular name: Thor's Helmet. The field here is ½° wide with north up. Photo: Daniel Verschatse / Observatorio Antilhue / Chile

First published March 2007

NGC 2367 is an open cluster that's often overlooked but deserves more attention. It sits 3.1° north of Tau. At 87x in my small scope, a dozen stars outline a scraggly fir tree. The tree is 5' tall with its top to the south. Someone at a star party once told me that he calls this Charlie Brown's Christmas tree. Anyone familiar with the animated Peanuts holiday program will immediately see the likeness. NGC 2367 is also about 5 million years young.

The lovely double star **h3945** lies about halfway along and 38' west of a line connecting NGC 2367 and Tau. (This double also carries the moniker 145 G Canis Majoris, though the designation is often incorrectly listed without the G. The G indicates it's from the 1879 *Uranometria Argentina* by Benjamin Apthrop Gould.) It's the brightest star in the area and sports 5.0- and 5.8-magnitude components 27" apart. Although striking in appearance, this is only an optical pair whose unrelated stars lie along the same line of sight. Astronomy author James Mullaney dubbed this duo the Winter Albireo for its resemblance to the famous gold and blue double in Cygnus. In a small scope, they seem gold and white to me. What colors do you see?

By Jiminy!

Bright Castor and Pollux draw attention to Gemini, among whose stars you'll find many treasures.

When wintry tempests o'er the savage sea
Are raging, and the sailors tremblingly
Call on the Twins of Jove with prayer and vow,
Gathered in fear upon the lofty prow.

— Homer, "Hymn to Castor and Pollux"

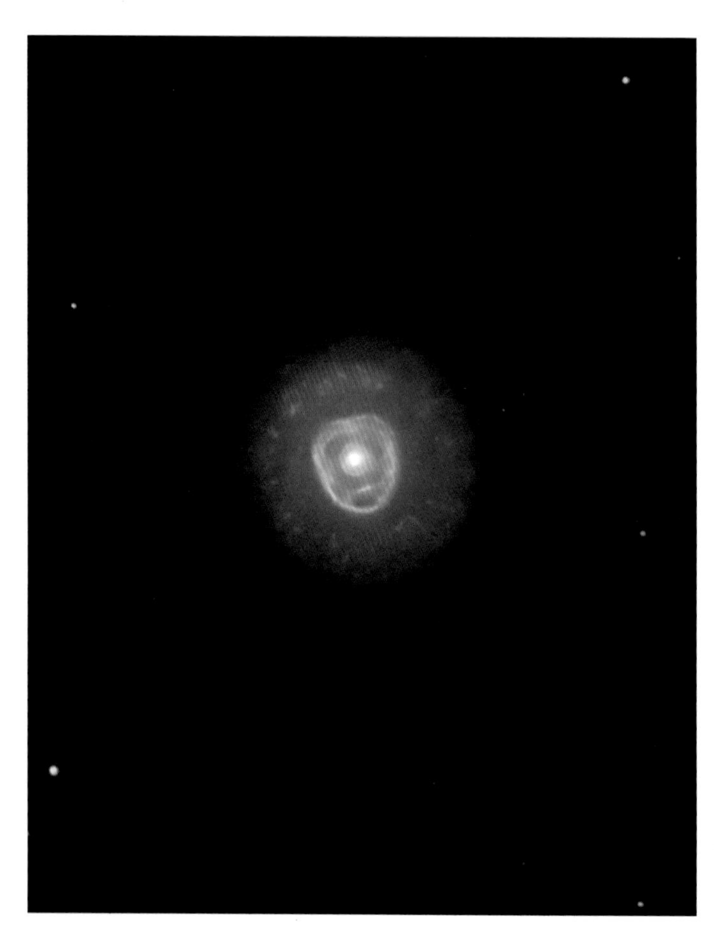

Castor and Pollux, the famed twins of Greco-Roman mythology, hold a place in the sky as the constellation Gemini. Once objects of adjuration, the twins were invoked with the oath "By Gemini," which eventually evolved into "By Jiminy."

There are many celestial wonders by Gemini, not the least of which is the amazing star **Castor**, or Alpha (α) Geminorum. Just 52 light-years away, Castor is a remarkable system of six suns engaged in an intricate gravitational dance. Through my 105mm refractor at a magnification of 87x, I see Castor A and Castor B cleanly split. These white-hot suns form a visual binary with an orbital period of 467 years. The 3rd-magnitude companion is 4.8 arcseconds (4.8") northeast of its 2nd-magnitude primary in 2011, and their separation will slowly widen to a maximum of 7.3" in the year 2085.

I also see Castor C to the south-southeast, a pale orange speck very widely separated from its bright companions. Its carroty color shows up much better in my 8-inch scope. Each member of this trio is actually a very close pair — too close to be split with any telescope. We know that they are paired stars in tight embraces only because their spectra show periodic wavelength shifts, the telltale indicators of orbital motion.

A Gemini showpiece, the planetary nebula NGC 2392 earned its Eskimo Nebula nickname from its likeness to a human face surrounded by a parka's fur-rimmed hood. The nebula began forming about 10,000 years ago when its central star started shedding material. Shown with south approximately up to emphasize the "face," the field is about 3 arcminutes (3') wide. Photo: Bernd Flach-Wilken / Volker Wendel

Our next stop is a planetary nebula whose two lobes have separate designations, **NGC 2371** for the southwestern extension and **NGC 2372** for the northeastern one. They were discovered in 1785 by William Herschel as he swept his English sky with an 18.7-inch reflector. His description reads: "Two, faint of an equal size, both small, within a minute of each other; each has a seeming nucleus, and their apparent atmospheres run into each other."

Deep yellow Iota (ι) Geminorum serves as a handy starting point for locating this fine bipolar planetary, sometimes called the Gemini Nebula for its twin nature. From Iota, climb 1.7° due north, where you'll find NGC 2371/72 hiding out in the eastern side of a 13' triangular patch of faint stars. My little refractor at 87x shows a small, oblong nebula. At 174x, I can distinguish the lobes, the southwestern one being brighter. In my 10-inch reflector at 166x, each half grows brighter toward an off-center patch, and the southwestern lobe holds a starlike spot.

Adding an oxygen III filter makes this spot stand out much better, indicating that it's not a star but, rather, a tiny, intense knot in the nebula. At 213x without a filter, I see faint haze between the lobes and in a thin envelope around them.

Outrigger arcs of nebulosity northwest and southeast double the apparent size of NGC 2371/72. Practiced observers have been able to see these with the help of oxygen III filters and scopes as small as 10 inches in aperture. The 14.8-magnitude central star that is nestled between the main lobes has been glimpsed through 11-inch and larger scopes at high magnification without a filter.

Another splendid planetary lies near Delta (δ) Geminorum. To find it, slide 1.8° east from Delta to a 7th-magnitude orange star, which forms the double star **South 548 AC** (S548) with a yellow-white, 9th-magnitude companion a wide 37" to the west. This double marks the pointy bottom of a skinny, upside-down kite formed with three brighter stars about 35' to its south.

Follow the line formed by the eastern side of the kite southward to what might, at low power, look like a north-south double star. Careful scrutiny will show the southern "star" to be the tiny bluish disk of **NGC 2392**. With my 105mm refractor at 153x, this bright, round, blue-gray planetary is mottled and contains an obvious central star. Gazing directly at the nebula through my 10-inch reflector at low power makes the central star stand out, but gazing a little off to one side (known as averted vision) makes the nebula appear more prominent. Boosting the magnification to 213x brings out

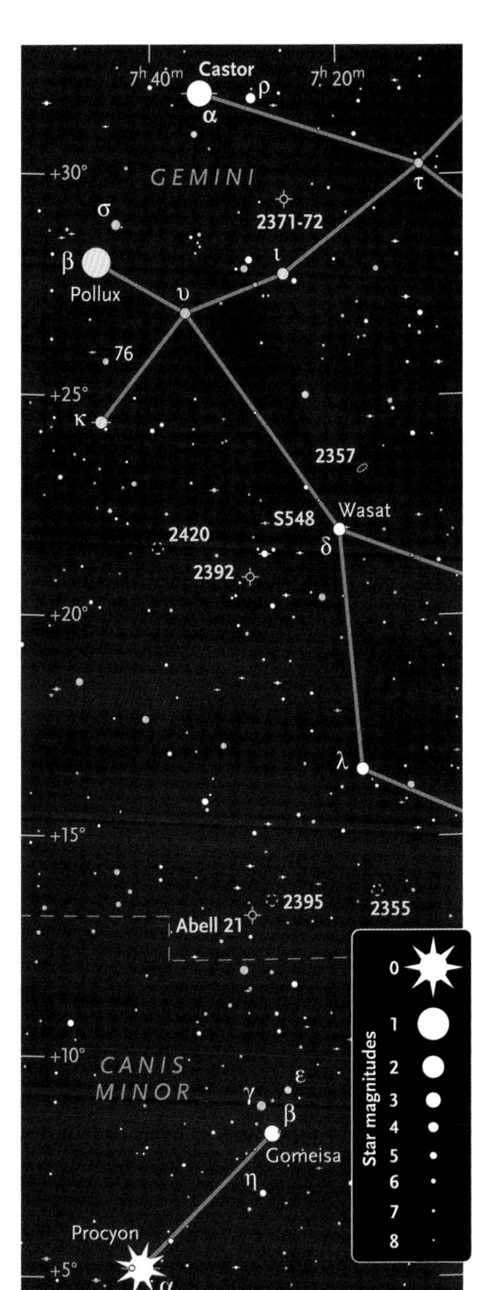

Believed by William Herschel to be two objects, and thus designated separately, the planetary nebula NGC 2371/72 is sometimes called the Gemini Nebula to link its association with the celestial Twins. North is up and the field is 3' wide. Photo: © 1999 Subaru Telescope / NAOJ

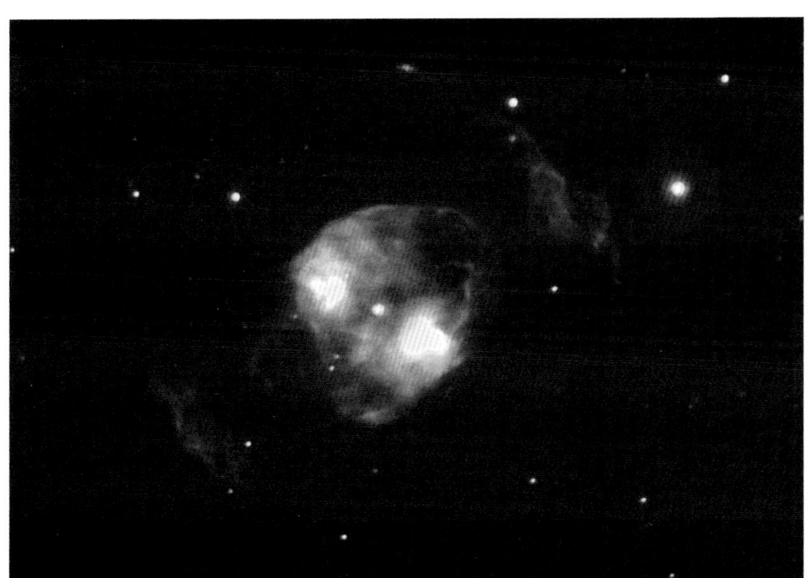

Some Telescopic Treasures in Gemini

Object	Type	Magnitude	Size/Sep.	Right Ascension	Declination	*MSA*	*U2*
Castor	Triple star	1.9, 3.0, 9.8	4.8", 71"	7^h 34.6^m	+31° 53'	130	57R
NGC 2371/72	Planetary nebula	11.2	55"	7^h 25.6^m	+29° 29'	130	57R
S548	Double star	7.0, 8.9	37"	7^h 27.7^m	+22° 08'	152	75R
NGC 2392	Planetary nebula	9.1	50"	7^h 29.2^m	+20° 55'	152	75R
NGC 2420	Open cluster	8.3	10'	7^h 38.4^m	+21° 34'	152	75R
NGC 2357	Galaxy	13.3	3.5" × 0.4'	7^h 17.7^m	+23° 21'	153	75R
NGC 2355	Open cluster	9.7	9'	7^h 17.0^m	+13° 45'	201	95L
NGC 2395	Open cluster	8.0	15'	7^h 27.2^m	+13° 37'	200	95L
Abell 21	Planetary nebula	10.3	12.4' × 8.5'	7^h 29.1^m	+13° 15'	200	95L

Angular sizes and separations are from recent catalogs. Visually, an object's size is often smaller than the cataloged value and varies according to the aperture and magnification of the viewing instrument. The columns headed *MSA* and *U2* give the appropriate chart numbers in the *Millennium Star Atlas* and *Uranometria 2000.0*, 2nd edition, respectively. All the objects this month are in the area of sky covered by Chart 25 in *Sky & Telescope's Pocket Sky Atlas*.

tantalizing features. The planetary presents a large, blotchy, inner oval surrounded by a wispy, round halo.

Some large-scale, deep images of NGC 2392 make the halo look like the furred rim of a parka's hood while the details within resemble a face. The likeness earned this planetary the nickname Eskimo Nebula. I had the great fortune of seeing the Eskimo in 2001 at the Winter Star Party, in the Florida Keys, through a 24-inch reflector owned by Georgia amateur Alex Langoussis. The magnification exceeded 1,000x, far beyond what atmospheric seeing usually allows, and the view was mind-blowing. This is the only time I've seen the complex detail shown in such photos.

For the next object, return to the asterism used to find the Eskimo. From the kite's top (its southern part), scan 2.5° east to the open cluster **NGC 2420**. It's an obvious patch of haze in my little refractor at 17x. At 87x, I see 15 stars within 8' and a little unresolved haze lingering in the center. The brightest star is an 11th-magnitude speck on the western side, and the rest are mostly 12th magnitude. This pretty group is rich in faint to very faint stars in my 10-inch scope. At 166x, I count 50 stars, many draped in little arcs, sprinkled across 10'.

Our next target is **NGC 2357**, a spiral galaxy included in the *Revised Flat Galaxy Catalogue* (Igor D. Karachentsev and others, 1999) for its highly elongated appearance.

Flat galaxies are spirals with little or no central bulge that are seen edge-on. NGC 2357 was discovered by Edouard Stephan in 1885 with the 31.5-inch reflector at Marseille Observatory, the first large telescope with a silver-on-glass mirror. Stephan described NGC 2357 as a "small spindle elongated southeast to northwest; length about 3'; excessively excessively faint; very thin; wrapped in several very small stars; a slightly brighter point near the center."

Don't let "excessively excessively faint" scare you; NGC 2357 is challenging, but it's visible in a 10-inch scope. It's exactly 1.5° north-northwest of Delta (δ) Geminorum, and a good star chart is a help for pinning down its position.

Rounding out the trio of planetary nebulae in Gemini, Abell 21 is a challenging object for backyard telescopes, though the author has viewed it with a 105mm refractor equipped with an oxygen III nebula filter. The field is 17' tall with north up. Photo: Chris Schur

Breathe carefully

Warm breath on a cold eyepiece can temporarily halt an observing session. Even moisture from your eye can fog an eyepiece's exposed lens on winter nights. You can prevent these problems by rotating a pair of frequently used eyepieces between a warm coat pocket and your telescope.

First published March 2008

Inspect the area for a pair of 9½-magnitude stars separated by 1.2' and aligned east-northeast to west-southwest. Extend a line from this duo through a third star of similar brightness 8.6' east-southeast. Continue for nearly the same distance again to reach NGC 2357.

You'll need medium to high power to see it. With my 10-inch scope at 299x, the galaxy is a ghostly sliver of light with a slightly brighter center. A 13th-magnitude star is poised above its northwestern end. NGC 2357 is an interesting subject for astrophotographers, who can try to capture its gently warped profile and superposed star.

Farther south, we find the rich open cluster **NGC 2355**. A line from Delta to Lambda (λ) Geminorum and extended for half that distance again indicates its position. In my 105mm scope at 17x, I see two faint stars in a small hazy spot just south-southwest of an 8th-magnitude star. At 87x, this is a pretty, raggedy-edged group with 20 stars in 10'. They are most densely crowded in the central 3½', where some haze remains.

Sweeping 2½° eastward brings us to the cluster **NGC 2395**. My little refractor at 87x reveals 20 stars loosely scattered across 15'. At 28x, it merely shows a granular-looking patch with two faint stars, but something remarkable happens when I add an oxygen III filter. Although completely invisible before, **Abell 21**, the Medusa Nebula, joins the scene ½° southeast of the cluster! I can see it with direct vision, but it shows up better with averted vision. This unusual planetary nebula is about 8' across, dented in on its northwest side, and brightest toward the northeast and southwest. With my 10-inch scope at 68x, I prefer viewing Abell 21 with a narrowband nebula filter (rather than the oxygen III filter), which shows this large, impressively detailed planetary to be very uneven in brightness.

In Starry Skies

Cancer is a telescopic store
of glimmering clusters and double stars.

In starry skies, long years ago,
I found my Science. Heart aglow
I watched each night unfold a maze
Of mystic suns and worlds ablaze,
That spoke: "Know us and wiser grow."
— Sterling Bunch, "In Starry Skies"

We stand poised at the turn of the seasons, and nowhere is this made clearer than in our view of the starry sky. West, we see the brilliant star patterns that kept us company throughout the winter, while east, the more subdued constellations of spring come to greet us. But no matter the time of year, the celestial realm holds wonders worthy of heed.

Cancer, the Crab, is one of the dimly lit patterns now taking center stage in the evening sky. As one of the zodiacal constellations, it is sometimes called the Dark Sign. Its starry nature lies not so much in the suns that outline the Crab as in the deep-sky wonders found within its realm.

The crowning glory of Cancer is **Messier 44** (M44), also known as Praesepe or the Beehive Cluster. In *Round the Year with the Stars*, Garrett P. Serviss described it as "a glimmering spot, a kind of starry cobweb" visible to the unaided eye in a moderately dark sky. I see it as a delicate, granular haze.

M44 is a glittering treasure for binoculars of all sizes. My 15x45 image-stabilized binoculars furnish a particularly fine view. Nine bright stars form a sideways V with a dozen more scattered around it. Four of the brightest boast a golden hue: one at the V's point, one in its northern arm, one north of the V, and one that the northern arm points toward. In total, I count 55 stars flung across 1½°.

You won't see the entire Beehive in your field of view with a large telescope, but you can use the labeled photo on page 72 to hunt down the swarm of small, faint galaxies that softly buzz in its background. I can spot three of them in my 10-inch reflector at 213x: NGC 2624 and NGC 2625 in the western side of the cluster, and NGC 2647 in the east. Images of NGC 2625 show a faint star wedded to the galaxy's western flank, but I've been unable to distinguish it. With my 14.5-inch reflector, I can also capture NGC 2643, IC 2388, UGC 4526, and one of the CGCG 89-56 galaxy pair. The Beehive hosts NGC 2637, one NGC galaxy that I've never seen with certainty. Can you spot it?

Sweeping 7° west-southwest of M44 brings us to **Zeta (ζ) Cancri**, a lovely trio of 5th- and 6th-magnitude stars that looks like a double at low magnifications. In my 105mm refractor at 47x, it's a light yellow primary with a dimmer yellow attendant to the east-northeast. At 174x, the brighter star looks pinched, both lobes having the same hue, while at 203x, it's a kissing couple, the fainter partner northeast. In my 10-inch scope, the close pair is split by a hair at 213x and well split at 299x.

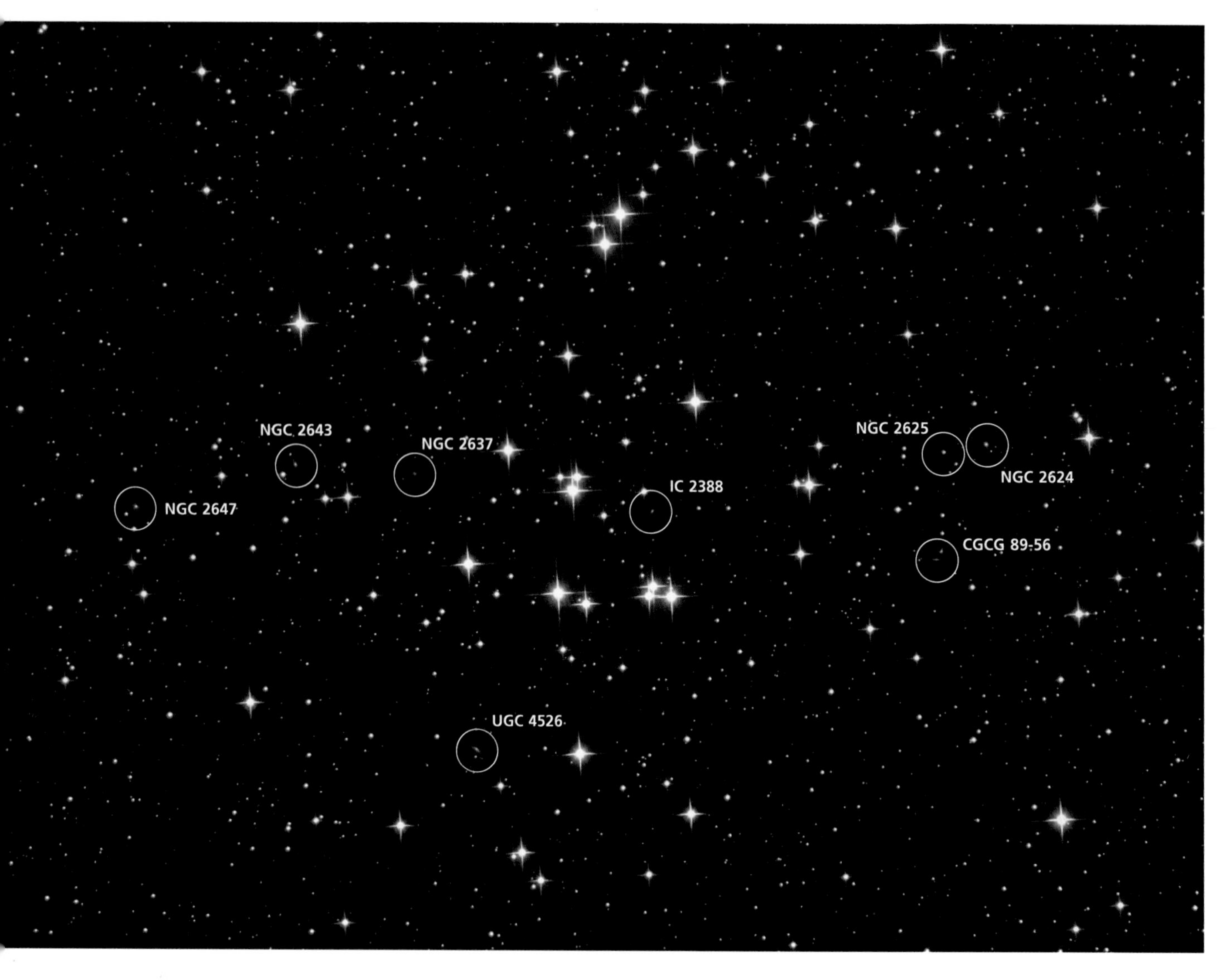

If you look closely at Robert Gendler's splendid photograph of the Beehive Cluster, you'll see many faint galaxies lurking within it. And you'll find the same thing if you examine the cluster carefully under dark skies with a 10-inch or larger telescope.

Zeta is a nearby star system, just 83 light-years away. From our vantage point, the components slowly change position and separation over the years. The wider pair has an orbital period of roughly 1,115 years, and the close pair's period is 60 years.

To admire a more strongly colored star, let's visit **X Cancri**, located 2° north-northwest of Omicron (ο) Cancri. X Cancri is a carbon star, a cool giant that has an exceptionally ruddy tint because most of its blue light is filtered out by carbon molecules and carbon compounds in its atmosphere. It's also a semiregular variable star whose brightness ranges between magnitude 5.6 and 7.5, well within the grasp of a small telescope.

Next, sweep 1.7° west from Alpha (α) Cancri to the only other confirmed cluster in Cancer, **Messier 67**. While M44 is about 610 light-years distant, M67 is nearly five times farther away and displays stars that look correspondingly dimmer. M67 is also much older: 2.6 billion years compared with 730 million for M44.

Through my little refractor at 17x, Alpha and M67 share the field of view. This stunning group of many barely resolved and densely packed stars is irregular in both concentration and outline. A considerably brighter, yellow-orange star adorns its northeastern edge. At 47x, the bright star is part of a 15' wedge of suns pointing east-northeast. A much more heavily populated tree of stars lies west of the wedge. The shining trunk and star-leafed branches curve up toward the west, and the tree is 11' tall. At 87x, I count 80 stars in this amazing cluster, which spans about 22'.

M67 inspires many comparisons. Irish amateur Kevin Berwick sees a "bright fountain of stars," with the "rusty orange" lucida at its top. Also noting this colorful star, English observer Dale Holt calls M67 the Golden-Eye Cluster and deems it "a little treasure." In Australia, Doug and Janet

Left: Dean Salman's 6½-hour exposure shows the faint nebulosity south-southeast of Abell 31's relatively bright ring. Inset: Here's Jaakko Saloranta's impression of Abell 31 as seen through an 80mm telescope equipped with an oxygen III filter.

Adams visualize M67 as a Pac-Man (from the video game by that name) and point out a "concave shape to the cluster, as if it's gobbling stars ahead of it." New York amateur Joe Bergeron remarks, "M67 has a loopy appearance and suggests many forms, including a crouching long-tailed monkey. I also often see it as a chalice." What flights of fancy does this magnificent cluster prompt for you?

The sizable planetary nebula **Abell 31**, labeled on many star charts as PK 219+31.1, rests 3° south-southeast of M67. In my semirural skies, Abell 31 is faintly visible at low power with a nebula filter through my 10-inch scope but makes a significantly better showing in my 14.5-inch. It spans about 12', with one 12th- and two 10th-magnitude stars superposed. The latter two form the southern side of a five-star, stick-figure house whose roof peaks to the east.

Unlike most planetary nebulae, Abell 31 is dominated by lines of doubly ionized oxygen (O III) only in its central region. Small scopes equipped with oxygen III filters often show just this area. Savvy observers under dark skies have nabbed parts of this gauzy net of gossamer in scopes as small as 80mm.

If the galaxies in the Beehive are too difficult for you, try Cancer's brightest galaxy, **NGC 2775**. It's located near the Cancer-Hydra border, 2.2° north-northeast of Omega (ω) Hydrae. Through my 105mm scope at 17x, the galaxy is a tiny but fairly bright spot, looking much like a planetary nebula with a brighter center. At 87x, NGC 2775 is a 3¼' x 2½' oval leaning north-northwest, wreathed by a handful of extremely faint stars.

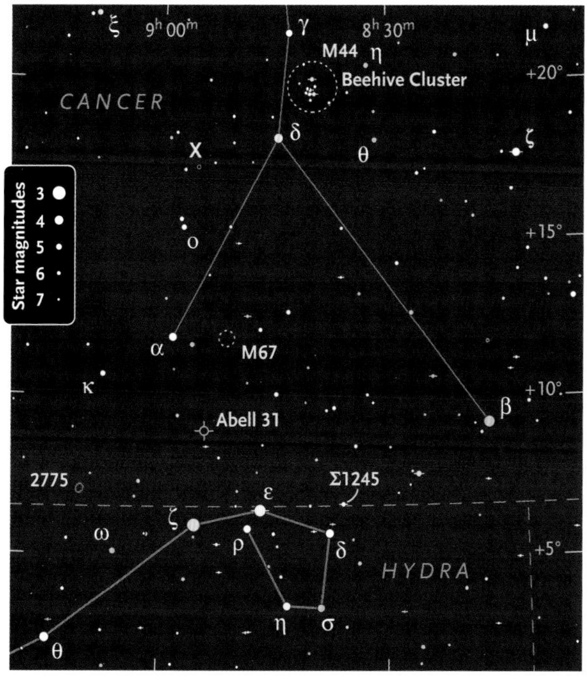

Messier 67 appears smaller and fainter than the Beehive because it's five times farther from us. But if both clusters were at the same distance, M67 would appear much more impressive than its counterpart. Photo: Robert Gendler

First published March 2009

NGC 2775 appears like an exotic hybrid of an elliptical galaxy in the center with a tightly wound spiral in its outer disk. Photo: Jeff Newton / Adam Block / NOAO / AURA / NSF

Its relatively large core clasps a minute, brighter nucleus.

Deep images of NGC 2775 are spectacular. A large portion of the galaxy's interior is essentially featureless, much like an elliptical galaxy. Then, quite abruptly, this smooth glow gives way to a picturesque, fleecy spiral structure. Adding depth to the scene, a distant clump of background galaxies can be seen through NGC 2775's outer halo.

Also hugging the Hydra border, the pretty double star **Σ1245** can be found 1° north-northwest of Delta (δ) Hydrae. It's nicely split even at 28x in my little refractor. The yellow, 6th-magnitude primary holds a deep yellow, 7th-magnitude companion 10" to the north-northeast — a pretty pair of gems with which to end our deep-sky tour.

The Crab's Collection

Object	Type	Magnitude	Size/Sep.	Right Ascension	Declination
M44	**Open cluster**	**3.1**	**1.5°**	**8^h 40.4^m**	**+19° 40'**
ζ Cancri	**Triple star**	**5.3, 6.3, 5.9**	**1.1", 5.9"**	**8^h 12.2^m**	**+17° 39'**
X Cancri	**Carbon star**	**5.6–7.5**	**—**	**8^h 55.4^m**	**+17° 14'**
M67	**Open cluster**	**6.9**	**25'**	**8^h 51.4^m**	**+11° 49'**
Abell 31	**Planetary nebula**	**12.0**	**16'**	**8^h 54.2^m**	**+8° 54'**
NGC 2775	**Galaxy**	**10.1**	**4.3' × 3.3'**	**9^h 10.3^m**	**+7° 02'**
Σ1245	**Double star**	**6.0, 7.2**	**10.2"**	**8^h 35.9^m**	**+6° 37'**

Angular sizes and separations are from recent catalogs. Visually, an object's size is often smaller than the cataloged value and varies according to the aperture and magnification of the viewing instrument.

Keeper of Secrets

Little-known Lynx hosts some exotic deep-sky objects.

To some native North Americans, the lynx was a totem animal regarded as a keeper of secrets. This is an apt description for the constellation Lynx, which bears no star brighter than 3rd magnitude and no deep-sky object that immediately springs to mind for the casual observer.

Lynx was devised by Johannes Hevelius and depicted in his 1687 atlas *Firmamentum Sobiescianum*. In her monumental work on celestial cartography, *The Sky Explored*, Deborah Jean Warner writes, "Lynx stands as a reminder of Hevelius' distrust of telescopic sights for stellar observations: to see the stars as a constellation, he wrote, you must be as sharp-sighted as a lynx."

Since Lynx isn't easy to pick out in the sky, let's start our tour of deep-sky wonders in nearby Cancer. Also a faint constellation, Cancer enjoys the advantage of having a distinctive, upside-down Y shape that is lodged between the bright patterns of Gemini and Leo.

Iota (ι) **Cancri**, which marks the base of the Y, is a beautiful double star. Its deep yellow primary and white companion are easily visible through any telescope at low power. Some observers see the companion star as bluish, a color-contrast illusion. Using enough magnification to put lots of space between the components will help you discern the true colors.

Working our way toward the border of Lynx, we find the triple star **57 Cancri** at the end of a curvy line of 7th- to 9th-magnitude stars trending northeast from Iota. The primary is a deep yellow, 6th-magnitude star. Low power shows a 9th-magnitude attendant a roomy 55" to the south-southwest. But you will need additional magnification to separate the primary from a companion barely fainter than itself. Through my 105mm refractor, the pair is elongated at 87x, kissing at 122x, split by a hair at 153x, and well split at 203x. The companion cuddles up northwest of its primary and gleams with a yellow-orange hue.

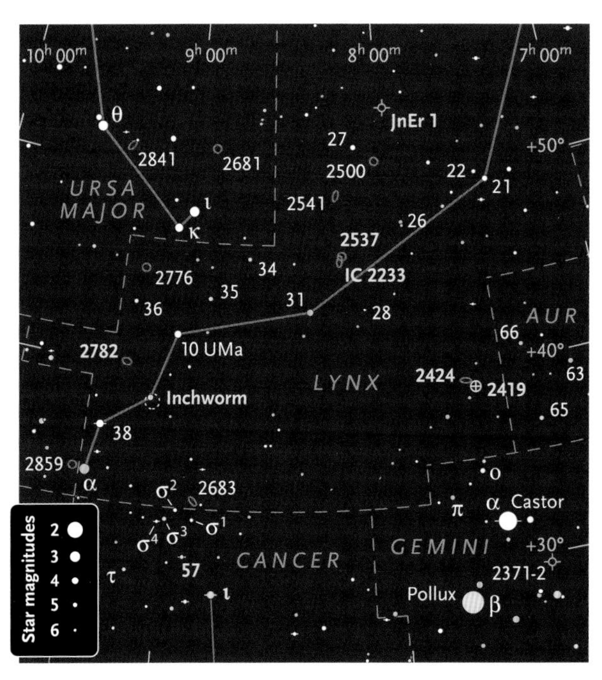

Below: The edge-on spiral NGC 2683 is the most spectacular galaxy in Lynx. Photo: Doug Matthews / Adam Block / NOAO / AURA / NSF

From 57 Cancri, hop 2° north to Sigma^1 (σ^1) Cancri and then 1° farther to Lynx's brightest galaxy. The elongated smudge of **NGC 2683** hangs out with the four members of the Sigma Cancri gang through my little refractor at 17x. NGC 2683, Sigma^1, and Sigma^2 mark the corners of an equilateral triangle, and a trapezium of four faint stars dangles 12' south of the galaxy. At 87x, NGC 2683 is a 6' x 1½' spindle, tilted to the northeast, that harbors a 3'-long core. A faint star lies off the galaxy's southeastern flank, and a very faint star sparkles on the galaxy's northern edge. The core looks mottled at 127x. In my 10-inch reflector at 213x, NGC 2683 is very pretty. The halo spans 7½', and the textured core grows more intense toward a very small, bright nucleus.

NGC 2683 is a spiral galaxy seen nearly edge-on. Relatively nearby, at 23 million-light years, this metropolis of stars presents us with such a striking profile that some observers call it the UFO Galaxy.

Puttering around in Lynx with my 105mm scope at 17x, I happened upon a cute asterism that I've christened the **Inchworm**. It's found 3° northwest of 38 Lyncis and contains the yellow, 4.6-magnitude star HD 77912. The Inchworm is 46' long with 10 stars down to 10th magnitude.

In the Lair of the Lynx

Object	Type	Magnitude	Size/Sep.	Right Ascension	Declination
ι Cancri	Double star	4.1, 6.0	31"	$8^h 46.7^m$	+28° 46'
57 Cancri	Triple star	6.1, 6.4, 9.2	1.5", 55"	$8^h 54.2^m$	+30° 35'
NGC 2683	Spindle galaxy	9.8	9.3' × 2.1'	$8^h 52.7^m$	+33° 25'
Inchworm	Asterism	4.3	46'	$9^h 05.9^m$	+38° 16'
NGC 2782	Tidal-tail galaxy	11.6	3.5' × 2.6'	$9^h 14.1^m$	+40° 07'
NGC 2419	Globular cluster	10.4	5.5'	$7^h 38.1^m$	+38° 53'
NGC 2424	Flat galaxy	12.6	3.8' × 0.5'	$7^h 40.7^m$	+39° 14'
NGC 2537	Dwarf galaxy	11.7	1.7' × 1.5'	$8^h 13.2^m$	+45° 59'
NGC 2537A	Face-on galaxy	15.4	0.6'	$8^h 13.7^m$	+46° 00'
IC 2233	Flat galaxy	12.6	4.7' × 0.5'	$8^h 14.0^m$	+45° 45'
JnEr 1	Planetary nebula	12.1	6.8' × 6.0'	$7^h 57.9^m$	+53° 25'

Angular sizes and separations are from recent catalogs. Visually, an object's size is often smaller than the cataloged value and varies according to the aperture and magnification of the viewing instrument.

The yellow star resides in the humped-up part of the Inchworm's body, and his glowing eyes are at the northwestern end.

Moving 2½° northeast of the Inchworm takes us to the peculiar spiral galaxy **NGC 2782**. In my 5.1-inch refractor at 23x, it shows up as a little spot of mist just north of a wide pair of very faint stars. At 102x, the galaxy is a 1.6' x 1.3' oval tipped a bit east of north, and its large core grows much brighter toward the center. My 10-inch reflector at 192x exposes a stellar nucleus and a detached haze 2' east-northeast, with a very faint star on its western edge. A diaphanous wisp weds the southern end of this haze to the rest of the galaxy.

The extended structure east of NGC 2782 is a tidal tail formed about 200 million years ago when a galaxy much like our own merged with a galaxy one-quarter as massive.

For a change of pace, let's visit the globular cluster **NGC 2419**, which garnishes a 3° arc of orange 5th- and 6th-magnitude stars near the Lynx-Auriga border. Through my 5.1-inch refractor at 63x, I see it as a softly glowing globe off the end of a curve of three progressively brighter stars, magnitudes 9 to 7. The center star holds a faint companion 24" to its north-northeast. The 4' cluster grows brighter toward the center and has several faint foreground stars scattered around the edges. The highly elongated galaxy **NGC 2424** is visible 36' northeast. At 102x, it appears about 2½' long, and it's tipped a bit north of east.

Only large-scope users can hope to glimpse NGC 2419's member stars, because the brightest feebly shine at 17th magnitude. The apparent dimness of the stars is due to the globular's astonishing distance of 275,000 light-years. Because of its remoteness, NGC 2419 was dubbed the Intergalactic Tramp, a nickname that has clung with amazing tenacity despite the fact that we've long known this globular is well within the gravitational thrall of our galaxy.

Our next stop is the Bear Paw Galaxy, **NGC 2537**, which sits 3.3° north-northwest of 31 Lyncis. My 5.1-inch scope at 102x shows a bright, 50" core that exhibits odd, large-scale patchiness. The halo is just a thin fringe around it. My 14.5-inch reflector at 170x reveals three lumps — the southward-pointing "toes" of the paw. The face-on spiral **NGC 2537A** joins the scene as a ghostly little orb floating 4½' east of the Bear Paw.

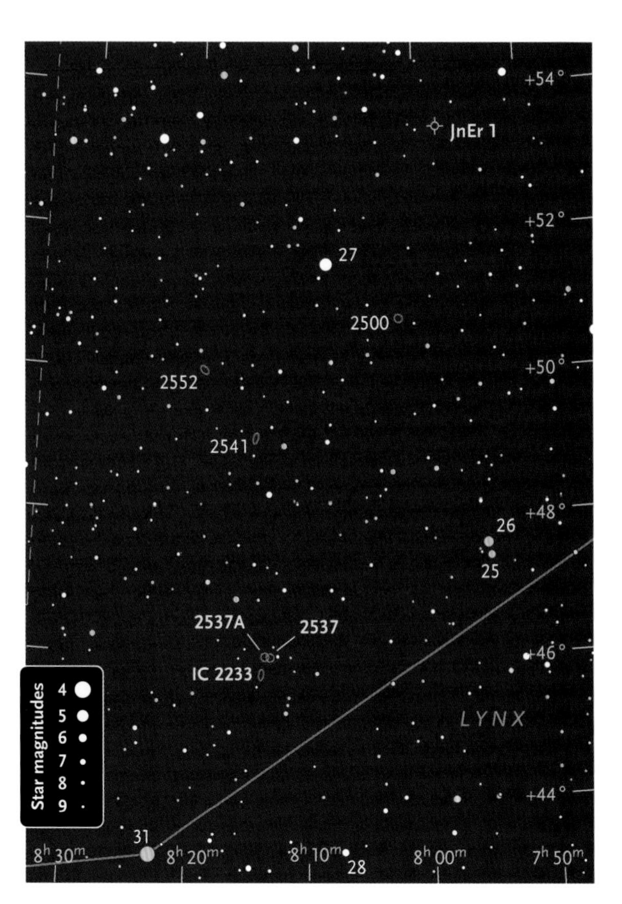

Above: The planetary nebula Jones-Emberson 1 is an elusive target through the eyepiece of a telescope.
Photo, above: Adam Block / NOAO / AURA / NSF.
Photo, right: POSS-II / Caltech / Palomar

IC 2233 is an exceptionally thin, edge-on galaxy 17' south-southeast of the Bear Paw. In my 5.1-inch refractor at 164x, it slashes the sky just west of a 10th-magnitude star with a dim companion. This ashen filament spans 1½', and its northern tip is pinned to the sky with a faint star. The view through my 10-inch reflector doubles the wafer-thin galaxy's length.

The Bear Paw is a compact dwarf galaxy patterned by vast star-forming complexes teeming with bright blue stars. It's about 26 million light-years away. NGC 2537A is perhaps 20 times more distant.

We'll wind up our tour with the huge planetary nebula **Jones-Emberson 1** (JnEr 1; PK 164+31.1). Elaine Osborne of Virginia and Dr. David Toth of Ohio joined me for views through my 5.1-inch scope at the 2009 Winter Star Party in Florida. The nebula is quite elusive at 63x. The brightest patch makes a right isosceles triangle with 11th-magnitude stars east and northeast. A dimmer patch is faintly visible 4' northwest. The two are tenuously connected by phantom arcs that convert the nebula into a 6' ring. Using a narrowband nebula filter or a higher magnification betters the view.

JnEr 1 was first reported in the Harvard College Observatory Bulletin in 1939 by Rebecca B. Jones and Richard Maury Emberson. Strangely, the bulletin states: "A faint nebular ring has been detected joining two condensations, NGC 2474, observed by Sir John Herschel, and NGC

2475." While the article's image does, indeed, show the nebula, NGC 2474 and NGC 2475 are a galaxy pair found ½° farther south. This misidentification was reproduced in many later references.

As you can see, our secretive Lynx holds many unusual treasures. I hope you'll find them worth uncovering.

First published March 2010

Spring

Gregarious Galaxies

Something Italian amateur Mirko Villi noticed in this galaxy group sparked intense interest among professionals.

Galaxies are social creatures. Most gather into groups that range in size from exclusive cliques with just a few individuals to vast flocks of several thousand. The relatively small M96 Group, or Leo I, is well placed for observing on spring evenings. This is the nearest galaxy cluster containing bright spirals and a bright elliptical galaxy. The M96 Group may be physically associated with the nearby Leo Triplet (or M66 Group).

First let's visit our cluster's namesake galaxy, **M96**. On the April all-sky star map on page 305, you'll see Leo's brightest star, Regulus, near the meridian (an imaginary line vertically dividing the map in half) and also quite close to the ecliptic. If you look a little eastward along the ecliptic, you will see a much dimmer blue-white star, Rho (ρ) Leonis. M96 lies about one-third of the way from Rho to Theta (θ) Leonis (the star at the right angle of the triangle that defines the Lion's hindquarters).

For more accurate pointing, use your finderscope and the chart on this page to locate 5.3-magnitude 53 Leonis 4.2° east-northeast of Rho; it is the brightest star in the area and forms a nice little equilateral triangle with a 7th- and an 8th-magnitude star nearby. Switching to your telescope and a low-power eyepiece, scan 1.4° north-northwest of 53 Leonis to look for the 9th-magnitude blur of M96.

Through my 14x70 binoculars, M96 is visible as a faint, fuzzy patch with some brightening toward the center. My 105mm refractor at 127x reveals only a little more detail. The galaxy is slightly oval, and the long dimension runs northwest to southeast. Its dim outer halo contains a large, bright core and a stellar nucleus.

In 1998, Italian amateur astronomer Mirko Villi discovered a Type Ia supernova in M96. Type Ia supernovae are considered good "standard candles" for determining the distance to remote galaxies. Such an object in a nearby galaxy — one whose distance had already been determined using other techniques — sparked intense interest among professional astronomers. Several recent studies, using improved values derived from this stellar outburst, have helped astronomers adjust the cosmic distance scale and refine the rate of expansion of the universe. The most recent work on Villi's star (designated Supernova 1998bu) yields a rate of about 70 kilometers per second per megaparsec. This is remarkably close to the weighted value of 72 found in the Hubble Space

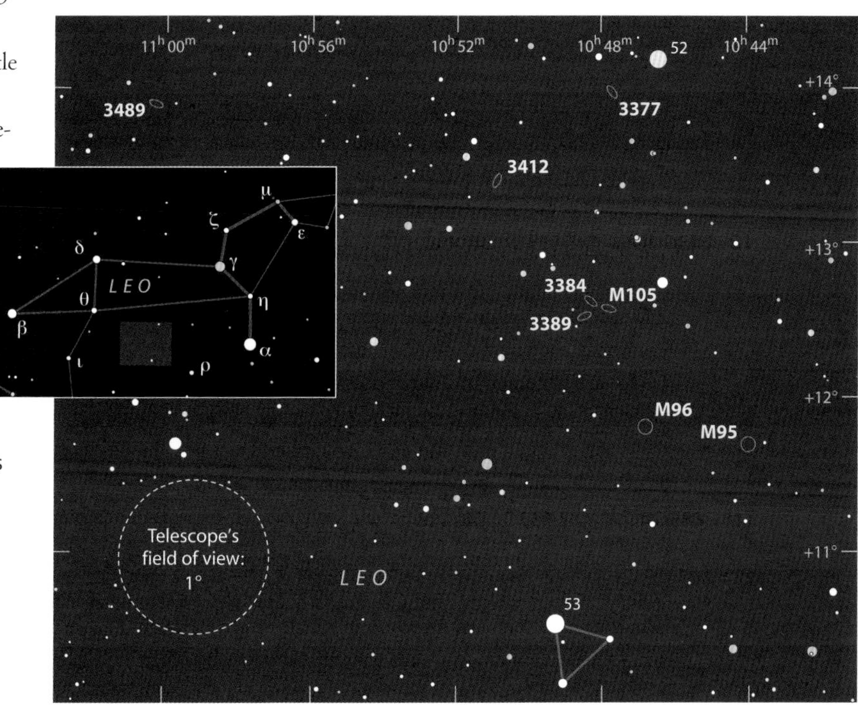

With the help of the inset chart, aim your telescope so that the stars 52 and 53 Leonis are both visible in the finder. Now you're ready to hunt down this month's galaxies, keeping in mind that the main scope has a much narrower field of view (roughly indicated at lower left). This chart goes to 11th magnitude and shows nearly all the stars visible in a 76mm telescope under the best conditions. Chart: Adapted from *Millennium Star Atlas* data

Insets: Taken with a Meade 16-inch telescope, these images of M95 (right) and M96 (below) bring out structural differences in the otherwise similar pair. M95 has a pronounced central bar with a spiral arm coming off each end. M96 is notable for its twisting dust lane. Telescopes seldom show these features visually. Inset photo, right: Bill and Sue Galloway / Adam Block / NOAO / AURA / NSF; Inset, below: Dan Stotz and Mike Ford / Adam Block / NOAO / AURA / NSF

Telescope Key Project. For currently favored cosmological models, these values imply that the universe's age is roughly 13.7 billion years.

Two other Messier galaxies belong to the M96 Group. **M95** lies 42' west of M96, and the two will fit together in the same low-power field. Through 14x70 binoculars, M95 is a very dim smudge. It has the lowest visual magnitude and surface brightness of the Messier trio. My 105mm scope at 127x shows a faint, roundish halo and a brighter core that intensifies toward a tiny bright nucleus. With his 4-inch refractor, the keen-eyed astronomy writer Stephen James O'Meara has been able to detect the little wings that extend beyond the core of this barred spiral as well as a faint outer ring running around the galaxy's edge.

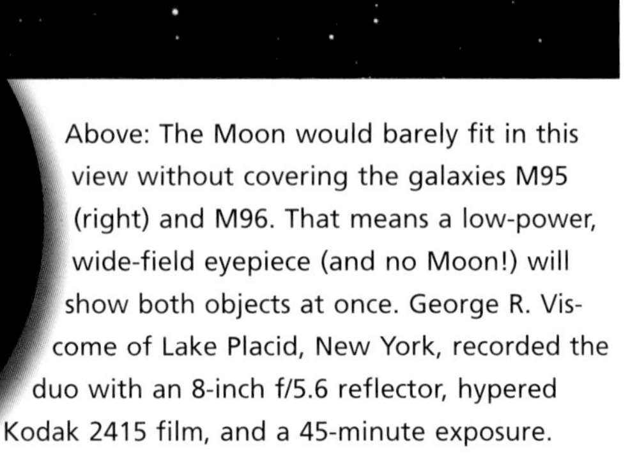

Above: The Moon would barely fit in this view without covering the galaxies M95 (right) and M96. That means a low-power, wide-field eyepiece (and no Moon!) will show both objects at once. George R. Viscome of Lake Placid, New York, recorded the duo with an 8-inch f/5.6 reflector, hypered Kodak 2415 film, and a 45-minute exposure.

The elliptical galaxy **M105** is located 48' north-northeast of M96. M105 has about the same visual magnitude as M96 but is smaller and therefore has a higher surface brightness. Through my 14x70 binoculars, this galaxy is small and faint and shares the field with M95, M96, and 53 Leonis. The four objects are arranged in a Y shape. A small telescope at 30x can encircle all three galaxies in the same field of view.

Counterclockwise from the big elliptical galaxy M105 at right are NGC 3384 (a lenticular galaxy) and NGC 3389. Martin C. Germano of Thousand Oaks, California, used his 14¼-inch reflector and SBIG ST-4 autoguider to take this 100-minute exposure on April 12, 1997.

In my 105mm scope at 87x, I can see M105 and another galaxy, **NGC 3384**, just to its east-northeast. Their centers are a mere 7' apart. On M105, the same instrument at medium to high power shows a tiny, bright nucleus embedded in a slightly oval glow that fades gradually toward the periphery. NGC 3384 looks like a smaller, dimmer version of M105. Close inspection reveals a very faint third galaxy, **NGC 3389**, only 10' east-southeast of M105. NGC 3384 is a member of the Leo I Group, but NGC 3389 is generally thought to be a background galaxy.

In 1781, French observer Pierre Méchain was first to notice the three main galaxies of the M96 Group, but he did not pass on his discovery of M105 in time for inclusion in Charles Messier's final catalog. M105 was added to the list much later, by Helen Sawyer Hogg in 1947.

At least three other Leo I galaxies are within the grasp of a small telescope, all

First published April 2002

Deep-Sky Denizens of the M96 Group (Leo I)

Object	Galaxy Type*	Magnitude	Size/Sep.	Right Ascension	Declination
M96	Sbp	9.3	7.1' × 5.1'	$10^h 46.8^m$	+11° 49'
M95	SBb	9.7	7.4' × 5.1'	$10^h 44.0^m$	+11° 42'
M105	E1	9.3	4.5' × 4.0'	$10^h 47.8^m$	+12° 35'
NGC 3384	SB0	9.9	5.9' × 2.6'	$10^h 48.3^m$	+12° 38'
NGC 3389	Sc	11.9	2.8' × 1.3'	$10^h 48.5^m$	+12° 32'
NGC 3377	E5	10.4	4.4' × 2.7'	$10^h 47.7^m$	+13° 59'
NGC 3412	E5	10.5	3.6' × 2.0'	$10^h 50.9^m$	+13° 25'
NGC 3489	E6	10.3	3.7' × 2.1'	$11^h 00.3^m$	+13° 54'

*Galaxies belong to two broad categories, those of spiral form (S) and the less-structured ellipticals (E).

lying within a finder field of M105. Each appears brighter than nonmember NGC 3389 but dimmer than NGC 3384. Search for **NGC 3377** 1.4° north of M105 and 23' southeast of 5.5-magnitude, yellowish 52 Leonis. It is small, not quite round, and has a tiny, bright nucleus. **NGC 3412** lies 1.1° northeast of M105 and 16' southwest of a pair of white and golden 8th-magnitude stars. It is faint, very small, and roundish, with a brighter, stellar nucleus. **NGC 3489** is the most difficult of the three to find. There are no bright stars or distinctive pairs to serve as guideposts, so use the star chart on page 81 to star-hop along the snaking chain of 8th- and 9th-magnitude stars that begins northeast of the M105 galaxy triplet. Alternatively, you can put the white and golden pair of stars near NGC 3412 in the southern part of a low-power field and then scan slowly eastward looking for the telltale glow of NGC 3489. It is a small oval comparable in brightness to NGC 3412, and it has a brighter nucleus.

The galaxies of Leo I are near enough to our own Milky Way that astronomers can study many of the Cepheid variable stars they contain. These well-known distance indicators have placed the cluster at about 38 million light-years, roughly two-thirds the distance to the great Virgo Galaxy Cluster.

Highlights of Hydra

Follow the twists and coils of this celestial snake for some interesting sights to see with your small telescope.

Hydra, the Water Snake, is our largest constellation both in area and in length. The she-serpent winds across more than a quarter of the heavens, and this is the only time of year when the evening sky will show you Hydra in her entirety. Her head is marked by an oval of stars south of Cancer, the Crab, while her tail abuts Libra, the Scales. With her enormous length, Hydra can keep you entertained throughout the evening as the passing hours bring each new wonder to its highest place in the sky.

Our first showpiece is **M48**, the brightest open star cluster in Hydra. As twilight ends in April, the cluster is already descending toward the west, so you should catch this one early in the evening. To locate M48, look for 3.9-magnitude C Hydrae 8° south-southwest of Hydra's head. C Hydrae is easy to recognize in a finder because two 5.6-magnitude stars flank it. Placing C near the northeastern edge of your finder's field will bring the hazy glow of M48 into view.

When you switch to a low-power telescopic view, M48 is a beautiful sight! My 105mm scope at 47x shows more than a hundred stars of 8th to 12th magnitude in nearly 1°. The central ½° is more densely populated, while the stars loosely cast around the periphery blend into the background sky.

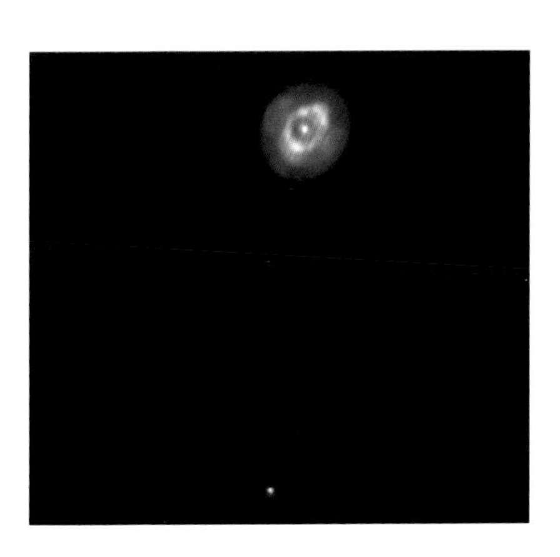

While it looks tiny, pale, and ghostly in small telescopes, the planetary nebula NGC 3242 (top) shows fine structure in this CCD image taken with an RC Optical Systems 12.5-inch reflector. The 11th-magnitude star at bottom lies just 2.4' south of the nebula. Photo: Richard D. Jacobs

Sights in Hydra

Object	Type	Magnitude	Size/Sep.	Distance (l-y)	Right Ascension	Declination	*MSA*	*U2*
M48	Open cluster	5.8	54'	2,500	$8^h\ 13.7^m$	–5° 45'	810	114R
NGC 3242	Planetary nebula	7.3	40" x 35"	2,900	$10^h\ 24.8^m$	–18° 39'	851	151R
V Hydrae	Carbon star	6–13	—	1,600	$10^h\ 51.6^m$	–21° 15'	850	151L
M68	Globular cluster	7.8	11'	33,000	$12^h\ 39.5^m$	–26° 45'	869	150L
M83	Spiral galaxy	7.5	13' × 11'	15,000,000	$13^h\ 37.0^m$	–29° 52'	889	149L

Angular sizes are from catalogs or photographs; most objects appear somewhat smaller when a telescope is used visually. Approximate distances are given in light-years based on recent research. The columns headed *MSA* and *U2* give the chart numbers of objects in the *Millennium Star Atlas* and *Uranometria 2000.0*, 2nd edition, respectively.

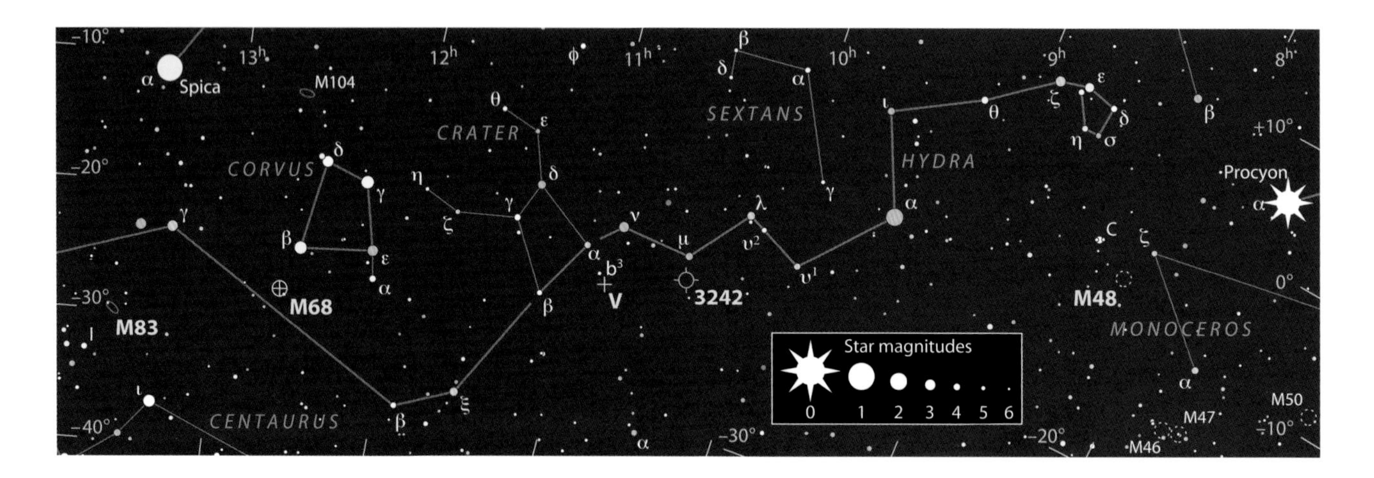

The group's brightest member is found at the southeastern edge of the central mass and shines with a yellowish hue.

M48 was once considered a missing Messier object, since there is no cluster at the coordinates Charles Messier gave in the late 18th century. Some older atlases plot M48 in the original, incorrect position — a fact I discovered the hard way when learning my way around the sky. Messier described M48 as a cluster of faint stars near the three stars that begin the Unicorn's tail. These stars are C Hydrae and its attendants, which no longer belong to the Unicorn (Monoceros), according to modern constellation boundaries. Since the cluster NGC 2548 fits Messier's description, newer atlases equate it with M48.

Our next target is Hydra's brightest planetary nebula, **NGC 3242**, or Caldwell 59, popularly known as the Ghost of Jupiter. In 1785, William Herschel discovered this nebula and was the first to compare it to that planet, saying that its light was the color of Jupiter. But it was Captain William Nobel in 1887 who first wrote that it looks "just like the ghost of Jupiter."

NGC 3242 is found 1.8° south and slightly west of 3.8-magnitude Mu (μ) Hydrae. The planetary is easy enough to see through a finder but looks like an 8th-magnitude star. You might be able to recognize it by its non-starlike color, which is usually described as blue or green. Magnifications of 50x or more will reveal its true nature.

Despite Herschel's claim, the color of NGC 3242 does not resemble Jupiter's to my eyes. With my little refractor, the planetary looks distinctly turquoise-blue. At 203x, it is roundish, nearly uniform in surface brightness, and slightly fuzzy around the edge. Those with better skies or larger telescopes should look for a northwest-to-southeast elongation with brighter patches at each end. You might also be able to spot the 12th-magnitude central star.

Now we'll move to the Water Snake's reddest star, **V Hydrae**. V Hydrae is a carbon star and, as such, has a number of carbon-bearing molecules in its atmosphere that act like a red filter. It varies in a period of roughly 17 months superposed on a grander cycle spanning 18 years. The brightest short-term cycles take the star from magnitude 6 to 9, roughly, while the dimmest ones carry it from magnitude 10 to 13.

To find V Hydrae, look first for 5th-magnitude b^3 Hydrae. It forms a right triangle with 4th-magnitude Alpha (α) Crateris and 3rd-magnitude Nu (ν) Hydrae, all visible within the same finder field. A low-power view through your telescope will show that b^3 Hydrae forms another right triangle with a 6.6-magnitude star to its southeast and a 7.1-magnitude star to its south. Drawing a gentle curve from b^3 Hydrae through the southern star and continuing for a little more than that distance again should bring you to V Hydrae. The star's color varies along with its brightness, from ruddy at minimum light to deep orange at maximum.

V Hydrae appears to be a dying red-giant star ready to form a planetary nebula. Studies indicate that planetary nebulae are largely shaped by high-speed bipolar outflows that cover a mere few hundred to a thousand years of a star's lifetime. V Hydrae is the first star to be caught in the act. A paper published in the November 20, 2003, issue of *Nature* reports its capture with the Hubble Space Telescope's imaging spectrograph. The research team involved suggests that the outflow may be driven by an unseen companion star or giant planet.

Hydra's brightest globular cluster is **M68**. The stars of Corvus, the Crow, serve as handy pointers for tracking it down. Draw an imaginary line from Delta (δ) through Beta (β) Corvi and continue for half that distance again. There, through a low-power eyepiece, you will see a 5th-magnitude star with a fuzzy spot to its northeast. My 105mm scope at 87x turns the fuzzy patch into a 9' ball of light with a large, bright, mottled core surrounded by a sparse and tattered halo that sparkles with faint stars. Increasing the magnification to 153x plucks out more of the cluster's stars right down to the center. The dappled face of M68 prompts some observers to picture dark lanes running through it.

Unlike a close-up image through a large telescope, this wide-field photograph of the galaxy M83 more nearly shows what observers can expect to see with small telescopes. This view, 3° wide, includes three stars on the finder chart opposite; north is up.
Photo: Akira Fujii

Our final target is the brightest galaxy in Hydra, **M83**, conveniently placed two-thirds of the way from 3rd-magnitude Gamma (γ) Hydrae to 4th-magnitude 1 Centauri. Northern skygazers may find the latter dimmed below naked-eye visibility when the sky is bright or the horizon hazy. If so, try star-hopping south-southeast from Gamma along a curvy line of 6th- and 7th-magnitude stars. M83 lies 1° east-southeast of the last star in the chain and can share the same low-power field.

At 127x, my little scope shows a small bright core and a fairly bright inner halo, 5' x 2', running east-northeast to west-southwest. This is surrounded by a dim oval halo about 8' across. I see some brighter patches but cannot trace M83's spiral structure. Three 10th-magnitude field stars form a tangent to the southeast side of the galaxy.

Seeking the highlights in the constellation Hydra should convey a sense of the colossal size of this constellation. Yet, amazingly, Hydra continues eastward beyond M83 for another 18½°.

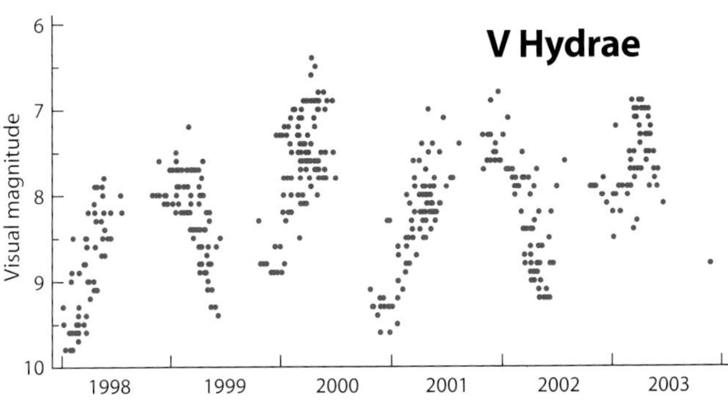

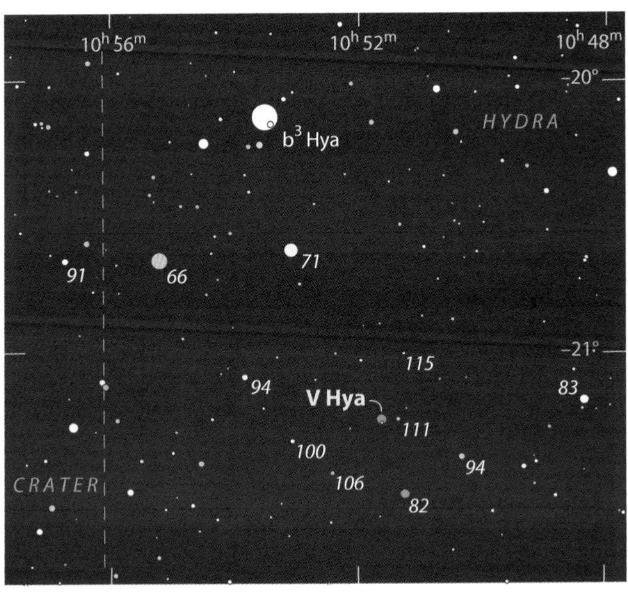

Left: For monitoring brightness changes in V Hydrae, this close-up chart gives the magnitudes of comparison stars to tenths, with the decimal point omitted. Top left: Based on observations collected by the American Association of Variable Star Observers, this graph shows how the magnitude of V Hydrae varied from 1998 through 2003. Gaps appear in the light curve around the time of the star's conjunction with the Sun, which occurs in early September each year.
Graph courtesy AAVSO

First published April 2004

A Bear's Bunch

Ursa Major offers many tantalizing telescopic targets.

He who would scan the figured skies
Their brightest gems to tell
Must first direct his mind's eye north
And learn the Bear's stars well.

— William Tyler Olcott, *Star Lore of All Ages*

The ancient origins of the Great Bear involve only the seven stars that we know as the Big Dipper, an eye-catching pattern that can point us toward many less conspicuous configurations in the sky. Later on, the figure of Ursa Major annexed a host of fainter stars, and her constellation eventually became the third largest, topped only by Hydra and Virgo.

In some modern renditions of the Great Bear, the star 24 Ursae Majoris (24 UMa) marks her ear, and it's here that we'll begin our bear hunt. A magician may pull a shiny coin from behind your ear, but the Bear's ear shields something far more astounding — two of the most beautiful galaxies in our northern sky.

Magnificent **M81** and **M82** are found 2° east of 24 UMa and are bright enough to be spotted in an 8x50 finder under fairly dark skies. I see the pair nicely in 14x70 binoculars. M81 is an oval glow that grows much brighter toward the center. To its north (below M81 in the evening sky at this time of the year), M82's spindle of light is smaller and fainter, with a uniform surface brightness.

This is a very pretty duo in my 105mm refractor at 17x. The oval of M81 runs north-northwest to south-southeast and displays a tiny, intense nucleus at its heart. When I stare directly at M81, it doesn't seem much larger than M82. Training my eye on M82, however, forces an averted-vision view of M81 that reveals a large faint halo and greatly increases the galaxy's apparent size. M82 is slightly blotchy, and its distinct cigar shape runs east-northeast to west-southwest. A 10th-magnitude star sits at the southern edge of its western end.

Bumping the telescope's magnification up to 28x makes M81's halo quite a bit more apparent. Two 11th-magnitude stars are superposed on the galaxy where the southern part of its outer core fades into the halo. At 47x, the mottled appearance of M82 is more obvious, and it looks particularly bright and patchy along the western two-thirds of its length. With a wide-angle eyepiece giving 87x, the galaxies still share a single

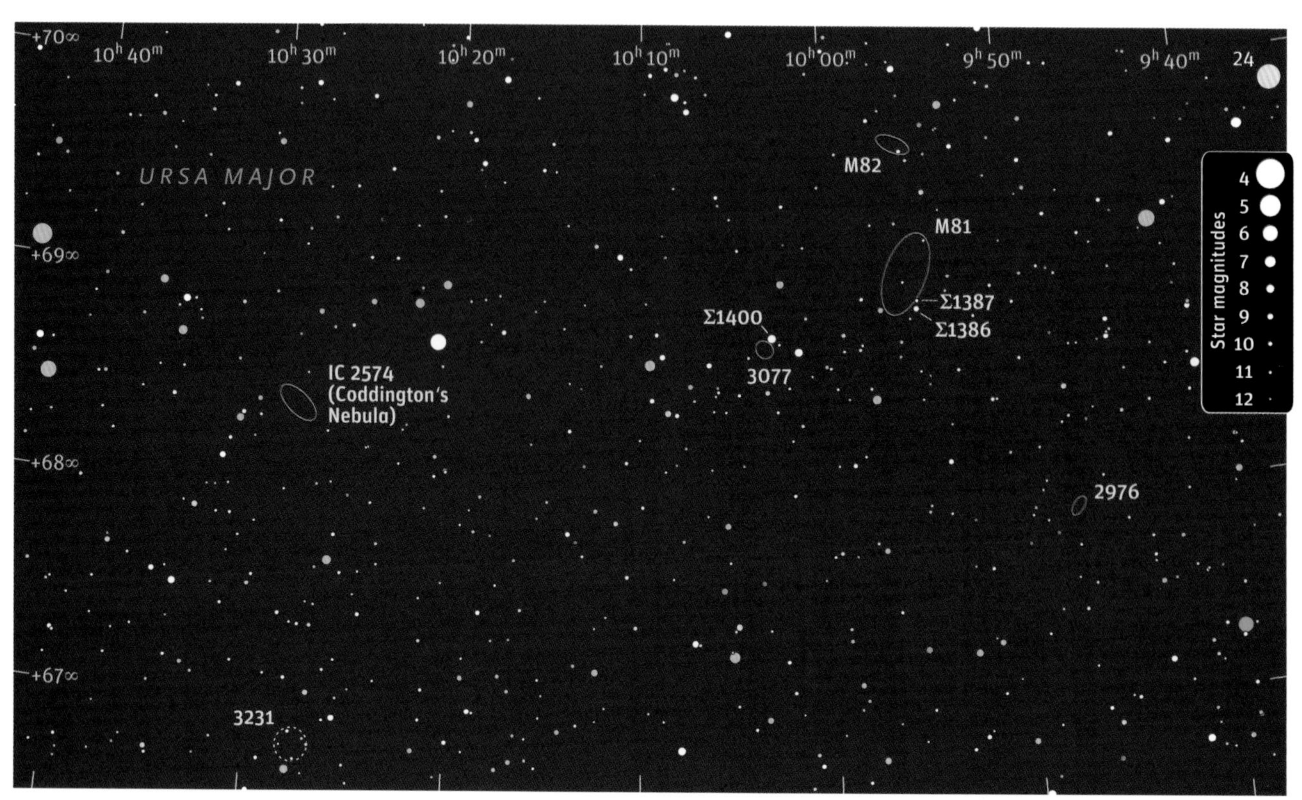

Chart adapted from *Tycho-2 Catalogue*

Considered by many to be the northern sky's most magnificent pairing of galaxies visible in a single eyepiece view, M81 (left) and M82 are ideally placed for observing during April evenings. While M81 has the classic appearance of a spiral galaxy, M82 has a chaotic look that is attributed to a burst of star formation that blew massive amounts of gas out of the galaxy's core. The field is 1° wide, with east up. Photo: Richard D. Jacobs

field of view, but without much room to spare. M81's inner core presents a slightly dappled face, while M82 appears quite intricate. A bright, wedge-shaped patch sits east of M82's center, and smaller patches strung westward from there give the galaxy a beaded look.

With my 10-inch reflector at 170x, M81 spans 12' x 7'. The two 11th-magnitude stars superposed on this galaxy point to a faint star on the other side of the galaxy near the edge of the outer core. A north-south pair of 9th- and 10th-magnitude stars lies south-southwest of M81, and both are double (**Σ1386** and **Σ1387**). This scope reveals the tapered form of M82 spanning 8' x 2' with three main dark patches. A broad one east of the galaxy's center cuts across M82, with the brightest portion of the galaxy lying to its east. A less distinct darkening borders the eastern side of the bright area, and an even subtler one sits west of center.

M81 is the largest member of a cluster of galaxies known as the M81 Group. At a distance of about 12 million light-years, it's one of the closest clusters to our own Local Group. Just as nearby star clusters often look larger on the sky than distant ones, nearby galaxy clusters also appear larger — in spades! Possible members of the M81 Group are flung across Ursa Major, Draco, and Camelopardalis.

We can see a few of them without ranging very far afield. In fact, **NGC 3077** can be spotted in a low-power field along with M81 and M82. To find it, look for an 8th-magnitude golden star about ½° east of M81's southern tip. Then drop about half that distance southward to a yellow-white star of similar brightness. NGC 3077 is just to the southeast.

My 105mm refractor at 47x nicely displays NGC 3077 together with its big brothers. NGC 3077 is a small, faint, northeast-southwest oval with a little brightening toward the center. The galaxy appears fairly bright in my 10-inch scope at 170x, which shows it with a large core surrounding an elusive starlike nucleus. The 8th-magnitude star off its northwest side, **Σ1400**, has a very close 10th-magnitude companion. Noted 19th-century British amateur astronomer William Henry Smyth was taken with the intense blackness of the sky surrounding this galaxy. He colorfully proclaimed that it "shows the nebula as if floating in awful and illimitable space, at an inconceivable distance."

In photographs, M82 looks highly disheveled, while NGC 3077 is only slightly mussed. Both are sibling-rivalry victims suffering from mutual gravitational encounters. As the most massive galaxy in the group, M81 has come out of these scuffles comparatively unscathed.

The dwarf spiral galaxy **IC 2574** is a dim but interesting member of the M81 Group. Astronomer Edwin Foster Coddington of the Lick Observatory in San Jose, California, discovered it in 1898 while examining photographs taken with the 6-inch Crocker telescope. Coddington and William Joseph Hussey followed up this discovery with a visual sight-

Celestial Baubles in the Great Bear

Object	Type	Magnitude	Size/Sep.	Right Ascension	Declination	*MSA*	*U2*
M81	Spiral galaxy	6.9	26.9' × 14.1'	9^h 55.6^m	+69° 04'	538	14L
M82	Irregular galaxy	8.4	11.2' × 4.3'	9^h 55.9^m	+69° 41'	538	14L
Σ1386	Double star	9.3, 9.3	2.1"	9^h 55.1^m	+68° 54'	538	14L
Σ1387	Double star	10.7, 10.7	8.9"	9^h 55.0^m	+68° 56'	538	14L
NGC 3077	Irregular galaxy	9.9	5.4' × 4.5'	10^h 03.4^m	+68° 44'	538	14L
Σ1400	Double star	8.0, 9.8	3.4"	10^h 02.9^m	+68° 47'	538	14L
IC 2574	Spiral galaxy	10.4	13.2' × 5.4'	10^h 28.4^m	+68° 25'	538	14L
NGC 3231	Open cluster	9.0	9.5'	10^h 27.5^m	+66° 48'	549	14L
NGC 2976	Spiral galaxy	10.2	5.9' × 2.7'	9^h 47.3^m	+67° 55'	550	14L

Angular sizes or separations are from recent catalogs. The columns headed *MSA* and *U2* give the chart numbers of objects in the *Millennium Star Atlas* and *Uranometria 2000.0*, 2nd edition, respectively.

Noted 20th-century deep-sky writer Walter Scott Houston often commented that celestial objects with popular names have a special appeal. As such, the dwarf spiral galaxy IC 2574 will likely entice backyard observers with its moniker, Coddington's Nebula. This ½°-tall view has north up. Photo: Martin C. Germano

ing through the observatory's 12-inch refractor. Coddington wrote: "The telescope shows it to be large, irregular, very faint, and composed of a number of condensations." IC 2574 is often called Coddington's Nebula.

Since this galaxy has a very low surface brightness, it will be helpful for you to accurately pinpoint its position when you're searching for it. Look for a 6th-magnitude star 1.6° east of NGC 3077. Coddington's Nebula is 45' east-southeast of this star, and both will fit within a low-power field.

The best view of IC 2574 through my little refractor comes at a modest 47x, where it is a gossamer northeast-southwest glow that extends for about 5'. Using averted vision helps, so look off to the side of the galaxy rather than directly at it. At 70x, through my 10-inch scope, Coddington's Nebula is a compelling sight and spans a generous 12' x 4'. Its low surface brightness lends it a tantalizing ghostly appearance. Several faint to very faint stars are sprinkled over this haunting galaxy, mostly around the edge.

The open cluster **NGC 3231** lies 1.6° south of IC 2574. Although it has nothing to do with the M81 Group, star clusters in Ursa Major are uncommon enough to make this a worthwhile side trip. Look for it just north of a yellow 8th-magnitude star. In my 105mm refractor at 17x, the group shows only six stars of around 11th magnitude. Moving up to 87x makes NGC 3231 appear more clusterlike. I count 15 stars within 10'. There is not much more to be seen in a larger scope, for my 10-inch shows only 17 stars gathered around a large central void.

Another member of the M81 Group, **NGC 2976**, is centered just 1.4° southwest of M81. In my little refractor at 28x, the galaxy's pale glow, which is weakly brighter in the center, is ensconced in a rough circlet of faint stars. NGC 2976 shows more clearly at 87x. It appears about 4' x 2'. A faint star is visible halfway along the southwest-facing side. Large scopes at high power give this galaxy a distinctly flocculent appearance, another relic of gravitational interactions within the M81 Group.

First published April 2005

Leo's 11th Hour

April evenings are a fine time to dip into the sea of galaxies in eastern Leo.

The constellation Leo, the Lion, now graces our evening sky. As the Lion slowly prowls westward during the night, the last part to linger in the sky is Leo's 11th hour of right ascension. This gives us plenty of time to dive into the sea of galaxies that enriches this area of the heavens.

Sky Atlas 2000.0, 2nd edition, plots 40 galaxies in Leo's 11th hour, while the *Millennium Star Atlas* maps a whopping 249. Let's pare down this rather overwhelming number of galaxies by concentrating on the Lion's hindquarters, where we find the NGC 3607 Group. This physically related collection of galaxies is about 75 million light-years away, but various sources assign different members to the group. My selection is simple. From the lists, I've chosen any galaxy that I've viewed with my 105mm refractor plus two that share the field in my 10-inch reflector.

Let's begin with **NGC 3607** itself, visually the brightest and largest member of the group. It lies about halfway along and ½° east of a line between Delta (δ) and Theta (θ) Leonis. With my little refractor at 28x, I see two small fuzzy spots. The southern one is NGC 3607. Zooming in on the scene at higher magnifications, I find my best view at 127x. NGC 3607 shows a stellar nucleus, a bright round core, and a faint oval halo that's 2' long northwest to southeast. A triangle of faint stars rests 4' southeast, and the star at the southern point hugs a close companion.

NGC 3608, the northern galaxy, occupies the same high-power field as NGC 3607. A smaller and fainter version of its companion, NGC 3608 runs east-northeast to west-southwest for 1½'. It grows brighter toward the center with a nearly round core and an elusive stellar nucleus. Two 12th-magnitude stars guard the galaxy's northern rim.

A third galaxy inhabits the field 3' southwest of NGC 3607's nucleus. While averted vision helps me spot **NGC 3605**, I can hold it with direct vision once I know exactly where it is. The galaxy is just a little smudge in the refractor, but my 10-inch reflector at 166x reveals an elongated 1' glow with a broadly brighter oval core.

This trio of galaxies sits at the heart of the NGC 3607 Group. The NASA/IPAC Extragalactic Database gives mean distances to galaxies based on data taken from astronomical literature. For NGC 3607, NGC 3608, and NGC 3605, the distances are 70 million, 77 million, and 73 million light-years, respectively. These galaxies are a bit too near for their radial velocities to yield reliable distances and a bit too far away for easy study of their stars. This makes it particularly

The lenticular galaxy NGC 3607 anchors a small collection of galaxies in Leo's hindquarters that are well within range of a 6-inch telescope under dark skies. The author has seen all the galaxies labeled at left with her 105mm refractor. The view here is 0.6° wide with north up. Photo: POSS-II / Caltech / Palomar

difficult to pin down how far away they are, and their actual distances may be off by as much as 25 percent from the figures quoted here.

If I drop the magnification in my 10-inch scope to 115x, I can fit the galaxy trio in the same field with **NGC 3599**, located ⅓° to the west in a barren field. I see a 1'-long oval sloped west-northwest to east-southeast and gently brightening toward the center, harboring a stellar nucleus. In my 105mm refractor, the galaxy is faintly visible with averted vision at 87x and with direct vision at 127x. It's estimated to be 65 million light-years distant.

Sweeping 3° westward takes us to **NGC 3507**, which exhibits low-level emissions from its nuclear region that are thought to arise from a combination of starburst and black-hole activity. In my little refractor at 47x, NGC 3507 is a faint glow enveloping an 11th-magnitude star. It's sandwiched between a slightly dimmer star 5' north-northeast and a 10th-magnitude star 3' south-southwest. At 127x, the galaxy is 2' long and slightly oval east-west. It has a brighter core that's difficult to make out because of interference from the superposed star just off its northeast side. With my 10-inch scope at 166x, I also see the flat (edge-on) galaxy **NGC 3501** lying 13' southwest. It appears very thin and 2' long. NGC 3501 has a low surface brightness, and placing the nearby 9th-magnitude star out of the field gives me a better view. NGC 3507 and NGC 3501 are 65 and 76 million light-years distant, respectively.

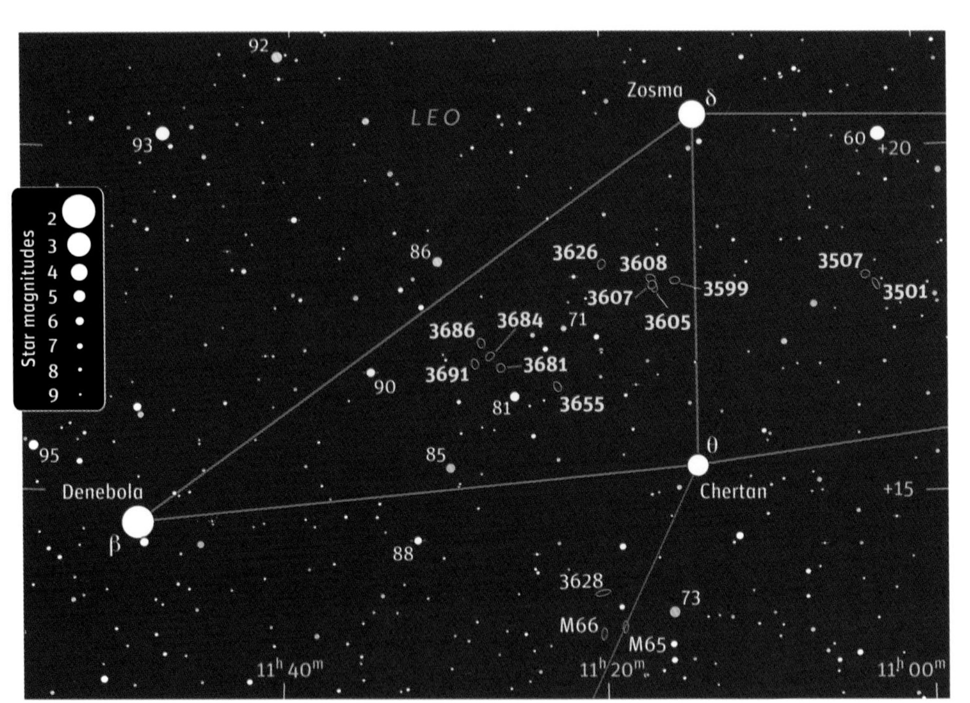

Let's return to NGC 3607 and then work our way eastward. In my small refractor at 47x, **NGC 3626** sits 48' east-northeast in the same field. It's easily visible as a little oval with a bright center. Boosting the magnification to 87x, I can tease out a stellar nucleus. In the 10-inch reflector at 115x, the galaxy's 2' halo is tipped a little west of north, and its tiny nucleus is quite intense. NGC 3626 is a strange galaxy whose gas and stars circle around its

Galaxies Galore

Object	Type	Magnitude	Size/Sep.	Right Ascension	Declination	*MSA*	*U2*
NGC 3607	**Lenticular galaxy**	**9.9**	**5.5' × 5.0'**	**11h 16.9m**	**+18° 03'**	**705**	**73L**
NGC 3608	**Elliptical galaxy**	**10.8**	**4.2' × 3.0'**	**11h 17.0m**	**+18° 09'**	**705**	**73L**
NGC 3605	**Elliptical galaxy**	**12.3**	**1.6' × 1.2'**	**11h 16.8m**	**+18° 01'**	**705**	**73L**
NGC 3599	**Lenticular galaxy**	**12.0**	**2.7' × 2.2'**	**11h 15.5m**	**+18° 07'**	**705**	**73L**
NGC 3507	**Barred spiral galaxy**	**10.9**	**4.6' × 3.7'**	**11h 03.4m**	**+18° 08'**	**705**	**73L**
NGC 3501	**Spiral galaxy**	**12.9**	**4.6' × 0.6'**	**11h 02.8m**	**+17° 59'**	**705**	**73L**
NGC 3626	**Spiral galaxy**	**11.0**	**3.2' × 2.3'**	**11h 20.1m**	**+18° 21'**	**705**	**73L**
NGC 3655	**Spiral galaxy**	**11.7**	**1.5' × 0.9'**	**11h 22.9m**	**+16° 35'**	**704**	**91R**
NGC 3686	**Barred spiral galaxy**	**11.3**	**3.2' × 2.4'**	**11h 27.7m**	**+17° 13'**	**704**	**91R**
NGC 3684	**Spiral galaxy**	**11.4**	**3.0' × 2.0'**	**11h 27.2m**	**+17° 02'**	**704**	**91R**
NGC 3681	**Barred spiral ring galaxy**	**11.2**	**2.0' × 2.0'**	**11h 26.5m**	**+16° 52'**	**704**	**91R**
NGC 3691	**Barred spiral galaxy**	**11.8**	**1.3' × 0.9'**	**11h 28.2m**	**+16° 55'**	**704**	**91R**

Angular sizes or separations are from recent catalogs. The visual impression of an object's size is often smaller than the cataloged value and varies according to the aperture and magnification of the viewing instrument. The columns headed *MSA* and *U2* give the chart numbers of objects in the *Millennium Star Atlas* and *Uranometria 2000.0*, 2nd edition, respectively.

First published April 2006

center in opposite directions. This may be evidence of a long-ago merger with a gas-rich dwarf galaxy. The NASA/IPAC Extragalactic Database places NGC 3626 at approximately the same distance as NGC 3507.

At 100 million light-years, **NGC 3655** is the most remote galaxy in our tour. To locate it, look for a right triangle of three 7th-magnitude stars about 1¼° southeast of NGC 3626. Then center the longer leg of the triangle in your field of view and drop ⅔° southward. This little galaxy is very faint in my 105mm scope. It displays a somewhat brighter center and a 13th-magnitude star 2.5' to the east-northeast. With my 10-inch at 166x, I see a ¾' oval leaning northeast, a brighter oval core, and a stellar nucleus.

Now center the shorter leg of the triangle in your field of view, and sweep 52' eastward. Here, we come to **NGC 3686**, the largest of a group of four galaxies clumped within a ½° patch of sky. In my little refractor at 68x, NGC 3686 is a small oval, tipped north-northeast, with a large, weakly brighter core. The other three galaxies are less obvious. **NGC 3684** sits 14' southwest and shows a faint northwest-southeast oval with a uniform surface brightness. Another 14' southwest, **NGC 3681** is a bit more apparent and forms a nice straight line with the galaxies above. It's small and round with a brighter center, and it makes a shallow curve with two nearby 12th-magnitude stars. With my 10-inch scope, the tiny galaxy **NGC 3691** joins the scene. Look for it 15' east-southeast of the center galaxy in the line. In the order presented here, these galaxies are 69, 74, 79, and 82 million light-years away.

This ½°-long line of 11th-magnitude galaxies begins (at lower right) about ½° north-northeast of the 5.6-magnitude star 81 Leonis. The field is 0.6° wide with north up. Photo: Digitized Sky Survey

The Lion's Lair

April is galaxy time.

Where yet the Lion climbs the ethereal plain,
And shakes the Summer from his radiant mane.
— Erasmus Darwin, *The Botanic Garden*, 1791

These lines, written by the grandfather of the renowned British naturalist Charles Darwin, refer to the time of year when the Sun treads the starry halls of Leo, the Lion. During the final weeks of northern summer, the Lion captures the golden rays of the Sun and seems to beam their warmth to us as though augmented by his tawny mane. Yet now in northern spring, Leo rules the evening sky and casts the light of far more distant suns our way.

Let's focus on the eastern side of Leo, starting with **M65** and **M66**, two galaxies from Charles Messier's famous catalog of 1781. M65 lies about halfway between Theta (θ) and Iota (ι) Leonis, with M66 just 20' to its east-southeast. Both are visible as small ovals through 14x70 binoculars, M66 appearing somewhat larger and brighter. In my 105mm refractor at a magnification of 68x, M65 is five times longer than wide and tipped a bit west of north. The galaxy shows a fairly bright oval core and sports a faint star on its southwestern edge. M66 has about the same orientation as its neighbor but looks only a bit more than two times longer than wide. The galaxy harbors a bright oval core, and a 10th-magnitude star decorates its northwestern edge. Placing M65 and M66 in the southern part of the eyepiece field adds **NGC 3628** to the scene. This edge-on galaxy is large and beautiful, though rather faint. Its slender profile is eight times longer than wide and tilts east-southeast. The ghostly galaxy brightens gently toward a greatly elongated core.

The trio — often called the Leo Triplet — reveals further secrets through my 10-inch reflector. At 68x, the galaxies still inhabit a single field of view. M65 covers a full 7.3' x 2'. The brightest area of its core takes on a mottled appearance and holds a starlike nucleus. M66 displays a north-south outer halo that measures 6' x 2'. The galaxy's misaligned core grows considerably brighter toward a tiny nucleus. Lanky NGC 3628 stretches for 13' but is a mere 2' wide. A shadowy dark lane is softly charcoaled across the galaxy's core. At 115x, I can no longer fit all three galaxies in the same field of view, but they are individually quite pretty. M65's nucleus becomes a small round spot with a bright knot north-northwest, and the core of M66 assumes a decidedly patchy facade. The dusky lane running through NGC 3628 is slanted so that it's

Deep-sky observers know that springtime in the Northern Hemisphere means galaxies, and Leo is home to many. Among the best is the pair of M65 and M66 observed by French comet hunter Charles Messier in 1780. NGC 3628, however, is far more challenging. William Herschel discovered it on April 8, 1784. The field is 0.8° wide with north up. Photo: Robert Gendler

higher (farther north) at the eastern end than the western.

The Leo Triplet is about 30 million light-years distant and is part of a larger galaxy group known as Leo I. The twisted dust lane of NGC 3628 and the distorted spiral arms of M66 are probably evidence of tidal interactions within the group. Studies indicate that NGC 3628 and M66 may have undergone a close encounter about a billion years ago.

Nearby **NGC 3593** is also part of the Leo I Group. Look for it 1.1° west-southwest of M65 and 34' south-southwest of the golden 5th-magnitude star 73 Leonis. My 105mm scope at 28x encompasses all four galaxies. NGC 3593 is a fairly small and faint, east-west oval. It sits at the western point of a 20' right triangle that it forms with two 7th-magnitude stars. At 87x, the galaxy appears 2½' long by two-thirds as wide and brightens inward. In my 10-inch scope at 43x, NGC 3593 shows a bright core, and M65 sits on the opposite side of the 1.5° field of view. At 166x, the galaxy stands alone. It spans 3' x 1½' and holds a large oval core with a tiny, slightly brighter spot at its heart.

Now let's take a closer look at **Iota (ι) Leonis**, an attractive but challenging double star. On a night of variable atmospheric stability, I could intermittently split the pair at 218x with my little refractor. The 4.1-magnitude primary looked pale yellow, and its 6.7-magnitude companion seemed to flicker with hints of blue or green. A magnification of 318x kept the stars apart, and the dimmer star simply looked white. On a night of slightly better seeing, my 10-inch reflector showed the pair just kissing at 171x and separated at 220x. The stars orbit each other with a period of 186 years. The separation in 2011 is 2.0". It will increase to 2.7" over the next few decades, as the companion crawls from east to east-northeast of its primary.

The galaxy **NGC 3705** lies exactly 2° southeast of Iota. I see it perched at the peak of a squat, stick-figure house to its north made with four stars, magnitudes 9 to 11. NGC 3705 is a nice little 3' spindle in my 105mm refractor at

Looking Around Leo

Object	Type	Magnitude	Size/Sep.	Right Ascension	Declination	*MSA*	*U2*
M65	Spiral galaxy	9.3	9.8' × 2.8'	$11^h 18.9^m$	+13° 06'	729	92L
M66	Spiral galaxy	8.9	9.1' × 4.1'	$11^h 20.3^m$	+12° 59'	729	92L
NGC 3628	Spiral galaxy	9.5	14.8' × 2.9'	$11^h 20.3^m$	+13° 35'	729	92L
NGC 3593	Lenticular galaxy	10.9	5.2' × 1.9'	$11^h 14.6^m$	+12° 49'	729	92L
ι Leonis	Double star	4.1, 6.7	1.9"	$11^h 23.9^m$	+10° 32'	728	91R
NGC 3705	Spiral galaxy	11.1	4.9' × 2.0'	$11^h 30.1^m$	+09° 17'	728	91R
τ Leonis	Double star	5.1, 7.5	89"	$11^h 27.9^m$	+02° 51'	776	112L
83 Leonis	Double star	6.6, 7.5	29"	$11^h 26.8^m$	+03° 01'	776	112L
NGC 3640	Elliptical galaxy	10.4	4.3' × 3.4'	$11^h 21.1^m$	+03° 14'	776	112L
NGC 3641	Elliptical galaxy	13.2	1.0' × 1.0'	$11^h 21.1^m$	+03° 12'	776	112L
NGC 3521	Spiral galaxy	9.0	11.0' × 7.1'	$11^h 05.8^m$	–00° 02'	777	112L

Angular sizes and separations are from recent catalogs. The visual impression of an object's size is often smaller than the cataloged value and varies according to the aperture and magnification of the viewing instrument. The columns headed *MSA* and *U2* give the appropriate chart numbers in the *Millennium Star Atlas* and *Uranometria 2000.0*, 2nd edition, respectively. All the objects this month are in the area of sky covered in Chart 34 in *Sky & Telescope's Pocket Sky Atlas.*

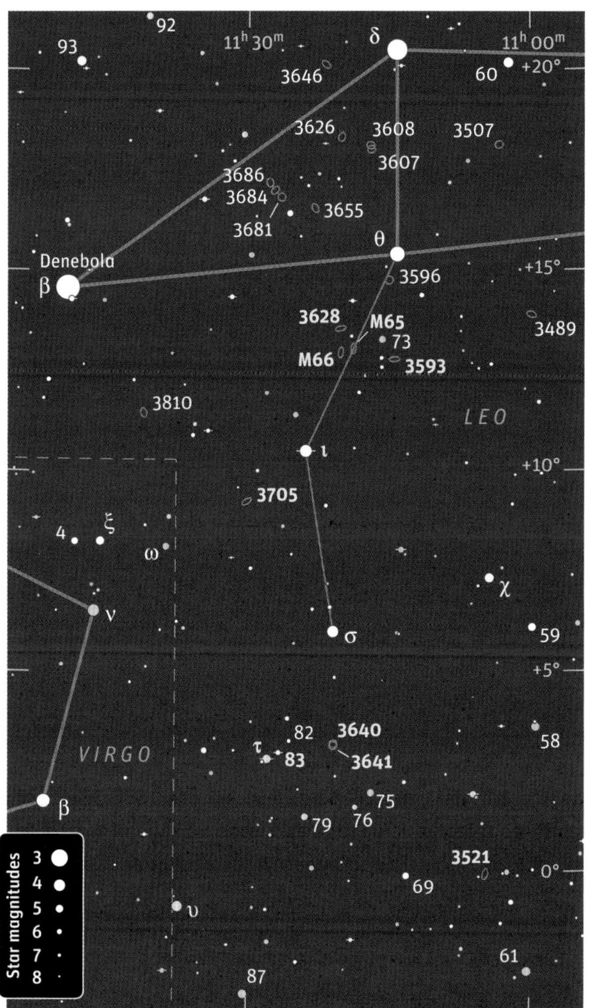

87x. It grows brighter toward the center and is canted east-southeast. In my 10-inch scope at 115x, this pretty galaxy bridges 4' and envelops an oval core with a bright stellar nucleus.

By dropping 6.4° southward, we come to a colorful pair of double stars, **Tau (τ)** and **83 Leonis**. The two are only 20' apart and easily share a low-power field. They both show nicely in my little refractor, even at 17x. Tau is a 5.1-magnitude, deep yellow primary with a yellow-white, 7.5-magnitude companion widely spaced to its south. The closer 83 Leonis pair presents a golden 6.6-magnitude primary and a pale orange, 7.5-magnitude secondary to the south-southeast.

Tau and 83 Leonis are evenly spaced along a shallow curve of stars with 82 Leonis (to the northwest). If you place 82 Leonis in a low-power field and sweep 1.1° westward, you'll find yourself at the doorstep of **NGC 3640**. This galaxy is small and moderately faint in my 105mm refractor at 87x. It appears slightly oval, brightens toward the center, and hosts a stellar nucleus. In my 10-inch reflector at 166x, NGC 3640 covers 2' x 1½' and runs nearly east-west. The galaxy **NGC 3641** is seen 2½' south-southeast, measured center to center. It's fairly faint, very small, and has a brighter center. I can spot this minuscule galaxy in my little refractor when conditions are right. To nab it, I use averted vision (directing my gaze a little off to the side of the object) and cover my head with a dark cloth to block stray light.

Our final stop is the showy galaxy **NGC 3521**. From NGC 3640, slip 1.6° south-southwest to 75 and 76 Leonis, and then continue for another 2.2° to 69 Leonis. NGC 3521 is precisely 2° due west of this 5th-magnitude star. In my 105mm scope at 28x, it's a bright and easy spindle with a stellar nucleus. It measures about 5' x 2' and slopes south-southeast. Boosting the magnification to 87x accents a small oval core. NGC 3521 is a lovely target for larger telescopes. In their *Observing Handbook and Catalogue of Deep-Sky Objects*, authors Christian Luginbuhl and Brian Skiff call the galaxy beautiful and describe the view through a 12-inch telescope: "The ends are ragged, the core and halo mottled. The oval core is roughly centered, but the brighter parts become progressively more eccentric to the [western] edge, where a dark lane 20" wide passes."

Photographs show that NGC 3521's spiral structure is choppy and disjointed. Systems with this type of fragmentation are known as flocculent galaxies. M63, the Sunflower Galaxy in Canes Venatici, has a similarly fleecy appearance. Deep images of NGC 3521 also show an unusually large and prominent halo.

Top: NGC 3521 was discovered by William Herschel on February 22, 1784. It's an interesting spiral galaxy that should be better known by observers with small telescopes. Those with larger apertures can look for a mottled appearance in the core. The field is 10' wide with north up. Photo: Robert Gendler

First published April 2007

Bear Toes

On April evenings, you'll find a fascinating array of telescopic sights by following a celestial beast's footsteps.

Ursa Major, the Great Bear, is most renowned for its dominant group of seven stars, widely known to skygazers as the Big Dipper. Among the brightest Bear stars outside the Dipper are the three pairs that mark her toes, a distinctive asterism in their own right. The Arabs saw these stars as the Three Leaps of the Gazelle, which, startled by the lion Leo, left these tracks as it bounded away.

The Bear's toes still carry reminders of the Arabian gazelle. The easternmost pair, **Nu (ν)** and **Xi (ξ) Ursae Majoris**, are known as Alula Borealis and Alula Australis. Alula is derived from an Arabic phrase meaning "the first leap," while Borealis and Australis are the familiar Latin distinctions for northern and southern. Both hoofprints are multiple stars. Nu is a double with a bright, golden-hued primary clasping a much fainter secondary 7.4 arcseconds (7.4") to its south-southeast. They are nicely split through my 105mm refractor at a magnification of 87x.

Easy to locate in small telescopes, the spiral galaxy NGC 2841 in Ursa Major is layered in subtle detail that unfolds when viewed with telescopes of increasing aperture. North is up in this ¼°-wide photograph. Photo: Jim Misti

The components of Xi are more closely matched in brightness but require more magnification to resolve. They appear to be kissing at 122x, are barely separated at 153x, and form a beautiful yet close pair at 174x. The primary is yellow-white, and its companion has a yellow tint.

Xi's stars have an orbital period of only 60 years. The pair are 1.6" apart in 2011, increasing to 3.1" in 2034. William Herschel discovered this double in 1780 and noted a "very extraordinary change in the angle of position" after he watched the fainter star move from southeast to east of its primary during a span of 22 years. In 1827, French astronomer Félix Savary made Xi the first binary to have its orbit determined.

But there is more to Xi than meets the eye. In 1905, Danish astronomer Niels Erik Nørlund found a discrepancy between the stars' observed and computed positions. He concluded that the most natural explanation was perturbations by an unseen companion. Indeed, later studies indicate that Xi's primary is itself a double, one star slightly hotter than our Sun and the other a red dwarf, with a maximum separation of 0.08" — unbearably close.

Xi's visible companion is an even closer binary composed of a star that's slightly cooler than the Sun mated to a dwarf with an orbital period of four days. This subsystem is sometimes listed as having a third component, but its existence is by no means certain. Complex and fascinating, Xi tantalizes us from about 27 light-years away.

The second most distant globular cluster in our galaxy sits 3½° southeast of Xi. **Palomar 4** is about 355,000 light-years away. (The Milky Way's most distant globular, at 402,000 light-years, is Arp-Madore 1 in the southern constellation Horologium.)

This remoteness makes Palomar 4 a challenging target even for skygazers with large telescopes. A good star chart (like the one on page 96) will increase your odds of spotting it, and here's how I hunt it down. I start by looking 2½° southeast of Xi for a 54'-long, westward-pointing triangle of 7th-magnitude stars. The two at the eastern end point southward to a widely spaced, nearly east-west pair of 8th- and 9th-magnitude stars. About ½° west of them, you'll see a 10th-magnitude star. Palomar 4 lies just 6' farther west, with a 13th-magnitude star off its east-northeastern side.

In my 15-inch reflector at 144x, the spectral glow of Palomar 4 is perhaps 1⅓' across and quite difficult to see. Even with averted vision, I can't hold it steadily in view. Virginia amateur astronomer Kent Blackwell has managed to nab this evasive cluster with a 10-inch scope — a very impressive feat.

The gazelle's next hoof-fall is marked by Lambda (λ) and Mu (μ) Ursae Majoris, known as Tania Borealis and Tania Australis (where Tania comes from an Arabic phrase meaning "the second leap"). In 7x50 binoculars, they make a lovely color-contrast pair, with Lambda appearing white and Mu distinctly orange. Mu is a red-giant star about 249 light-years distant, while Lambda is a subgiant only half as far away.

Two noteworthy spiral galaxies are found nearby. **NGC 3184** is 46' west of Mu and 11' east-southeast of a 6½-magnitude, pale orange star. It's visible as a small round patch of mist in my little refractor at 17x, and at 47x, I see a 12th-magnitude star pinned to its northern edge. Boosting the magnification to 87x uncovers a small, slightly brighter core within the 4½' glow. In my 10-inch reflector at 118x, subtle traces of uneven illumination indicate a pair of spiral arms that seem to emanate from north and south of the galaxy's core and then wrap counterclockwise. At 171x, I see brighter patches in the halo: one northwest, one southwest, and a very elusive one to the east. The first two bear their own designations, NGC 3180 and NGC 3181.

The second galaxy, **NGC 3198**, is also easy to locate. Start at Lambda and look for a 6½-magnitude yellow star 18' to its east-northeast. Then hop 1° north to a star of similar color

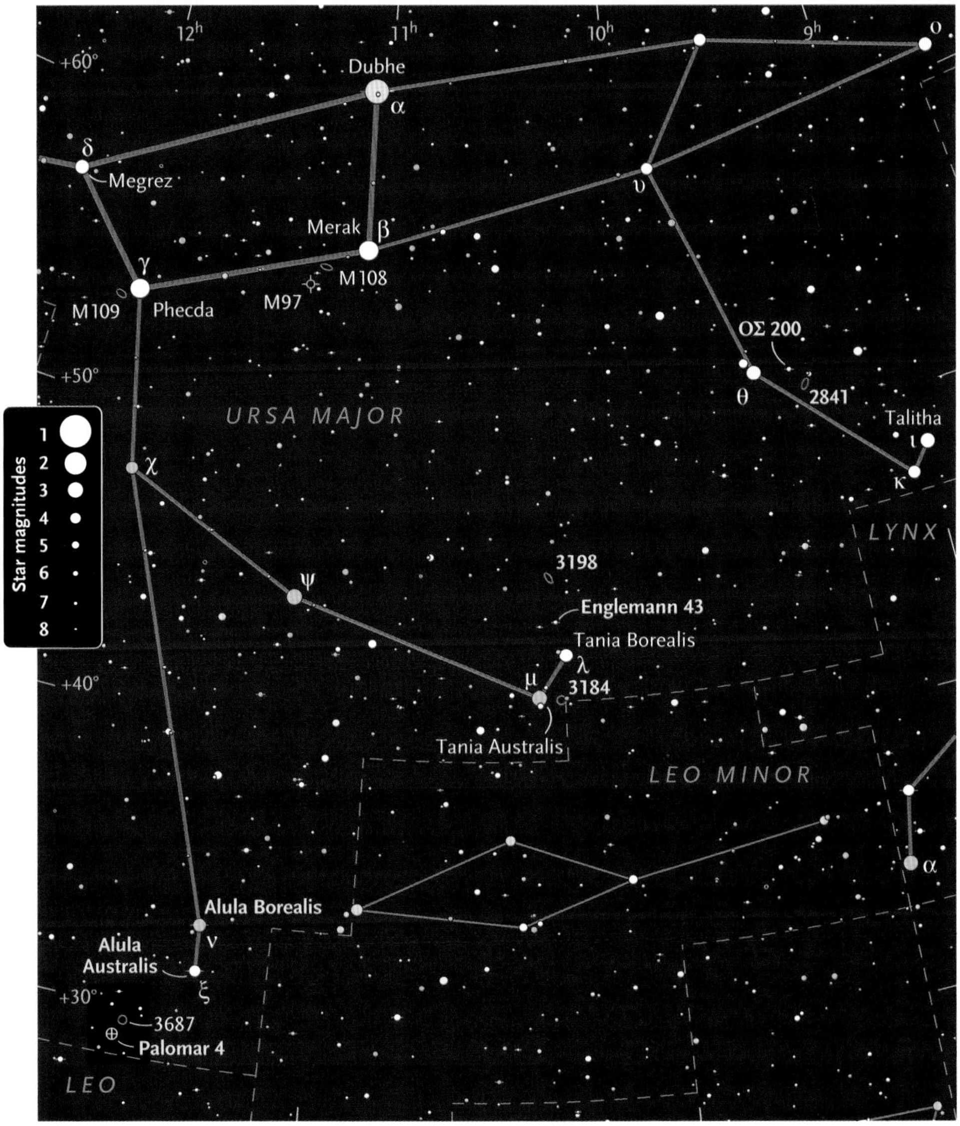

Deep-space search

If thoughts of hunting down distant globular clusters catch your fancy, check out Barbara Wilson's listing on the Web at www.astronomy-mall.com/Adventures.In.Deep.Space/obscure2.htm.

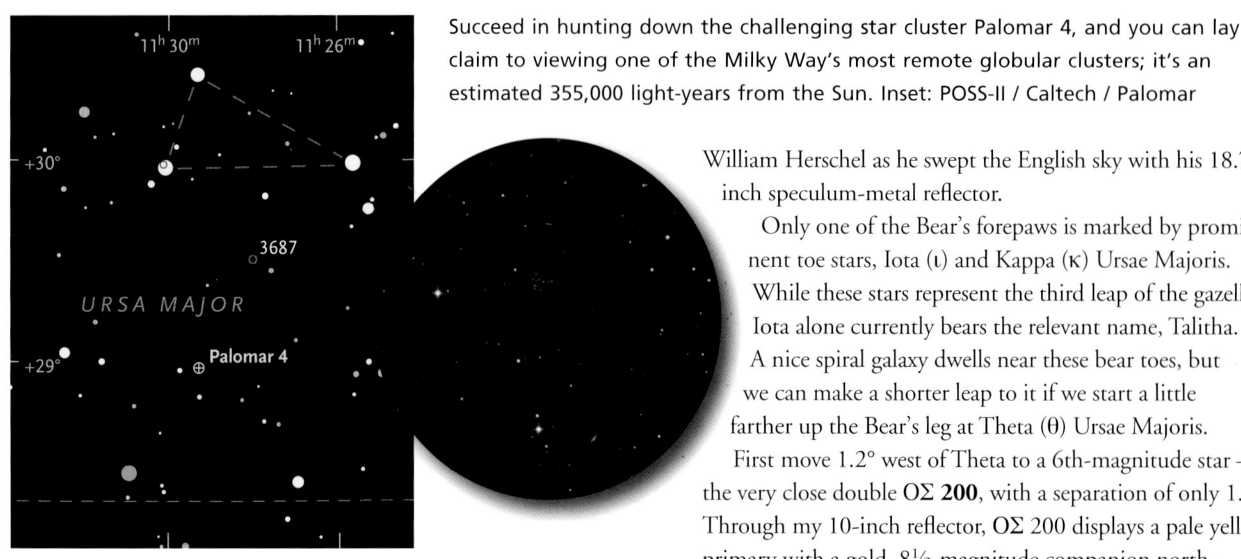

Succeed in hunting down the challenging star cluster Palomar 4, and you can lay claim to viewing one of the Milky Way's most remote globular clusters; it's an estimated 355,000 light-years from the Sun. Inset: POSS-II / Caltech / Palomar

and brightness. This is the double star **Englemann 43**, which shows an extremely wide, 9th-magnitude companion to the east. Draw a line from the first yellow star to Englemann 43, and then continue the line 1½ times that distance again to NGC 3198.

Acquiring the area with my 105mm scope at 17x, I see NGC 3198 as a small smudge sitting in a long curvy line of 9th- and 10th-magnitude stars. A magnification of 87x reveals a gossamer oval 5½' long and one-fourth as wide. An 11th-magnitude star hovers north of the galaxy's northeastern tip. At 153x, NGC 3198 takes on a slightly mottled appearance. The galaxy's woolly facade is nicely displayed in my 10-inch scope at 171x, where it covers 7' x 2' and shows a few faint stars cradling its southwestern half.

The Spitzer Infrared Nearby Galaxy Survey (SINGS) places NGC 3198 and NGC 3184 at distances of 28 and 32 million light-years, respectively. Both galaxies were discovered in the 1780s by William Herschel as he swept the English sky with his 18.7-inch speculum-metal reflector.

Only one of the Bear's forepaws is marked by prominent toe stars, Iota (ι) and Kappa (κ) Ursae Majoris. While these stars represent the third leap of the gazelle, Iota alone currently bears the relevant name, Talitha. A nice spiral galaxy dwells near these bear toes, but we can make a shorter leap to it if we start a little farther up the Bear's leg at Theta (θ) Ursae Majoris.

First move 1.2° west of Theta to a 6th-magnitude star — the very close double **OΣ 200**, with a separation of only 1.3". Through my 10-inch reflector, OΣ 200 displays a pale yellow primary with a gold, 8½-magnitude companion north-northwest. At 213x, separation comes and goes with variations in atmospheric turbulence, but I can split the pair consistently at 311x.

From OΣ 200, scan 43' west-southwest to a star of similar brightness. **NGC 2841** rests 21' southeast and forms a right triangle with the two 6th-magnitude stars. Through my little refractor at 17x, the galaxy is an easily visible oval that leans northwest with an 8½-magnitude golden star to the east. At 87x, NGC 2841 holds a small, round, bright core, and an 11th-magnitude star lies just off the galaxy's northern tip. At 127x, the galaxy appears 4' long and one-third as wide.

In my 10-inch reflector at 171x, NGC 2841 engulfs the northern star, and I see a faint star deeper in the halo 1.8' north-northwest of the galaxy's tiny nucleus. The halo and the outer part of the oval core are slightly mottled. The nucleus plus the brightest part of the inner halo look almost barlike, and there's a subtle darkening along their eastern side that hints at a dusty lane adorning the galaxy's spiral arms. SINGS places this galaxy, also discovered by William Herschel, at the same distance as NGC 3198.

Where the Great Bear Walks

Object	Type	Magnitude	Size/Sep.	Right Ascension	Declination	*MSA*	*U2*
ν Ursae Majoris	**Double star**	**3.5, 10.1**	**7.4"**	**11^h 18.5^m**	**+33° 06'**	**657**	**54L**
ξ Ursae Majoris	**Double star**	**4.3, 4.8**	**1.6"**	**11^h 18.2^m**	**+31° 32'**	**657**	**54L**
Palomar 4	**Globular cluster**	**14.2**	**1.3'**	**11^h 29.3^m**	**+28° 58'**	**657**	**54L**
NGC 3184	**Spiral galaxy**	**9.8**	**7.4' × 6.9'**	**10^h 18.3^m**	**+41° 25'**	**617**	**39L**
NGC 3198	**Spiral galaxy**	**10.3**	**8.5' × 3.3'**	**10^h 19.9^m**	**+45° 33'**	**617**	**39L**
Englemann 43	**Double star**	**6.7, 9.4**	**145"**	**10^h 18.9^m**	**+44° 03'**	**617**	**39L**
OΣ 200	**Double star**	**6.5, 8.6**	**1.3"**	**9^h 24.9^m**	**+51° 34'**	**580**	**39R**
NGC 2841	**Spiral galaxy**	**9.2**	**8.1' × 3.5'**	**9^h 22.0^m**	**+50° 59'**	**580**	**39R**

Angular sizes and separations are from recent catalogs. Visually, an object's size is often smaller than the cataloged value and varies according to the aperture and magnification of the viewing instrument. The columns headed *MSA* and *U2* give the appropriate chart numbers in the *Millennium Star Atlas* and *Uranometria 2000.0*, 2nd edition, respectively. All the objects this month are in the area of sky covered by Charts 32 and 33 in *Sky & Telescope's Pocket Sky Atlas.*

First published April 2008

The Dry Bear

Every corner of Ursa Major is chock-a-block with galaxies.

The Bear, that sees star setting after star
In the blue brine, descends not to the deep.
— William Cullen Bryant, "The Order of Nature"

This ursine verse refers to the circumpolar nature of the Great Bear, ever visible as she nightly prowls around the North Star. But you have to be pretty far north before Ursa Major does not at least wet her toes in the oceans of the world. At my latitude of 43° north, only her northernmost toe remains forever dry. To keep her southernmost toe from splashing in the briny blue, you'd have to be north of 58.5°. During the evening at this time of the year, however, the Bear snoozes high in the north. She looks as though she's lying on her back with her toes brushing the zenith — a perfect time for Bear watching.

Let's start at Upsilon (υ) Ursae Majoris, the star at the top of the Bear's foreleg. Through my 105mm refractor at 28x, Upsilon shares the field of view with the galaxy **NGC 2950**, located 1.1° west and 11' south of the star. At such a low power, the galaxy is very small and faint, but it displays a brighter center. At 87x, I can see that NGC 2950 presents us with an oval profile canted northwest, and its core harbors a bright, starlike nucleus.

With my 10-inch reflector at 213x, NGC 2950 covers 1¼' x ¾' and shows a 14.9-magnitude star near the midpoint of its southwestern flank. This star sits at the south-southeastern end of a straight, ¾'-long line that it makes with two others, magnitudes 15.4 and 15.5. But I didn't spot them. Can you? If so, the 15.0-magnitude star ½' west-northwest of the northernmost star will turn the line into a stubby hockey stick.

The simple appearance of this galaxy through the eyepiece belies its complicated structure. NGC 2950 is a double-barred lenticular (lens-shaped) galaxy. The two bars rotate at different speeds, and recent studies suggest that the small inner bar may rotate in the opposite direction from the large outer bar and the galaxy's large-scale disk. It seems unlikely

Austrian astrophotographer Bernhard Hubl's image captures all four galaxies in the group Holmberg 124.

SPRING
April

Left: Can you see both of NGC 2950's bars in this image from the Sloan Digital Sky Survey? Right: This deep image, also from the Sloan Digital Sky Survey, shows NGC 2681 to be a subtle but beautiful barred lenticular galaxy that's aimed face-on to Earth. North is to the lower left.

that such a bar would survive very long unless the galaxy's inner disk is also counterrotating.

Next, we'll visit **NGC 2768**, which lies halfway between Upsilon and Omicron (ο) Ursae Majoris, the star that marks the Great Bear's nose. This galaxy is faintly visible in my little refractor even at 17x. At 47x, its east-west oval grows brighter toward a stellar nucleus. A magnification of 87x gives a prettier view, with a 10th-magnitude star off the galaxy's west-northwestern edge and a faint star close to its east-northeastern edge. If I push my scope northwest, **NGC 2742** shares the 53' field of view with a little room to spare. It's similar to NGC 2768 in shape and orientation but is smaller and has a more uniform surface brightness. NGC 2742 is accompanied by a deep yellow star northwest and a little triangle of faint stars southwest.

NGC 2950, NGC 2768, and NGC 2742 lie at similar distances of roughly 65 million light-years.

Dropping about a degree south of NGC 2742 takes you to a 6th-magnitude star, from which you can hop 1.8° west to another. **NGC 2685**, the Helix Galaxy, rests 27' southeast of the second star. It's a very faint northeast-southwest oval through my 105mm refractor at 28x, and a faint star roosts north of its northeastern tip. At 87x, it's 1½' long and grows slightly brighter toward its center.

At 45 million light-years, the Helix (pictured below) is one of the nearest polar-ring galaxies. A polar-ring galaxy exhibits a disk (or ring) of gas, dust, and stars that's nearly perpendicular to the disk of the host galaxy. The ring is thought to arise as the result of a merger with or an accretion of material from a neighboring galaxy. Through my 10-inch reflector at high power, the only suggestion of the Helix Galaxy's polar ring is a somewhat greater central width than I might expect otherwise. My 14.5-inch scope very faintly shows stubby protrusions tipped a bit clockwise from perpendicular to the main galaxy, with the stub on the northwestern side more obvious.

Climbing northward takes

Far left: Ken Crawford's amazingly detailed image of NGC 2685 shows why this is known as the Helix Galaxy. Left: Through a modest 8-inch telescope, Finnish stargazer Jere Kahanpää was able to see and sketch the essential needle-and-cocoon shape of the Helix Galaxy.

Nine Galaxies and One Double Star

Object	Type	Magnitude	Size/Sep.	Right Ascension	Declination
NGC 2950	Galaxy	10.9	2.7' × 1.8'	9^h 42.6^m	+58° 51'
NGC 2768	Galaxy	9.9	6.4' × 3.0'	9^h 11.6^m	+60° 02'
NGC 2742	Galaxy	11.4	3.0' × 1.5'	9^h 07.6^m	+60° 29'
NGC 2685	Galaxy	11.3	4.6' × 2.5'	8^h 55.6^m	+58° 44'
NGC 2805	Galaxy	11.0	6.3' × 4.8'	9^h 20.3^m	+64° 06'
NGC 2814	Galaxy	13.7	1.2' × 0.3'	9^h 21.2^m	+64° 15'
NGC 2820	Flat galaxy	12.8	4.3' × 0.5'	9^h 21.8^m	+64° 15'
IC 2458	Galaxy	15.0	0.5' × 0.2'	9^h 21.5^m	+64° 14'
Σ1321	Double star	7.8, 7.9	17"	9^h 14.4^m	+52° 41'
NGC 2681	Galaxy	10.3	3.6' × 3.3'	8^h 53.5^m	+51° 19'

Angular sizes and separations are from recent catalogs. Visually, an object's size is often smaller than the cataloged value and varies according to the aperture and magnification of the viewing instrument.

us to an interesting quartet of galaxies, Holmberg 124 (see photo on page 97). The largest member of the group is **NGC 2805**, 1.2° east-northeast of Tau (τ) Ursae Majoris. This face-on spiral has very low surface brightness, and observers armed with small telescopes may only be able to spot its tiny, brighter core. Through my 10-inch reflector at 170x, this ghostly galaxy appears irregularly round and 4' to 5' across. It's bracketed by a pair of fairly bright stars, magnitudes 9 and 10, while a 12th-magnitude star guards its northwestern edge.

Sharing the field of view, **NGC 2814** is perched 10.5' north-northeast of its bigger companion and has an 11th-magnitude star near its southern tip. This edge-on spiral has a significantly dimmer total magnitude than NGC 2805 does, but its light is confined to a much smaller area, resulting in considerably higher surface brightness. My scope shows a very faint, north-south slash nearly 1' long.

NGC 2820 is a bit more obvious than NGC 2814 and centered only 3.7' to its east. NGC 2820 is another edge-on spiral, thin enough to be listed in the *Revised Flat Galaxy Catalogue* (Igor D. Karachentsev and colleagues, 1999). Entries in this catalog are disklike galaxies with little or no central bulge that are seen edge-on and appear at least seven times longer than wide. Through my 10-inch scope, the slender form of NGC 2820 cuts east-northeast to west-southwest for 2½' and joins the field with its neighbors.

The Holmberg galaxies are the most distant in our tour, approximately 76 million light-years away. I missed the final member of the group, faint little **IC 2458**, when surveying the area with my 10-inch scope. Can you spot this galaxy dangling from the western tip of NGC 2820?

Now we'll plunge southward to 15 Ursae Majoris, the trailhead for our final two deep-sky wonders. The double star **Struve 1321** (Σ1321) lies halfway between and a shade east of a line connecting 15 and 18 Ursae Majoris, which fit in the same field of view through a finderscope. Its matched components are split at 17x in my little refractor, which shows a lovely east-west pair of orange and gold suns.

Next, place 15 Ursae Majoris a little north of center in your telescope's field of view and sweep 2.4° west to the galaxy **NGC 2681**. My 105mm scope at 17x shows a fairly small, round, faint bit of mist with a brighter center. At 47x, I see a pair of faint stars just off the west-northwestern edge. Through my 10-inch scope at 220x, the galaxy's starlike nucleus gleams in a slightly lumpy core, both cloaked in a soft gray mantle 3' across with a faint star nuzzling its eastern border.

Sitting only 38 million light-years away, NGC 2681 is the nearest galaxy in our deep-sky tour of the "dry Bear."

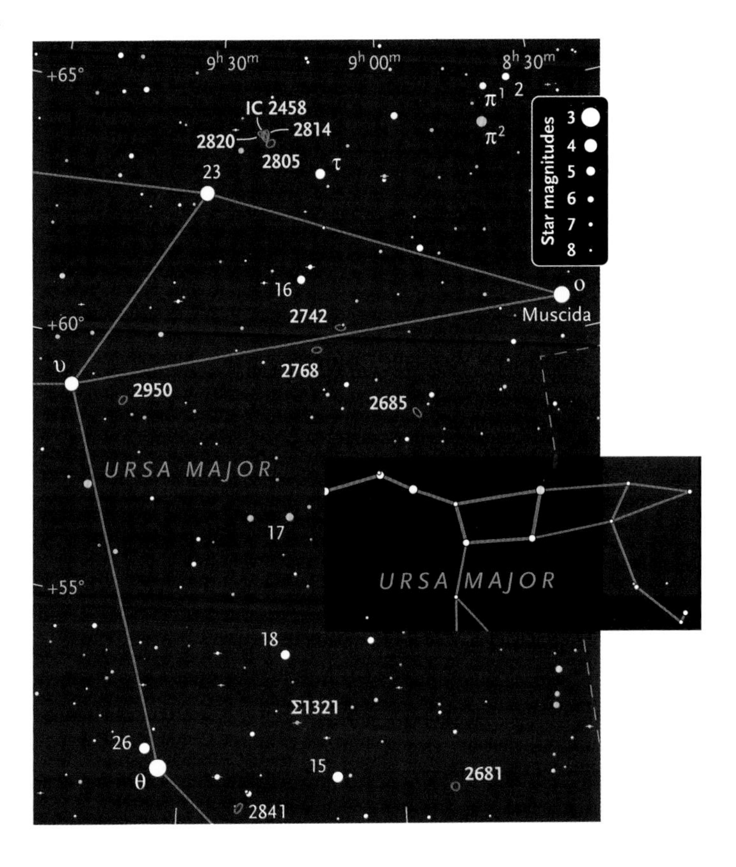

First published April 2009

Between the Bears

The area north of the Great Bear is worthy of study but often neglected.

Here the huge Snake in many a volume glides,
Winds like a stream, and either Bear divides.
— Virgil, *Georgics*, Book I

Parting the constellations of Ursa Major and Ursa Minor is the tail of Draco, the serpentine Dragon of the sky. The region between the Bears owns no particularly bright stars, but it does possess a memorable assortment of deep-sky wonders.

Let's begin our journey at 4th-magnitude Kappa (κ) Draconis. As plotted on the April all-sky chart on page 305, it's the second star from the tip of Draco's tail and makes a readily recognizable triangle with Megrez (δ) and Dubhe (α) in the Big Dipper. Kappa is unevenly bracketed by a pair of 5th-magnitude attendants that share a low-power telescopic field of view. They make a colorful stellar trio shining orange, blue-white, and gold, from south to north.

The barred spiral galaxy **NGC 4236** lies 1.5° west and a bit south of Kappa. With a total integrated magnitude of 9.6, this is the brightest galaxy in the area. In other words, NGC 4236 would shine as brightly as a 9.6-magnitude star if all its light were gathered into a single point. However, this light is actually smeared over an oval roughly 22' long and 7' wide, so it's highly attenuated. The galaxy's average

NGC 4236 is a barred spiral galaxy with unusually prominent star-forming regions. The brightest of these is designated VII Zw 446. Photo: Brian Lula

magnitude per square arcminute, known as surface brightness, is only 15.0. Seeing a tiny galaxy with such a low surface brightness would be a hopeless task with a small telescope, but our eyes are better at detecting dim objects when they're large. Another factor in our favor is that the light of NGC 4236 isn't evenly spread. The galaxy is brighter than average, and therefore easier to see, across a large portion of its interior.

In my semirural skies, I notice NGC 4236 easily through my 105mm refractor at 47x. Its oval form leans north-northwest and is sheltered by a distinctive pattern of stars that helps pinpoint its exact position. The galaxy is lovely at 87x and shows a fair amount of detail.

NGC 4236 appears large in our sky because it's relatively nearby — only 14 million light-years away. Its proximity allows observers with large telescopes to glimpse the star-forming regions that spangle its disk. A bright but tiny spot 4.5' south-southeast of the galaxy's center bears the designation VII Zw 446, signifying its place in a catalog of compact galaxies by Fritz and Margrit Zwicky. But in deep, color images of NGC 4236, this spot appears to be a region of ionized hydrogen (H II), and it stands out better when using a narrowband nebula filter with my 14.5-inch scope at 100x. Nebula filters may also tease out other H II regions within the galaxy.

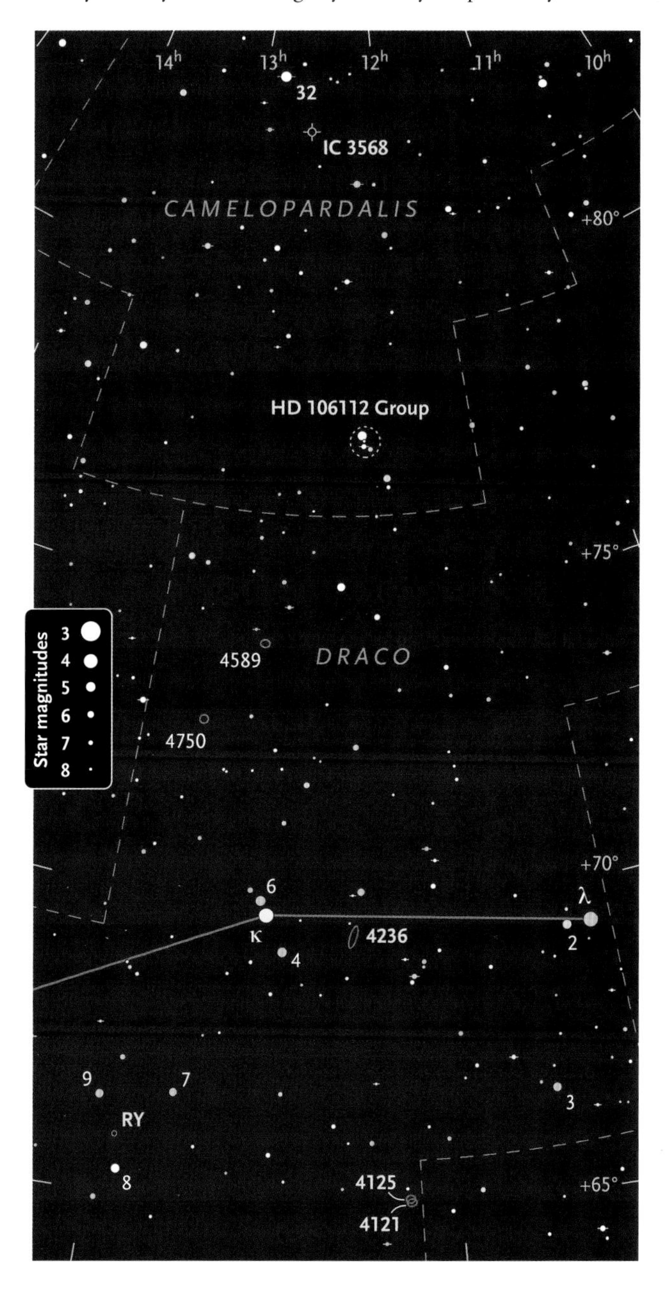

Let's compare NGC 4236 with the elliptical galaxy **NGC 4125**, located 4.4° south-southwest, where it sits sentinel at the border of Ursa Major. At magnitude 9.7, NGC 4125 has nearly the same total brightness as NGC 4236, but its light is concentrated into a much smaller area. Its 6' x 3' oval averages magnitude 12.9 per square arcminute, a much higher surface brightness than its large neighbor has. As a result, NGC 4125 is significantly easier to spot.

NGC 4125 is about 75 million light-years away. If placed at the same distance as NGC 4236, it would appear one-third again as large as NGC 4236 and shine at 6th magnitude.

Through my 105mm refractor at 47x, NGC 4125 is a moderately bright oval tipped a bit north of east. It grows considerably brighter toward the center, and a 10th-magnitude star embosses its eastern edge. The tiny companion galaxy **NGC 4121** rests 3.8' south-southwest and is visible most of the time with averted vision. I can hold NGC 4121 steadily in view at 87x. The galaxy marks the southern corner of a right triangle that it makes with the star and the center of NGC 4125. At this magnification, I estimate NGC 4125's visible size as 3¼' x 1½'.

Sweeping 5° eastward, we find the carbon star **RY Draconis** along the eastern side of an isosceles triangle made by the 5th-magnitude stars 7, 8, and 9 Draconis. RY is a semiregular variable with poorly known periodicity. It appears to have long, superimposed cycles that undergo slow changes. The star is generally found between magnitude 6 and 8 and is an impressively deep reddish orange when seen through my 5.1-inch refractor. It nicely contrasts with the triangle stars, of which the northern two glow yellow-orange and the southern one white.

Draco isn't the only constellation between the Bears. The head of Camelopardalis, the Giraffe, also divides them. With unaided eyes, I see a little fuzzy patch in Camelopardalis that lies about one-third of the way from Kappa Draconis to Polaris. The two stars at the end of the Little Dipper's bowl roughly point to it. Aiming my 5.1-inch scope toward this smudge at low power, I discovered an eye-catching little asterism. Two

SPRING April

Deep-Sky Treats in Draco and Camelopardalis

Object	Type	Magnitude	Size/Sep.	Right Ascension	Declination
NGC 4236	**Galaxy**	**9.6**	**21.9' × 7.2'**	**$12^h\ 16.7^m$**	**+69° 28'**
NGC 4125	**Galaxy**	**9.7**	**5.8' × 3.2'**	**$12^h\ 08.1^m$**	**+65° 10'**
NGC 4121	**Galaxy**	**13.5**	**0.6' × 0.6'**	**$12^h\ 07.9^m$**	**+65° 07'**
RY Draconis	**Carbon star**	**6–8**	**—**	**$12^h\ 56.4^m$**	**+66° 00'**
HD 106112 Group	**Asterism**	**4.7**	**17'**	**$12^h\ 11.3^m$**	**+77° 29'**
32 Camelopardalis	**Double star**	**5.3, 5.7**	**21"**	**$12^h\ 49.2^m$**	**+83° 25'**
IC 3568	**Planetary nebula**	**10.6**	**18"**	**$12^h\ 33.1^m$**	**+82° 34'**

Angular sizes and separations are from recent catalogs. Visually, an object's size is often smaller than the cataloged value and varies according to the aperture and magnification of the viewing instrument.

slightly diverging arcs of three stars each are nested one beside the other and topped by a 5th-magnitude star. The arc stars are magnitudes 6.8 to 9.7, and the southernmost star in the western arc gleams yellow.

Sure that others must have noticed this asterism, I looked it up in the Deep Sky Hunters (a Yahoo group) database, where the asterism is listed as the **HD 106112 Group**. The designation stands for the brightest star's *Henry Draper Catalogue* number.

The 5th-magnitude star **32 Camelopardalis** (32 Cam) dwells 6° north of HD 106112. Similar designations for most of the bright northern stars represent their number in the 1721 unauthorized version of John Flamsteed's star catalogue. For example, the three stars in the triangle bearing RY Draconis have Flamsteed numbers. But if you look at a star chart that labels numbered stars, you'll find that 32 Cam is completely out of order in the eastward progression of Flamsteed numbers across the constellation. That's because the number did not come from Flamsteed's catalog but, rather, from *Prodromus Astronomiae*, the 1690 catalog of Johannes Hevelius. Most Hevelius numbers have fallen by the wayside, but 32 Cam lingers on many star charts, to the recurrent puzzlement of stargazers.

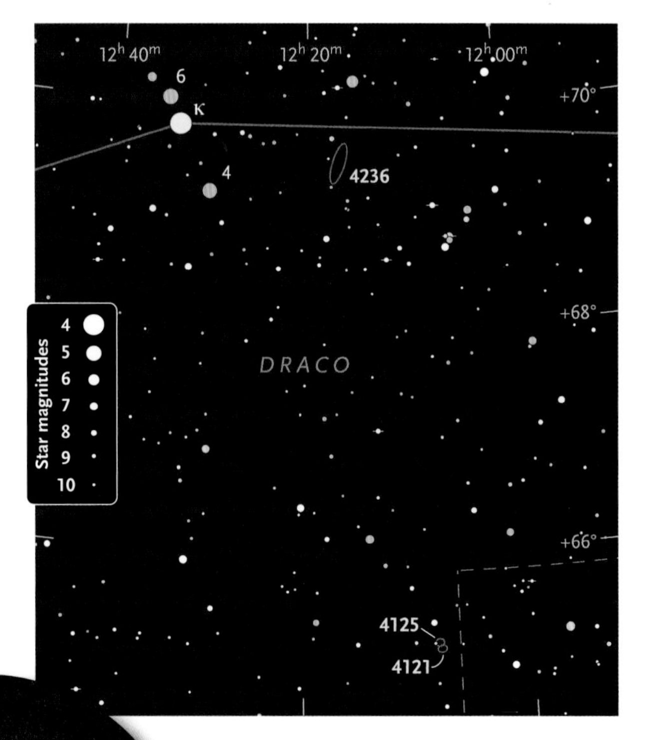

This 1997 Hubble Space Telescope photo gave the planetary nebula IC 3568 its nickname the Lemon Slice.

Telescopically, 32 Cam is a nearly matched pair of white suns with the companion star northwest of its primary. At a generous separation of 21", the stars are easily split at low power.

The planetary nebula **IC 3568** sits 1° south-southwest of 32 Cam. It has been called the Lemon Slice because of the radial structure and yellow color displayed in its 1997 Hubble Space Telescope image. Amateur astronomer Jay McNeil nicknamed it the Baby Eskimo for its resemblance to the Eskimo Nebula (NGC 2392).

IC 3568 is fairly bright through my 5.1-inch refractor at 63x. It could be mistaken for a star at first glance, but it doesn't look as sharp as a star and shows a decidedly nonstellar, bluish gray hue. I see a very bright center surrounded by a faint, bluish fringe at 102x, while at 164x, the fringe loses its color but the core becomes slightly uneven in brightness. With my 10-inch scope at high power, I see a 13th-magnitude star nuzzling the planetary's western edge. My 15-inch reflector helps bring out details in the brilliant core and the elusive star cloaked in its bright center. With its patterned core and fainter fringe, IC 3568 truly does look like a miniature of the Eskimo Nebula.

First published April 2010

A Toehold in the Virgo Cluster

Once you've familiarized yourself with the area, you can star-hop, or even galaxy-hop, to new sights.

The hazy band of the Milky Way runs through the constellations that hug the horizon on May's all-sky star map on page 306. With the plane of our galaxy and its obscuring dust clouds riding so low, we can gaze up into an unobstructed sky and peer far into deep space. Our clearest view is toward the North Galactic Pole in the constellation Coma Berenices, where we just happen to find the nearest large cluster of galaxies.

Some 60 million light-years away, the Virgo Galaxy Cluster splays out across the borders of Virgo and Coma. Its 2,000 member galaxies form the core of the Local Supercluster, with our own Local Group as an outlying member. The tremendous mass of the Virgo Cluster acts on the galaxy groups around it, slowing the recession they would otherwise have as part of the universe's overall expansion. The Milky Way Galaxy and other members of the Local Group may eventually fall into and be swallowed by the Virgo Cluster.

About 50 Virgo Cluster galaxies, including 16 Messier objects, are bright enough to be seen through a small telescope. With such an abundance of galaxies in an area devoid of bright stars, novice skygazers easily become lost in this realm. In such a case, it is wise to remember the old adage: Well begun is half done. We need a toehold in the Virgo Cluster — a familiar base from which to start and then carry our explorations farther afield.

If your sky is dark enough for you to spot the 5.1-magnitude star 6 Comae Berenices, you're already on your way. If not, begin at the bright star Denebola in the tail of Leo, the Lion. Scan eastward from Denebola for 6½° to find 6 Comae. It is the brightest star along the way and the westernmost star in a distinctive T-shaped asterism with four other stars ranging from magnitudes 6.4 to 6.9.

This group of stars will be our home base in the Virgo Cluster. A finder will easily encompass the entire T. The group will also fit within the field of your main scope, if you have a low-power eyepiece that gives you a true field of 2° or more.

First, start at 6 Comae and use it to find the galaxy **M98**. This nearly edge-on spiral lies ½° due west of 6 Comae and shares the field at magnifications under about 70x. Although M98 has a low surface brightness, it can be seen in a 60mm scope under dark skies. Through a

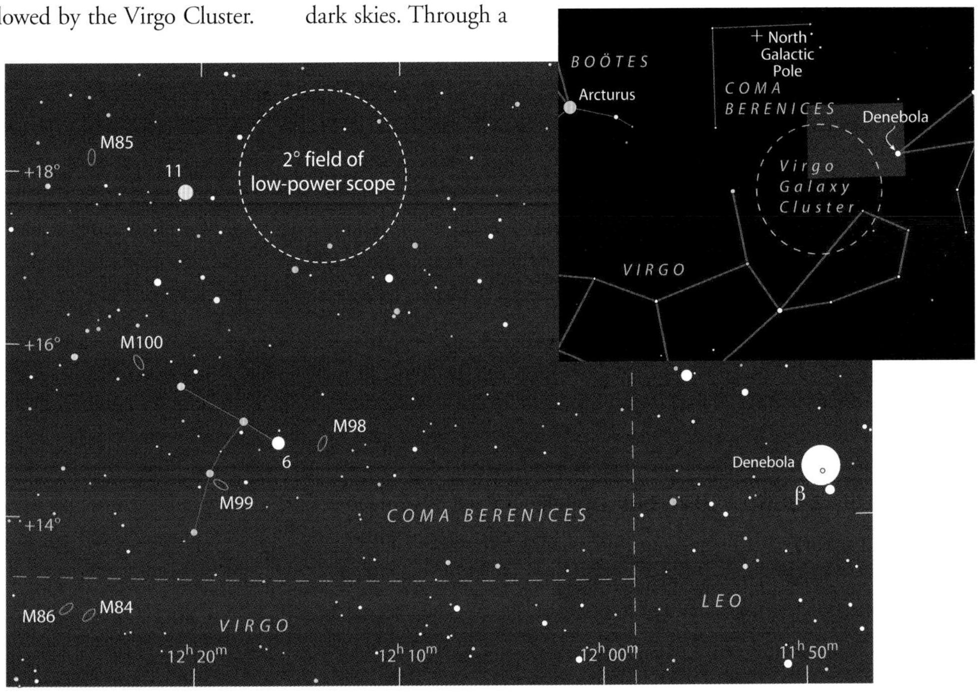

Easy Galaxies in the Virgo Cluster

Galaxy	Type	Magnitude	Dimensions	Right Ascension	Declination	Constellation
M98	Spiral	10.1	9' × 3'	12^h 13.8^m	+14° 54'	Coma Berenices
M99	Spiral	9.8	5'	12^h 18.8^m	+14° 25'	Coma Berenices
M100	Spiral	9.4	7' × 6'	12^h 22.9^m	+15° 49'	Coma Berenices
M85	Elliptical	9.2	7' × 5'	12^h 25.4^m	+18° 11'	Coma Berenices
M86	Elliptical	9.2	7' × 6'	12^h 26.2^m	+12° 57'	Virgo
M84	Elliptical	9.3	5' × 4'	12^h 25.1^m	+12° 53'	Virgo

Recent estimates place the center of the Virgo Galaxy Cluster 40 to 75 million light-years away; distances to individual member galaxies are not known. Catalog dimensions, listed here, are greater than those seen with small telescopes.

105mm scope at around 100x, the galaxy is about 6' x 2', elongated north-northwest to south-southeast. It contains a brighter, extended, patchy core and an off-center, nearly stellar nucleus. Through his 4-inch refractor, noted observer Stephen James O'Meara sees the brighter areas of M98 forming a Klingon vessel from "Star Trek." Less experienced observers will probably need a larger telescope to see the arcing spiral arms that give M98 this appearance.

M98 is approaching us at 125 kilometers per second. Since the Virgo Cluster as a whole is receding from us at about 1,100 kilometers per second, M98 must be moving in our direction at a rate of 1,225 kilometers per second with respect to the center of its cluster. The immense mass of the Virgo Cluster gravitationally accelerates many of its members to high individual velocities so that the light from some, like M98, actually shows us an approaching blueshift instead of the more usual redshift of cosmological expansion.

If you have trouble spotting M98, take heart. Our next two targets are a little easier. This time, we'll start from the 6.5-magnitude star in the center of our T's upright. The face-on spiral galaxy **M99** lies just 10' southwest of this star and will even fit in the same high-power field of view. A 105mm scope at around 100x shows a 3' x 2' oval core that brightens toward the center and is surrounded by a very faint, oval halo. The core's elongation runs east to west, while that of the halo runs northeast to southwest. Under dark skies, a 6-inch scope can start to show hints of spiral structure. Knots of slightly brighter haze emerge from the eastern side of the galaxy and wrap around the south.

M99's spiral structure is asymmetric, probably from arm-wrenching gravitational interactions with other Virgo Cluster members. M99 also has a high velocity of its own. Its measured redshift, unusually large for a galaxy at M99's distance, indicates it is hurtling away from us at 2,324 kilometers per second. That's twice the recession rate of the Virgo Cluster as a whole!

To locate our next galaxy, we'll start at the 6.5-magnitude star at the eastern side of our T's crossbar. The face-on spiral galaxy **M100** lies just 35' east-northeast of this star. Through a 105mm scope at around 100x, this galaxy appears slightly oval, about 4' x 3', and

The symmetry of M100's spiral arms makes it a showpiece on long-exposure images like these. Note the many fuzzy objects in the immediate vicinity of M100. They're more galaxies! But only NGC 4312, the needlelike object ⅓° to the south-southwest, is within the reach of a small scope visually. The bright star at lower right in the wide view is one of those in the T-shaped asterism plotted on the chart shown on page 103. Photo, left: Martin C. Germano; Inset: Robert Gendler

First published May 2001

Left: The most nearly edge-on of this month's galaxies, M98 is also a little fainter than the others. The star 6 Comae Berenices lies just outside the left edge of this close-up view. Photo: George R. Viscome. Right: M99 completes a triangle with stars of 6th and 9th magnitude that shine brightly in the same telescopic field. This galaxy's miniature pinwheel, so striking on photographs, requires a large scope to be detected visually. Photo: Martin Germano

is tipped east-southeast to west-southwest. Its halo is fairly uniform and contains a small, round, brighter nucleus. A 6-inch scope can start to show, within the halo, slightly brighter patches that define the galaxy's spiral arms. Photographs of M100 reveal the beautiful structure that astronomers label a *grand-design spiral* (one having two principal arms).

The T of stars and the galaxies M98, M99, and M100 give us our toehold in the Virgo Cluster, but there's no need to stop here. Once you've familiarized yourself with the area, use a good atlas to star-hop, or even galaxy-hop, to new sights. For example, from M100 you can scan 45' east to a 6.7-magnitude star. From there, it is 2.3° north to **M85** or 2.9° south to **M86**. M86 lies right next to **M84**, and this pair marks one end of a long, distinctive curve of galaxies known as Markarian's Chain.

Take it a little at a time, and eventually, you'll master the brighter galaxies of the Virgo Galaxy Cluster. Then, someday soon, perhaps you'll be the one showing a novice how to navigate them.

Coma Squared

This region's main star cluster is so large that a telescope is of little use.

Coma Berenices (Berenice's Hair) flows high across the evening skies of spring for observers at midnorthern latitudes. This area is spangled with faint stars when seen under a very dark sky. But only three of these stars are bright enough to make our May all-sky star map on page 306, where they look like a rotated L just above Virgo. Filling in the blank corner of Coma with the galaxy M85 gives us a square about 10° on each side, about as wide as your fist at arm's length. In and near this celestial square, deep-sky sights abound.

The constellation Coma Berenices is named for a real person, Queen Berenice II, who lived in the 3rd century BC and was the wife of Ptolemy III, king of Egypt. Shortly after their marriage, Ptolemy went off to war in response to diplomatic intrigues that included the murder of his sister (also named Berenice).

Out of fear for her husband, her king, and (no doubt) her position, Queen Berenice offered her royal tresses to the gods as ransom to secure Ptolemy's safety. When he returned victorious, Berenice made good her vow. Her amber locks were shorn and placed in the temple of Aphrodite, but they vanished shortly thereafter.

The royal couple's outrage was quelled by the palace astronomer, Conon of Samos. Conon wisely claimed that the gods were so pleased with Berenice's lustrous gift that they placed it in the sky for all to see, now shining as a

Left: This amazing sliver of light — the almost perfectly edge-on spiral galaxy NGC 4565 — is a fairly easy catch for any small telescope under dark skies. But the tiny smudge ¼° toward lower right, the galaxy NGC 4562, is a tough challenge visually for 12-inch telescopes. Bob and Janice Fera combined three 90-minute exposures made with their 11-inch Celestron on Mount Pinos, California. Right: The paired galaxies M85 (right) and fainter NGC 4394 hug the center of this view with north up. Photo: POSS-II / Caltech / Palomar

delicate tracery of glittering stars:

Who scans the bright machinery of the skies
and plots the hours of star-set and star-rise...
By you this soft effulgence first was seen
*who knew at once the ringlets of the Queen.**

We'll start our sky tour at the constellation's Alpha (α) star, which is sometimes called Diadem but is always pictured at the wrong end of the queen's hair to mark her jeweled crown. The globular cluster **M53** lies 1° northeast of this pale topaz gem. M53 looks nearly stellar through a small finder, but 14x70 binoculars reveal a moderately bright, round, fuzzy patch that brightens toward the center. With my 105mm refractor, M53 begins to look grainy at 153x. At 203x, I can see a few elusive pinpricks of light on a steady night.

If M53 seems easy, try for the very dim globular **NGC 5053** just 1° to the east-southeast. Stephen James O'Meara, who has glimpsed this in his 4-inch refractor, delightfully describes NGC 5053 in *The Messier Objects* as "the departed soul of its more brilliant neighbor."

Now let's move to the diagonally opposite corner of the Coma square marked by golden Gamma (γ). Scattered over a region 5° across, with Gamma at the northern edge, is the nearby open cluster known as the Coma Star Cluster, or **Melotte 111**. Because this grouping is so large, a regular telescope is of little use here; binoculars or a finderscope work much better. In 8x40 binoculars, I can count a dozen bright stars and twice as many fainter ones. The cluster includes the very wide, bright double star **17 Comae Berenices**. A small scope at low power shows a subtle color contrast between the 5.2-magni-

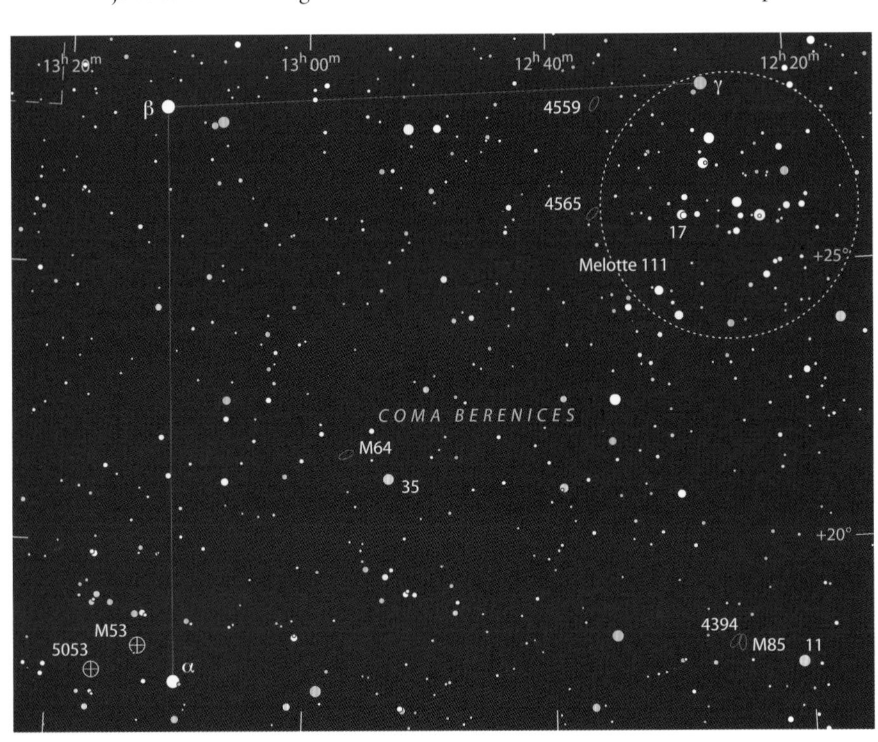

Zooming in on the three main stars of Coma Berenices, this chart includes fainter stars to magnitude 9.5 for help in locating the objects described by the author. Most finderscopes have a field as wide as the dotted circle at top right, which encloses the loose star cluster Melotte 111. (Your main scope, even at low power, shows a much smaller area of sky.)

*From *The Poems of Catullus, A Bilingual Edition* (Berkeley and Los Angeles: University of California Press, 1969).

Deep-Sky Riches of Coma Berenices

Object	Type	Magnitude	Size/Sep.	Distance (l-y)	Right Ascension	Declination
M53	Globular cluster	7.6	12'	60,000	13^h 12.9^m	+18° 10'
NGC 5053	Globular cluster	9.5	8'	54,000	13^h 16.5^m	+17° 42'
Melotte 111	Open cluster	1.8	5°	290	12^h 25.1^m	+26° 07'
17 Comae	Double star	5.2, 6.6	145"	270	12^h 28.9^m	+25° 55'
NGC 4565	Galaxy	9.6	16' x 3'	32 million	12^h 36.3^m	+25° 59'
NGC 4559	Galaxy	10.0	11' x 5'	32 million	12^h 36.0^m	+27° 58'
35 Comae	Double star	5.0, 9.8	29"	320	12^h 53.3^m	+21° 15'
M64	Galaxy	8.5	9' x 5'	19 million	12^h 56.7^m	+21° 41'
M85	Galaxy	9.1	7' x 5'	60 million	12^h 25.4^m	+18° 11'
NGC 4394	Galaxy	10.9	3'	60 million	12^h 25.9^m	+18° 13'

tude primary and the 6.6-magnitude secondary. They are actually blue-white and white, respectively, but some observers see the companion star as blue-green.

NGC 4565 lies 1.7° east of 17 Comae. It is one of the most impressive galaxies in the *Revised Flat Galaxy Catalogue* (Igor D. Karachentsev and colleagues, 1999), which lists thousands of highly elongated, edge-on spirals. My little refractor at 47x shows NGC 4565 about 7' long and very thin with a small, brighter bulge in the center. Switching to 87x brings out some mottling across the core, hinting at the dark lane that runs along the galaxy's length.

Another nice galaxy can be seen 2° north of NGC 4565. **NGC 4559** is also elongated, but it appears a little shorter and fatter than its neighbor to the south. Two faint stars lie on either side of the galaxy's southeastern end.

Nearly two-thirds of the way from 17 Comae to Alpha, we find the 4.9-magnitude golden star **35 Comae Berenices.** It is easy to recognize as the brightest star in the area, and it has a 9.8-magnitude companion to the southeast that is well separated at 17x.

We can use 35 Comae to find **M64**, the Black Eye Galaxy, 1° to the northeast. The Black Eye gets its name from the prominent dust lane that looks so stunning in photographs. Observers have been able to glimpse this feature in telescopes as small as 60mm. In my 105mm scope at 127x, the galaxy appears about 6' x 3'.

Now we'll move to the southwestern corner of Coma's square, marked by the galaxy **M85**. If your sky is dark enough, you can try to spot nearby 4.7-magnitude 11 Comae with the unaided eye (see the chart on the facing page). It looks yellow-orange through the eyepiece, and you'll find M85 only 1.2° (one low-power field) to the east-northeast. M85 is the northernmost Messier galaxy in the Virgo Galaxy Cluster and one of the brightest. It is easily visible in my 14x70 binoculars as a small, oval patch of light. My little refractor at 87x shows a 2' tapered oval that grows smoothly brighter toward the center. A very faint star can be seen on the northern edge, and a 10th-magnitude star lies close to the southeast. In dark skies, it's possible to glimpse the much fainter galaxy **NGC 4394** east of M85 — their centers are separated by a mere 7.5'. It appears as a small oval with a bright, stellar nucleus.

These are just a handful of the deep-sky riches tangled in the strands of Berenice's Hair. With the help of a good star atlas, you'll find many more to while away the hours on these warming nights of spring.

It's no mystery how M64, the Black Eye Galaxy, got its name. The photographic image, taken with a Celestron 11, spans about ¼°. The author's pencil drawing, made at 127x with her Astro-Physics Traveler 105mm refractor, shows a much wider field. Both views are reproduced with north up and east to the left. Photo: Bob and Janice Fera

First published May 2002

Markarian's Chain

What region of the sky shows the most galaxies in a single telescopic view?

Strung along a 1½° arc that straddles the Virgo–Coma Berenices border are eight galaxies known as Markarian's Chain. Two of its members (M84 and M86) were spotted in 1781 by Charles Messier, while the rest are best known by their numbers from Johan L. E. Dreyer's 1888 *New General Catalogue* (NGC 4435, 4438, 4458, 4461, 4473, and 4477). But the moniker for the whole group arises from a paper titled "Physical Chain of Galaxies in the Virgo Cluster and Its Dynamic Instability" by Russian astrophysicist Benjamin E. Markarian (*Astronomical Journal*: December 1961, page 555).

Markarian's Chain has long been one of my favorite areas for playing a little observational game of seeing how many galaxies I can spot in one field of view. There are many places elsewhere in the sky where quite a few galaxies will fit in a relatively small eyepiece field, but these are generally in the purview of large telescopes. Here in the heart of the Virgo Cluster of galaxies, even small scopes can play the game.

There are trade-offs in this counting pastime. A large-aperture telescope can show fainter galaxies, but most small scopes can achieve a wider field and encompass more galaxies. By increasing the magnification, we can make a faint galaxy easier to see (our eyes are better at seeing a large dim object than a small dim one), but this decreases the field of view and spreads its neighbors farther apart.

I think my most appealing view of the area is through a 10-inch reflector with an eyepiece that gives me 43x and an 89' field. I can capture all of Markarian's Chain plus five or six other galaxies in a single look. Although a low-power view with its wealth of galaxies is captivating, my descriptions were made at moderate powers of 75x to 100x to help tease out more detail. These galaxies are also visible in my 105mm refractor, but some are challenging to pick out.

Let's visit each in turn.

Our first target will be **M84**, which sits at the western end of Markarian's Chain, halfway between Vindemiatrix (ε Virginis) and Denebola (β Leonis). My 105mm scope shows a slightly oval glow, aligned northwest to southeast, enveloping an intense round core. The 10-inch reflector exposes a tiny bright nucleus and reveals more of the halo before it fades into the background sky.

M86 lies just 17' to the east. With my little refractor, it appears larger and slightly dimmer than M84. Aligned in the same direction as its companion, this more elongated galaxy brightens

Star magnitudes
7
8
9
10
11
12

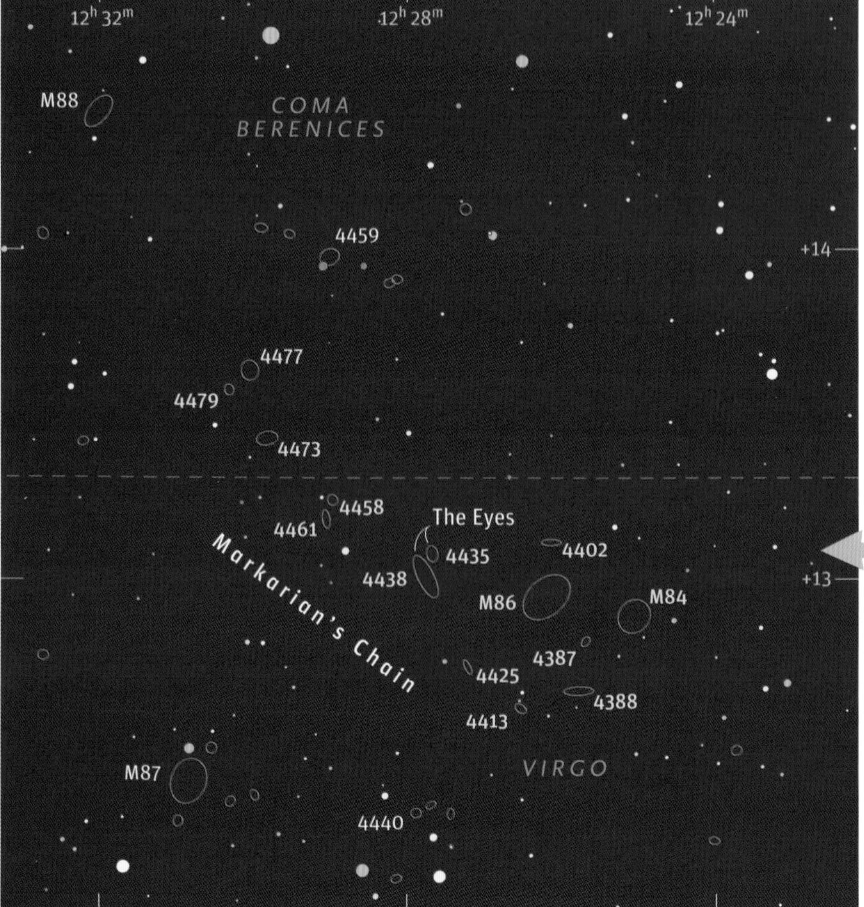

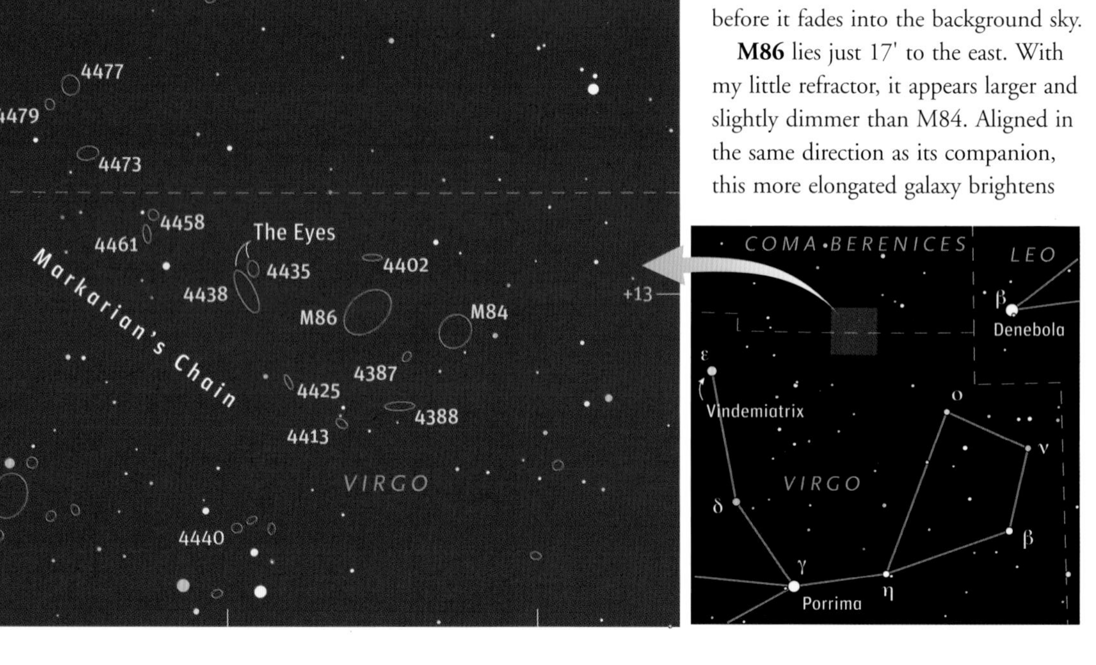

Two galaxies known as the Eyes lie practically in the middle of Markarian's Chain. NGC 4435 sits 5' north-northwest of larger NGC 4438, which shows distortions in its symmetry due to the gravitational interaction between the pair. For this close-up, Adam Block used a 20-inch RC Optical Systems telescope and an SBIG ST-10XME camera on Kitt Peak in Arizona. Photo: NOAO / AURA / NSF

considerably toward its center. The view through the 10-inch adds a stellar nucleus.

Several galaxies that are not part of Markarian's Chain lie in the immediate area. **NGC 4388** is south of M84 and M86 and forms an equilateral triangle with them. This misty east-west slash is fairly easy in the refractor, but it's nearly featureless and shows only a subtly brighter, extended core. The 10-inch picks out a slight bulge in the core, offset to the west of center, and a faint stellar nucleus.

NGC 4387 marks the middle of the triangle formed by the above galaxies. Through the 105mm, it is a minuscule bit of fuzz, but the 10-inch turns it into a small oval with a brighter heart. Together, these galaxies make up what I think of as the Great Galactic Face. M84 and M86 are its eyes, NGC 4388 forms the mouth, and NGC 4387 indicates the nose. Our face even seems to have one raised eyebrow.

That eyebrow is **NGC 4402**, a highly elongated spindle similar to NGC 4388 but considerably dimmer. It is a very difficult catch for the little refractor. The larger reflector shows a nearly uniform glow with barely detectable hints of mottling. This eyebrow gives our face a quizzical look, as though its owner is wondering who shaved off the other one.

Two additional galaxies in the vicinity could be mistaken for faint stars in the 105mm. **NGC 4413** is the more difficult of the pair. East of NGC 4388, look for a 10.9-magnitude star with an 11.5-magnitude star 1.6' to its south-southeast. NGC 4413 is centered due south of the second star by about the same distance and looks like a dimmer fuzzy star. **NGC 4425** is a little easier. Look for it 4' west-southwest of the brightest star just east of the face. The 10-inch scope helps coax out the galactic nature of these dim objects, and both appear oval with brighter centers.

Returning to members of Markarian's Chain, we'll wander eastward from M86 to **NGC 4435** and **NGC 4438**. In February 1955, *Sky & Telescope* carried an article called "Adventuring in the Virgo Cloud" by Leland S. Copeland, who had been this magazine's first Deep-Sky Wonders columnist in the 1940s. Copeland made a chart, which he called Coma-Virgo Land, and labeled it with fanciful names for patterns of stars and galaxies. He dubbed this pair of galaxies the Eyes, and the name stuck.

Both are easy targets for the small refractor. NGC 4435 is a small oval aligned almost north-south, and it harbors a bright stellar nucleus. NGC 4438 is larger, more highly elongated, and tipped more to the north-northeast. A small bright core lies at its heart. The 10-inch increases the apparent size of the galaxies and adds a faint stellar nucleus to NGC 4438.

Recent observations with the Chandra X-ray Observatory indicate that this galaxy pair underwent a high-speed collision about 100 million years before our current view of them. This distorted NGC 4438 and ejected much of the hot gas that formerly belonged to NGC 4435. NGC 4438 may also have an active galactic nucleus, wreaking further havoc.

If you move 21' east-northeast of this pair, you'll find another galactic duo, **NGC 4458** and **NGC 4461**. With

Galaxies abound in and around Markarian's Chain, which stretches from M84 (right) to NGC 4477 (upper left). North is up in this 2°-wide field. Photo: Robert Gendler

First published May 2004

Galaxies of Markarian's Chain

Galaxy	Magnitude	Right Ascension	Declination
M84	9.1	12ʰ 25.1ᵐ	+12° 53'
M86	8.9	12ʰ 26.2ᵐ	+12° 57'
NGC 4388	11.0	12ʰ 25.8ᵐ	+12° 40'
NGC 4387	12.1	12ʰ 25.7ᵐ	+12° 49'
NGC 4402	11.7	12ʰ 26.1ᵐ	+13° 07'
NGC 4413	12.3	12ʰ 26.5ᵐ	+12° 37'
NGC 4425	11.8	12ʰ 27.2ᵐ	+12° 44'
NGC 4435	10.8	12ʰ 27.7ᵐ	+13° 05'
NGC 4438	10.2	12ʰ 27.8ᵐ	+13° 01'
NGC 4458	12.1	12ʰ 29.0ᵐ	+13° 15'
NGC 4461	11.2	12ʰ 29.1ᵐ	+13° 11'
NGC 4473	10.2	12ʰ 29.8ᵐ	+13° 26'
NGC 4477	10.4	12ʰ 30.0ᵐ	+13° 38'
NGC 4479	12.4	12ʰ 30.3ᵐ	+13° 35'

the 105mm, NGC 4458 appears small, round, and quite faint. NGC 4461 is brighter and elongated nearly north-south. The 10-inch makes NGC 4458 easier to spot and shows a brightish core within a faint halo. NGC 4461 displays a stellar nucleus within a small, round core.

Crossing the constellation border from Virgo into Coma Berenices, we next come to **NGC 4473**. A very easy target for the 105mm, this galaxy is an east-west oval of mist with a starlike nucleus. The 10-inch increases the apparent size of the galaxy and shows that it fades gradually outward from the center.

The final galaxy in Markarian's Chain is **NGC 4477**. It is another easy capture for the small refractor, which shows it to be small and round with a brighter center. Both the fringe and the core of the galaxy look slightly oval in the 10-inch, and a stellar nucleus sits within.

To fit both NGC 4477 and M84 in the same low-power field of the 10-inch, I must put them on opposite edges of the field with part of their halos cut off. With some careful maneuvering, I can also lay **NGC 4479** along the edge. But this is a faint galaxy, hard to pick out at the edge of a low-power field unless the sky is reasonably dark. NGC 4479 is a difficult little smudge in the 105mm and easier, but not much more impressive, in the 10-inch.

There are many variations of the galaxy-counting game. Observers with small telescopes capable of giving a field a few degrees across will lose some of the dim galaxies, but they'll gain bright Messier galaxies nearby. Large-scope users working with a more restricted field of view can sacrifice the eastern end of the chain but pick up several faint NGC and IC galaxies.

The reality of Markarian's Chain as a true physical system is still debatable, but that surely won't keep skygazers from enjoying this wonderfully rich area of the sky. So let me leave you with these words from another article by Leland S. Copeland:

The earth's great shadow sweeps around
And ends the fulgent day;
The mountains doff their twilight blue,
The lowlands lose their gray.
But overhead the stars return,
The cloven Milky Way,
And eyes that look through telescopes
Amazing things survey.

Bowled Over in Ursa Major

The region around the Big Dipper's Bowl is rich in deep-sky wonders.

Only avid stargazers are familiar with the complete figure of Ursa Major, the Great Bear, but seven of her stars are among the best known in all the sky. In North America, we fashion them into the Big Dipper, with three stars in its bent handle and four in its bowl. Here, we'll tour the area around the bowl, a region rich in deep-sky wonders.

Let's begin with another seven-starred pattern 1½° west of Beta (β) Ursae Majoris, or Merak. In 1993, Barlow Pepin wrote about this asterism in the *Journal of the British Astronomical Association*. Pepin credited its discovery to Johannes Sachariassen, whose telescopic exploits were chronicled by Pierre Borel in 1655-56. Borel styled Sachariassen's stars as *Uniti Belgii*, a bundle of seven arrows representing the United Dutch Provinces. If Pepin correctly identified the group near Merak as Sachariassen's discovery, we are seeing, when we view this group, one of the first telescopic asterisms ever recorded.

Many observers picture the **Seven Arrows** as an incomplete loop of 7th- to 10th-magnitude stars forming a 15' oval. Because of its battered appearance, astronomy writer Philip S. Harrington calls this asterism the Broken Engagement Ring and pictures the brightest star as its diamond. I've always seen two additional stars east of the ring as part of this pattern. This gives me five stars in a 24' bar sloping east-northeast to west-southwest with four stars arcing

Bopping Around the Bowl

Object	Type	Magnitude	Size/Sep.	Right Ascension	Declination	*MSA*	*U2*
Seven Arrows	Asterism	6.8	15' × 9'	10h 50.6m	+56° 08'	577	25L
M108	Spiral galaxy	10.0	8.7' × 2.2'	11h 11.5m	+55° 40'	576	24R
M97	Planetary nebula	9.9	3.4'	11h 14.8m	+55° 01'	576	24R
NGC 3718	Spiral galaxy	10.8	9.2' × 4.4'	11h 32.6m	+53° 04'	575	24R
NGC 3729	Spiral galaxy	11.4	3.0' × 2.2'	11h 33.8m	+53° 08'	575	24R
h2574	Double star	11.8, 11.9	33"	11h 32.5m	+53° 02'	575	24R
Hickson 56	Galaxy group	14.1 (total)	2.4' (total)	11h 32.6m	+52° 57'	575	24R
M109	Spiral galaxy	9.8	7.6' × 4.6'	11h 57.6m	+53° 22'	575	24R
NGC 4026	Lenticular galaxy	10.8	5.2' × 1.4'	11h 59.4m	+50° 58'	575	24R
M40	Double star	9.7, 10.2	53"	12h 22.2m	+58° 05'	559	24L
NGC 4290	Spiral galaxy	11.8	2.3' × 1.5'	12h 20.8m	+58° 06'	559	24L
NGC 4284	Spiral galaxy	13.5	2.5' × 1.1'	12h 20.2m	+58° 06'	559	24L

Angular sizes or separations are from recent catalogs. The columns headed *MSA* and *U2* give the chart numbers of objects in the *Millennium Star Atlas* and *Uranometria 2000.0*, 2nd edition, respectively.

northward from its center. Sometimes, I imagine it as a flat boat with a billowing sail, perhaps a Chinese junk. At other times, it reminds me of the Coathanger asterism (Collinder 399) in Vulpecula or the Mini-Coathanger in Ursa Minor. Some things, like socks, always seem to go missing, while others multiply. Coathangers appear to be among those magically reproducing items — even in the sky.

Some better-known deep-sky denizens are also found near Merak. The spiral galaxy **M108** sits 1½° east-southeast of the star. A 7th-magnitude golden star lies a little less than halfway from Merak to M108 and serves as a good pointer for finding the galaxy. Through my 105mm refractor at 47x, M108 is a mottled 5' x 1.1' spindle tipped a bit north of east. A fairly bright star is superposed just west of center, and a 12th-magnitude star sits off the western tip. The galaxy is impressively patchy at 166x through my 10-inch reflector, which also reveals an extremely faint star along the south following edge. Photos of M108 display a chaotic mix of obscuring dust clouds and bright star-forming regions.

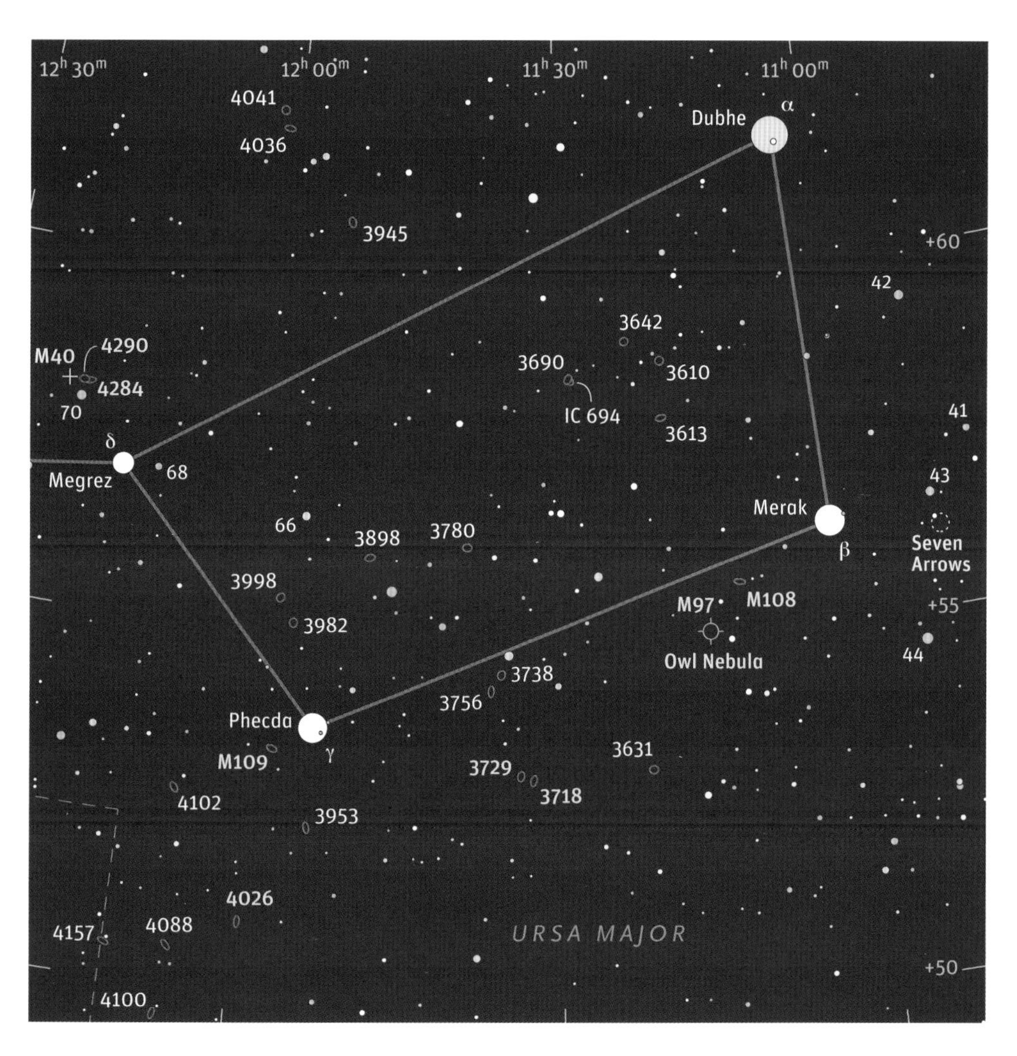

The planetary nebula **M97** is 48' southeast of M108 and

shares the same low-power field of view. With my little refractor at 47x, M97 is the brighter of the two. It's about 3½' across and has a 12th-magnitude star off its north-northeast side. At 127x, the nebula looks very slightly oval. Careful study reveals two subtly dark patches that earn this planetary its nickname, the Owl Nebula. These dusky eyes are placed along M97's long axis. The narrow strip between the eyes is brighter, as are the sides lined up with it. A little pip of brightness is visible at the nebula's heart, but don't mistake this for the 16th-magnitude central star, which is well beyond the grasp of small telescopes.

The Owl Nebula was named by the 19th-century Irish observer Lord Rosse, whose sketch of the object has a whimsical mien. Although we can embrace the Owl and M108 within the same telescopic field, they are nowhere near each other in space. M97 resides within our own galaxy at a distance of around 2,000 light-years, while M108 is estimated to be 46 million light-years away.

A golden 5.6-magnitude star lies three-fifths of the way from Merak to Gamma (γ) Ursae Majoris, also known as Phecda. If you place the golden star in a low-power field and drop 1.7° southward, you'll come to the interacting galaxy pair **NGC 3718** and **NGC 3729**. Both are faintly visible in my 105mm scope at 87x. NGC 3718 is a 2½'-long oval just north-northeast of a matched pair of 12th-magnitude suns (the double star **h2574**). NGC 3729 is a smaller smudge with a 12th-magnitude star at its south-southwest edge.

This is a pretty galaxy pair in my 10-inch scope at 166x. NGC 3718 has a small nucleus surrounded by a large, bright, patchy oval core running north-northwest to south-southeast. Short, faint smears arc away from each end of this oval, giving it a shallow, mirrored S curve and a 5½' x 2½' halo. NGC 3729 is smaller but has a higher surface brightness and is tipped a little west of north. Its large core is surrounded by a thin atmosphere of faint haze for an overall size of about 2' x 1'.

Photographs show the ultracompact galaxy group **Hickson 56** a few arcminutes south of NGC 3718. It has been seen under very dark skies in apertures as small as 12 inches. With five galaxies crammed into 2½' of sky, it's a challenge to distinguish them as separate knots. High magnification is required. Here, too, we have a study in contrasts. NGC 3718 is about 50 million light-years distant, while Hickson 56 is believed to be more than 10 times farther away.

Moving eastward, we find the galaxy **M109** just 39' east-southeast of Phecda. The star is a hindrance when we observe with binoculars, but the galaxy is easily visible at 47x through my small refractor, despite Phecda's glare. At 87x, M109 is an oval running east-northeast to west-south-west with a brighter core. A faint star is superposed north of the core, and another lies off the galaxy's easterly tip. With my 10-inch reflector, the galaxy appears about 5½' x 3', and the oval core is tipped with respect to the rest of the galaxy. It extends 1½' northeast to south-west and grows brighter toward the center.

Now we can look for a 9th-magnitude star ⅓° east of M109. It marks the northern end of a 2⅓° north-south line of four very unevenly spaced stars of similar brightness and color. The pretty spindle-shaped galaxy **NGC 4026** sits 7' south-southwest of the chain's southernmost star. My little refractor at 87x shows a moderately bright, 3½' north-south streak with a brighter

This visually interesting pair of interacting galaxies, NGC 3718 (right) and NGC 3729, is a study in contrasts with the compact galaxy cluster Hickson 56 located below NGC 3718. While modern estimates place the interacting pair roughly 50 million light-years distant, the Hickson group is perhaps 10 times farther, a value intuitively appreciated from the relative size and brightness of the objects. North is up in this 22'-wide field. Photo: Robert Gendler

First published May 2005

After only a glance at his sketch (below), it's easy to understand why 19th-century Irish astronomer Lord Rosse bestowed the moniker "Owl Nebula" on the planetary M97 in Ursa Major. While photographs often emphasize subtle features that are difficult to detect at the eyepiece, the Owl's dark eyes can be seen visually in modest-aperture telescopes. North is up and the field 5' wide. Photo: Gary White / Verlenne Monroe / Adam Block / NOAO / AURA / NSF; Sketch: Sci. Papers; W. Parsons, 3rd Earl of Rosse, 1800-1867

oval core and a small round nucleus. NGC 4026 and M109 belong to a group of galaxies about 60 million light-years away.

Our next object is **M40**, a double star near Delta (δ) Ursae Majoris, or Megrez. M40 was once considered a missing Messier object. In his 1784 catalog, Charles Messier described his 40th object as "two stars very close to one another and very small, placed at the root of the great Bear's tail." Messier's position corresponds to just such a star pair, so there's no reason to consider anything "missing." He noted this double while searching for a "nebulous star" reported by Johannes Hevelius. Messier assumed that Hevelius mistook these two stars for a nebula, but the latter was actually referring to a different star — also without nebulosity.

To locate M40, start at Megrez and hop 1.1° northeast to 5.5-magnitude 70 Ursae Majoris. Continue that line for ¼° to arrive at M40. My 105mm scope at 28x reveals an east-west pair of 10th-magnitude stars, with the western one slightly brighter. Through my 10-inch scope, I see the bright one as yellow-orange and its companion as yellow-white. Two galaxies share the field at 118x. **NGC 4290** is a small northeast-southwest oval, and **NGC 4284** is a tiny faint spot forming a 1½' triangle with two 13th-magnitude stars. The galaxies are about 140 and 190 million light-years away, respectively. Although the distances to its stars are poorly known, M40 is probably an optical (unrelated) pair.

Independently discovered in 1863 by German astronomer Friedrich August Theodor Winnecke, M40 carries the double-star designation Winnecke 4. Winnecke is also the original discoverer of eight NGC objects, and there are 10 comets that bear his name. Periodic Comet 7P/Pons-Winnecke is the parent body supplying debris for the June Boötids, a highly unpredictable meteor shower with peak rates as high as 100 meteors per hour.

Ode to Joy

The celestial Hunting Dogs help us flush out galaxies in May's gentle evening sky.

The constellation Canes Venatici, the Hunting Dogs, is usually portrayed as a pair of hounds whose leashes are held by Boötes, the Herdsman. Johannes Hevelius introduced our present-day depiction in his 1687 atlas, *Firmamentum Sobiescianum*, but the celestial woofers were popular on much earlier star maps. They seem to have made their debut on a 1493 globe by Johannes Stoeffler, where two dogs are placed between Boötes and Serpens. They moved to the opposite side of the Herdsman on Peter Apian's 1533 planisphere but were found just above Virgo's northern arm. The 1602 globe of Willem Blaeu brought a small version of the dogs north, where the area's two brightest stars adorned their collars.

Hevelius changed the hunting dogs to their present size and position, and he awarded them the status of a separate constellation. On his atlas, the northern dog is labeled Asterion (Starry) and the southern dog Chara (Joy). Chara lays claim to the constellation's brightest stars, Alpha (α) and Beta (β) Canum Venaticorum, which will guide us to M94,

The nearly face-on spiral galaxy M94 sports a ring of intense star formation around its tiny, bright nucleus. Both features are visible in small telescopes and have inspired at least one veteran observer to liken the view to a cosmic eye staring back at us across 16 million light-years of space. The field is 10' wide with north up. Photo: Jim Misti; Processing: Robert Gendler

the first in a string of fascinating galaxies that inhabit this region of the sky.

M94 makes a squat isosceles triangle with Alpha and Beta. Placing Beta in a low-power field and sweeping 3.2° eastward will take you right to it. My 105mm refractor at 28x shows a fairly bright oval halo enclosing a small intense core and tiny nucleus. A gossamer outer halo can be faintly glimpsed. A 9½-magnitude star sits north of M94, and an 11th-magnitude star is anchored off its western edge. The pretty gold and yellow double star Espin 2643 shares the field. Look for it 1° north-northwest of M94 in an arrow-shaped asterism of 8th-magnitude stars. Upping the magnification to 87x enhances the beauty of the scene by drawing out several faint stars decorating the periphery of M94. Another power boost to 127x reveals some patchiness in the galaxy's core.

Roughly 16 million light-years away, M94 is the closest starburst-ring galaxy that we see nearly face-on. A band of intense star formation encircles the galaxy's center at a distance of about 3,800 light-years. Interaction with M94's small barred nucleus is thought to drive the star-formation process.

According to the HyperLeda extragalactic database, four NGC galaxies are part of a physical group associated with M94. An interacting pair lies about halfway between M94 and Beta. The brighter member is **NGC 4618**. My little refractor at 87x displays an oval glow 2' long and two-thirds as wide sloped north-northeast. It harbors an offset barlike core tipped east-northeast. Nearby in the same medium-power field of view, **NGC 4625** is a small bit of fluff with a faint star off its western side.

Hunting (Dog) Galaxies

Object	Type	Magnitude	Size/Sep.	Right Ascension	Declination	*MSA*	*U2*
M94	Ringed spiral galaxy	8.2	14.3' × 12.1'	$12^h 50.9^m$	+41° 07'	610	37L
NGC 4618	Magellanic spiral galaxy	10.8	4.2' × 3.4'	$12^h 41.5^m$	+41° 09'	611	37R
NGC 4625	Magellanic spiral galaxy	12.4	1.6' × 1.4'	$12^h 41.9^m$	+41° 16'	611	37R
NGC 4490	Barred spiral galaxy	9.8	6.3' × 2.7'	$12^h 30.6^m$	+41° 39'	611	37R
NGC 4485	Magellanic irregular galaxy	11.9	2.3' × 1.6'	$12^h 30.5^m$	+41° 42'	611	37R
NGC 4449	Magellanic irregular galaxy	9.6	6.1' × 4.3'	$12^h 28.2^m$	+44° 06'	611	37R
NGC 4460	Barred lenticular galaxy	11.3	4.0' × 1.2'	$12^h 28.8^m$	+44° 52'	611	37R
NGC 4346	Barred lenticular galaxy	11.2	3.3' × 1.3'	$12^h 23.5^m$	+47° 00'	591	37R
M106	Barred spiral galaxy	8.4	18.8' × 7.3'	$12^h 19.0^m$	+47° 18'	592	37R
NGC 4248	Irregular galaxy	12.5	3.1' × 1.1'	$12^h 17.8^m$	+47° 25'	592	37R
NGC 4232	Barred spiral galaxy	13.6	1.4' × 0.7'	$12^h 16.8^m$	+47° 26'	592	37R
NGC 4231	Lenticular galaxy	13.3	1.1'	$12^h 16.8^m$	+47° 27'	592	37R
NGC 4217	Spiral galaxy	11.2	5.7' × 1.6'	$12^h 15.8^m$	+47° 06'	592	37R
NGC 4226	Spiral galaxy	13.5	1.0' × 0.5'	$12^h 16.4^m$	+47° 02'	592	37R

Angular sizes or separations are from recent catalogs. The visual impression of an object's size is often smaller than the cataloged value and varies according to the aperture and magnification of the viewing instrument. The columns headed *MSA* and *U2* give the chart numbers of objects in the *Millennium Star Atlas* and *Uranometria 2000.0*, 2nd edition, respectively.

Both galaxies are Magellanic spirals, which often exhibit a single prominent spiral arm springing from one end of an off-center stellar bar. Gravitational interactions may enhance this unusual structure, but such lopsidedness is also found in Magellanic spirals without physical companions. If you have a 10-inch or larger scope, try to spot NGC 4618's spiral arm, which has regions of patchy brightness that start east of the core and wrap southward.

Another interacting pair associated with M94 lies 40' west-northwest of Beta. In my 105mm scope at 87x, **NGC 4490** appears 4½' long, one-third as wide, and fatter at its southeastern end. A wide, irregularly bright bar runs along the galaxy's long axis. Hovering north of its companion's narrow end, **NGC 4485** is a small, round glow with a slightly brighter core.

Let's move on to a different group of related galaxies, starting with **NGC 4449**. If you place NGC 4490 in the eastern side of a low-power field and scan 2.5° northward, this moderately bright galaxy should be easy to spot. In my 105mm scope at 87x, it's elongated northeast-southwest with a broadly brighter core and a thin halo. The core looks wider on the northeast end and is a bit patchy. NGC 4449 is a Magellanic irregular galaxy, noted for multiple extended regions of star formation. With his 4-inch refractor at 189x, skilled observer Stephen James O'Meara has seen several of these regions as brighter spots within NGC 4449.

Sweeping 42' northward takes us to the pale yellow and deep yellow double star Σ1645. The pair is widely split at 87x and shares the field with **NGC 4460**, located 8' east-northeast. This faint galaxy appears about 2' long and one-quarter as wide, elongated northeast to southwest.

A 6th-magnitude orange star 2.6° north-northwest of Σ1645 will serve as a jumping-off point for our next two galaxies. It's the brightest star in the area, and **NGC 4346** lies just 19' toward the southeast. In my small scope at 47x, this nearly east-west galaxy appears 1½' long and one-third as wide with a small, considerably brighter core.

M106, located ½° west-northwest of the orange star, is the dominant member of our second galaxy group. It's big and beautiful in my little refractor at 87x. The faint halo spreads 10' x 3½' with a dim star superposed on its north-northwestern end. Its much brighter, oval core is mottled and about half as long. The small, round, intensely bright nucleus nestled in its heart is thought to be powered by a black hole of 35 million solar masses. The companion galaxy **NGC 4248** is faintly visible as an elongated smudge west-northwest of M106 and pointing toward its northern end.

In my 10-inch reflector at 118x, the core of M106 forms a fat but shallow S curve betraying its spiral nature. NGC 4248 is dominated by an unusual barlike core and encompasses a faint star in the western end of its thin halo. The faint side-by-side galaxies **NGC 4232** and **NGC 4231** are also visible, 11' west-northwest of NGC 4248. Increasing the magnification to 170x, I see NGC 4232 as small, elongated north-northwest to south-southeast, and having a brighter center. Just to its north, NGC 4231 is small and round. The galaxies make a little isosceles triangle with a 14th-magnitude star to their west. NGC 4232 and NGC 4231 are not part of the M106 group, since M106 is about 25 million light-years away and the pair is 14 times more distant.

Two additional field galaxies share a medium-power view with M106. With my 105mm refractor at 87x, **NGC 4217** is faintly visible 34' west-southwest of M106 and sports an 11.6-magnitude star on its northern side. Although this is not a particularly faint galaxy, the view is hindered by the glare of the 9th-magnitude star just above its northeastern tip. Observers with large telescopes can look for the dark dust lane that runs the length of this edge-on galaxy. While my 10-inch scope doesn't show me the dust lane, it does pick out **NGC 4226**, located 7' to the southeast. This small oval galaxy is weakly brighter toward the center and tilted east-southeast. NGC 4217 is about twice as far away as M106, while NGC 4226 appears to be a distant galaxy associated with NGC 4232 and NGC 4231.

In June, we'll continue our trek among the rich assortment of galaxies populating this part of the sky.

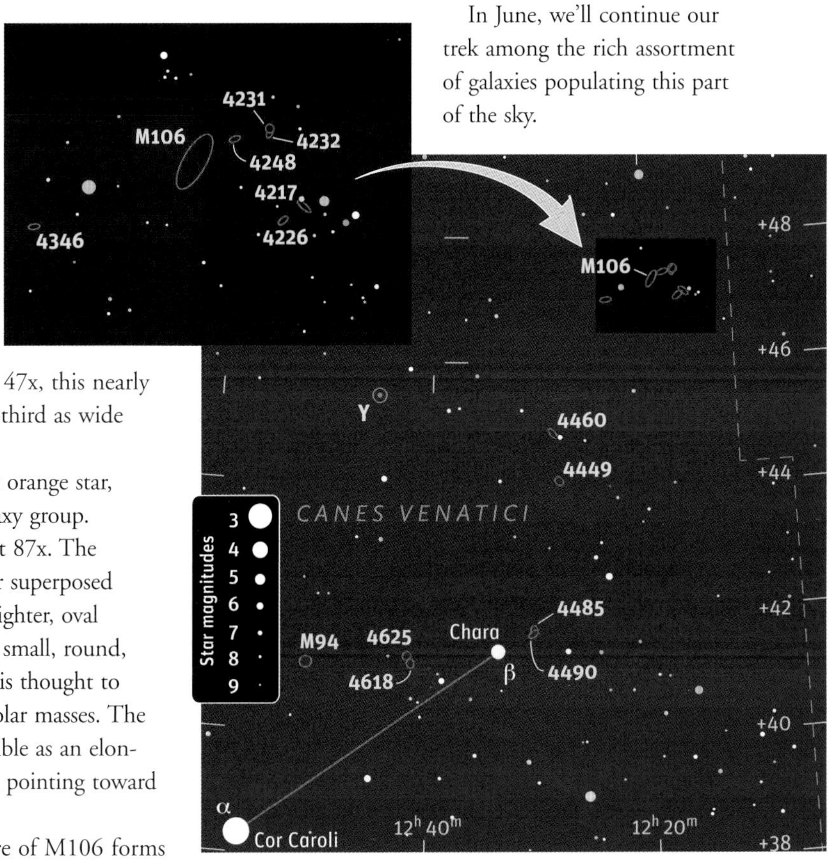

SPRING

May

First published May 2006

V Is for Virgo

One small corner of Virgo holds celestial sights near and far.

The brightest stars in western Virgo form a distinctive and very wide V that always helps me find my way in this galaxy-rich area of the sky. From east to west, the stars forming the V are Epsilon (ε), Delta (δ), Gamma (γ), Eta (η), and Beta (β) Virginis. These stars held a special place in ancient Arabian astronomy, as they delineated one of the *manâzil al-qamar*, mansions of the Moon. In the course of its monthly journey around Earth, the Moon appears to move through the starry backdrop of the constellations and visits one of these 28 lunar stations each night. The mansion embraced by Virgo's V is al-ʿAwwaʾ. The meaning is uncertain, but it is often associated with a dog kennel and translated as "the Barker."

Gamma Virginis, also known as **Porrima**, marks the point of Virgo's V. Porrima is a visual binary with a period of 169 years. In May 2005, its twin yellow-white suns reached their smallest apparent separation, an excruciatingly close 0.4". In 2011, the stars widened to 1.6", a challenge for a 76mm telescope at high power under steady skies. If you can't split the twosome, look for a nearly north-south elongation to their combined light. They will continue to widen over the next several decades, reaching a maximum separation of 6.0" in 2088.

Porrima serves as the trailhead for a star-hop to the galaxy **NGC 4517**. First, jump about 1° northwest to a 7.2-magnitude star, the brightest in the area. Then make an equal hop in the same direction to an 8.6-magnitude golden star adorning the western edge of a smattering of dimmer gems. Two of the brightest and this saffron spark form a nice arrowhead. Following this handy road sign the same distance yet again takes us to a spot halfway between an 8th-magnitude star and NGC 4517, both of which are visible in the same field of view.

In my 105mm refractor at a magnification of 47x, NGC 4517 is a faint slash with an 11th-magnitude star nestled against its northern flank. The galaxy's highly elongated profile leans slightly north of east. My 10-inch reflector at 118x shows NGC 4517's slender form extending 8½' and brighter along the southern side. The positively ghostly galaxy **NGC 4517A** joins the scene in my 15-inch reflector at 153x. It rests 17' north-northwest of its companion and makes a 4' triangle with two stars northwest and west, magnitudes 10.0 and 11.5, respectively. NGC 4517A appears slightly oval and very faint. Averted vision (that is, directing your gaze a little off to one side of the object) makes it easier to spot. Despite the designation, NGC 4517A doesn't form a physical pair with its neighbor but, rather, lies 20 million light-years beyond it.

NGC 4517 is a "flat galaxy." The visible portion of a spiral galaxy is a flattened disk that looks roundish when seen face-on. When viewed edge-on, it appears very elongated, and if the galaxy doesn't have a large central bulge, it looks quite flat. More than seven times longer than wide, NGC 4517 has earned a place in the 1999 *Revised Flat Galaxy Catalogue* by

Virgo is filled with galaxies, and one of the nicest for small telescopes is the face-on spiral M61, which has a sharply defined stellar core. Far more challenging to see is the faint companion galaxy NGC 4303A located 10' to the northeast (upper left in this 14'-wide view). Photo: Jim Misti / Robert Gendler

Spotting the sky's brightest quasar, 3C 273, gives you bragging rights to having seen light that originated nearly 2 billion years ago, before advanced life forms appeared on Earth. The area covered by the inset finder is indicated by the dark square on the chart. The visual magnitudes of selected stars are given to tenths with the decimal points omitted (so that 101 means magnitude 10.1), and the quasar varies irregularly between magnitude 12.3 and 13.0. North is up.

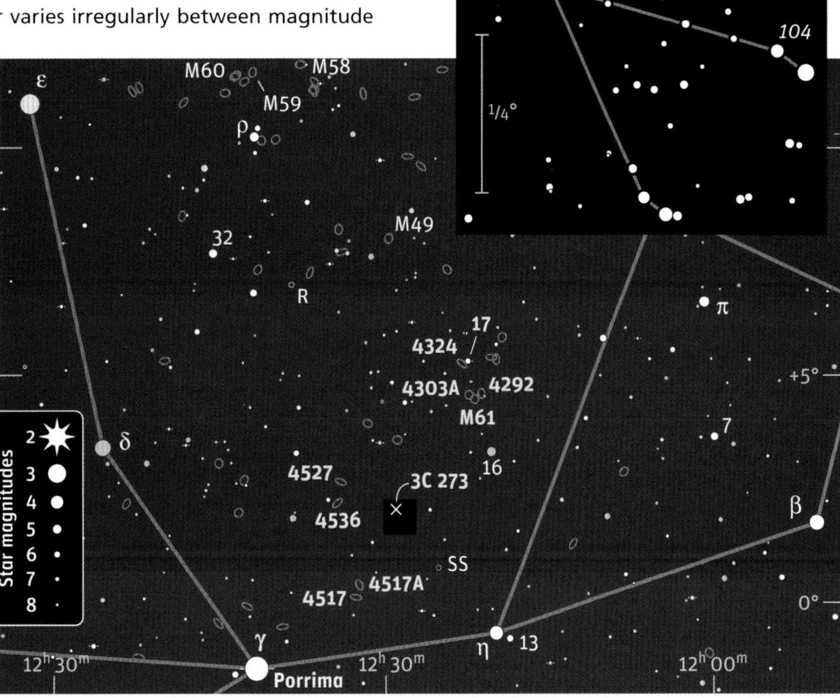

Igor D. Karachentsev and his colleagues.

A brighter galaxy pair awaits us just a short leap to the north. Rise 1° north from NGC 4517 to a yellow 8th-magnitude star. From there, move 1.4° north-northeast to a north-south pair of 9th-magnitude stars, which lie halfway between **NGC 4527** and **NGC 4536**. In my little refractor at 47x, both galaxies are visible as faint ovals. NGC 4536 is larger but has a lower surface brightness. NGC 4527 has a brighter core and seems to grow when viewed with averted vision. The two galaxies still share the field of view at 87x. NGC 4536 is 3½' long, one-third as wide, and leans southeast. NGC 4527 is about the same length, one-quarter as wide, and tips east-northeast. With my 10-inch scope at 118x, NGC 4536 reveals a large, patchy, irregular core with a stellar nucleus. The lens-shaped center has projections that give the galaxy a shallow S curve. NGC 4527 bares a tiny nucleus and grows considerably brighter toward its long axis.

The brightest quasar in the sky, **3C 273**, is 1.3° west of NGC 4536. Stellar-looking sources of vast energy, quasars are the active cores of distant galaxies. I never thought of 3C 273 as a small-scope target until Arizona astronomer Brian Skiff reported seeing it in his 70mm refractor. While my upstate New York observing site is considerably more

A Galaxy Romp in Northwestern Virgo

Object	Type	Magnitude	Size/Sep.	Right Ascension	Declination	*MSA*	*U2*
Porrima	**Double star**	**3.5, 3.5**	**≥1.6"**	**12h 41.7m**	**−01° 27'**	**772**	**111L**
NGC 4517	**Galaxy**	**10.4**	**11.2' × 1.5'**	**12h 32.7m**	**+00° 07'**	**773**	**111L**
NGC 4517A	**Galaxy**	**12.5**	**5.1' × 3.4'**	**12h 32.5m**	**+00° 23'**	**773**	**111L**
NGC 4527	**Galaxy**	**10.5**	**6.9' × 2.4'**	**12h 34.1m**	**+02° 39'**	**773**	**111L**
NGC 4536	**Galaxy**	**10.6**	**8.4' × 3.2'**	**12h 34.4m**	**+02° 11'**	**773**	**111L**
3C 273	**Quasar**	**12.3–13.0**	**—**	**12h 29.1m**	**+02° 03'**	**773**	**111L**
M61	**Galaxy**	**9.7**	**6.5' × 5.7'**	**12h 21.9m**	**+04° 28'**	**749**	**111L**
NGC 4292	**Galaxy**	**12.2**	**1.7' × 1.2'**	**12h 21.3m**	**+04° 36'**	**749**	**111L**
NGC 4303A	**Galaxy**	**13.0**	**1.5' × 1.2'**	**12h 22.5m**	**+04° 34'**	**749**	**111L**
17 Virginis	**Double star**	**6.6, 10.5**	**21.4"**	**12h 22.5m**	**+05° 18'**	**749**	**111L**
NGC 4324	**Galaxy**	**11.6**	**3.1' × 1.3'**	**12h 23.1m**	**+05° 15'**	**749**	**111**

Angular sizes and separations are from recent catalogs. The visual impression of an object's size is often smaller than the cataloged value and varies according to the aperture and magnification of the viewing instrument. The columns headed *MSA* and *U2* give the appropriate chart numbers in the *Millennium Star Atlas* and *Uranometria 2000.0*, 2nd edition, respectively. All the objects this month are in the area of sky covered by Chart 45 in *Sky & Telescope's Pocket Sky Atlas*.

First published May 2007

light-polluted than Anderson Mesa, Arizona, I thought it was worth a try with the 90mm and 105mm refractors my husband Alan and I use. We both succeeded. The larger scope at 87x showed 3C 273 and a 13.6-magnitude star next to it, the quasar being significantly brighter. The smaller scope at 113x showed the quasar with direct vision, but the star could be spotted only with averted vision. After we accomplished our feat, Finnish amateur Jaakko Saloranta told me that he's seen 3C 273 in a refractor stopped down to just 40mm in aperture.

When Alan and I saw 3C 273, it was shining at magnitude 12.6. This quasar varies irregularly between about 12th and 13th magnitude, so the ease of spotting it varies as well. But when you nab 3C 273, you can brag that you've seen something an incredible 2 billion light-years away. The light that bathes us now began its journey when oxygen was just becoming a significant component of Earth's atmosphere.

Now we'll work our way over to the galaxy **M61**, starting from Eta Virginis in Virgo's V. With a finderscope, I look for a long, skinny triangle formed by Eta, 13, and 16 Virginis. After centering 16, I switch to a low-power telescopic view, which includes three 8th- and 9th-magnitude stars in a straight line. Scanning from 16 through the northernmost star in the line and continuing for almost twice this distance again takes me close enough to M61 to see it in the field. Zeroing in for a better look, I see a fairly large, oval glow with my 105mm scope at 87x. The galaxy seems very patchy with hints of structure, and it harbors a small, brighter core. The galaxy **NGC 4292** is visible as a small, very faint spot 12' northwest of M61 and 1.3' south-southeast of a 10th-magnitude star. At 127x, a stellar nucleus gleams at M61's heart.

In my 10-inch reflector at 70x, M61 displays a bright, 1½' ring enveloped by a faint halo that gently fades into the background sky. At 171x, the ring breaks up into patches that suggest a pair of spiral arms opening counterclockwise. A faint star guards their western edge. NGC 4292 and **NGC 4303A** share the field of view. NGC 4292 is a north-south oval with a brighter center. NGC 4303A is a small, very faint glow 10' northeast of M61 and 2½' east of a 13th-magnitude star.

The pretty double star **17 Virginis** lies ¾° north of NGC 4303A. It's widely split in my little refractor at 47x and boasts a light yellow primary star accompanied by a much dimmer orange attendant to the north-northwest. The galaxy **NGC 4324** is faintly visible 9' east-southeast. Inspecting the galaxy at 87x reveals an oval glow, tilted east-northeast, with a brighter core and stellar nucleus. The view gives us a wonderful study in contrasts, since 17 Virginis is only 97 light-years away while NGC 4324 is nearly a million times more distant.

The Immortal Beast

The Bear is a great place to hunt for unusual galaxies and stars.

Looking at Ursa Major, the Great Bear, on the May all-sky star map on page 306, you can "imagine that the immortal beast is descending the slopes of heaven with majestic tread," as described by Peter Lum in *The Stars in Our Heaven*. The chart shows our starry Bear's foreleg joined to its body at the star Upsilon (υ), where we'll begin our ursine tour of deep-sky wonders.

Yellow and gold 6th-magnitude stars 42 arcminutes (42') apart rest about 2° south-southeast of Upsilon and make a nearly straight line with it. Extending that line 1.5° will bring you to our first target, a colorful triple star. Its A and B components bear the designation **Σ1402**. This name indicates that the pair was discovered by the 19th-century German-Russian astronomer Friedrich Georg Wilhelm von Struve. My 105mm refractor at 28x shows the gold, 8th-magnitude primary holding a wide, yellow, 9th-magnitude companion to the east-southeast.

The third star belongs to the pair **GIR 2** AC, where GIR stands for Pierre Girard, the British amateur astronomer who reported it in the Webb Society's Double Star Section Circulars in 1996. The C component is a deep yellow 10th-magnitude star lounging a lavish 2.2' south of the primary. According to the *Tycho Double Star Catalogue* (published by Claus Fabricius and others in

The edge-on spiral galaxy NGC 3079 dwarfs its neighbor NGC 3073. The third-brightest member of this grouping, designated CGCG 265-55, looks like a small, very faint smudge through the author's 14.5-inch reflector. Photo: Robert Gendler

Hunting the Bear in May

Object	Type	Magnitude	Size/Sep.	Right Ascension	Declination
Σ1402/GIR 2	Multiple star	7.7, 8.9, 9.6	32", 134"	10^h 04.9^m	+55° 29'
NGC 3079	Galaxy	10.9	7.9' × 1.4'	10^h 02.0^m	+55° 41'
NGC 3073	Galaxy	13.4	1.3' × 1.2'	10^h 00.9^m	+55° 37'
W Ursae Majoris	Variable star	7.8–8.5	—	9^h 43.8^m	+55° 57'
Spade	Asterism	—	1.1°	9^h 42.6^m	+53° 17'
UGC 5459	Galaxy	12.6	4.0' × 0.5'	10^h 08.2^m	+53° 05'

Angular sizes and separations are from recent catalogs. Visually, an object's size is often smaller than the cataloged value and varies according to the aperture and magnification of the viewing instrument.

2002), this saffron gem is an excruciatingly close pair of matched suns measured just once — 0.4" apart in 1991. There's even a chance that this star isn't really double. Arizona astronomer Brian Skiff recommends using a 16-inch or larger scope under very steady skies.

The lovely galaxy **NGC 3079** sits 28' west-northwest of Σ1402. Its southern tip grazes a triangle of one 8th- and two 9½-magnitude stars. In my little refractor at 47x, I see a nice spindle about seven times longer than wide. It tilts a little west of north and harbors a large, highly elongated core. The tiny companion galaxy **NGC 3073** joins the scene at 87x, making a perfect playing-card diamond with three field stars. This galaxy appears faint and round when I use averted vision but disappears when I look straight at it. At 127x, a very faint star is pinned to the northern tip of NGC 3079.

In my 10-inch reflector at 202x, NGC 3079 seems flat along its western flank but bulges on the opposite side. The core is slightly mottled and contains a bright elliptical center engulfing a nearly stellar nucleus. NGC 3073 also hosts a tiny, brighter heart.

This galaxy pair is about 50 million light-years distant. NGC 3079 has an active nucleus that emits high-energy jets of particles powered by a supermassive black hole. According to a 2007 study, these jets are encountering dense clouds of interstellar matter and blowing a lumpy, bipolar bubble of hot gas. Images recorded in the light emitted by ionized hydrogen and nitrogen show the bubble extending 3,500 light-years from the galaxy's core.

Now let's visit the fascinating binary star **W Ursae Majoris** (UMa), just 162 light-years away. Its components are so close together that they're pulled into teardrop shapes that actually touch! Both stars are yellow dwarfs, one a bit larger than our Sun and the other a little smaller. The conjoined stars whirl madly around each other every 8 hours, undergoing two eclipses 4 hours apart as they alternately pass in front of each other.

What makes this spinning dumbbell so appealing is its continuous variation in brightness as it presents an ever-changing face to inquisitive skygazers. It takes only 2 hours for the pair to go from maximum to

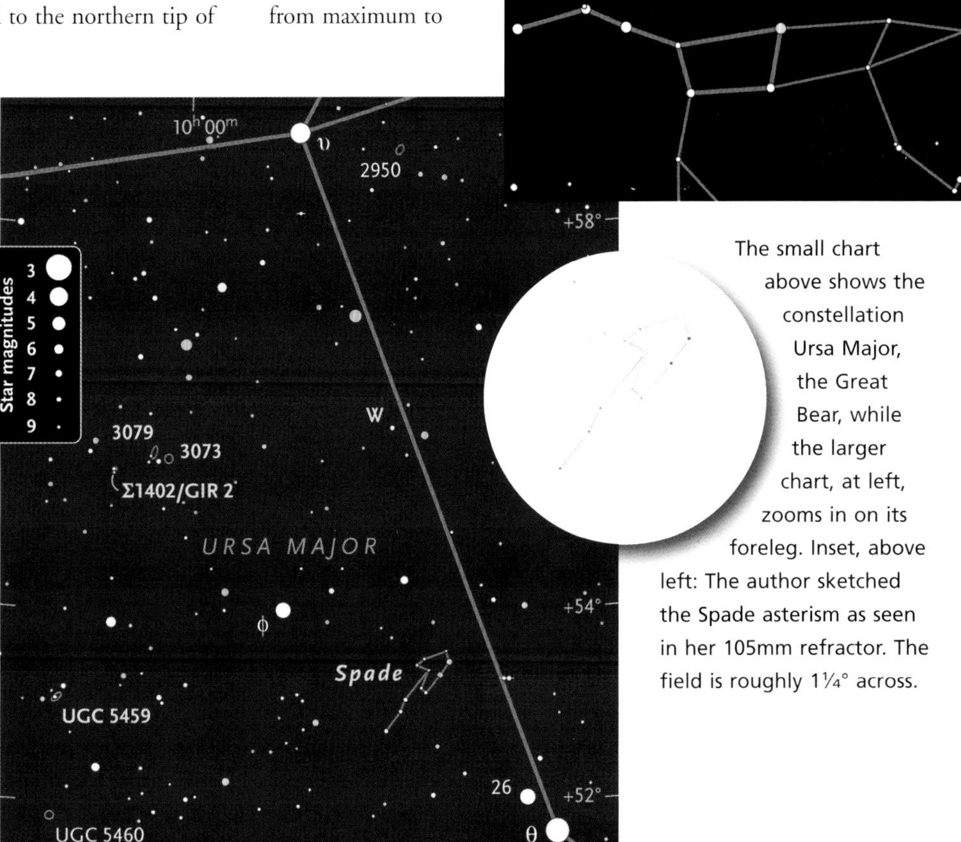

The small chart above shows the constellation Ursa Major, the Great Bear, while the larger chart, at left, zooms in on its foreleg. Inset, above left: The author sketched the Spade asterism as seen in her 105mm refractor. The field is roughly 1¼° across.

SPRING

May

First published May 2008

minimum brightness or vice versa. W UMa peaks at about magnitude 7.8 and dips to 8.5, the two minima being nearly equal.

To locate W UMa, sweep 1.2° west from NGC 3079 to an 8.0-magnitude star. Then hop 1.7° farther west to a 6.5-magnitude star, the westernmost and brightest in a parallelogram of stars measuring ½° north to south. The northern and southern stars in the figure are magnitude 8.9, a little fainter than W UMa at its minimum. W UMa is the eastern star. If you check from time to time, you should be able to catch these stars-that-kiss in their quick-change act.

Next, we'll start to dig a little deeper, burrowing down to an asterism that John Chiravalle calls the **Spade** in his book *Pattern Asterisms*. A good target for a small telescope or large binoculars, it is conveniently located 1.6° southwest of Phi (φ) Ursae Majoris. The celestial shovel runs southeast to northwest and spans 1.1°. My 105mm scope at 28x displays three stars in the handle and eight in the head.

Starting once again at Phi, slip 2.6° east-southeast to a pair of 7.8-magnitude stars with a 9.6-magnitude star halfway between them. The trio forms a slightly curved, southeast-northwest line 16' long. The flat galaxy **UGC 5459** parallels this line 5' to the south. An 8.7-magnitude star tries to hide its southeastern tip. In my 10-inch reflector at 213x, the galaxy appears 2½' long and very thin. Two faint stars sit 1½' west of the northwestern tip.

Flat galaxies are nearly bulgeless spirals that appear thin because we view them virtually edge-on. Many of these galaxies have very low rates of star formation and often do not show the dusty dark lanes that characterize their more active cousins, such as the magnificent edge-on spiral NGC 4565 in Coma Berenices.

As the Crow Flies

Exotic galaxies and multiple-star systems adorn this southerly constellation.

The distinctive trapezium of stars marking Corvus, the Crow, takes wing across the southern sky at this time of year. Although the pattern in no way resembles a bird, the ancient Greeks were not the only culture to make an avian association. The Chinese pictured it as a cart riding the wind in the astrological palace of the Red Bird. Corvus may have been the Euphratean Messenger of the Evil Wind, or Great Storm Bird, and in Brazil, it was seen as a Heron.

Corvus houses no Messier objects, but snug against its northern border, we find **M104**, the Sombrero Galaxy, which is well worth a visit while we're in the neighborhood. The Sombrero resides in Virgo, 3.6° south of Chi (χ) Virginis, and is readily visible as an elongated glow even in my 12x36 image-stabilized binoculars. Through my 105mm refractor at 47x, M104 stretches east-west for 6¾' and harbors a bright oval core 2¼' wide with a small, intense center. Black as a crow, a narrow dust lane runs the length of the galaxy, defining the brim of the Sombrero. A 10th-magnitude star dangles like a bobble from the Sombrero's western tip. Most of the core sits north of the dusky rim and constitutes the crown of our celestial headgear. Higher magnifications render the impressive dust lane more prominent.

Unlike the dust wafting around your home, the dust that encircles the Sombrero Galaxy is no lightweight. It totals 16 million times the mass of our Sun.

Through my little refractor at 17x, this majestic galaxy shares the 3.6° field of view with an amazing asterism and a superb sextuple star. First, look 24' west-northwest of M104 for a little group of four 8th- and 9th-magnitude stars. This is the toothy grin of a ½°-long asterism that astronomy author Philip S. Harrington calls **Jaws**. I see six fainter stars reaching north and then curving east to trace the shark's scrawny body, and one star west of the body that marks his dorsal fin. Don't swim too close; he looks really starved.

Irish amateur Kevin Berwick sees the shark's grin as a small version of the constellation Sagitta, the Arrow. It's a colorful sight through my 10-inch reflector. One of the stars shines yellow, and two are yellow-orange. Our shark should brush his teeth.

Next, soar 1.1° west-southwest from M104 to the multiple star **Struve 1659** (Σ1659). In my 105mm scope at 47x, I see a 5' isosceles triangle of stars with a smaller isosceles triangle centered inside it. My 10-inch scope shows color in all but the dimmest of the six stars. In order of decreasing brightness, I see them as yellow-white, deep yellow, pale yellow, yellow, and gold.

I was first introduced to this sextuplet by John Wagoner in the 1980s at the Texas Star Party. He dubbed the group Stargate, because it reminded him of the hyperspace stargate used by the hero Buck Rogers in the 1979-1981 television series. Wagoner created and ran the Astronomical League's Bulletin Board Service, which he named Stargate. Not surprisingly, others have discovered this group as well.

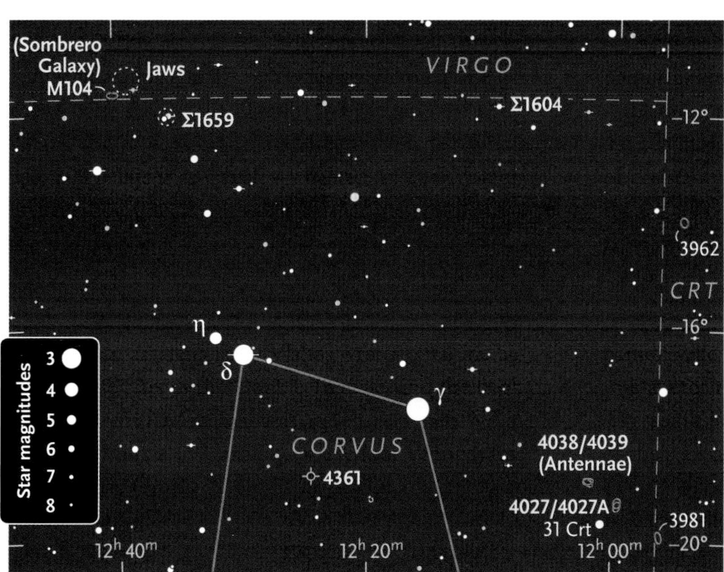

In photographs, the most striking feature of Messier 104 is its wide, dark, equatorial dust lane. The section of the galaxy north of the dust lane — the only part that's visible at first glance through a telescope — gives M104 its nickname, the Sombrero. Photo: Martin Pugh

Australian amateur Perry Vlahos wrote to tell me he knows these stars as the Double Triangle, and the book *Star Clusters* (Brent A. Archinal and Steven J. Hynes; Willmann-Bell, 2003) lists it as Canali — named for Pennsylvania amateur Eric Canali, who calls it "that pretty little triangle-asterism-thingy."

While I know of no other triangle-in-triangle multiples, triple stars that form triangles are common. After all, any three stars not in a straight line make some type of triangle. It's rarer to find a triple that outlines a nearly equilateral triangle, yet one happens to lie just 6.4° west of Σ1659. As displayed through my little refractor at 47x, the primary star of **Struve 1604** (Σ1604) appears yellow, its brighter companion yellow-white, and the dimmer one orange.

Data for the companions are mangled in various sources. The confusion ensues from the large relative proper motion of the stars and the fact that the C component is brighter than B. (The members of most triplets are designated A, B, C, in order of decreasing brightness.) In our table,

Corvus Rides the Wind Across the Southern Sky

Object	Type	Magnitude	Size/Sep.	Right Ascension	Declination
M104	Galaxy	8.0	8.7' × 3.5'	12^h 40.0^m	–11° 37'
Jaws	Asterism	6.4	30'	12^h 38.9^m	–11° 21'
Σ1659	Sextuple star	6.6–11.0	6'	12^h 35.7^m	–12° 02'
Σ1604	Triple star	6.6, 9.4, 8.1	9.1" E, 10.2" NNE	12^h 09.5^m	–11° 51'
NGC 4361	Planetary nebula	10.9	2.1'	12^h 24.5^m	–18° 47'
NGC 4038	Galaxy	10.5	3.4' × 1.7'	12^h 01.9^m	–18° 52'
NGC 4039	Galaxy	11.2	3.1 ' × 1.6'	12^h 01.9^m	–18° 53'
NGC 4027	Galaxy	11.1	3.2' × 2.1'	11^h 59.5^m	–19° 16'
NGC 4027A	Galaxy	14.5	0.9' × 0.6'	11^h 59.5^m	–19° 20'

Angular sizes and separations are from recent catalogs. Visually, an object's size is often smaller than the cataloged value and varies according to the aperture and magnification of the viewing instrument.

separation and direction to the companion from its primary are for the year 2010.

Swooping deeper into Corvus, we come to Delta (δ) and Gamma (γ) Corvi at the top of the trapezium. The unusual planetary nebula **NGC 4361** makes a nearly isosceles triangle with these stars and sits 2.4° east-southeast of Gamma.

In my 105mm scope at 87x, NGC 4361 is a fairly bright, roundish planetary with a faint central star. My 10-inch reflector reveals considerable detail. At 115x, I see a faint, round halo enfolding a brighter, northeast-southwest oval spanned by a bright, slightly oval core that's tipped a bit south of east. Viewing at 166x, I estimate the halo as 2' across and the larger oval as 1½' long. There's a dimmer patch in the core, south-southwest of the central star. The east end of the core has a short extension reaching southwest, and the west end has a fainter, stubby one opposed. A narrowband nebula filter accentuates some features at the lower magnification, but I prefer a filterless view at 166x. Through my 15-inch reflector at a magnification of 133x, the extensions lengthen and curve out counterclockwise, making the nebula look for all the world like a spiral galaxy.

Let's conclude our tour of Corvus with a quartet of quirky galaxies north of 31 Crateris. (This star was considered to be part of the constellation Crater when it was named but lies in Corvus, according to modern definitions.) **NGC 4038** and **NGC 4039** form a well-known pair of dramatically merging galaxies nicknamed the Antennae. Even my little refractor shows hints of their strangeness. At 47x, I merely see a faint but sizable glow that's brighter in the north. But at 87x, NGC 4038 is an east-west oval, while fainter NGC 4039 has a bright patch where it blends into the eastern side of its companion and stretches southwest from there. Their combined glow covers about 3' north-south and 2½' east-west.

The Antennae galaxies are quite striking through my 10-inch scope at 213x. Together, they form a fetal shape, concave west, and show much large-scale patchiness. The galaxy pair discloses progressively more detail through increasing apertures. I was once fortunate enough to view it with a 36-inch reflector, through which it was wonderfully complex. The bright patches mark the cores of the galaxies

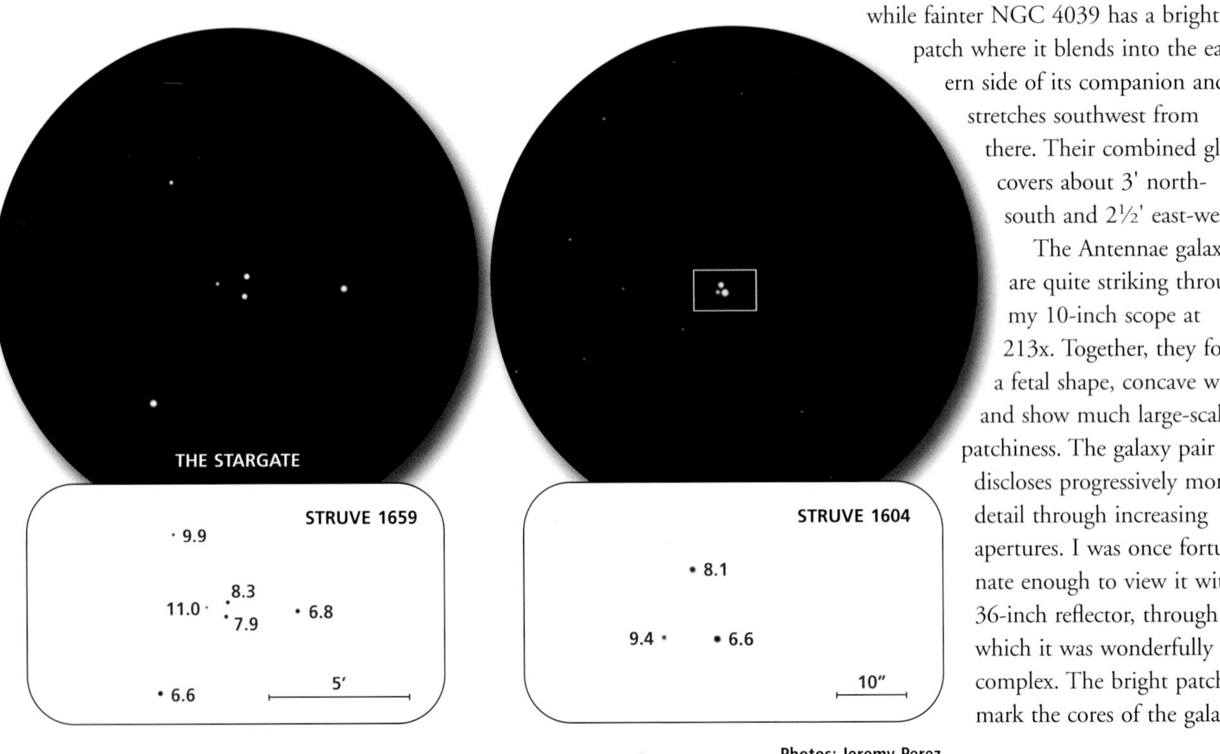

Photos: Jeremy Perez

The mouth of the Jaws asterism points directly at Messier 104, and the six bright stars of the shark's body arc around that galaxy. Photo: POSS-II / Caltech / Palomar

The collision of the Antennae has caused an extraordinary burst of star formation. This Hubble photo shows dozens of regions glowing blue with the light of hot, young stars and pink from hydrogen-alpha emissions. Contrast these colors with the yellow light of the old stars in the galaxy cores. North is to the upper right. Photo: NASA / ESA / Hubble Heritage Team / STScI / AURA

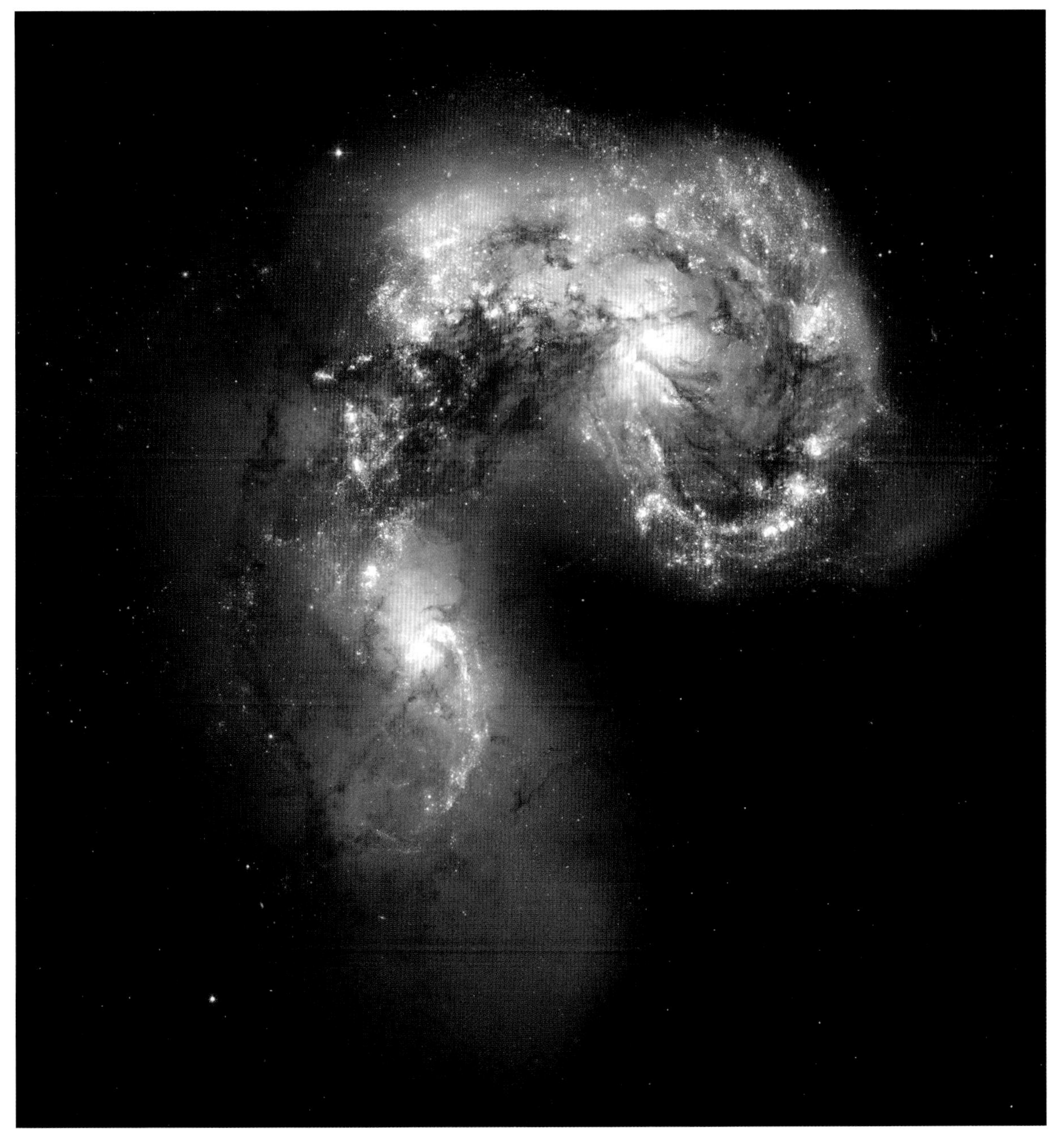

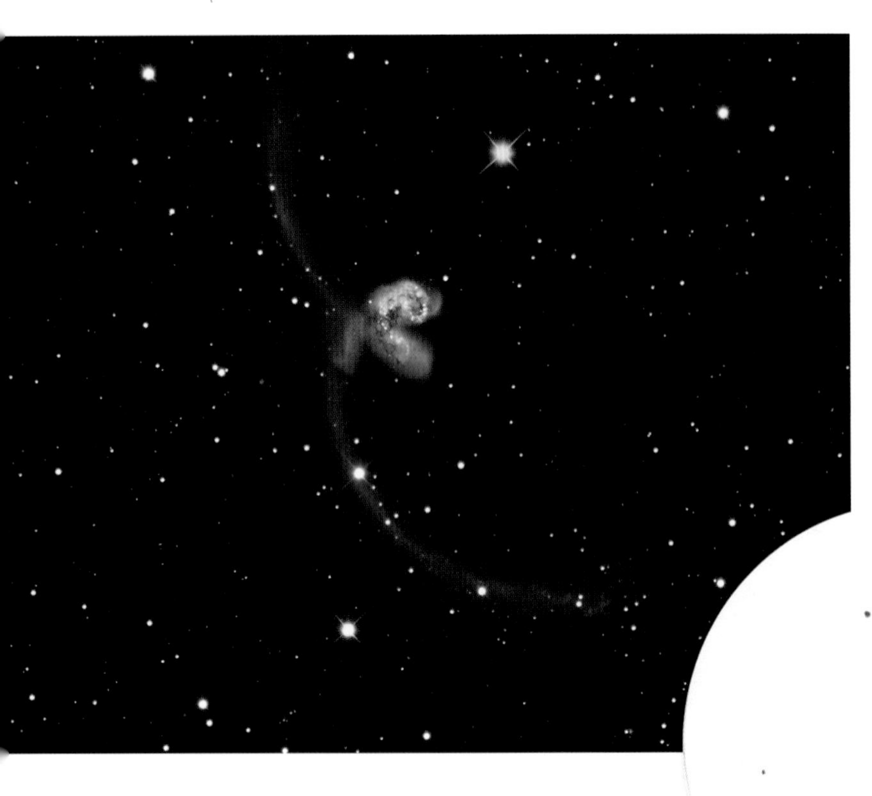

and regions of intense star formation triggered by the ongoing collision.

The Antennae pair takes its name from the long, gossamer tails of gas and stars revealed in deep images. Caused by tidal interactions, these magnificent arcs sweep across a half-million light-years. Some amateurs have managed to glimpse traces of them with scopes as small as 12.5 inches in aperture, but the tails can be quite challenging even in a 20-inch scope.

The second pair of interacting galaxies rests 41' southwest of the Antennae and is often overlooked, despite the fact that its brighter galaxy is an exotic, one-armed spiral. Though just a faint glow in my small refractor at moderate power, **NGC 4027** bares both its peculiar nature and its diminutive companion, **NGC 4027A**, through my 10-inch scope at 213x.

On star-filled spring evenings, spend some time exploring Corvus. If you feather your nest with the shiny trinkets it holds, you'll surely have something to crow about.

Above: The collision of NGC 4038 (top) and NGC 4039, a galaxy pair often called the Antennae, has strewn tails of gas across a huge swath of space. Photo: Martin Pugh. Right: The author sketched the exotic one-armed spiral galaxy NGC 4027 and its faint companion NGC 4027A as seen at 213x through her 10-inch reflector.

The Gossamers of Coma Berenices

The constellation north of Virgo is littered with galaxies.

There was an old woman tossed up in a blanket,
Seventeen times as high as the moon;
Where she was going I could not but ask it,
For in her hand she carried a broom.
"Old woman, old woman, old woman," quoth I;
"O whither, O whither, O whither so high?"
"To sweep the cobwebs from the sky,
And I'll be with you by-and-by!"

— Anonymous

Garrett P. Serviss cites this nursery rhyme in his book *Astronomy with an Opera-Glass* when describing the constellation Coma Berenices, Berenice's Hair. He writes, "Nearly on a line between Denebola and Arcturus, and somewhat nearer to the former, you will perceive a curious twinkling, as if gossamers spangled with dew-drops were entangled there. One might think the old woman of the nursery rhyme who went to sweep the cobwebs out of the sky had skipped this corner, or else that its delicate beauty had preserved it even from her housewifely instincts."

Serviss's gossamers comprise Melotte 111, the huge cluster of stars that crowns Berenice's Hair. Gazing up at a clear dark sky, I can see several of its glittering gems entwined in

Illustration by Anne Anderson

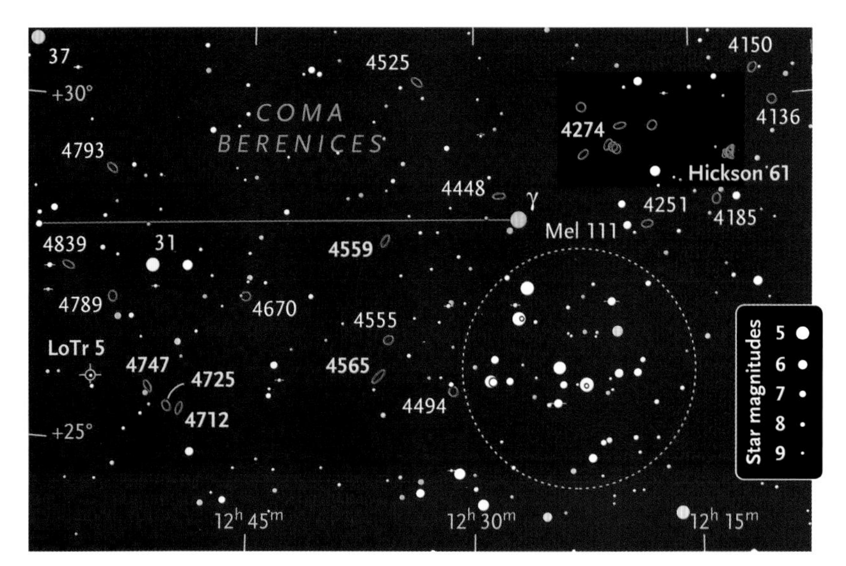

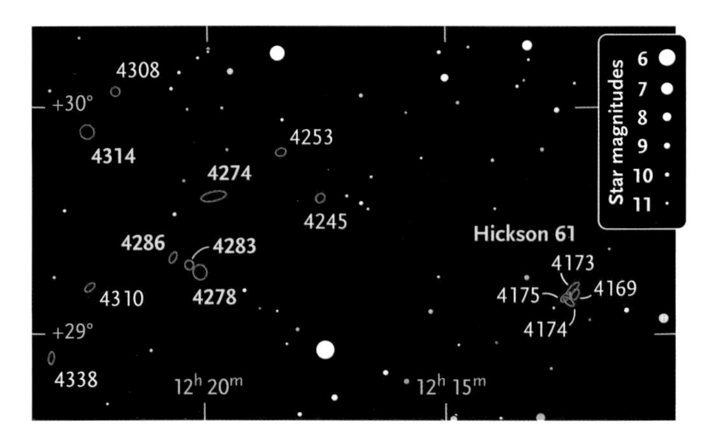

Most of the galaxies shown on these charts belong to the Virgo Galaxy Cluster, roughly 60 million light-years distant. Melotte 111, the Coma Star Cluster, lies just 280 light-years from Earth.

SPRING

May

her tresses, and dozens of stars spring forth in binoculars.

Through a telescope, we see more evidence that the old woman left this region of the sky unswept, for it's rife with dust-bunny galaxies. One of the most obvious bits of fluff is **NGC 4559**. It lies 2° east and a bit south of yellow-orange Gamma (γ) Comae, the bright foreground star that caps Melotte 111.

Through my 105mm refractor at 28x, NGC 4559 is an oval glow with faint stars hugging each side of its southeastern end. At 76x, a dimmer star pops out at the southeastern tip. The galaxy covers about 4' x 1½', has a brighter core, and is slightly uneven in brightness.

NGC 4559 is very pretty through my 10-inch reflector at 192x and spans 6½' x 2½'. Hazy wisps reach out toward the middle and easternmost of the three stars arcing across the galaxy. A fainter strand starting north of the galaxy's center trends north-northwest. An elusive, starlike nucleus rests at the heart of NGC 4559, and brighter spots ornament the galaxy's face.

The magnificent showpiece galaxy **NGC 4565** slashes the sky 2° south of NGC 4559. Its slender profile earned this edge-on spiral the nickname of the Needle Galaxy, while some observers fondly call it Berenice's Hair Clip.

As the brightest of the galaxies appearing at least seven times longer than wide, NGC 4565 is one of the best "flat" galaxies for a small telescope. In my 5.1-inch at 37x, its 9'-long streak contains a brighter area half as long with a small central bulge. At 102x, the core is a flattened oval, with a stellar nucleus and a faint star hovering over its northeastern pole. A dusky lane is subtly charcoaled across the core and a bit beyond, skimming just northeast of the nucleus.

In my 10-inch reflector at 192x, NGC 4565 is a beautiful sight. It bridges an incredible 15' of sky, and I can trace its dark lane for about 3'.

Moving 3.2° east and a little south takes us to a galaxy with remarkable structure, **NGC 4725**. A sharply peaked chain of several stars points toward it from the east. The galaxy is fairly large and bright in my 105mm scope at 28x. It features a small core

The barred spiral galaxy NGC 4559 is similar in structure to our own Milky Way. Photo: Jeff Hapeman / Adam Block / NOAO / AURA / NSF

Far right: NGC 4725, a barred spiral, and NGC 4712 are tilted roughly halfway between edge-on and face-on to Earth. But NGC 4712 is more than three times farther, so it appears much smaller. Photo: Sheldon Faworski / Sean Walker. Right: NGC 4565 is often cited as the most spectacular edge-on spiral galaxy in the sky. Photo: Johannes Schedler

with extensions northeast and southwest, all enveloped in a faint halo. A faint star dangles beneath the galaxy's southern edge. At 87x, the extensions become part of an oval haze, uneven in brightness, surrounding the core. The core itself is slightly oval and harbors a tiny, bright nucleus. NGC 4725 spans about 6½' x 4½' with an extremely faint star pinning its edge, north of the nucleus.

My 10-inch scope at 192x nicely displays details within NGC 4725 and reveals two nearby galaxies. NGC 4725 juts into one side of a roughly 10' rhombus of 12th-magnitude stars. The moderately faint galaxy **NGC 4712** lies along the opposite side in the same field of view. Its 2' x ¾' oval holds a small oval core. Farther afield, **NGC 4747** is 6' north of the brightest star in the chain. This very faint streak is 2' long with a somewhat brighter center.

Those with large scopes, dark skies, and a touch of masochism might like to try for the planetary nebula **Longmore-Tritton 5** (LoTr 5). It rests 52' east by north of NGC 4747, where it surrounds an 8.9-magnitude binary star whose fainter component is the nebula's progenitor. This planetary has extremely low surface brightness and is among the largest in our sky. It has a diameter of 8.8' and a structure reminiscent of the Helix Nebula (NGC 7293). To catch sight of it, try low magnification and an oxygen III nebula filter. I've gazed at this phantom planetary a few times with 14- to 15-inch scopes and could log it only as a "maybe." Observers with darker skies have reported success with scopes as small as 16 inches in aperture.

Now let's move 2.1° northwest of Gamma Comae to **NGC 4274**. It shares the field of view with **NGC 4278** and **NGC 4314** through my 105mm scope at 28x. NGC 4274 is a fairly bright oval, NGC 4278 is smaller and round with a very bright nucleus, and NGC 4314 is a faint smudge. At 87x, NGC 4274 is about 5½' x 1¾' and leans south of east. It enfolds a large oval core with a small, round, bright center. A close, unequal star pair lies 6' south. NGC 4278 grows much brighter toward the center. A very faint star floats 5' north, while the very small but fairly bright, round galaxy **NGC 4283** keeps it company 3½' east-northeast. NGC 4314 shows a bright, 2½' spindle with a round central bulge clasping a stellar nucleus. The spindle has a very faint star near its northwestern tip and is enclosed in a tenuous halo.

Both the inner arms and the outer halo of the spiral galaxy NGC 4274 appear to form ring structures around the galaxy's core. The bright inner ring is often likened to Saturn's rings. Photo: Steve and Sherry Bushey / Adam Block / NOAO / AURA / NSF

NGC 4274 looks rather strange through my 10-inch scope at 192x. The oval halo seems disconnected from the sides of the core by darker ears. It's like looking at a ghostly Saturn! There's a diaphanous envelope around the whole thing that quickly fades outward. **NGC 4286** emerges 5' east-northeast of NGC 4283. Its faint oval glow has a small brighter core and a dim star off its south-southeastern tip.

Our final stop is **Hickson 61**, resting 1.7° west of NGC

4278. Commonly known as The Box, this compact group consists of four galaxies tightly packed into 6' of sky. My 105mm refractor at 87x shows three of them. NGC 4169 is the brightest. It's tilted north-northwest and has a large oval core. NGC 4174 and NGC 4175 are very faint. The former is a small fuzz-spot tipped northeast, while the latter is larger and leans northwest. The fourth member, NGC 4173, is elusive even in my 10-inch scope. Its long, highly elongated profile is in line with NGC 4175 and best seen with averted vision.

Odd though it may seem, NGC 4173 is a foreground galaxy not associated with the other three. We see evidence of this in photos, where NGC 4173 looks larger, more detailed, and bluer than its chance companions on the sky. The center of Virgo Cluster is 60 million light-years away, but most of its individual galaxies are closer to or farther from us. The other Hickson 61 galaxies and NGC 4712 are about three times more distant.

Dust Motes in the Queen's Hair

Object	Type	Magnitude	Size/Sep.	Right Ascension	Declination
NGC 4559	Galaxy	10.0	10.7' × 4.4'	12^h 36.0^m	+27° 58'
NGC 4565	Galaxy	9.6	15.9' × 1.9'	12^h 36.3^m	+25° 59'
NGC 4725	Galaxy	9.4	10.7' × 7.6'	12^h 50.4^m	+25° 30'
LoTr 5	Planetary nebula	—	8.8'	12^h 55.6^m	+25° 54'
NGC 4274	Galaxy	10.4	6.8' × 2.5'	12^h 19.8^m	+29° 37'
Hickson 61	Galaxy group	12.2–13.4	6'	12^h 12.4^m	+29° 12'

Angular sizes and separations are from recent catalogs. Visually, an object's size is often smaller than the cataloged value and varies according to the aperture and magnification of the viewing instrument.

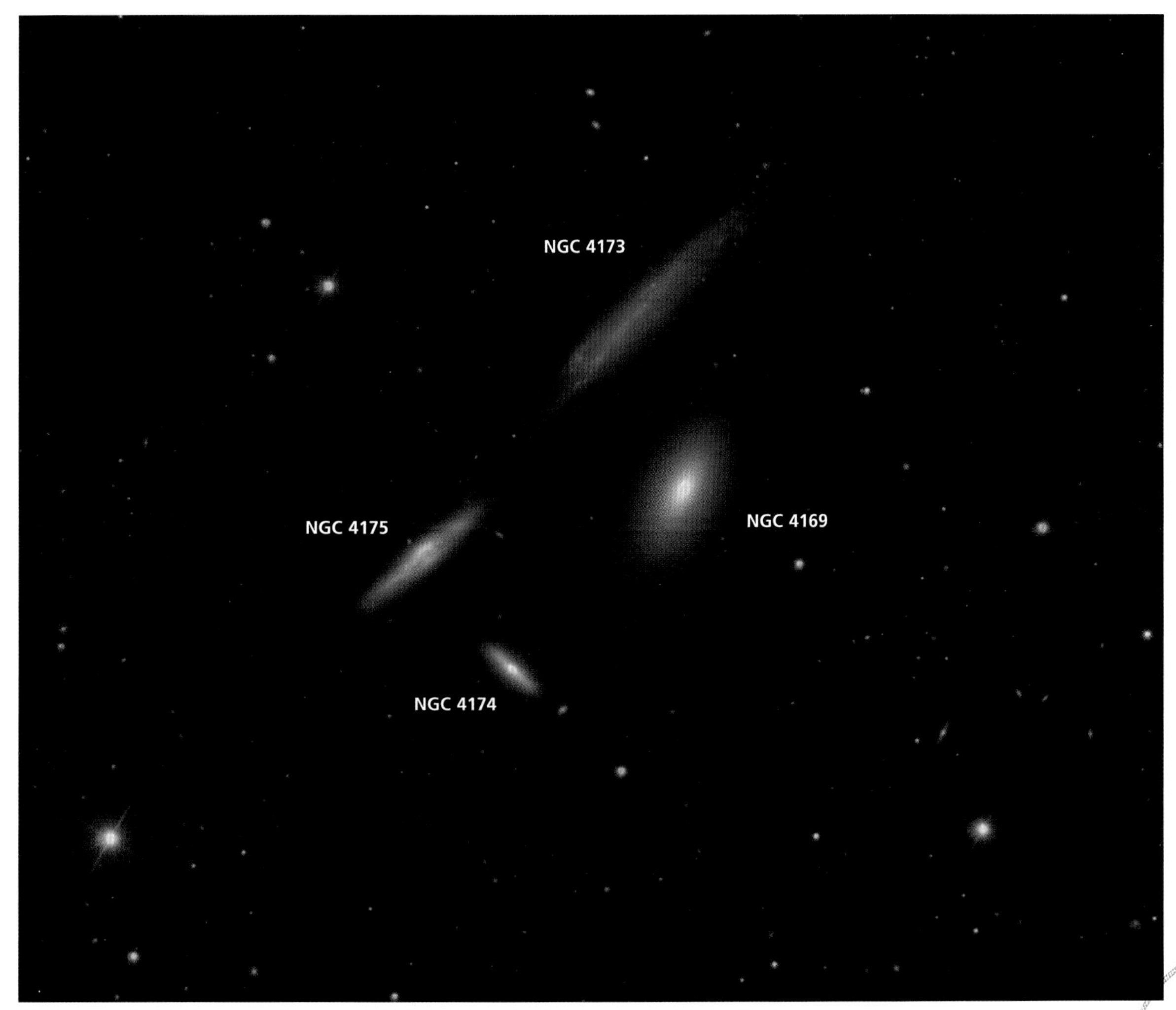

Sloan Digital Sky Survey

SPRING

May

First published May 2010

Dazzling Doubles, Glittering Globulars

A not-so-well-known constellation yields up a host of delights for small-scope users.

The constellation Serpens Caput, the Serpent's Head, winds its way across the early-summer skies between Ophiuchus (aptly called the Serpent Bearer) and Boötes, the Herdsman. Hidden among the stars of the Serpent is one of the finest — but often overlooked — globular clusters in our sky. Let's begin our tour with this isolated object. From there, we'll hop along a bridge of double stars to a close pair of globulars in nearby Ophiuchus.

M5 is the brightest globular cluster north of the celestial equator. It is located 7.7° southwest of Serpens' brightest star, Alpha (α) Serpentis. At magnitude 5.8, it is easily visible in a 60mm telescope as a small, round glow with a fainter, granular halo. A 5th-magnitude yellow star, 5 Serpentis (Σ1930), lies just 0.3° to its south-southeast in the same field of view. A 4-inch scope at 150x resolves many stars around the cluster's fringes. The core appears as a bright blaze intensifying toward a starlike nucleus. The cluster looks slightly elliptical, with the long dimension oriented northeast-southwest. M5 is 13 billion years old — one of the most ancient globular clusters known. It is also among the largest, at 130 light-years across. It was first spotted in

5 Serpentis (Σ1930)

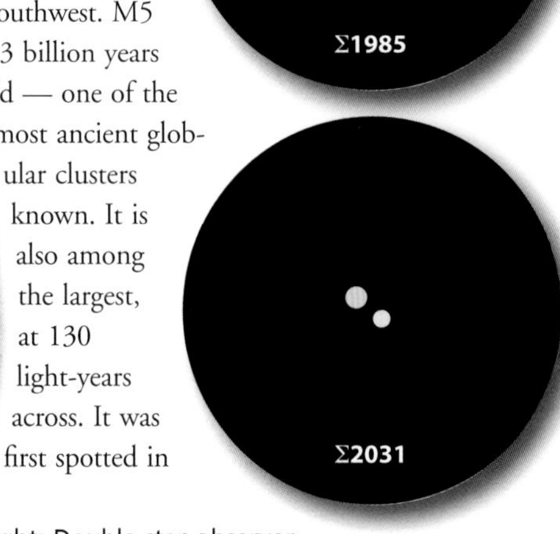

Right: Double-star observer Sissy Haas sketched 5 Serpentis (Σ1930), Σ1985, and Σ2031. These three figures suggest how the stars appeared in her small refractor. North is up. Left: A pair of globular clusters lie at the heart of Ophiuchus. In this image by Akira Fujii, M12 is at the upper right and M10 is to the lower left. North is up.

First published June 2000

Sky-Hopping for Globular Clusters

Object	Type	Magnitude	Distance (l-y)	Right Ascension	Declination
M5	Globular cluster	5.8	25,000	15^h 18.6^m	+2° 05'
V42	Variable star	10.6–12.1	25,000	15^h 18.6^m	+2° 05'
5 Serpentis (Σ1930)	Double star	5.0,10	81	15^h 19.3^m	+1° 45'
Σ1985	Double star	7.0, 8.1	123	15^h 56.0^m	–2° 10'
Σ2031	Double star	7.0, 11	152	16^h 16.3^m	–1° 39'
M12	Globular cluster	6.6	18,000	16^h 47.2^m	–1° 57'
M10	Globular cluster	6.6	14,000	16^h 57.1^m	–4° 06'

1702 by German astronomer Gottfried Kirche. M5 is a gravitationally bound horde of hundreds of thousands of stars, about 25,000 light-years away.

V42. One of M5's many variable stars can be found in the cluster's halo 3' southwest of the nucleus. It may be a challenge to find, but it's worth the hunt. V42 pulses in brightness from visual magnitude 10.6 to 12.1 and back over a period of 25.7 days. At maximum, it is the brightest star in the halo, while at minimum, it may fade from view through a small telescope.

5 Serpentis. Go back to that bright star to the southeast of M5. It's actually a double. The 5th-magnitude primary is a sunny yellow star. Its 10th-magnitude companion lies 12" to the northeast and has an orange hue that may be difficult to discern through a small telescope. The two are cleanly split at 80x. This double star is also known as Σ1930. The Greek letter Σ (Sigma), found in many double-star designations, indicates that the object is listed in the *Micrometric Measurement of Double and Multiple Stars* catalog compiled by double-star pioneer Friedrich Georg Wilhelm von Struve and published in 1837.

Σ1985. Another yellow-orange pair lies just 2° northeast of the 3.5-magnitude star Mu (μ) Serpentis. This pair features a 7th-magnitude primary with an 8th-magnitude secondary 6" to the north. It takes around 120x to comfortably split the two.

Σ2031. About 5° east of Σ1985, we find a

Serpens Caput extends from just below Corona Borealis through the outstretched hand of Ophiuchus, the Serpent Bearer. On this chart, north is up and east is left. To find which way is north in your eyepiece, nudge your telescope slightly toward Polaris; new sky enters the view from the north edge. (If you're using a right-angle star diagonal at the eyepiece, it probably gives a mirror image. You can take it out to see a correct image that will match the map.)

The glittering globular cluster M5 rides the serpent's back in solitary splendor. The variable star V42 lies 3' southwest of the cluster's center. The field is 20' wide, and north is up. Photo: Robert Gendler

third Struve double, near the naked-eye stars Delta (δ) and Epsilon (ε) Ophiuchi, also known as Yed Prior and Yed Posterior, respectively. Yed comes from the Arabic word for "hand"; these stars mark the left hand of Ophiuchus as he grasps the Serpent. Yed Prior is the leading (western) star of the hand, while Yed Posterior follows it in the stars' unceasing march across the sky.

To find Σ2031, look 2.1° north-northeast of Yed Prior. This wide double is easily separated at 50x, but it may prove challenging since the companion is very faint and nearly colorless through a small telescope. Look for it 21" to the southwest of the 7th-magnitude orange primary.

M12. Our Struve stepping-stones to the globulars of Ophiuchus have swept us progressively farther from home. Σ1930, Σ1985, and Σ2031 lie at 81, 123, and 152 light-years, respectively. Our jump to the globular cluster M12 takes us about 18,000 light-years. M12 is located 7.7° east of Σ2031. It is visible through binoculars or a finder as a tiny patch of fuzz. A 4-inch scope at high powers can resolve some of the stars in the outer halo, while the core remains a mottled patch of mist. At 200x, a 6-inch scope can reveal several dozen stars, with many in the halo arranged in meandering chains framing dark, starless voids.

M10. Just 3.3° southeast of M12, we find a similar cluster, which may be easier to locate than its companion because it is only 1° west of the orange 4.8-magnitude 30 Ophiuchi. If you have trouble sweeping up M12, look for M10 first and proceed from there. The twosome can be seen together through a finder.

M10 appears slightly brighter than its companion and a little more concentrated toward its center. Through a 4-inch scope at high powers, the center looks granular and some stars are resolved around the edges. A 6-inch can show dozens of stars strewn across a hazy background. M10 is 14,000 light-years away, making our two Ophiuchus globulars relatively close neighbors. If the distances are correct, each would be about a 4th-magnitude object in the other's sky.

Polar Nights

Ursa Minor may lack good deep-sky objects, but have you ever sorted through its stars?

The sad and solemn night
Hath yet her multitude of cheerful fires;
The glorious host of light
Walk the dark hemisphere till she retires....
And thou dost see them rise,
Star of the Pole! and thou dost see them set.
Alone, in thy cold skies,
Thou keep'st thy old unmoving station yet.

— William Cullen Bryant
"Hymn to the North Star"

Dim though it may be at 2nd magnitude, the North Star is the most famous in the sky. Polaris stands essentially still above your landscape all night and all year (assuming you're in the Northern Hemisphere), holding its station while all others revolve around it. **Polaris** marks not only true north (to better than ¾°) but also your latitude on Earth — which nearly equals Polaris's altitude above your horizon.

Observationally, Polaris has more to offer us than marking the way north. A small telescope at low power will show that it's part of a 40' circlet of mostly 6th- to 9th-magnitude stars extending in the direction toward Perseus. This asterism is sometimes called the **Engagement Ring**, with Polaris as its sparkling diamond. Despite its name, however, the Engagement Ring seems to have suffered some rough

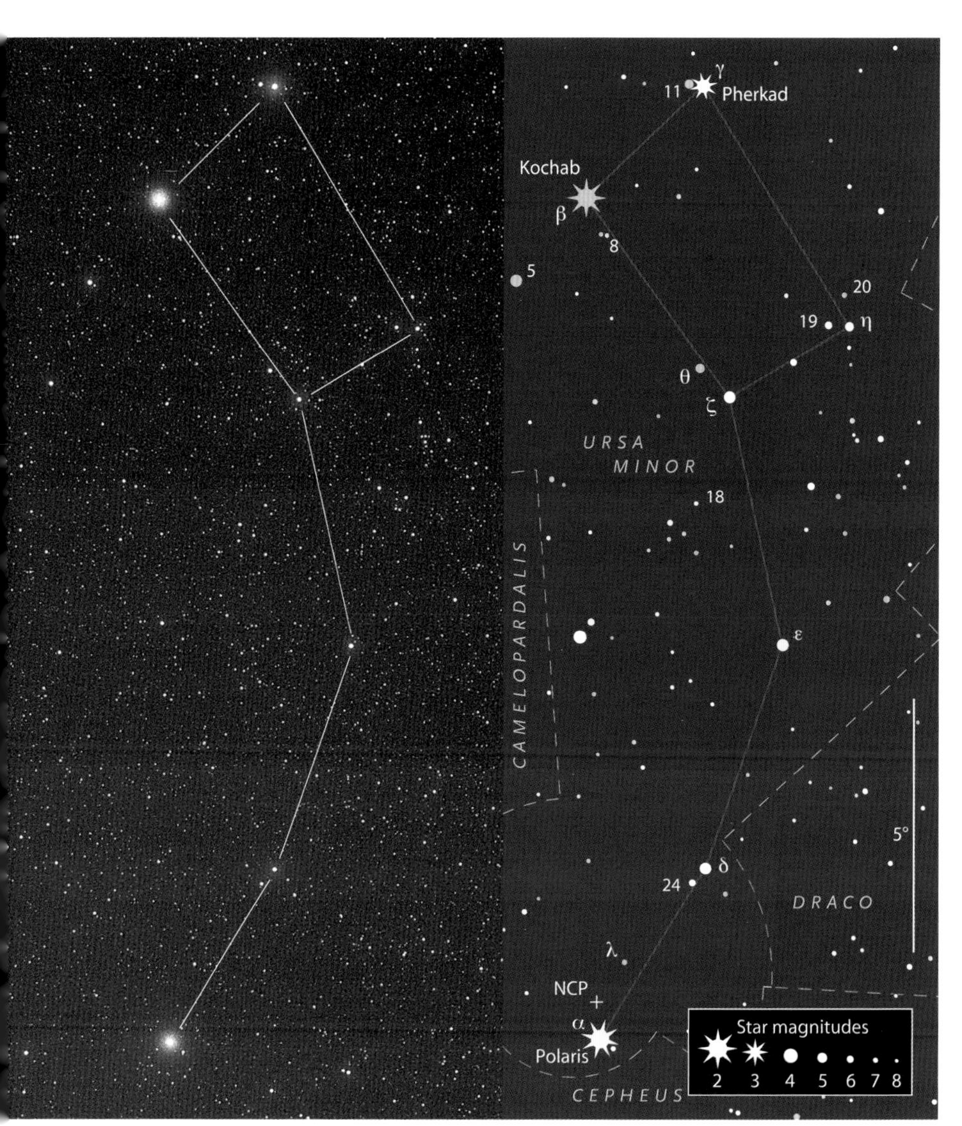

The Little Dipper extends from Polaris to the Guardians of the Pole, Beta (β) and Gamma (γ) Ursae Minoris — a span of nearly 20°, or about three binocular fields of view. All illustrations here are oriented with north down, to match how the Little Dipper stands on late-spring evenings. Chart: Adapted from *Millennium Star Atlas* data.

treatment, being oval and badly dented on its south side.

Polaris itself is a double star; the 2nd-magnitude primary has a 9th-magnitude companion spark a generous 18" away. It seems our abused Engagement Ring also has a chip off its diamond. This would be an easy low-power split if the stars were more alike, but their difference of seven magnitudes (a difference of about 600 times in brightness) means the companion is easily overpowered by the brighter star's glare. Try a magnification of at least 80x to bring it out. Polaris is pale yellowish. The faint companion looks pale blue by contrast, but measurements show that their colors are actually very similar.

Polaris enjoys its special position because the Earth's axis happens to point toward it. In other words, it stands almost directly above the North Pole. As the world turns, the geographic poles are the only places that face the same direction all night long. Consequently, the star straight above the pole is the only one that seems to stand still.

Yet Polaris is not quite the unmoving cynosure described by poets. The axis of our planet misses it slightly; the true north celestial pole is currently about ¾° away, in the direction from Polaris toward the Little Dipper's other 2nd-magnitude star, Kochab, or **Beta (β) Ursae Minoris** (UMi). Our North Star travels in a little circle around the true pole each day.

The circle will shrink to just under ½° radius in the coming century. Why? The Sun and Moon tug on the equatorial bulge of the tipped Earth and try to pull it toward the ecliptic plane. The resulting torque on the spinning planet causes precession, a slow change in the orientation of the Earth's axis with respect to the stars. Precession carries the celestial pole around the sky in a large near-circle that takes some 26,000 years to complete.

So when will Polaris be closest to the north celestial pole? The answer isn't simple. Variations in the gravitational effects of the Moon add a tiny, 18.6-year nodding motion (nutation) to the celestial poles as they tour the precessional circle. Superimposed on nutation is a tiny annual shift in the apparent positions of all stars due to Earth's 30-kilometer-per-second orbital motion with respect to their incoming light (annual aberration). And

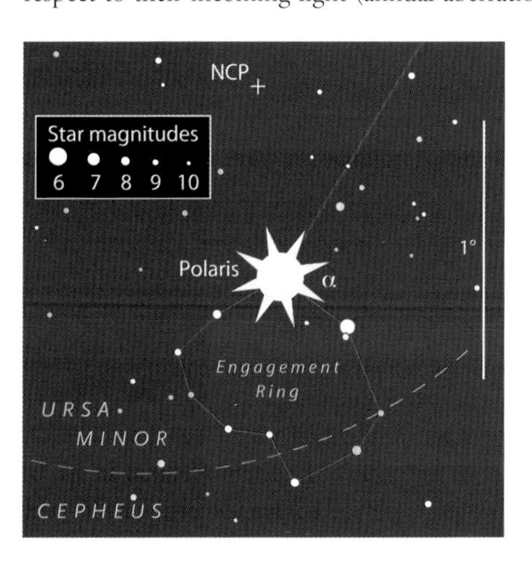

How many observers know the faint Engagement Ring asterism that holds Polaris as its jewel? The ring is ⅔° across, so use your telescope's lowest-power, widest-field eyepiece.

The Mini-Coathanger is a ⅓°-long asterism floating near Epsilon Ursae Minoris, between the Little Dipper's handle and bowl.

Polaris changes its own position (proper motion) because of its sideways motion with respect to the Sun in space.

Taking all this into account, computational wizard Jean Meeus finds that Polaris will be closest to the north celestial pole on March 24, 2100, with an apparent separation of 27' 09". While this is probably not the final word on Polaris as the polestar, it shows how a simple question has an ever more complicated answer, depending on how precise you want the answer to be.

Among the other stars making up the Little Dipper are four very wide pairs. None is a true binary, but two of them offer attractive color-contrast duos for binoculars or a small telescope.

One pair is formed by **Zeta** (ζ) and **Theta** (θ) UMi, where the Little Dipper's handle joins its bowl. Fourth-magnitude Zeta UMi is white; 5th-magnitude Theta is orange. Separated by 49', they will fit together only in a low-power, wide-angle view.

In the opposite corner of the Little Dipper's bowl are 3rd-magnitude Pherkad and 5th-magnitude Pherkad Minor, **Gamma** (γ) and **11** UMi, respectively. They're 17' apart and have hues similar to our previous pair, but the dimmer star is very slightly paler orange, with a color index of 1.4 compared with Theta's 1.6. (A color index of +0.2 is pure white; lower and negative values are slightly bluish, while larger values denote yellow, orange, and red.) Use your lowest power to keep each pair as close together as possible to heighten apparent color contrasts.

The corner of the bowl opposite bright Kochab is formed by 5th-magnitude **Eta** (η) and **19** UMi. They're 26' apart and tinted, respectively, white with the barest hint of yellow and white with a trace of blue. Can you detect the difference? Forming an isosceles triangle with them is 6th-magnitude **20** UMi, which is yellow-orange.

The fourth pair is **Delta** (δ) and **24** UMi, the first stars down the handle from Polaris. They're magnitudes 4 and 6, separated by 23', and both white.

North of the Little Dipper's bowl is a pretty asterism discovered by Pennsylvania amateur Tom Whiting. He dubbed it the **Mini-Coathanger** for its likeness to the familiar Coathanger pattern in Vulpecula. It is composed of ten 9th- to 11th-magnitude stars 1.9° south-southwest of **Epsilon** (ε) UMi. Like its larger cousin, the Mini-Coathanger resembles an old-fashioned coathanger with a straight wooden bar and metal hook. Seven stars make up the bar, which is about 17' long. Three faint stars comprise the hook.

Ursa Minor owes most of its recognition to the fact that it contains the North Star. But there are subtler treasures to be found here, and best of all, they can be seen by most readers all year long.

Stars in the Little Dipper

Star	Magnitude	Spectral Type	Color Index	Distance (l-y)	Right Ascension	Declination
Polaris A	**1.9–2.1**	**F5–8 Ib**	**+0.6**	**430**	**2^h 31.8^m**	**+89° 16'**
Polaris B	**9.0**	**F3 V**	**+0.4**	**430**	**2^h 31.8^m**	**+89° 16'**
β UMi	**2.0**	**K4 III**	**+1.5**	**125**	**14^h 50.7^m**	**+74° 09'**
ζ UMi	**4.3**	**A3 V**	**0.0**	**375**	**15^h 44.1^m**	**+77° 48'**
θ UMi	**5.0**	**K5 III**	**+1.6**	**800**	**15^h 31.6^m**	**+77° 21'**
γ UMi	**3.0**	**A3 III**	**+0.1**	**480**	**15^h 20.7^m**	**+71° 50'**
11 UMi	**5.0**	**K4 III**	**+1.4**	**390**	**15^h 17.1^m**	**+71° 49'**
η UMi	**5.0**	**F5 V**	**+0.4**	**97**	**16^h 17.5^m**	**+75° 45'**
19 UMi	**5.5**	**B8 V**	**–0.1**	**660**	**16^h 10.8^m**	**+75° 53'**
20 UMi	**6.4**	**K2 IV**	**+1.2**	**760**	**16^h 12.5^m**	**+75° 13'**
δ UMi	**4.4**	**A1 V**	**0.0**	**180**	**17^h 32.2^m**	**+86° 35'**
24 UMi	**5.8**	**A2**	**+0.2**	**155**	**17^h 30.8^m**	**+86° 58'**
ε UMi	**4.2**	**G5 III**	**+0.9**	**350**	**16^h 46.0^m**	**+82° 02'**

Follow the Arc

Boötes is not as devoid of deep-sky sights as many amateurs believe.

Boötes, the Herdsman, is an easy constellation to locate. The Big Dipper's familiar pattern leads us to it if we remember the saying, "Follow the arc to Arcturus." Extending the curve of the Big Dipper's handle takes us right to Arcturus, the brightest star in Boötes. The skinny kite-shaped figure of Boötes stretches northeast from this brilliant golden star. One of the Herdsman's arms reaches out toward the handle of the Dipper as if he is clutching the Great Bear by the tail. Perhaps this is how the Great Bear's tail has been stretched as she flees before Boötes in their nightly chase around the polestar.

Although simple to find, Boötes is a constellation lacking deep-sky wonders — or so you might think. I asked several active observers in my astronomy club whether they could think of any deep-sky objects in Boötes, and the only suggestion was Izar, also known as Epsilon (ε) Boötis, a beautiful gold and white double star. Yet Boötes does hold other sights worth seeking. Let's visit a few.

We'll start with the **Arcturus Group**. I learned of this gathering of 5th- to 9th-magnitude stars in the fabulous book *Star Clusters* by Brent A. Archinal and Steven J. Hynes. The authors unearthed what seems to be the first mention of the group in the 1856 edition of *The Geography of the Heavens* by Hiram Mattison and Elijah H. Burritt. There, it is described as "a rich group of stars in the vicinity of Arcturus, and surrounding that star. May be seen with small telescopes." The figure from the companion atlas shows 48 stars, but neither scale nor directions are indicated. Binoculars reveal a roughly

Visible at right in this photograph, NGC 5466 looks dim and almost too sparse for a globular star cluster. Yet it can be seen in almost any telescope. The 7th-magnitude star at left lies just ⅓° east. George R. Viscome took this 22-minute exposure on 3M 1000 film with his 14.5-inch f/6 Newtonian reflector at Lake Placid, New York.

Buried Booty of Boötes

Object	Type	Size	Magnitude	SB	Right Ascension	Declination	*MSA*	*U2*
Arcturus Group	Asterism	5° or so	—	—	$14^h 19.5^m$	+19° 04'	696	70R
Picot 1	Asterism	20' × 7'	—	—	$14^h 14.9^m$	+18° 34'	696	70R
NGC 5466	Globular cluster	9'	9.0	—	$14^h 05.5^m$	+28° 32'	650	70R
M3	Globular cluster	18'	6.2	—	$13^h 42.2^m$	+28° 23'	651	71L
NGC 5529	Spiral galaxy	6.4' × 0.7'	11.9	13.5	$14^h 15.6^m$	+36° 14'	627	52R
NGC 5557	Elliptical galaxy	3.6' × 3.2'	11.0	12.6	$14^h 18.4^m$	+36° 30'	627	52R
NGC 5676	Spiral galaxy	4.0' × 1.9'	11.2	13.2	$14^h 32.8^m$	+49° 27'	586	36L
NGC 5689	Spiral galaxy	4.0' × 1.1'	11.9	13.0	$14^h 35.5^m$	+48° 45'	586	36L

Angular sizes are from catalogs or photographs; most objects appear somewhat smaller when a telescope is used visually. SB is the average surface brightness expressed as the equivalent visual magnitude per square arcminute. The columns headed *MSA* and *U2* give the chart numbers of objects in the *Millennium Star Atlas* and *Uranometria 2000.0*, 2nd edition, respectively.

corresponding collection of stars. There is a core group about 3½° x 2° with the long dimension running east-west, and outliers expand the group to perhaps 5° x 3°. A small telescope will bring out hints of yellow and orange in some of the members, but few scopes can achieve a wide enough field of view to encompass them all. The Arcturus Group is not a true star cluster but, rather, an asterism of unrelated stars that happen to lie along the same line of sight.

A much smaller asterism lies in the southern edge of the Arcturus Group. It is called **Picot 1** in the online catalog at http://tech.groups.yahoo.com/group/deepskyhunters. Picot 1

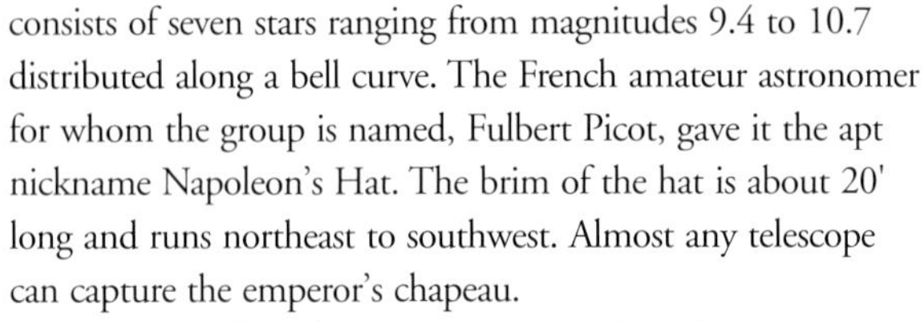

consists of seven stars ranging from magnitudes 9.4 to 10.7 distributed along a bell curve. The French amateur astronomer for whom the group is named, Fulbert Picot, gave it the apt nickname Napoleon's Hat. The brim of the hat is about 20' long and runs northeast to southwest. Almost any telescope can capture the emperor's chapeau.

Moving northward, we come to one of my favorite targets in Boötes, **NGC 5466**, a ghostly globular star cluster with a low surface brightness. Despite this, it can be glimpsed in a dark sky through binoculars. Look for it 6° west-southwest of Rho (ρ) Boötis next to a 7th-magnitude orange star.

Through a small telescope, NGC 5466 is a weak, round glow about 5' across. My 105mm refractor at 87x can pick out only a few of the cluster's brightest stars, but a 6- to 8-inch scope might catch a dozen. Larger apertures will let you boost the magnification without making the cluster fade away. With my 10-inch reflector at 213x, NGC 5466 is 6' across and very slightly elongated east-west. More than two dozen faint to very faint stars shimmer over unresolved haze, the brightest strewn mostly around the edges.

One of the most spectacular globular clusters in the northern sky lies 5° due west of NGC 5466. Although a resident of Canes Venatici, not Boötes, **M3** is well worth the side trip. The cluster is bright enough to be seen in binoculars, and some observers have spotted it with the unaided eye. Through my 105mm scope at 127x, M3 appears sparse around the edges with a much brighter core that intensifies toward the brilliant center. Some stars are resolved right down to the cluster's heart. The core, elongated northwest-southeast,

French observer Charles Messier discovered the globular cluster M3 while tracking Comet Bode of 1779, and he made it the third entry in his famous deep-sky catalog. Yet he saw only a glow here, not any stars. The author resolves individual stars with her 105mm refractor. To avoid overexposing the center, George R. Viscome limited this exposure to 10 minutes. North is up and the field is 0.4° wide.

Edge-on galaxies are among the sky's most startling sights on a clear, moonless night. NGC 5529 was captured with a 16-inch telescope on Kitt Peak in Arizona. The field is ¼° wide with north up. Photo: Bill Kelly / Sean Kelly / Adam Block / NOAO / AURA / NSF

is buried in a halo 12' across. A 10th-magnitude star shines at the northwest edge, and an 8th-magnitude star lies outside the halo to the south-southeast. This globular takes magnification well, so boost the power to help bring out its fainter stars.

Larger telescopes resolve M3 into a glorious blaze teeming with countless stars. Outliers radiate from the central mass in meandering streams and glittering rays. M3 appears more impressive than its neighbor NGC 5466 because it is intrinsically six times brighter and only two-thirds as far away.

Many of the deep-sky denizens of Boötes are galaxies, none terribly bright. One of the most interesting is **NGC 5529**, a remarkably flat galaxy. NGC 5529 is actually a spiral whose thin disk, seen edge-on, looks highly elongated, or "flat." It is one of the entries in the *Revised Flat Galaxy Catalogue* (Igor D. Karachentsev and colleagues, 1999), a compilation of 4,236 galaxies that appear at least seven times longer than wide. The vast majority of those galaxies are not visible in the average backyard telescope, but NGC 5529 is one exception.

To find it, move 2° west-southwest of Gamma (γ) Boötis to a 7th-magnitude star, the brightest in the area, and then extend your sweep for that distance again. Although NGC 5529 has been seen with instruments as small as 6 inches in aperture, I have never looked at it in anything smaller than my 10-inch Newtonian. At 115x, I see its needlelike sliver running west-northwest to east-southeast. The galaxy is a little wider and brighter in the center, and it harbors an extremely faint stellar nucleus. It is flanked by two concentric arcs of three stars each, magnitudes 10 to 12. NGC 5529 is closer to the eastern curve, and its southeastern tip is near the arc's southernmost star.

Photographically, NGC 5529 is about 6.4' x 0.7' and bisected by a dark lane of dust. I'd be interested to know whether anyone has been able to see this lane visually.

A much rounder galaxy is found 38' east-northeast of NGC 5529 and can share the same low-power field. **NGC 5557** has a higher surface brightness and should be easier to see. My 105mm scope at 17x displays a small circular patch that is brighter in the center. At 87x, it appears slightly oval with a thin halo surrounding the brighter core. Boosting the magnification to 127x uncovers an elusive, starlike nucleus. Views through my 10-inch scope enlarge the galaxy a little by revealing more of its halo.

A small group of galaxies lies beside the Herdsman's upraised arm, at least two of them bright enough to be seen in a small telescope. First, look for a 5.7-magnitude reddish orange star 3° south-southeast of Theta (θ) Boötis. From there, our first stop will be 19' west-northwest, where we'll find **NGC 5676**. It is visible through my little refractor at 47x as an extended smudge running northeast to southwest. At 87x, it maintains a nearly uniform surface brightness. A 10-inch shows a faint halo three times longer than wide enclosing a large oval core. A brighter patch ornaments the eastern end of the core, a clue that this is a spiral galaxy.

Although slightly fainter, **NGC 5689** is also visible at 47x through the small refractor. It lies 38' south-southeast of the ruddy star mentioned above and shows an east-west elongation. At 87x, I can see that it grows brighter toward the center. A 10-inch at moderate to high power shows that the core is slightly mottled and harbors a stellar nucleus. NGC 5689 is a barred spiral galaxy viewed nearly edge-on.

We can see that Boötes does, indeed, boast cosmic wonders worth pursuit. So when some star-filled night beckons, be sure to *follow the arc* and herd up a few for yourself.

First published June 2004

Embracing Edasich

There are galaxies galore in the northern spring sky.

This month's exploration of the deep sky covers a patch of the heavens centered on Iota (ι) Draconis, or **Edasich**, an interesting star in its own right. Due to the precessional wobble of the Earth's axis, Edasich was our polestar around 6,400 years ago, though it never sat as close to the sky's north pole as Polaris does now.

Edasich is also parked near the radiants of two meteor showers, January's Quadrantids and the June Boötids (also known as the Iota Draconids). Peak rates for the latter shower range from zero to 100 meteors per hour, but only four significant displays have been recorded, in 1916, 1921, 1927, and 1998. Predictions for the upcoming peak of the Quadrantid meteor shower are given on the web page of the International Meteor Organization at www.imo.net, where viewing conditions for this and other showers are discussed.

Beaming at us from a distance of about 100 light-years, Edasich is a K2 orange giant star. It's the first giant star found to have a substellar companion, now called Iota Draconis b. The companion is believed to have 8.9 to 19.8 times the mass of Jupiter and thus could be either a giant planet or a brown-dwarf star. The companion travels in a highly eccentric orbit that takes it from 0.4 to 2.2 astronomical units from Edasich.

Edasich is a cinch to locate. Nearby Alpha (α) Draconis, or Thuban, lies halfway between the end of the Little Dipper's bowl and the star at the bend of the Big Dipper's handle. Edasich is 3rd magnitude and is the next star eastward from Alpha in Draco's sinuous body. It is easily recognized through a telescope by its golden color. It forms a very wide optical pair (BUP 162) with a reddish 9th-magnitude star to its northeast.

A host of noteworthy galaxies surrounds Edasich. The brightest is **NGC 5866**, found 4° to the star's southwest. This nearly edge-on lenticular galaxy is small but bright through my 105mm refractor at 28x. It sits outside the faintest corner of a 40' right triangle formed by three 7th- and 8th-magnitude field stars. A bright yellow star lies 1.2° south of the galaxy, and an 11th-magnitude star nestles against its northwestern flank. Upping the magnification to 87x, I also see a dimmer star along the opposite side of NGC 5866. The galaxy appears spindle-shaped, about

One glance reveals why the edge-on spiral NGC 5907 in Draco was often called the Splinter Galaxy by former *Sky & Telescope* columnist Walter Scott Houston. North is up in this 12'-wide field. Photo: Robert Gendler

2½' long and a third as wide, with a bright elongated core. A tiny nucleus intermittently shows at 127x. Observers with large-aperture telescopes (say, 12 inches or more) should look for a dark dust lane running the length of the galaxy and thin wings of light stretching outward from each end of the core.

Although the identification is controversial, NGC 5866 is often plotted on atlases or included on lists with the Messier catalog designation M102. (See *Sky & Telescope*, March 2005, page 78, and www.maa.clell.de/Messier/E/m102d.html for contrasting articles.)

NGC 5866 is the brightest member of a group of galaxies roughly 50 million light-years distant. The second-brightest is **NGC 5907**, located just 1.4° east-northeast of NGC 5866. As you sweep toward NGC 5907, take note of a ¼° arc of three 8th-magnitude stars. Following the star chain and extending it for another arc's length will lead you to the galaxy.

The spiral disk of NGC 5907 is viewed almost perfectly edge-on and appears much flatter than its neighbor. It's sometimes called the Splinter Galaxy, a tag that seems to have originated with former *Sky & Telescope* columnist Walter Scott Houston, who used the term several times dating back as far as the June 1970 issue. My 105mm scope at 87x shows a fairly faint, but very nice, long needle running south-southeast to north-northwest. It appears 7' long and grows brighter toward both its long axis and its center. My 10-inch reflector at 115x stretches this gorgeous knife-edge galaxy to 9' and shows a mottled core. A faint star cuddles up to the galaxy's western edge, and several more are sprinkled east of the galaxy's northern half.

Dropping a degree or so south from NGC 5907 brings us to the galaxies **NGC 5905** and **NGC 5908**. They share the field of view in my small refractor at 87x. NGC 5908, another nearly edge-on galaxy, is the brighter of the two. It appears about 2' long and a quarter as wide. The galaxy brightens considerably toward its long axis and slightly more toward its center. NGC 5905, 13' to the west-northwest, is just a small featureless smudge.

My 10-inch telescope at 115x still displays both galaxies together, and while this scope adds only to the apparent brightness of NGC 5908, the view of NGC 5905 is much improved. The smudge seen in the 105mm becomes the core of a much larger galaxy with an oval halo tipped northwest and a bright stellar nucleus. A close pair of 11th-magnitude stars (**Wirtz 13**) lies 2½' south of the galaxy's southeastern tip. Large-scope users should look for the bar in the core of NGC 5908 and the dust lane in NGC 5905. Can you spot them?

NGC 5905 and NGC 5908 aren't part of the NGC 5866 group but, rather, are an interacting pair of background galaxies lying three times farther away. We can, however, find the third-brightest member of the group by going back to NGC 5907 and then moving a degree northwest. Here, you'll find the spiral galaxy **NGC 5879** with a yellow-white 7th-magnitude star 7' to its north-northwest. My small refractor at 87x reveals a 1½' x ½' oval aligned nearly north-south. It grows considerably brighter toward the center and sports a pair of faint stars a few arcminutes to the east. My 10-inch scope grows the galaxy to a 3' x 1' oval and unveils a stellar nucleus.

We can now return to Edasich and sweep 1.8° east-northeast to one of the most stunning galaxy trios in the sky. **NGC 5981**, **NGC 5982**, and **NGC 5985** form a physical group 100 million light-years away, yet each appears strikingly different. Here, we see an edge-on spiral, an elliptical, and a spiral tilted to proffer a view of its arms.

The elliptical galaxy, NGC 5982, is easiest to spot and can be seen in my 105mm scope at 28x. The best view of

SPRING

June

Engaging Galaxies Near Edasich

Object	Type	Magnitude	Size/Sep.	Right Ascension	Declination	*MSA*	*U2*
Edasich	**Double star**	3.4, 8.9	255"	$15^h\ 24.9^m$	+58° 58'	553	22R
NGC 5866	**Lenticular galaxy**	9.9	6.4' × 2.8'	$15^h\ 06.5^m$	+55° 46'	568	22R
NGC 5907	Spiral galaxy	10.3	12.9' × 1.3'	$15^h\ 15.9^m$	+56° 20'	568	22R
NGC 5905	Spiral galaxy	11.7	4.7' × 3.6'	$15^h\ 15.4^m$	+55° 31'	568	22R
NGC 5908	Spiral galaxy	11.8	3.2' × 1.6'	$15^h\ 16.7^m$	+55° 25'	568	22R
Wirtz 13	Double star	11.1, 11.5	9"	$15^h\ 15.6^m$	+55° 27'	568	22R
NGC 5879	Spiral galaxy	11.6	4.2' × 1.4'	$15^h\ 09.8^m$	+57° 00'	568	22R
NGC 5981	Spiral galaxy	13.0	3.1' × 0.6'	$15^h\ 37.9^m$	+59° 23'	553	22R
NGC 5982	Elliptical galaxy	11.1	2.5' × 1.8'	$15^h\ 38.7^m$	+59° 21'	553	22R
NGC 5985	Spiral galaxy	11.1	5.5' × 2.9'	$15^h\ 39.6^m$	+59° 20'	553	22R
NGC 6015	Spiral galaxy	11.1	5.4' × 2.1'	$15^h\ 51.4^m$	+62° 19'	553	22R

Angular sizes or separations are from recent catalogs. The columns headed *MSA* and *U2* give the chart numbers of objects in the *Millennium Star Atlas* and *Uranometria 2000.0*, 2nd edition, respectively. All the objects in the table except for the galaxy NGC 6015 are plotted on the chart on page 138.

Compact enough to fit within a single field of view of most telescopes, this trio of galaxies in Draco is especially noteworthy for its visual variety. From left, they are NGC 5985, NGC 5982, and NGC 5981. North is up in this ¼°-wide field. Photo: Robert Gendler

the trio, however, comes at a magnification of 102x and shows all three galaxies evenly spaced along a shallow curve. NGC 5982's oval is tipped a bit to the south of east and harbors a brighter core and stellar nucleus. NGC 5985 is much larger but has a lower surface brightness. Its nearly north-south oval has a fairly large, slightly brighter core, and a 12th-magnitude star punctuates the galaxy's northern tip. NGC 5981 is a very faint streak pointing toward an 11th-magnitude star to its north-northwest.

My 10-inch reflector at 118x also embraces this triplet within one field of view. NGC 5981 appears 1.7' long and very thin, NGC 5982 has a 1.5' x 1' halo enveloping a core half that size, and NGC 5985 measures 3' x 2'. Under good conditions, a 10-inch scope begins to show mottling that hints at NGC 5985's spiral arms. If you have a large telescope, look for the tiny nucleus whose emissions place this spiral in the family of active galaxies known as Seyferts.

Our final stop will be the flocculent galaxy **NGC 6015**, located 3.3° to the north-northeast, off our chart. You'll find it 37' east-southeast of a 5th-magnitude star. My little refractor at 87x shows a faint 3½' oval tipped north-northeast that grows brighter toward the center. An 11th-magnitude star sits off the western side, and a very faint pair of stars lies off the southern tip. In my 10-inch scope at 170x, I judge the size to be 3' x 1.2'. The galaxy's large, bright core and inner halo look intriguingly patchy. A 13th-magnitude star is perched on the edge of the galaxy at the eastern side of the southern tip. NGC 6015 is an isolated galaxy about 50 million light-years away.

This ennead of engaging galaxies is well placed for midnorthern observers throughout the summer, and you never have to wander more than a finder's field from Edasich to track them down.

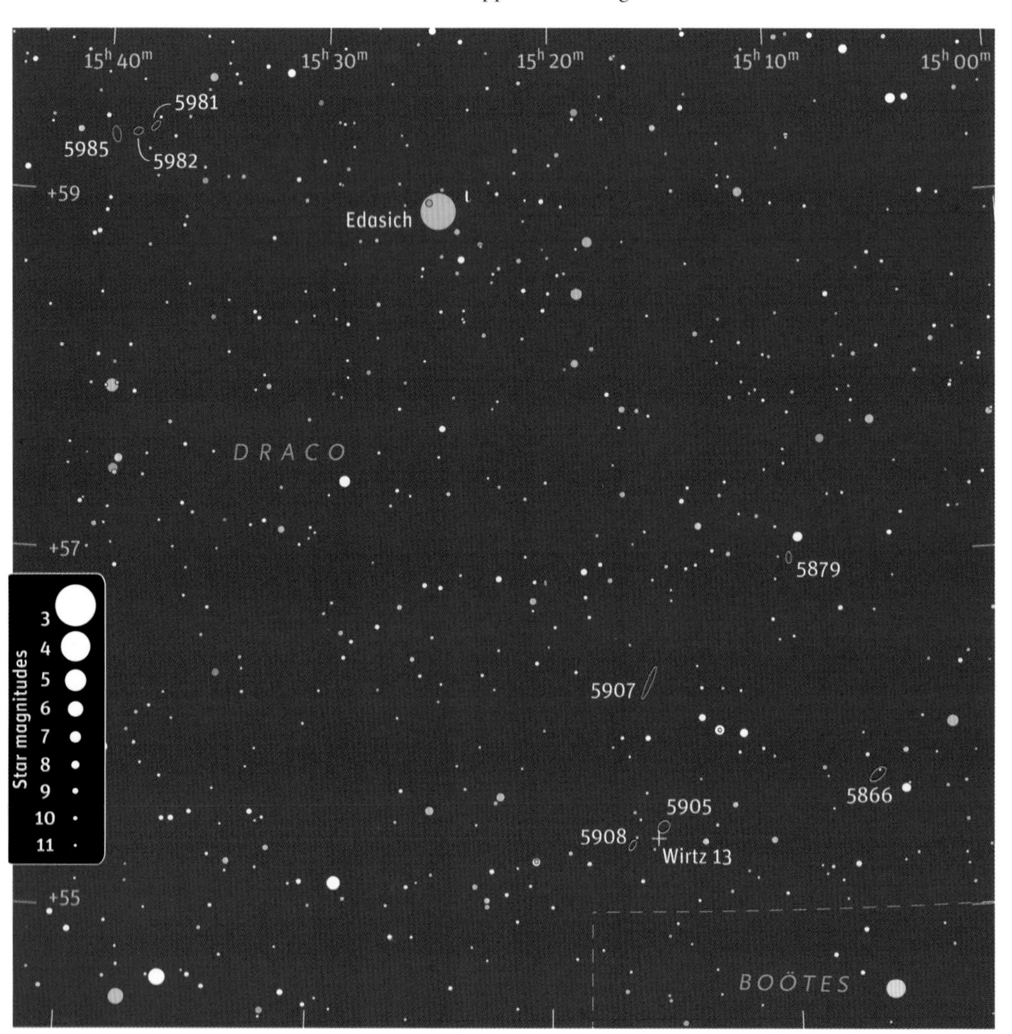

First published June 2005

Canes Redux

June's evening sky holds a bevy of interesting galaxies for deep-sky hunters.

In "Ode to Joy," on page 113, I surveyed a celestial string of deep-sky wonders draped from M94 to M106 in Canes Venatici. Let's explore a region farther south, starting just inside neighboring Ursa Major and working our way toward two of my favorite galaxies.

Our first stop is the lovely multiple star **67 Ursae Majoris** (UMa). If your sky is fairly dark, you should be able to see its 5th-magnitude primary with your unaided eye, especially if you use Alpha (α) and Beta (β) Canum Venaticorum as guide stars that roughly point to 67 UMa. Even in an 8x50 finder, 67 UMa is easy to recognize by its three attendants, while a telescope will draw out the colors of this widely spaced quartet. The primary is white, but its companions shine with various shades of yellow. The 7th-magnitude secondary, a roomy 4.6' east-northeast, is deep yellow. An 8th-magnitude companion 6.2' north-northeast of the primary glows yellow-orange, and the comparably distant 9th-magnitude star to the west has a pale yellow hue.

From 67 UMa, you can sweep 54' due east to **NGC 4111** in Canes Venatici. This galaxy looks like a fuzzy star in my 105mm refractor at low power. But at 87x, I see a thin slash 2½' long and tipped north-northwest. It shows an intense stellar nucleus, a bright elongated core, and a faint halo. A mere 3.8' east-northeast, the double star **h2596** consists of a deep yellow primary with a dim secondary on the side facing the galaxy. With my 10-inch reflector at 171x, another galaxy becomes visible on the opposite side of h2596. **NGC 4117** is about 50" long, slopes north-northeast, and has a weakly brighter core.

My 105mm scope at 28x shows two little smudges in the field with NGC 4111. The brighter one is **NGC 4143**, found 43' southeast near a yellow 8th-magnitude star. At 87x, this oval galaxy displays three distinct brightness steps: a faint oval halo running southeast to northwest, a small bright core, and a tiny nucleus. In my 10-inch reflector, the core is shaped like a convex lens and exhibits traces of structure with slightly brighter areas along the northeast and southwest sides. **NGC 4138** is located 46' northeast of NGC 4111. With my 105mm scope at 87x, I see a softly radiant oval that grows gently brighter toward the center and has a faint stellar nucleus.

Sweeping 41' east with my 10-inch scope brings me to **NGC 4183**, whose highly elongated form has earned it a place in the *Revised Flat Galaxy Catalogue*. This 1999 catalog by Igor D. Karachentsev and four colleagues lists edge-on spiral galaxies that appear at least seven times longer than they are wide. The view at 118x reveals a 4½' streak tipped north-northwest and enclosing a dappled 1½' core with a faint star at its southern tip.

The galaxies I've mentioned thus far belong to a physi-

SPRING

June

A Midyear Galaxy Hunt

Object	Type	Magnitude	Size/Sep.	Right Ascension	Declination	*MSA*	*U2*
67 UMa	Quadruple star	5.2, 6.6, 8.3, 8.9	4.6', 6.2', 6.1'	12h 02.1m	+43° 03'	612	37R
NGC 4111	Edge-on lenticular galaxy	10.7	5.2' × 1.2'	12h 07.0m	+43° 04'	612	37R
h2596	Double star	8.2, 11.6	34"	12h 07.3m	+43° 06'	612	37R
NGC 4117	Edge-on lenticular galaxy	13.0	2.1' × 0.9'	12h 07.8m	+43° 08'	612	37R
NGC 4143	Barred lenticular galaxy	10.7	2.4' × 1.8'	12h 09.6m	+42° 32'	612	37R
NGC 4138	Lenticular galaxy	11.3	3.0' × 2.4'	12h 09.5m	+43° 41'	612	37R
NGC 4183	Flat edge-on spiral galaxy	12.3	6.3' × 0.8'	12h 13.3m	+43° 42'	612	37R
NGC 4244	Flat edge-on spiral galaxy	10.4	17.7' × 1.9'	12h 17.5m	+37° 48'	633	54L
NGC 4214	Magellanic irregular galaxy	9.8	7.4' × 6.5'	12h 15.7m	+36° 20'	633	54L
Upgren 1	Asterism	6.2	20'	12h 35.3m	+36° 17'	632	54L
NGC 4631	Barred (?) edge-on spiral	9.2	15.4' × 2.6'	12h 42.1m	+32° 32'	632	54L
NGC 4656	Peculiar edge-on spiral	10.5	9.1' × 1.7'	12h 44.0m	+32° 10'	654	54L
NGC 4627	Peculiar elliptical galaxy	12.4	2.6' × 1.8'	12h 42.0m	+32° 34'	654	54L

Angular sizes or separations are from recent catalogs. The visual impression of an object's size is often smaller than the cataloged value and varies according to the aperture and magnification of the viewing instrument. The columns headed *MSA* and *U2* give the chart numbers of objects in the *Millennium Star Atlas* and *Uranometria 2000.0*, 2nd edition, respectively.

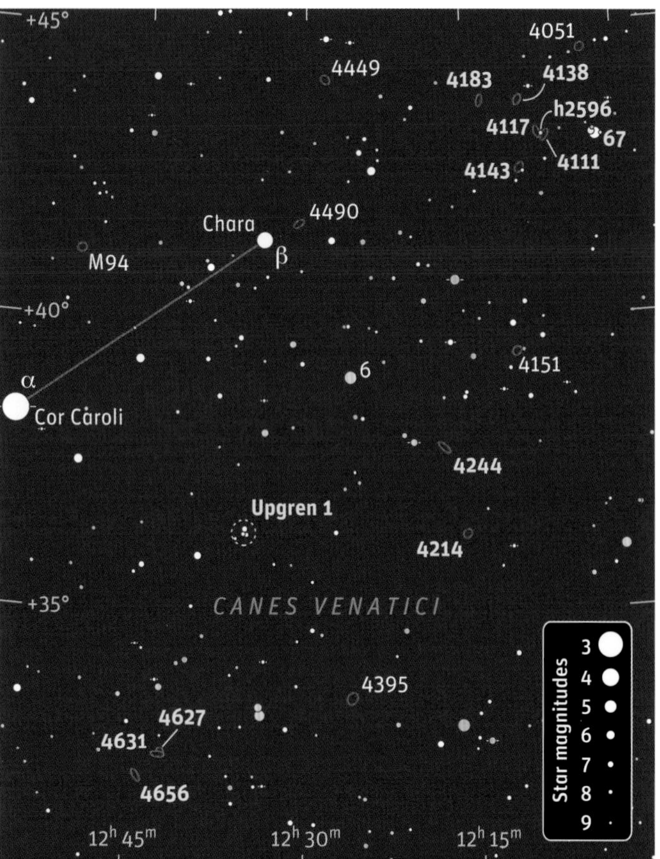
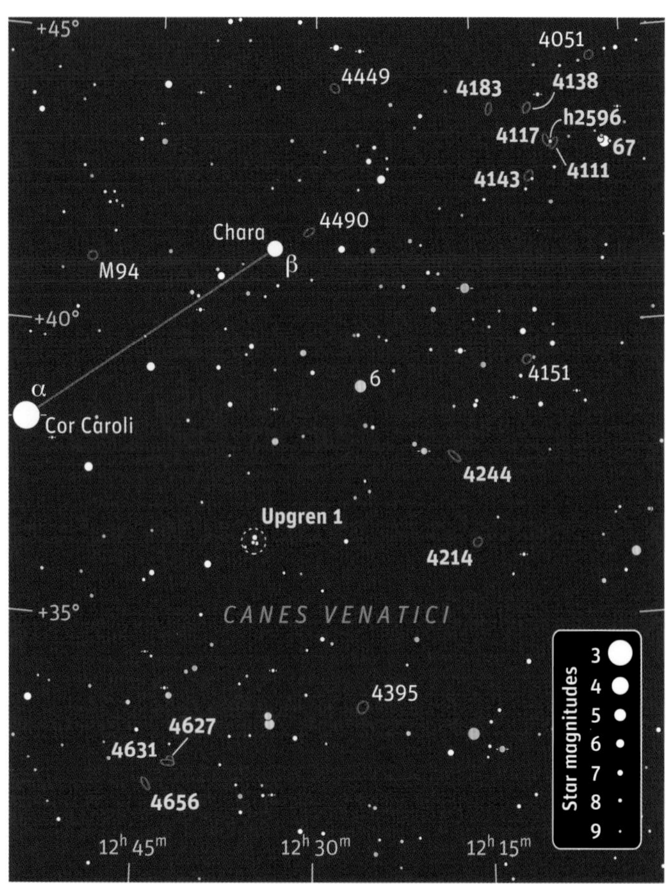

cally related system 50 million light-years away that's just a small part of the Ursa Major South galaxy group. Our next galaxy is only 14 million light-years distant and, consequently, appears much larger. To find **NGC 4244**, start at Chara and hop 2.8° south-southwest to 6 Canum Venaticorum, a 5th-magnitude yellow star that makes a right triangle with Alpha and Beta. From there, slide 2° southwest. NGC 4244, another flat galaxy, is suspended like a silvery sliver of light in my 105mm scope at 28x. Its needlelike form stretches northeast to southwest for 12' and gradually intensifies toward its long axis. In my 10-inch, this wafer-thin galaxy extends for 16' and is delicately mottled.

NGC 4244 is particularly interesting to astronomers because many of its stars can be resolved in large telescopes. This, combined with the edge-on orientation of the galaxy, helps astronomers analyze the structure and evolution of spiral galaxies. According to a recent study, NGC 4244 has a thin star-forming disk surrounded by a thick disk more than twice as wide containing older stars and a low-density, flattened halo nearly seven times as wide.

Now drop 1½° south-southwest to the irregular galaxy **NGC 4214**. My little refractor at 127x reveals a misty round glow growing sharply brighter toward an unevenly lit, barlike core that's wider at its northwestern end. Its misshapen facade is due to the myriad massive regions of ongoing star formation that riddle the galaxy. Several can be spotted with large backyard scopes, particularly in the southeastern part of the core. NGC 4214 is about 13 million light-years away.

Canes Venatici contains one open cluster — well, sort of. **Upgren 1** is a "compact grouping of F-type stars" brought to the attention of astronomers in 1965 by Arthur Upgren and Vera Rubin. Their research indicated that it might be the core of one of the oldest and closest open clusters. They suggested that the group had already lost its gravitational hold on fainter members and was in its final stages of dissolution. Although this paints an enchanting pic-

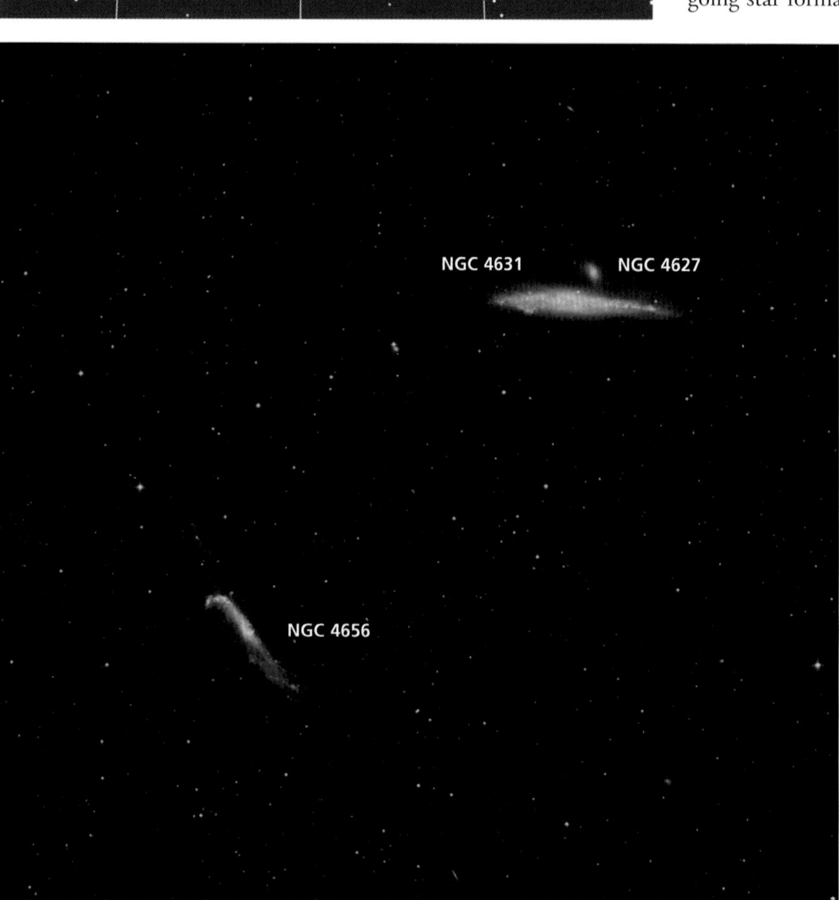

Astronomers think that intense tidal interactions among the large edge-on galaxies NGC 4631 and NGC 4656 and a smaller nearby companion, NGC 4627, are responsible for the unusual appearance of this trio, which lies about 25 million light-years away in the constellation Canes Venatici. The field is 1° wide with north up. Photo: POSS-II / Caltech / Palomar

First published June 2006

ture, later studies analyzing the motions of stars in the area provided evidence that Upgren 1 isn't a true cluster after all.

The seven stars studied by Upgren and Rubin are easy targets for any telescope. Look for the 20' asterism 4° east of NGC 4214. Three of its stars form an east-west line that is flanked by two stars north and two more southeast.

We'll finish our tour this month with two intriguing galaxies, **NGC 4631** and **NGC 4656**. Drop 3° south from Upgren 1 to a nearly north-south pair of bright, orange stars. Then scan 1.9° east-southeast, where you'll encounter beautiful NGC 4631. This long, thin, east-west spindle measures 8' x 1¼' in my little refractor at 87x. Its fleecy surface is rich in subtle detail, and a faint star is anchored off the center of its northern side. Just ½° southeast in the same field of view, the slender galaxy NGC 4656 stretches southeast to northwest for 6'. It's considerably fainter than its neighbor.

The bright knots and dark patches that decorate these galaxies become much more obvious in my 10-inch reflector. NGC 4631's eastern half is broad, but the western end tapers to a narrow point. A small bright spot a bit east of center coincides with a complex region of intense star formation that may mark part of a central ring or the end of a bar. A little companion galaxy, **NGC 4627**, hugs NGC 4631's northern side.

NGC 4656 is surprisingly faint southwest of its lumpy core. A pronounced hook that inspired the galaxy's nickname, the Hockey Stick, crazily distorts its northeastern end. The hook bends eastward and bears its own designation, NGC 4657. The chaotic appearance of NGC 4631 and NGC 4656 is due to fierce tidal interactions with each other and nearby NGC 4627. This turbulent trio lies approximately 25 million light-years away.

Where Virgo Walks

As we approach midyear, the rich galaxy fields of eastern Virgo beckon.

Virgo, the Maiden, is the second-largest constellation, surpassed only by lengthy Hydra, the Water Snake. From head to toe, statuesque Virgo spans 53° of sky with her feet in the constellation's eastern compass. As veteran observers know, the sky where Virgo walks is rich in galaxies, but it contains a few surprises as well.

Let's begin with the lovely edge-on spiral galaxy **NGC 5746**, conveniently located 20' west of the naked-eye star 109 Virginis. My 105mm refractor at a magnification of 47x shows a thin streak with a bright sliver of a core. A curve of four stars arcs over the galaxy's northern tip, the two brightest glowing like orange embers. When viewed at 87x, this splintery galaxy appears about 4' long. With my 10-inch reflector, NGC 5746 looks like a half-size version of the well-known galaxy NGC 4565 in Coma Berenices. At 166x, a dust lane lines the eastern side of the galaxy's patchy core.

NGC 5746 is about 96 million light-years distant and is part of the isolated galaxy pair KPG 434. Its smaller and fainter companion, **NGC 5740**, lies 18' south-southwest and shares the field of view through my 10-inch scope at 115x. NGC 5740's oval form is two times longer than wide and contains a broadly brighter core.

NGC 5746 is too far from its companion to suffer significant tidal interactions, and it doesn't show signs of starburst activity or an active galactic nucleus. Combined with our edge-on view, these traits make NGC 5746 a perfect test object for an irksome mystery in prevailing theories of galaxy formation. Spiral galaxies are believed to form from the col-

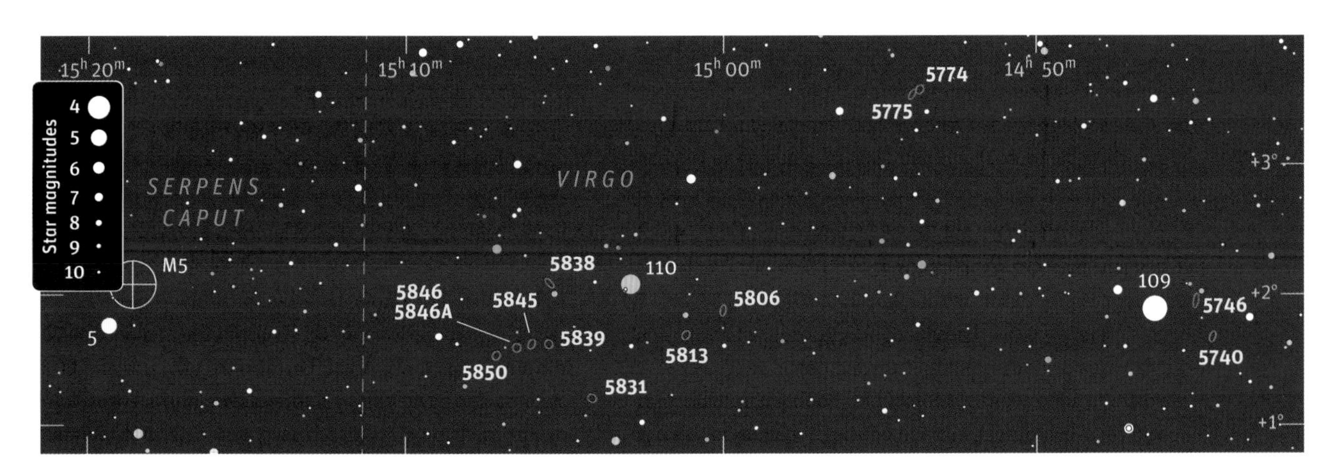

lapse of immense intergalactic clouds, and as such, they should be immersed in hot remnant halos of slowly infalling gas. Hot halos of gas *ejected* from active galaxies have been discovered, but no infalling halo was known until the orbiting Chandra X-ray Observatory turned its gaze toward NGC 5746. Chandra found an extensive halo reaching out to at least 65,000 light-years on each side of the galaxy's disk. It seems that the predicted halos do, indeed, exist, but detection awaited the scrutiny of Chandra's powerful X-ray eye.

Another interesting galaxy pair lies 2.5° northeast of 109 Virginis. KPG 440 consists of the face-on spiral **NGC 5774** and the edge-on spiral **NGC 5775**. Only the latter is visible in my little refractor. At 47x, I see a uniformly lit spindle, tipped approximately northwest, making a squat isosceles triangle with two 9th-magnitude stars to the south. Closer examination reveals no further detail. However, in my 10-inch reflector at 118x, NGC 5775 becomes a little patchy and covers 4' x 1'. Just a few arcminutes west-northwest, NGC 5774 is perceptible as a much fainter, slightly oval glow, which appears about half as long as its neighbor.

Studies indicate that gas and stars are flowing from NGC 5774 to NGC 5775. Many of the stars are too young to have formed in their parent galaxy and must have been born in the vast space in between. This remarkable pair is about 87 million light-years from Earth.

Now we'll move southeastward to 110 Virginis for a quick jaunt through a collection of galaxies outlining a 1.8°-long arrow. **NGC 5806**, the point of the arrow, sits 45' west-southwest of the star. In my 105mm scope at 87x, this galaxy appears 2' long and one-third as wide, tipped a bit west of north. It shares the field with **NGC 5813**, 21' east-southeast along the southern side of the arrowhead. NGC 5813 is a brighter and fatter oval held captive in a little box composed of one 13th- and three 12th-magnitude stars. The galaxy is about the same length as its neighbor and leans northwest.

NGC 5831 marks the southern corner of the arrowhead. You can reach it by scanning from NGC 5806 to NGC 5813 and continuing for 2½ times that distance. In my little refractor at 87x, NGC 5831 looks small and round with a slightly brighter center. The northern corner of the arrowhead, 38' east of 110 Virginis, is marked by **NGC 5838**. At the same power, this galaxy displays a faint 3' x 1' halo and a bright, round 1' core. An orange 8th-magnitude star with a faint companion sits 5' south-southwest.

The stem of the arrow is formed by a chain of galaxies with **NGC 5850** at its eastern end. This galaxy is quite faint in my 105mm scope at 87x. Its elusive halo tips west-northwest and encloses a small brighter core. Lying 10' west-northwest in the same field, bright and round **NGC 5846** is much more obvious. It spans 2' and grows more intense toward its center. My 10-inch scope at 231x shows a little companion within its halo, south of center. This is **NGC 5846A**, and it appears as a starlike spot with a tiny aureole of haze around it. Nudging the reflector 12' west reveals two more galaxies in the arrow's stem. **NGC 5845** is a small but bright oval with a brighter center. **NGC 5839** is half again as large but fainter, with a nearly stellar nucleus. The stem galaxies all belong to a phys-

Strolling Through Virgo

Object	Type	Magnitude	Size/Sep.	Right Ascension	Declination	*MSA*	*U2*
NGC 5746	Spiral galaxy	10.3	7.5' × 1.3'	14h 44.9m	+01° 57'	766	109L
NGC 5740	Spiral galaxy	11.9	3.0' × 1.5'	14h 44.4m	+01° 41'	766	109L
NGC 5774	Spiral galaxy	12.1	3.0' × 2.4'	14h 53.7m	+03° 35'	766	109L
NGC 5775	Spiral galaxy	11.4	4.2' × 1.0'	14h 54.0m	+03° 33'	766	109L
NGC 5806	Spiral galaxy	11.7	3.0' × 1.5'	15h 00.0m	+01° 53'	765	108R
NGC 5813	Elliptical galaxy	10.5	4.1' × 2.9'	15h 01.2m	+01° 42'	765	108R
NGC 5831	Elliptical galaxy	11.5	2.0' × 1.7'	15h 04.1m	+01° 13'	765	108R
NGC 5838	Lenticular galaxy	10.9	4.1' × 1.4'	15h 05.4m	+02° 06'	765	108R
NGC 5850	Spiral galaxy	10.8	4.6' × 4.1'	15h 07.1m	+01° 33'	765	108R
NGC 5846	Elliptical galaxy	10.0	3.5' × 3.5'	15h 06.5m	+01° 36'	765	108R
NGC 5846A	Elliptical galaxy	12.8	0.4' × 0.3'	15h 06.5m	+01° 36'	765	108R
NGC 5845	Elliptical galaxy	12.5	0.8' × 0.5'	15h 06.0m	+01° 38'	765	108R
NGC 5839	Lenticular galaxy	12.7	1.3' × 1.1'	15h 05.5m	+01° 38'	765	108R
NGC 5634	Globular cluster	9.5	5.5'	14h 29.6m	–05° 59'	791	129L
IC 972	Planetary nebula	13.6	43" × 40"	14h 04.4m	–17° 14'	840	129R
Abell 36	Planetary nebula	11.8	8.0' × 4.7'	13h 40.7m	–19° 53'	841	149L

Angular sizes are from recent catalogs. The visual impression of an object's size is often smaller than the cataloged value and varies according to the aperture and magnification of the viewing instrument. The columns headed *MSA* and *U2* give the appropriate chart numbers in the *Millennium Star Atlas* and *Uranometria 2000.0*, 2nd edition, respectively. All the objects this month are in the area of sky covered by Chart 46 in *Sky & Telescope's Pocket Sky Atlas*.

First published June 2007

At first glance, it's easy to mistake the edge-on spiral galaxy NGC 5746 in eastern Virgo for iconic NGC 4565 in Coma Berenices. While NGC 5746 is smaller and slightly fainter than its look-alike, it's much easier to find, since it's located only ⅓° west of the naked-eye star 109 Virginis. The view here has north at left, and the bright star is not 109 Virginis but, rather, an 8th-magnitude field star north-northwest of the galaxy. Photo: Robert Gendler

ically related group centered 70 million light-years from us.

Now for something completely different. **NGC 5634**, Virgo's only globular cluster, perches 8' south of the midpoint between Mu (μ) and Iota (ι) Virginis. My little refractor at medium powers shows a nice little cotton ball at the northeastern end of a straight line of one 8th- and two 10th-magnitude stars, the brightest embedded in the eastern edge of the cluster. NGC 5634 appears about 3' across, a bit patchy, and brighter toward the center. A 12th-magnitude star sits at its northwestern edge.

Our galaxy may have appropriated this globular from the Sagittarius Dwarf Spheroidal Galaxy. It seems that 2 billion years ago, this satellite galaxy of the Milky Way passed near the Large Magellanic Cloud and was deflected into a smaller elliptical orbit around the Milky Way. Our galaxy is slowly cannibalizing the dwarf galaxy and laying claim to several of its globular clusters. NGC 5634 could be one of them. It lies high above our galactic plane and 82,000 light-years away. At such an immense distance, I wouldn't expect to see any of this cluster's stars, yet the revised edition of *Hartung's Astronomical Objects for Southern Telescopes* claims that NGC 5634 is "clearly resolved into faint stars, just perceptible with 20 cm [8 inches of aperture]." Curious, I checked further and found that the brightest star belonging to the globular is magnitude 14.5. A total of five stars are brighter than magnitude 15.0, 13 brighter than 15.5, and 27 brighter than 16.0. Very dark, transparent, steady skies would be needed to resolve many of this cluster's stars in an 8-inch scope. In semirural skies with a 10-inch, high-power views show me only a few stars scattered around the periphery and a little fuzzy spot with an occasional star glint at the western edge. What do you see?

Only two objects in Virgo are positively identified as planetary nebulae, the glowing remains of Sunlike stars. The easier target is **IC 972**, located 21' east-northeast of the golden 6th-magnitude star HD 122577. It appears round, faint, and fairly small through my 10-inch scope at 171x. Although **Abell 36** has a greater total brightness than IC 972, its light is spread over a much larger area, making it quite a challenge. Its relatively bright, 11.5-magnitude central star is plotted on the *Millennium Star Atlas*; the symbol for the planetary nebula (labeled PK318+41.1) is plotted slightly off and should be shown centered on the star. I needed my 14.5-inch reflector to capture this planetary. At 63x with an oxygen III nebula filter, Abell 36 is faintly mottled and about 5' across.

Flying off the Handle

Many fine sights lie on or near the Big Dipper's handle.

We often associate flying off the handle with losing one's temper, yet the first definition in the Oxford English Dictionary is "to be carried away by excitement." There's a lot to be excited about when we fly off the handle of the Big Dipper, which is nestled amid some of the sky's most remarkable deep-sky wonders.

Our anchor point is **Mizar**, the star at the handle's bend. If your sky isn't severely light-polluted, you should need no optical aid to see Alcor, Mizar's 4th-magnitude companion, close by to the east-northeast. It's not certain whether Alcor is actually orbiting Mizar.

Nonetheless, Mizar is a celebrated double in its own right — the first physical pair discovered with a telescope. In January 1617, Benedetto Castelli wrote to his friend Galileo urging him to visit and have a look at Mizar, which he called "one of the most beautiful things in the sky." Galileo complied later that month, and his notes describe a pair of unequal stars 15 arcseconds (15") apart.

The observation is easy to duplicate with any modern

Near the Big Dipper's Handle

Object	Type	Magnitude	Size/Sep.	Right Ascension	Declination
Mizar	**Double star**	**2.3, 3.9**	**14.3"**	**13h 23.9m**	**+54° 56'**
Ferrero 6	**Asterism**	**7.0**	**28' × 20'**	**13h 10.0m**	**+57° 31'**
Latyshev 2	**Moving cluster**	**3.8**	**5°**	**13h 44.4m**	**+53° 30'**
M101	**Galaxy**	**7.9**	**29' × 27'**	**14h 03.2m**	**+54° 21'**
NGC 5474	**Galaxy**	**10.8**	**4.7'**	**14h 05.0m**	**+53° 40'**

Angular sizes and separations are from recent catalogs. Visually, an object's size is often smaller than the cataloged value and varies according to the aperture and magnification of the viewing instrument.

telescope. My 105mm refractor at a magnification of 47x shows Mizar and Alcor in the same field of view. Mizar's companion lies south-southeast and is about the same brightness as Alcor. Although all three are white, the two dimmer stars are a slightly yellower white to my eye.

Now that our anchor is established, let's fly off the handle to the charming asterism **Ferrero 6**, named for French amateur Laurent Ferrero. It's 3.3° northwest of Mizar and sits at the right angle of the triangle it makes with Mizar and Epsilon (ε) Ursae Majoris, also known as Alioth. My little refractor at 47x reveals 16 stars in a shape that makes me imagine the Eiffel Tower or perhaps a teepee, door and all, with a lightning rod sticking out the top. The figure is 28' tall with its tip pointing south-southwest, and its stars range from magnitudes 8 to 12.

The stars 81, 83, 84, and 86 Ursae Majoris form a slightly wavy line that starts east of Mizar and trends east-southeast for 3.3°. The brightest star is magnitude 4.6, and the rest are a magnitude fainter — barely bright enough to see with unaided eyes from my semi-rural home. But the formation shows nicely through binoculars or a finderscope. These are the brightest members of a possible moving cluster known as **Latyshev 2**. In 1977, astronomer Ivan Latyshev showed that these stars share the same motion through space and may have a common origin.

We can soar eastward along Latyshev 2 to find our next target, the magnificent spiral galaxy **M101**, sometimes called the Pinwheel Galaxy. A line of three 7th-magnitude stars stretches 1½° north to south ½° east of 86 Ursae Majoris. M101 is 52' east of the two northern stars and makes an isosceles triangle with them.

M101 is a face-on spiral with a very low surface brightness, yet it shows up well in my little refractor at 28x. The ashen halo spans ¼° and harbors a fairly small, slightly brighter core. At 87x, the galaxy is subtly dappled and a faint star is superposed a bit north of

Messier 101 (NGC 5457) is noted for the strikingly bright knots of light, many with NGC designations of their own, that stud its elegant spiral arms. Photo: Bernhard Hubl

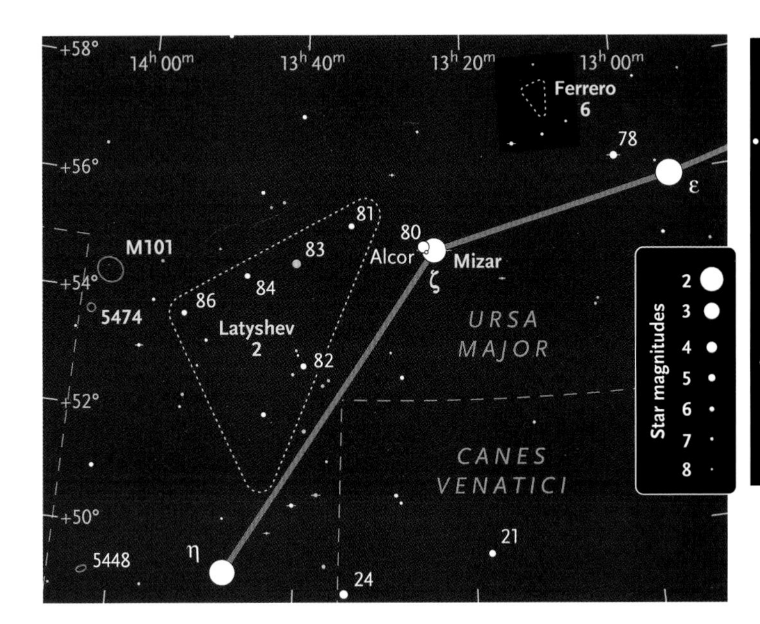

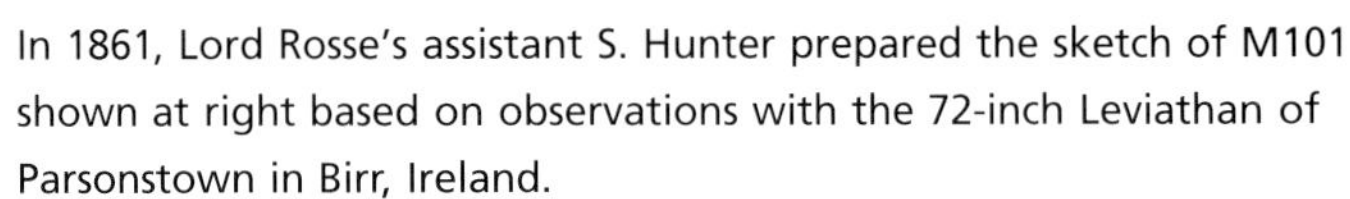
In 1861, Lord Rosse's assistant S. Hunter prepared the sketch of M101 shown at right based on observations with the 72-inch Leviathan of Parsonstown in Birr, Ireland.

the core. And at 127x, I see a small nucleus at the Pinwheel's heart and can detect the galaxy's diaphanous spiral arms.

Through my 10-inch reflector at 115x, M101 is a grand Catherine-wheel firework shedding multiple arm segments. Two arms north of the core unwrap to the west. One spirals tightly around the core and then joins a bright section on the east that reaches northward. The other wraps more loosely and can be followed to the southwest. A branch along the south and southeast contains the bright star-forming region NGC 5461. The arm fades just beyond this but brightens again at NGC 5462.

NGC 5471, a disconnected star-forming region often mistaken for a galaxy, sits 5.6' east-northeast of NGC 5462. This small roundish smudge marks the northeastern corner of the 3' trapezoid it forms with three field stars. Another arm segment, well west of M101's center and mostly composed of very faint patches, bears the easily visible, combined glow of NGC 5447 and NGC 5450, punctuated by a minute star immediately to their north. With careful study, I can also spot these star-forming regions as little islands of fluff through my 105mm scope. But my 10-inch scope at 170x discloses two more. NGC 5458 is a ghostly wisp 3' south of M101's nucleus. NGC 5455 is a very small, relatively bright spot that sits 3.8' farther south-southwest. It makes an isosceles triangle with two faint stars 2.3' northeast and north-northwest.

The captivating galaxy **NGC 5474** dwells 44' south-southeast of M101 at the eastern corner of a 10' trapezoid that it forms with three 11th-magnitude stars. My little refractor at 47x captures a soft round glow, while at 122x the galaxy appears 2' across and hints at being brightest in the north. The view gets really interesting through my 10-inch scope. At 213x, NGC 5474 delightfully reminds me of a quahog clamshell. Its peculiar facade is adorned with a bright hub in the north and then fans out toward the south. A faint star rests less than 1' northeast of the bright area.

M101 is interacting with NGC 5474, and it looks as though the larger galaxy is trying to rip the nucleus out of its companion! However, that's not the case. NGC 5474 is apparently undergoing libration due to tidal forces or mass infall. Astronomers say that the galaxy is being "sloshed," and it's only a matter of time before it sloshes the other way.

The asymmetric spiral galaxy NGC 5474 (at bottom) seems to be shying away from its huge companion, M101. Photo: *Sky & Telescope* / Dennis di Cicco

First published June 2008

Whorls of Stars

Canes Venatici is home to some of the northern sky's finest galaxies.

White whorls of stars slow turning in the sky,
Across the borders of the measured night,
On, on beyond the treasures of the eye,
Where only magic lenses garner sight:
Transparent reapers of dim sheaves of light.
— George Brewster Gallup, "Andromeda"

In the spring of 1845, William Parsons, Ireland's 3rd Earl of Rosse, became the first person to behold spiral structure in a "nebula." The target of his formidable 72-inch reflector was **Messier 51** (M51), which became known as the Spiral Nebula, or the Whirlpool Nebula. Other nebulae of peculiar structure were found, and Lord Rosse commented, "The discovery of these strange forms may be calculated to excite our curiosity, and to awaken an intense desire to learn something of the laws which give order to these wonderful systems."

Now aware that this elegant spiral is a vast and distant city of stars resembling our own Milky Way, we call it the Whirlpool Galaxy. Equipped with modern telescopes and the knowledge of what to expect, we need not have a Leviathan to detect its galactic whorls.

In his famous catalog, Charles Messier described Messier 51 as a double nebula with touching atmospheres. Thus the designation M51 belongs not only to the magnificent spiral, NGC 5194, but also to its interacting companion, NGC 5195. The pair sits 3.6° southwest of Alkaid, the star at the end of the Big Dipper's handle, and makes a squat isosceles triangle with Alkaid and 24 Canum Venaticorum.

On a good night at my semirural home, M51 is visible

Along with the Horsehead Nebula, the spiral galaxy M51 in Canes Venatici is one of astronomy's most iconic images. Although long-exposure photographs typically show more than can be seen in the eyepiece of a backyard telescope, many of the large-scale features within M51 and its companion galaxy, NGC 5195, can be identified with modest-aperture telescopes. North is at left. Photo: Johannes Schedler

Galaxies of the Hunting Dogs

Object	Type	Magnitude	Size/Sep.	Right Ascension	Declination
M51	Galaxy pair	8.4, 9.6	11' × 6'	$13^h 29.9^m$	+47° 14'
NGC 5198	Galaxy	11.8	2.1' × 1.8'	$13^h 30.2^m$	+46° 40'
NGC 5173	Galaxy	12.2	1.0' × 0.9'	$13^h 28.4^m$	+46° 36'
NGC 5169	Galaxy	13.5	2.2' × 0.7'	$13^h 28.2^m$	+46° 40'
M63	Galaxy	8.6	12.6' × 7.2'	$13^h 15.8^m$	+42° 02'
NGC 5023	Flat galaxy	12.3	6.0' × 0.8'	$13^h 12.2^m$	+44° 02'
Hickson 68	Galaxy group	11.0–13.6	10'	$13^h 53.7^m$	+40° 19'
NGC 5371	Galaxy	10.6	4.4' × 3.5'	$13^h 55.7^m$	+40° 28'
The Heron	Galaxy pair	11.4, 13.0	4' × 2'	$13^h 58.6^m$	+37° 26'

Angular sizes are from recent catalogs. Visually, an object's size is often smaller than the cataloged value and varies according to the aperture and magnification of the viewing instrument.

in a 50mm finderscope, but a hazy sky veils it. Even at a magnification of 17x, my 105mm refractor shows these galaxies as a distinct and beautiful pair, with the companion hugging the northern edge of its neighbor. NGC 5194 is a small, fairly bright, 6'-long oval with a small bright core, while NGC 5195, about one-fifth as large, is round and holds a tiny, intense heart. At 87x, the halo of the larger galaxy leans a little east of north and displays enough patchiness to suggest the galaxy's spiral structure. The core grows brighter toward the center and is slightly mottled. NGC 5195 now looks a bit larger with respect to its cohort and shows a bright, stellar nucleus.

NGC 5194 is very pretty through my 10-inch scope at 115x. Two spiral arms unwind clockwise and are flattened at their northern and southern extremes. The brightest region of NGC 5195 is elongated north-south, while the halo runs east-west and covers about 2' x 2½'. The distorted galaxies reach out to each other in the east, but their connecting bridge fades in the middle. The Whirlpool's spiral arms are very uneven in brightness and quite lovely at 166x. A particularly prominent, brightly mottled bar 2' northeast of the nucleus points toward the western side of NGC 5195. A similar enhancement is symmetrically placed on the opposite side of the galaxy, and a faint star nuzzles its inner edge. The first bar's spiral arm bears a smoother bright arc coreward and a small bright patch outward.

The double galaxy M51 is only about 25 million light-years distant, but nudging my 10-inch scope ½° southward brings into view a trio of galaxies more than four times farther away. **NGC 5198**, almost due south of M51, is the largest and brightest of the three. It appears round with a brighter core and starlike nucleus. **NGC 5173**, 19' west-southwest of NGC 5198, is faint and round with a much brighter center. Just 5½' north-northwest of NGC 5173, **NGC 5169** is difficult to hold with direct vision but grows a little brighter toward the center.

M63, another offering from Messier's catalog, is perched 1.2° north of the 6th-magnitude star 19 Canum Venaticorum. Sometimes called the Sunflower Galaxy, M63 is a flocculent spiral whose fleecy pattern is reminiscent of the arrangement of seeds in its namesake's blossom.

Through 12x36 image-stabilized binoculars, I see the Sunflower as a little hazy patch with a star at its western end. It shares the field of view with four bright stars, including 19 Canum Venaticorum, that form a checkmark. My 105mm refractor at 17x shows a fairly bright oval tipped west-northwest. At 87x, the galaxy covers 7' x 3' with a 3' x 1½' outer core, a small round inner core, and a faint

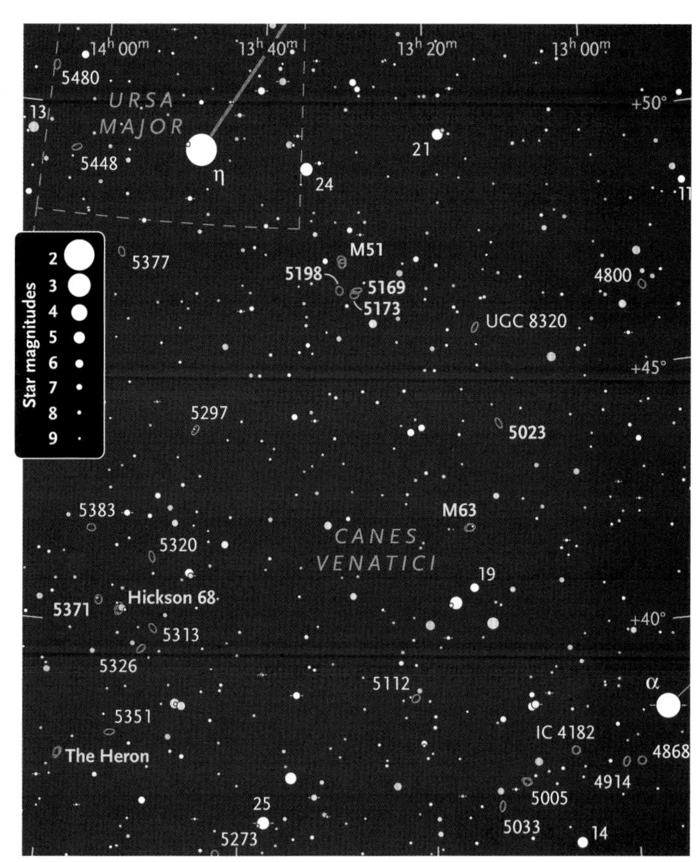

stellar nucleus. I've been unable to see M63's woolly appearance with this refractor, but Stephen James O'Meara has captured it with a similar instrument from his dark, high-altitude site in Hawaii.

Climbing 2.1° north-northwest brings us to the flat galaxy **NGC 5023**, which looks exceptionally thin because it's a bulgeless spiral seen edge-on. Through the 105mm scope at 87x, I see its slender form sloped north-northeast. Very wide pairs of stars lie several arcminutes off each flank, and the galaxy's elongated core sits on a line between the fainter stars of each pair. At first blush, this sliver of light appears about 2½' long, but with careful study, I can trace it to 3½'.

Our next stop is the compact galaxy cluster **Hickson 68** in far eastern Canes Venatici. Many of the Hickson groups are game only for moderate to large amateur telescopes. But Hickson 68 is one of the exceptions, and the 6.5-magnitude golden star at its western edge helps us locate the grouping.

Even at 47x, my 105mm refractor reveals three of its members. The combined glow of NGC 5353 and NGC 5354 is easiest to spot. Their halos blend together, but they faintly show separate cores. To their north, NGC 5350 is a northeast-southwest oval. Eastward, the non-Hickson galaxy **NGC 5371** shares the field of view. Larger than that of its neighbors, NGC 5371's oval glow is aligned north-south. NGC 5353 and NGC 5354 are easily differentiated at 127x, the southern one brighter and tilted northwest. With averted vision and patience, I see NGC 5355 as a very small, extremely faint smudge. NGC 5371 has a uniform surface brightness but hosts an elusive nucleus. Although my 105mm scope shows me the close, faint star pair south-southwest of NGC 5358, our final Hickson member, spotting the galaxy requires the additional light-gathering power of my 10-inch reflector. At 166x, its vague blur harbors a faint starlike center and merely hints at being elongated.

Popularly know as the Sunflower Galaxy, M63 can be spotted with modest binoculars, and observers with 4-inch telescopes have glimpsed some of the mottling readily visible in photographs. Photo: Johannes Schedler

Many Hickson galaxy groups prove challenging for small telescopes, but Hickson 68 in eastern Canes Venatici is an exception since most of its members are within range of a 4-inch-aperture scope. The group is also easy to locate because of its proximity to a prominent 6.5-magnitude golden-hued star. Photo: Bernhard Hubl

Our last target this month is the interacting pair of galaxies NGC 5395 and NGC 5394. Together, they are known as **The Heron** for their striking photographic resemblance to that waterbird.

Lying 3° south-southeast of Hickson 68, only NGC 5395 is visible in my 105mm refractor. At 87x, I see a north-south oval of uniform surface brightness, about twice as long as it is wide. Through my 10-inch scope at 171x, NGC 5395 has a faint, 2½' x 1' halo with a 1½' x ¾' core that grows a little brighter toward the center. The core exhibits brightness variations, particularly in the north. A 13.7-magnitude star rests just off the galaxy's southern tip. NGC 5394 is a small oval, slanted northeast, with about the same surface brightness as its companion's core.

Examining NGC 5395 with my 15-inch reflector at 192x, I see a long, faint arm starting at the galaxy's southern tip and reaching north along its western flank. The northern end of the galaxy has a short hook that bends south along the eastern flank. A faint star hovers over NGC 5395's northern tip, while another guards its western side. I've been unable to see the graceful, tidal arms of NGC 5394. Can you?

First published June 2009

Southern Skies

Scorpius and its neighbors host an unparalleled visual feast.

Southern skies, have you ever noticed southern skies?
Its precious beauty lies deep beyond the eye,
And it goes rushing through your soul
like the stories told of old.
— Allen Toussaint, "Southern Nights"

The southern sky is magnificent at this time of the year, and those who probe its depths with telescopes greet a wealth of celestial treasures. But observers at midnorthern latitudes are limited in just how far south they can survey. With that in mind, I've chosen a small sample of deep-sky wonders from the southern sky, beginning just below the celestial equator and ending with a marvelous sight that hugs my horizon in upstate New York.

Let's start our tour with the obscure planetary nebula **Shane 1** (PK 13+32.1), lying 1.3° south of Sigma (σ) Serpentis. Through my 6-inch reflector at low power, I can easily pinpoint Shane 1 in a V of 10th- to 12th-magnitude stars. The V is 18' tall, very slender, and its western side is slightly curved, as shown at right. A small triangle of stars marks the southern point. The middles of its sides are marked by Shane 1 (east) and an 11.6-magnitude star (west).

Shane 1 is a tiny thing, measuring only 6" x 5". Despite their size, diminutive planetaries often have a certain something that whispers, "I am not a star." They may not appear sharp enough or quite the right color for a star. To me, Shane 1 looks like the ghost of a star. A narrowband or an oxygen III nebula filter betrays the planetary's nature, for when viewed through such a filter, Shane 1 outshines the star to its west. My 10-inch reflector at high power adds a faint central star, a blue-gray hue, and some dimension to the nebula. Shane 1 is listed at magnitude 12.8, but it seems about a half magnitude brighter to my eye.

Charles Donald Shane, then director of Lick Observatory, near San Jose, California, discovered Shane 1 while examining the Lick Northern Proper Motion survey's first-epoch photographic plates, taken in the 1940s and 1950s. A recent catalog places Shane 1 roughly 31,000 light-years away.

Now drop down to **NGC 5897**, which is shown on page 150. This pretty globular cluster sits 1.7° southeast of Iota (ι) Librae. In my 6-inch scope at 38x, it's visible as a very soft glow cradled by a shallow arc of three 8th-magnitude stars. At 112x, NGC 5897 is a lovely sight. It spans about 7½' and clasps a large, somewhat brighter core. At least a dozen stars of mixed brightness sparkle within the cluster, and an

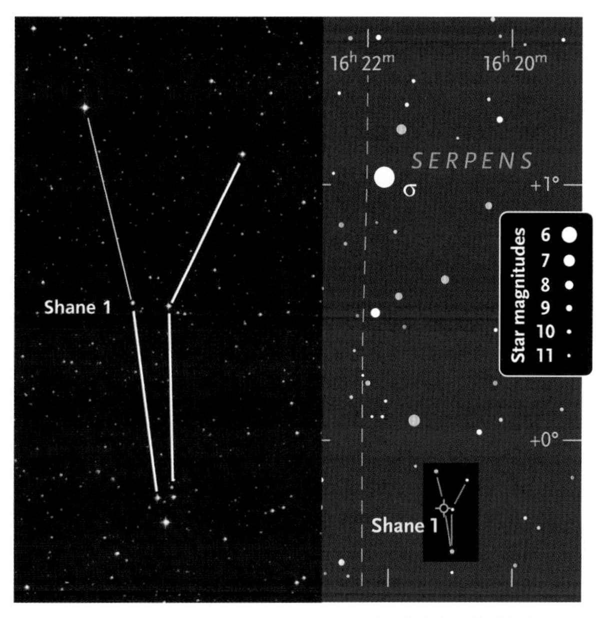

POSS-II / Caltech / Palomar

A Summer Smorgasbord

Object	Type	Constellation	Magnitude	Size/Sep.	Right Ascension	Declination
Shane 1	Planetary nebula	Serpens	12.8	6" × 5"	$16^h 21.1^m$	–00° 16'
NGC 5897	Globular cluster	Libra	8.5	11.0'	$15^h 17.4^m$	–21° 01'
54 Hydrae	Double star	Hydra	5.1, 7.3	8.3"	$14^h 46.0^m$	–25° 27'
NGC 5694	Globular cluster	Hydra	10.2	4.3'	$14^h 39.6^m$	–26° 32'
NGC 5986	Globular cluster	Lupus	7.5	8.0'	$15^h 46.1^m$	–37° 47'
NGC 6072	Planetary nebula	Scorpius	11.7	98" × 72"	$16^h 13.0^m$	–36° 14'
NGC 6231	Open cluster	Scorpius	2.6	14.0'	$16^h 54.2^m$	–41° 50'

Angular sizes and separations are listed from recent catalogs. Visually, an object's size is often smaller than the cataloged value and varies according to the aperture and magnification of the viewing instrument.

Photo: Daniel Verschatse / Observatorio Antilhue, Chile

11th- and 12th-magnitude star pair guards its north-northwestern edge. In my 14.5-inch reflector at 170x, NGC 5897 looks much like a rich open cluster. It covers about 10', with several of its brightest stars sprinkled across its eastern half.

If you have trouble locating NGC 5897 or NGC 5694 (discussed below) using the chart above, these objects are shown on page 57 of *Sky & Telescope's Pocket Sky Atlas.* This page and several others are available as free downloads at www.SkyandTelescope.com/psa.

Our next two targets are in Hydra, the Water Snake, though it's more fun to think of them as occupying Noctua, the Night Owl. This reputedly wise raptor was first depicted on Alexander Jamieson's 1822 celestial atlas. Flown from modern star charts, little Noctua once stood upon Hydra's tail and occupied some of the territory belonging to Libra and Virgo.

Our first owlish stop is the colorful double star **54 Hydrae**. The pair is delightfully split with my 5.1-inch refractor at 63x. The yellow-white primary holds a golden companion east-southeast. William Herschel discovered this double in 1783 and described the companion as bluish red. What color do you see?

Also in Noctua, the globular cluster **NGC 5694** sits along the western side of a 53' zigzag of four 7th-magnitude stars. Through my 6-inch reflector at 38x, NGC 5694 is a little fuzzball making a 2½' curve with a north-south pair of stars, magnitude 10½. The cluster bares no stars even at 176x, but it's rather attractive, with an intense, almost starlike center. The globular is fairly bright to a diameter of about 1', and a faint halo doubles its size.

NGC 5694 is intrinsically brighter and larger than NGC 5897, but it looks smaller and fainter because it's much farther away — 113,000 versus 40,000 light-years. Yet the loose appearance of NGC 5897 is not solely due to its proximity. The stars of NGC 5694 are 400 times more densely packed in the center.

Next, swoop down to **NGC 5986**, the brightest globular cluster in Lupus. My 105mm refractor at 87x reveals a mottled glow with hints of resolution. A brighter foreground star garnishes its eastern side. In my 10-inch reflector at 187x, NGC 5986 is a granular haze flecked with stars across much of its 5' face. My 14.5-inch scope shows stars right to the center.

At a distance of 34,000 light-years, NGC 5986 is the

nearest of our three globulars. This is enough to make it appear brighter in our sky than NGC 5897, despite being more obscured by interstellar dust and inherently a bit less luminous.

Just 1.4° east-northeast of Theta (θ) Lupi, the planetary nebula **NGC 6072** resides in neighboring Scorpius. Through my 6-inch scope at 112x, it appears roundish with dimmer patches inside. A yellow, 8.6-magnitude star hovers north. My 10-inch scope at 118x shows an east-west oval with fainter ends, a suggestion of annularity, and a dim star superposed on the northeastern side. The nebula stands out nicely with a narrowband or an oxygen III nebula filter. At 220x, it's about 1¼' long, two-thirds as wide, and harbors a brighter spot near the center.

NGC 5986 and NGC 6072 are too far south for those who are observing in the northern reaches of Canada and Europe, but they're approximately 10° above my horizon in upstate New York. By contrast, the **False Comet** tickles my horizon and cannot be seen in its entirety from any place much farther north.

Canadian amateur Alan Whitman pointed out this striking "cometlike Milky Way patch" while attending the Texas Star Party in 1983. With his unaided eyes, Whitman saw **NGC 6231** as the coma of the comet and Zeta^1 (ζ^1) and Zeta^2 (ζ^2) Scorpii as a nucleus at its leading edge. The comet tail arcs gently northward, formed mostly by the misty glow of Trumpler 24.

The False Comet's splendor isn't obvious to the unaided eye from my latitude, but its components are a lovely sight in my husband's 15x50 image-stabilized binoculars — even from our family camp, where the comet's nucleus crests a mere 3.3° above the horizon. The Zeta stars shine in nicely contrasting hues of white and gold, with the speckled glow of NGC 6231 to their north. Above this, a large arc of loosely scattered stars fans to the north-northeast for about 1¾°. Some sources identify this as **Trumpler 24** (Tr 24), but in his 1931 doctoral dissertation, Swedish astronomer Per Arne Collinder defined a larger group, Collinder 316, overlapping Trumpler 24 and centered farther southwest. Perhaps too scattered to rate open-cluster status, these groups are part of a youthful association of stars known as Scorpius OB1.

The False Comet is a beautiful naked-eye sight from the Winter Star Party in the Florida Keys. Viewed from Florida through New York amateur Betsy Whitlock's 105mm refractor, NGC 6231 is a stunning ¼° collection of 80 mixed bright and faint stars, while the comet's tail is a conspicuous gathering of well over a hundred stars.

My 10-inch reflector turns up more comet pieces. Within Trumpler 24, **van den Bergh-Hagen 205** (vdB-Ha 205) is a little knot of six stars, **Ruprecht 122** (Ru 122) shows a small misty sprinkling of faint suns, and the emission nebula **IC 4628** is a large east-west glow. A few tempting targets fence the comet's tail. **NGC 6242** is a prominent 9' collection of 25 stars, most confined to a bar tipped north-northwest. Dimmer and smaller, **NGC 6268** has many of its 25 stars arranged in two parallel lines. The dark nebula **Barnard 48** (B48) is a 35' x 10' star-poor strip leaning northeast, and **van den Bergh-Hagen 211** is a dim group dominated by an east-west line of stars.

The False Comet is truly a precious beauty in our southern skies. Enjoy!

The region north of Zeta Scorpii, sometimes called the False Comet, is one of the most magnificent fields in the entire sky. Photo: Adam Block / NOAO / AURA / NSF

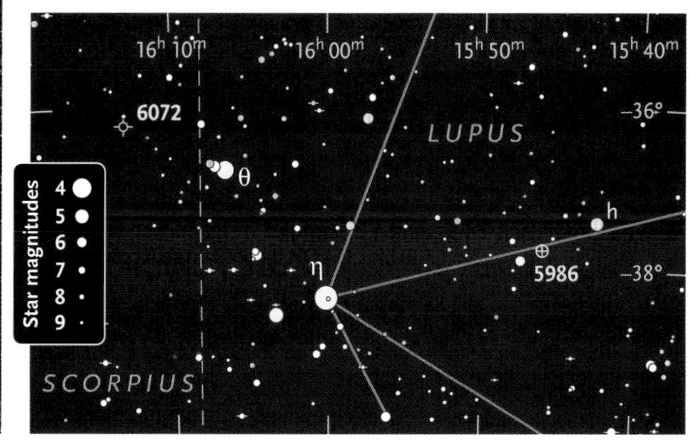

First published June 2010

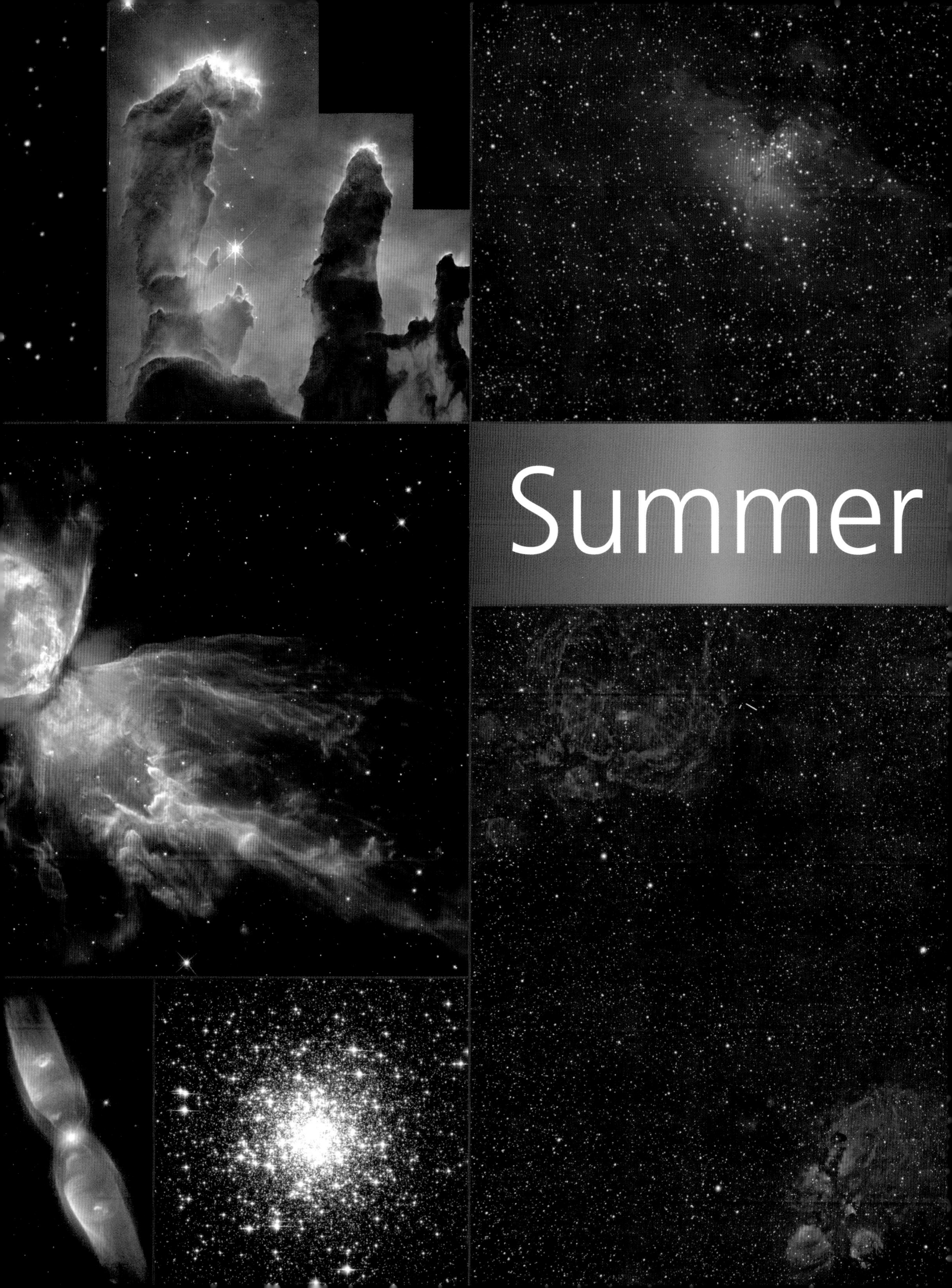

Summer

Around the Bend in Draco

An interesting part of this circumpolar constellation is high on July evenings.

Draco, the Dragon, twists tortuously through the northern sky, wrapping himself around neighboring polar star patterns. You can pick up the tip of the Dragon's tail between the bowls of the Big and Little Dippers and follow along as he curls around the Little Dipper and then takes a sharp bend toward the south. A number of noteworthy sights hug the bend of Draco, making them fairly simple to ferret out.

We'll begin at the quadrilateral that marks Draco's head. Its faintest star is **Nu (ν) Draconis**, a double that is wide enough to split in binoculars. My 105mm refractor at 17x shows matched suns aligned northwest (ν^1) to southeast (ν^2). Both stars appear white to me, but there's just a hint of yellow in ν^2.

The star joining Draco's head to his sinuous body is Xi (ξ) Draconis, whose popular name Grumium ("jaw") sounds as if it came from a Dr. Seuss book. Drawing a line from Nu Draconis through Grumium and continuing for 1⅓ times that distance again will bring you to 5th-magnitude **39 Draconis**. At 17x, I see an 8th-magnitude companion to the north-northeast, well split from the white primary. Boosting the magnification to 153x reveals a third component barely separated from the primary. Both companions are the same brightness, but the outlying one seems colorless, while the inner looks blue.

Moving 3.5° east-northeast of 39 Draconis takes us to slightly brighter **Omicron (ο) Draconis**. Omicron is an unequal pair of orange stars easily seen at 17x. The 5th-magnitude primary harbors an 8th-magnitude secondary to the northwest. In his classic 1844 *Bedford Catalogue*, William H. Smyth described the pair as orange-yellow and lilac. Other observers have seen the companion as blue-green, but these hues must be a contrast illusion, since the star has a spectral type of K3 (orange).

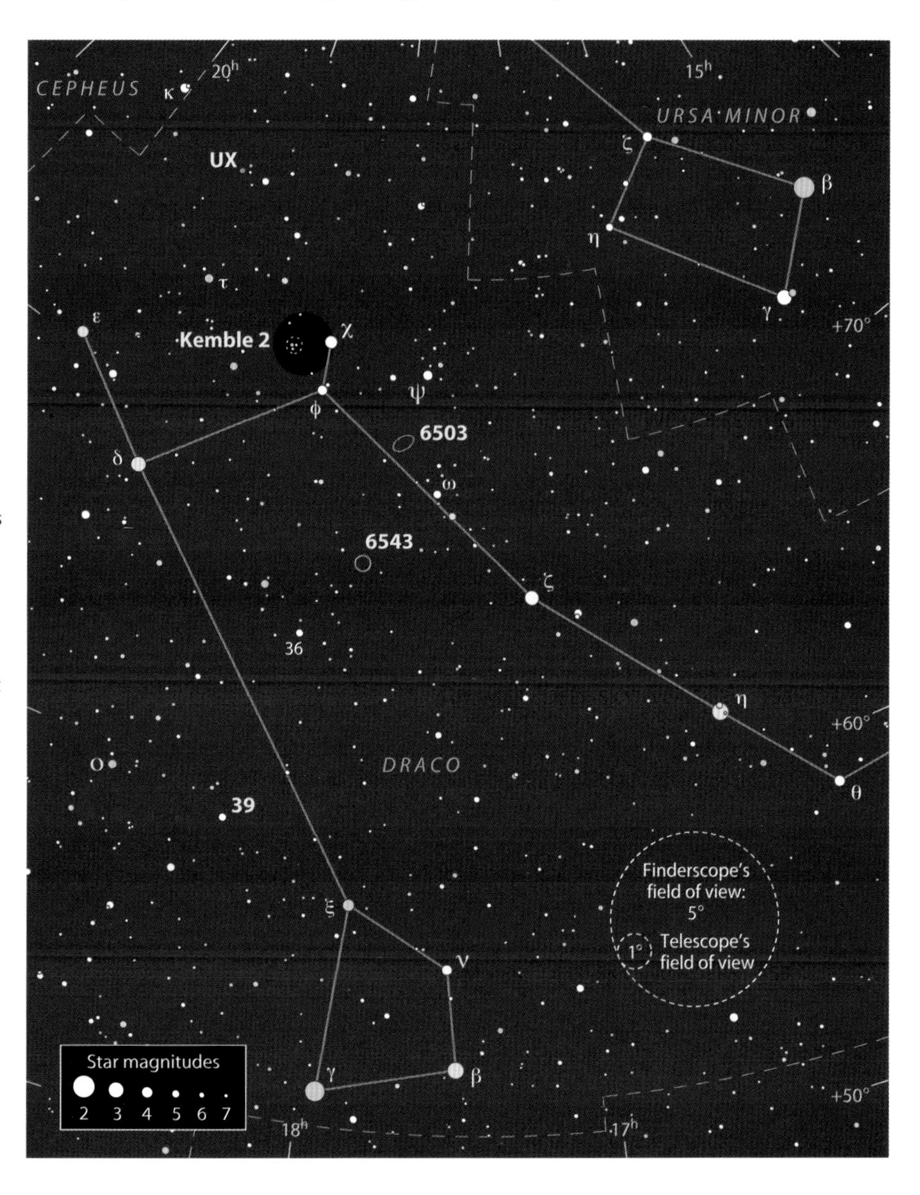

The Dragon's Deep-Sky Delights

Object	Type	Magnitude	Size/Sep.	Distance (l-y)	Right Ascension	Declination	*MSA*	*U2*
ν Draconis	Double star	4.9, 4.9	62"	99	$17^h\ 32.2^m$	+55° 11'	1095	21R
39 Draconis	Triple star	5.1, 8.0, 8.1	90", 3.7"	190	$18^h\ 23.9^m$	+58° 48'	1078	21L
ο Draconis	Double star	4.8, 8.3	37"	320	$18^h\ 51.2^m$	+59° 23'	1078	21L
UX Draconis	Carbon star	5.9–7.1	—	2,000	$19^h\ 21.6^m$	+76° 34'	1043	3R
Kemble 2	Asterism	5.6	20'	—	$18^h\ 35.0^m$	+72° 23'	1053	10R
ψ Draconis	Double star	4.6, 5.6	30"	72	$17^h\ 41.9^m$	+72° 09'	1054	11L
NGC 6503	Galaxy	10.2	7.1' × 2.4'	17 million	$17^h\ 49.5^m$	+70° 09'	1054	11L
NGC 6543	Planetary nebula	8.1	22" × 19"	3,000	$17^h\ 58.6^m$	+66° 38'	1066	11L

The columns headed *MSA* and *U2* give the chart numbers on which the objects may be found in the *Millennium Star Atlas* and *Uranometria 2000.0*, 2nd edition, respectively. Sizes and separations of objects are catalog values.

Up near the top of Draco's bend is Tau (τ) Draconis, an orange 4th-magnitude star. Sweeping 3.2° north from Tau, we come to a little clump of stars, one of which looks markedly reddish. This is the carbon star **UX Draconis**, one of the reddest stars in the sky. One measure of a star's color is its color index — the redder the star, the higher the number. The white star Vega (in Lyra) has a color index of 0.0, while reddish orange Antares (in Scorpius) has a color index of 1.8. Yet UX Draconis is much ruddier than Antares, with a color index near 2.7. It is a semiregular variable that shows changes in both color and brightness. UX Draconis goes from around magnitude 5.9 to 7.1 and back over a cycle of about 168 days.

Next, we'll seek out the charming asterism **Kemble 2**. This handful of 7th- to 9th-magnitude stars is very easy to locate, about 20' across and centered just 1.1° east-southeast of 3.6-magnitude Chi (χ) Draconis. The name Kemble 2, which appears in the second edition of *Uranometria 2000.0* and the software atlas *MegaStar 5.0*, honors Lucian J. Kemble (1922-1999), a well-known Canadian amateur and Franciscan friar who had written about the asterism in an article he never published. The description was brought to the attention of observers by his Norwegian friend Arild Moland, who dubbed it the Mini-Cas asterism. In *The Deep Sky: An Introduction*, Philip S. Harrington calls this striking bunch the Little Queen.

These alternate names echo what is so striking about Kemble 2, an uncanny resemblance to the familiar W of the five bright stars of Cassiopeia, the Queen. There's even a sixth star corresponding to Eta (η) Cassiopeiae, albeit slightly out of place. The shape is nicely displayed in the *Millennium Star Atlas* but not labeled. Through my little refractor at 28x, the brightest stars appear golden, while the northernmost looks orange. Arresting though it may be, Kemble 2 is a chance alignment of stars lying at different distances and moving in different directions through space.

Scanning 4° westward, we come to another wide double, **Psi (ψ) Draconis**. It is well split through my little scope at 17x. Both appear pale yellow, with the secondary north-northeast of the primary.

Psi Draconis makes a nice jumping-off point for this month's only extragalactic object, the dwarf spiral galaxy

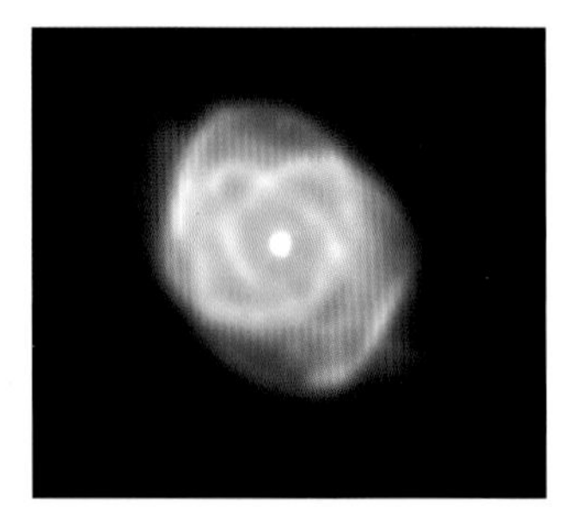

Below right: The Cat's Eye lives up to its name in this highly magnified close-up with a Meade 16-inch telescope. When viewed in a small instrument at low power, it resembles an out-of-focus star. The nebula's overall magnitude is about 8, while that of the central star is 11. Bottom: A spiral galaxy seen nearly edge-on, NGC 6503 looks more like a spindle than a spinning wheel. While it is bright enough to be spotted in small scopes, you won't see the fine structure visible in this image taken on Kitt Peak in Arizona. North is up. Photos: Adam Block / NOAO / AURA / NSF

First published July 2003

When viewed in a low-power telescope, Kemble 2 in Draco bears a marked similarity to the much larger W of Cassiopeia. The pattern is seen with bright Chi (χ) Draconis in this 2° field, which includes stars to magnitude 11.
Chart: Adapted from *Millennium Star Atlas* data

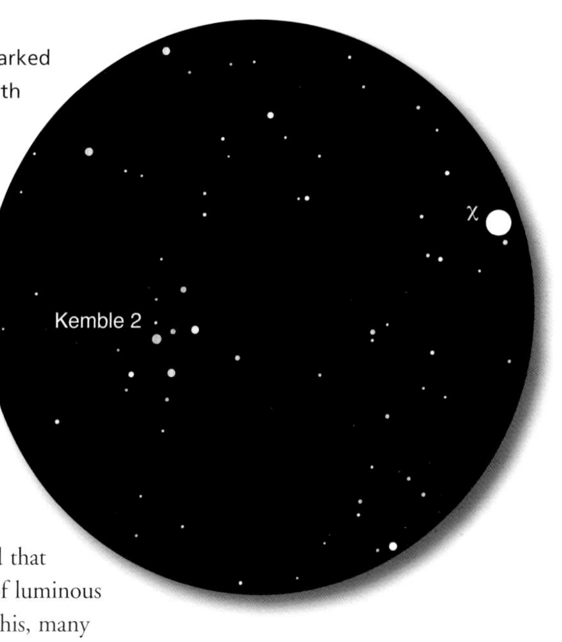

Kemble 2

χ

NGC 6503. To locate it, put Psi at the western edge of a low-power field and sweep 2.1° south. When I viewed it low in a slightly hazy sky, my 105mm scope at 68x showed a lens-shaped glow that brightened slightly toward the long axis (running west-northwest to east-southeast). The galaxy appeared about 3' long and one-quarter as wide. An 8.6-magnitude star lies just east of the southern end. This high-surface-brightness galaxy was discovered in 1854 by a German university student, Arthur von Auwers, with his 2.6-inch refractor. Auwers later became a professional astronomer, and a lunar crater bears his name.

Last, but not least, we'll visit **NGC 6543**, the Cat's Eye. This minute planetary nebula sits 3.6° south and a little east of NGC 6503. You can find it halfway between the 5th-magnitude stars Omega (ω) and 36 Draconis. At 17x, it appears almost stellar, but its telltale robin's-egg blue color gives it away. The nebula's central star can be distinguished even at this low power. A 10th-magnitude star just 2.7' to the west-northwest keeps it company. At 153x, the Cat's Eye looks slightly oval (elongated north-northeast to south-southwest) and perhaps a little darker in the very center. The outer edge seems to fuzz away into a thin and faint ring. The nebula takes magnification well if the air is steady.

In 1864, English amateur astronomer William Huggins examined NGC 6543 with a spectroscope. It was the first planetary nebula to be analyzed in this manner, and its spectrum proved that it is composed of luminous gases. (Prior to this, many astronomers had believed that with powerful enough instruments, all nebulae could be resolved into stars.) Huggins could not identify the bright spectral emission lines that give the nebula its blue-green color, so they were attributed to a hypothetical new element named nebulium.

Later studies showed that these were the "forbidden lines" of doubly ionized oxygen. This type of emission is virtually nonexistent on Earth, but it can take place within the rarefied gases of some nebulae. Thus, as Robert Burnham Jr. says in *Burnham's Celestial Handbook*, "The forbidden lines, then, are not truly 'forbidden' at all; they are merely, so to speak, frowned upon severely."

Stung by Beauty

The tail of the Scorpion can stun you in more ways than one.

Scorpius, the Scorpion, rides low in the sky for stargazers at midnorthern latitudes. Yet it beckons us with a wealth of deep-sky treasures that are hard to ignore, urging us to plumb these lower reaches of the sky. The area near the Scorpion's tail is an exceptionally fertile hunting ground. Indeed, you can hardly find a place to point your telescope without ensnaring an object or two.

Lambda (λ) and Upsilon (υ) Scorpii form the Scorpion's stinger and were once collectively referred to as Shaula, from the Arabic *al shaulah*, "the sting." Over time, the name migrated to Lambda, the brighter star of the pair, while Upsilon picked up the moniker Lesath, which comes from the Arabic word *las'a*, "stinging." But the name wasn't connected with this star in early Arabic astronomy texts.

The name Lesath followed a tortuous path to Upsilon, as is explained by star-name expert Paul Kunitzsch in *A Dictionary of Modern Star Names*. Ptolemy's *Almagest*, written in the 2nd century AD, mentions a "foggy conglomeration" following the sting of Scorpius. The Greek term for this was translated into Arabic as *al-latkha*, "the spot." This evolved into *alascha*, a Latin corruption used in astrological works relating it to the Scorpion's tail. In 1600, the great classical scholar Joseph Justus Scaliger wrongly assumed that *alascha* was derived from *las'a*, which then seemed a logical choice for the Scorpion's other sting star.

Ptolemy's foggy spot is now generally regarded as the earliest extant record of the open star cluster **M7**. At magnitude 3.3, it is clearly visible to my unaided eyes, despite being near my horizon. You could spend an entire evening exploring the treasures in and around M7.

Some Stunning Sights Near the Scorpion's Tail

Object	Type	Magnitude	Size/Sep.	Right Ascension	Declination	*MSA*	*U2*
M7	Open cluster	3.3	80'	17h 53.9m	–34° 47'	1437	164L
Tr 30	Open cluster	8.8	20'	17h 56.4m	–35° 19'	1437	164L
NGC 6444	Open cluster	—	12'	17h 49.5m	–34° 48'	1437	164L
NGC 6453	Globular cluster	10.2	3.5'	17h 50.9m	–34° 36'	1437	164L
NGC 6455	Starcloud	—	1°	17h 51.8m	–35° 11'	1437	164L
B287	Dark nebula	—	25' × 15'	17h 54.4m	–35° 12'	1437	164L
Cn 2-1	Planetary nebula	12.2	2"	17h 54.5m	–34° 22'	1437	164L
Hf 2-1	Planetary nebula	14.0	9"	17h 51.2m	–34° 55'	1437	164L
M 1-30	Planetary nebula	14.7	5"	17h 53.0m	–34° 38'	1437	164L
M6	Open cluster	4.2	30'	17h 40.3m	–32° 16'	1416	164L

Angular sizes are from recent catalogs; most objects appear somewhat smaller when a telescope is used visually. The columns headed *MSA* and *U2* give the chart numbers of objects in the *Millennium Star Atlas* and *Uranometria 2000.0*, 2nd edition, respectively.

M7 itself is a glorious prize in any instrument. My 15x45 image-stabilized binoculars show a central pattern of 10 stars in the shape of the Greek letter Chi (χ). Fifty additional bright and faint stars stretch the group out to 1¼°. Through a small scope, the inner ⅓° of the cluster is a knot of 30 stars that reminds me of the symbol # with every other extension cut off. Other observers have imagined a crushed box or the letter K. The group remains fairly rich to a diameter of 1°, while stragglers grow it to 1⅓° and almost 100 stars. Several stars are bright enough to show colors, mostly shades of blue-white. Three distinctly golden stars decorate the cluster. One is at the southwestern edge of the central knot, another is in the northwestern section of M7, and the third is on the cluster's north-northwestern border. M7's sparkling gems sometimes flash spectacularly in the poor seeing conditions I often encounter so low in the sky.

The small cluster **Trumpler 30** (Tr 30), also known as Collinder 355, kisses M7's southeast border. With 4- to 6-inch scopes, I see a 10' triangular group of 30 stars over haze. An 8th-magnitude star punctuates its northern point, and the rest of the stars shine at magnitudes 10 to 12. My 10-inch reflector draws out more stars (mostly southwest of the triangle) that swell the cluster to 20'. The brightest members seem to fill a five-pointed star, or perhaps they form a stick-figure man.

Open cluster **NGC 6444** lies just beyond M7's western edge and could easily be overlooked if you weren't searching for it. In a small scope, I see two dozen 11th- and 12th-magnitude stars loosely scattered across 12'. The brightest are gathered into an east-west bar spanning the group. In my 14.5-inch reflector, this is a very pretty cluster, rich in faint stars with many gathered into small bunches.

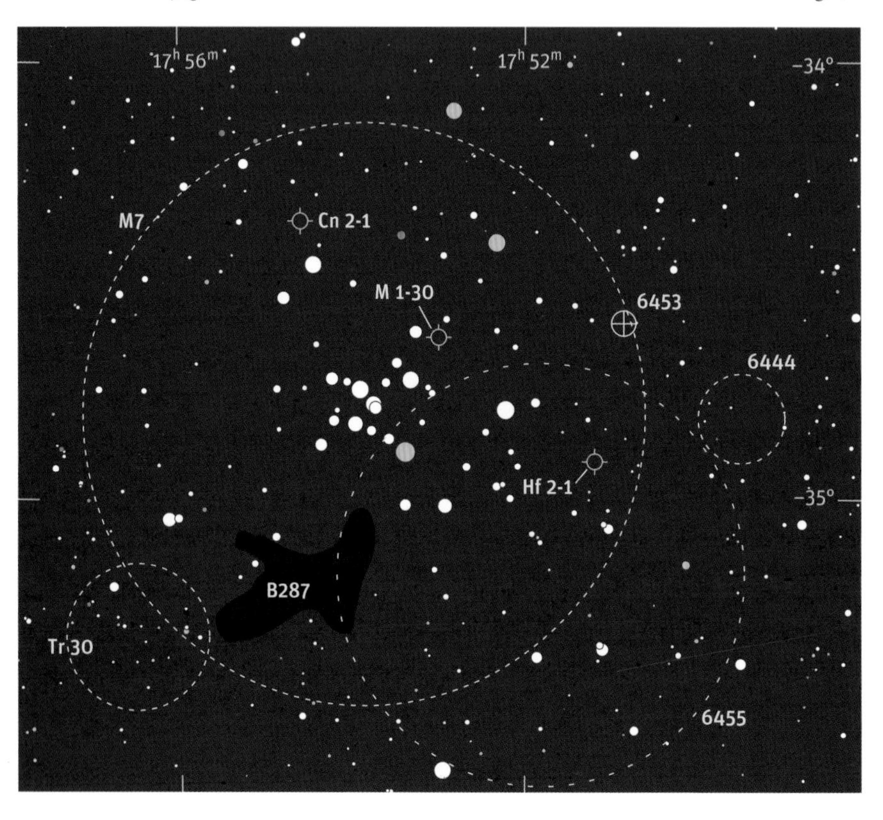

In and around the bright open star cluster M7 lie eight more objects that the author describes here. Hardest among them to spot are the three planetary nebulae, which require a fairly large amateur telescope, high magnification, and a very clear sky. This chart shows stars to magnitude 11. M7 is located between Scorpius and Sagittarius.

The globular star cluster **NGC 6453** seems attached to the west-northwestern fringe of M7, but it is actually 50 times more distant. It is round and looks like a faint little nebula through a small telescope. An 11th-magnitude star lies just to its west, and a 10th-magnitude star is somewhat farther away to its east. In my large reflector, NGC 6453 appears patchy. A few sparkles of light are visible, but considering the richness of the field, these may be foreground stars rather than cluster members.

NGC 6455 was discovered in the early 1830s by John Herschel, who was then observing from South Africa. He described it as "a very extensive nebulous clustering mass of the milky way. The stars [are] of excessive smallness, and infinite in number." That's an impressive description for an object listed as nonexistent in some modern sources. The first authors to ban NGC 6455 to the realm of fable were Jack W. Sulentic and William G. Tifft in *The Revised New General Catalogue of Nonstellar Astronomical Objects* (University of Arizona Press, 1976). Sulentic had inspected images from the *National Geographic Society–Palomar Observatory Sky Survey* and determined that there was no cluster in the designated position. Indeed, you won't see a cluster here but, instead, a large, bright cloud of Milky Way stars that is roughly 1° across and heavily overlaps M7. Its center is close to M7's southwest rim, and it extends into the near edge of M7's bright core. The enhancement is subtle at best, but dark nebulae threading their way through the area help define its boundaries. Your best chance for detection comes with a small, wide-field telescope or large binoculars. A southerly observing site is also a definite plus.

Several dark nebulae adorn this area of the sky, one within M7. **Barnard 287** (B287) wanders through the southeastern reaches of the cluster. I see it arranged in three dark arms totaling 20' across. A few stars are visible scattered across it. To spot the nebula, you must be able to see the backdrop of the Milky Way against which it lies.

Observers at lower latitudes (where M7 sits higher in the sky) may be able to pick out B287 with a small scope, while northerly observers may require a larger instrument.

Although they are actually background objects, three planetary nebulae appear to reside within M7. The brightest, **Cannon 2-1** (Cn 2-1; PN G356.2–04.4), has been spotted in scopes as small as 8-inch aperture. With my 14.5-inch reflector it is bright but tiny, even at 315x. It responds well to an oxygen III (green) filter. **Hoffleit 2-1** (Hf 2-1; PN G355.4–04.0) also has been seen with an 8-inch scope, but I find it considerably more difficult. My 14.5-inch at 315x with an oxygen III filter shows a small, faint, and roundish disk. I could glimpse it only occasionally without the filter. Despite repeated attempts, I have never nabbed the third planetary with any certainty. **Minkowski 1-30** (M 1-30; PN G355.9–04.2) has been detected by other observers with 10-inch or larger scopes.

Let's leave the realm of the very faint and jump 3.8° northwest of M7 to the delightful cluster **M6**. Under good sky conditions, 4.2-magnitude M6 is visible

Top: The core of the bright star cluster M7 occupies the center of this 1⅔°-wide field. The patchy dark nebula Barnard 287 lies below the field center. To the upper right is the globular cluster NGC 6453, partially resolved. This is a 30-minute exposure on Fujicolor SG400, taken with a 300mm f/6 telescope; north is up. Above: The Butterfly Cluster, M6, gets its name from the strings of stars that suggest the outlines of wings. The brightest orange star in this 1⅔°-wide view is the variable BM Scorpii. North is up. Photos: Akira Fujii

First published July 2004

to the unaided eye. My 15x45 binoculars show 20 stars in a 20' east-northeast to west-southwest oblong. My 6-inch reflector swells the star tally to about 50 stars in ½°, while my 10-inch scope reaches 90. The bright golden star gleaming in the eastern edge is the long-period, semi-regular variable BM Scorpii.

Through almost any telescope, many observers think M6 resembles a butterfly. This comparison may have first been made in *Splendour of the Heavens*, the 1923 two-volume compendium of popular astronomy by Theodore E. R. Phillips and William H. Steavenson. It says, "Somewhat irregular in shape, with a central rib of stars and resembles a butterfly with open wings." The group is now commonly known as the Butterfly Cluster. Oddly, not everyone outlines the butterfly in the same way. Most see a northwestward flying butterfly, but some arrange the stars into one that is winging its way northeast. How do you see it?

The region around the Scorpion's sting is a wonderful neighborhood for deep-sky hunting. If you take some time to scan the area with your telescope, I'm sure you'll uncover even more of the riches it holds.

Scuttling Around the Scorpion

Among the oldest of constellations, Scorpius is a wonderland for deep-sky observers.

There is a place above, where Scorpio bent,
In tail and arms surrounds a vast extent;
In a wide circuit of the heavens he shines,
And fills the space of two celestial signs.
— Ovid, *Metamorphoses*

In the above passage, the Roman poet Ovid describes Scorpius as portrayed by the ancient Greeks. This early version of the celestial Scorpion included the stars we now assign to Libra. The Greeks called these stars the Claws and considered them an asterism in its own right. Thus the ancient Scorpion took the place of two signs of the zodiac. Our modern Scorpion has been forced to pull his claws in toward his upper body, and it's here that we'll mount our July hunt for deep-sky wonders.

We'll begin at **Delta (δ) Scorpii**, or Dschubba, which suddenly became a variable star in June 2000, as the fast-spinning star developed a circumstellar disk. It brightened by nearly a magnitude, fluctuating irregularly and taking years to return to its original brightness. The star has a tight companion that approaches it every 10.7 years, and interactions between this companion and the disk may continue to make Dschubba an interesting star. Right before our eyes, Delta Scorpii has become an irregular eruptive vari-

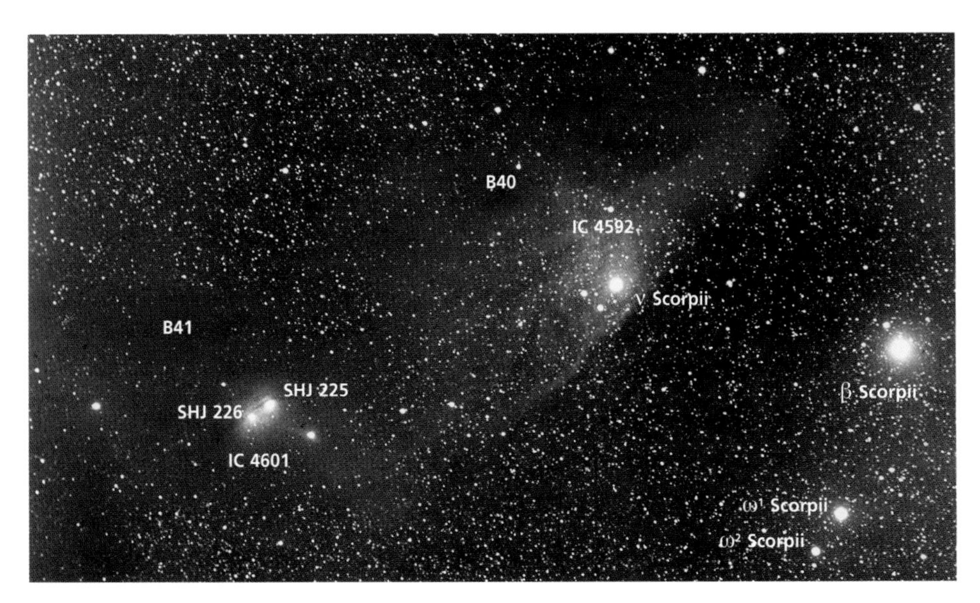

A century ago, famed astronomer Edward Emerson Barnard was at California's infant Mount Wilson Observatory shooting pictures that ultimately ended up in his *Photographic Atlas of Selected Regions of the Milky Way*. Because Barnard's photographs recorded only blue light, the apparent brightness of objects is somewhat distorted compared with what the human eye sees. The field is 5° wide with north up. Photo: Observatories of the Carnegie Institution of Washington

This 10°-wide photograph is centered on the compact globular cluster M80. Labels identify some of the objects featured in the accompanying photographs. Photo: Akira Fujii

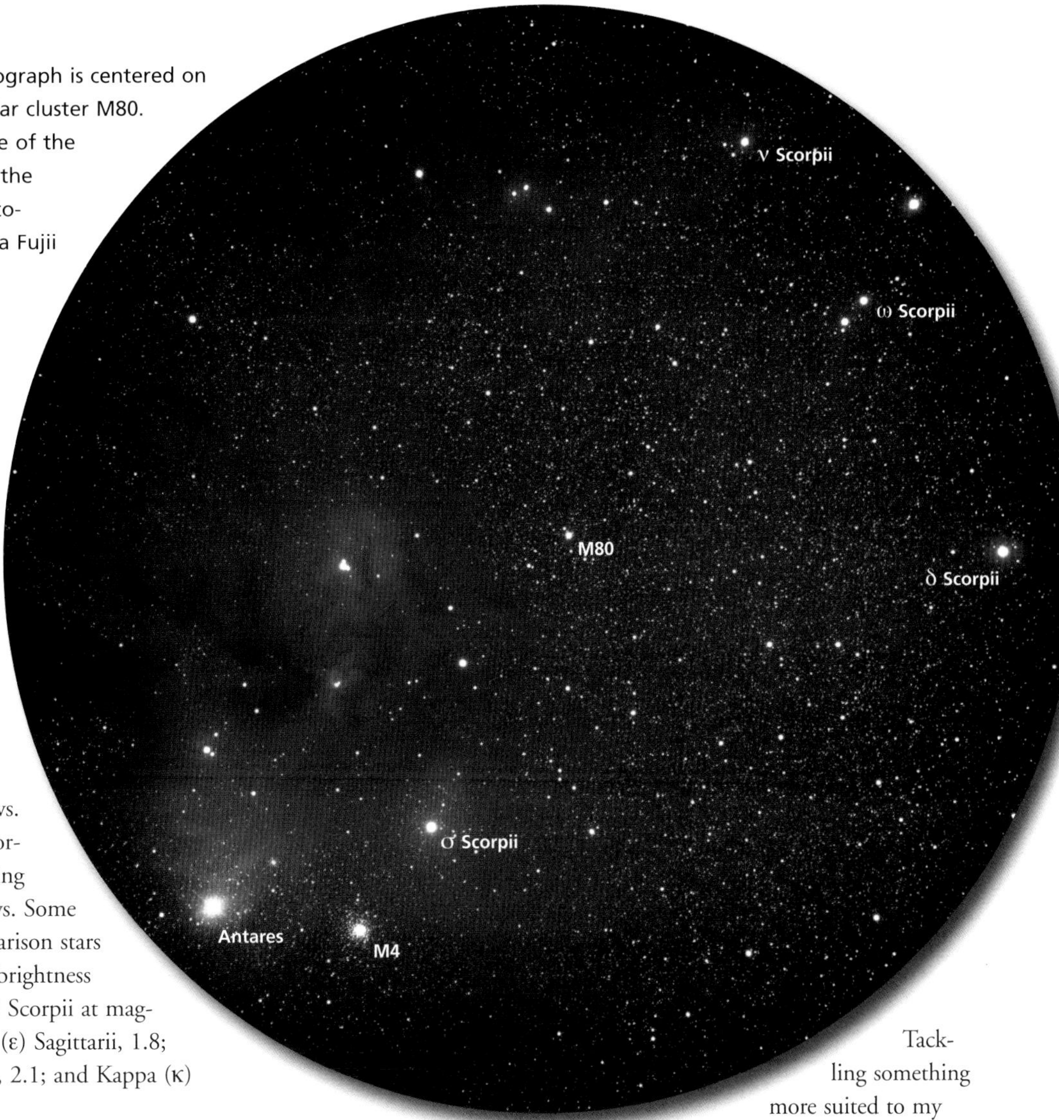

able. The prototype of its class is Gamma Cassiopeiae, which took 29 years to settle into its preoutburst magnitude. These variables are rapidly rotating blue-white stars that shed material from their equatorial regions. Temporary fading often follows such outflows. What will Delta Scorpii do in its upcoming years? No one knows. Some useful nearby comparison stars for judging Delta's brightness include Lambda (λ) Scorpii at magnitude 1.6; Epsilon (ε) Sagittarii, 1.8; Sigma (σ) Sagittarii, 2.1; and Kappa (κ) Scorpii, 2.4.

Moving 2.4° northeast of Delta Scorpii, we come to the pretty optical pair **Omega^1** (ω^1) and **Omega^2** (ω^2) **Scorpii**. Although they can be resolved with the unaided eye, these 4th-magnitude stars are best viewed in binoculars or a very low-power telescope to show off their contrasting blue and yellow colors.

Continuing northeast for 1.7° will take us to 4th-magnitude **Nu (ν) Scorpii**, a lovely double-double reminiscent of Epsilon Lyrae. My 105mm refractor at 17x shows the 4.2-magnitude primary with a 6.6-magnitude companion 41" to the north-northwest. At 87x, the fainter component splits and reveals a 7.2-magnitude companion 2.4" to its northeast. Both appear white. A much higher magnification of 203x picks out a 5.3-magnitude star kissing the northern edge of the primary, both bluish white.

While at the Winter Star Party in the Florida Keys in February 2005, I decided to have a look at some of the nebulae around Nu Scorpii. Nu is buried in the reflection nebula **IC 4592**, which is much too large to fit within the field of the 10-inch reflector I had with me. I look forward to viewing it with my little refractor sometime.

Tackling something more suited to my scope's abilities, I panned east of Nu to pick up a smaller reflection nebula and two dark nebulae. One of the latter, **Barnard 40** (B40), lies 1° east-northeast of Nu, where I spotted a tall, thick Z of darkness. Later, while perusing *A Photographic Atlas of Selected Regions of the Milky Way*, I saw that author Edward Emerson Barnard charted only the blackest, northernmost 15' of this figure as B40, which pares it down to a sideways V. Moving 1½° east-southeast, I looked for **Barnard 41** (B41). My scope showed a 40' patch that appeared to be merely one of the inkiest parts of a large complex of dark nebulae in the area. The faint reflection nebula **IC 4601** was visible off the southwest edge of B41. It covers about 10' x 15', northwest to southeast, surrounding two bright double stars, **SHJ 225** and **SHJ 226**, each blue and white. These observations were all made with an eyepiece giving 44x and a true field of 87'. I haven't had a chance to tackle these with a smaller scope yet. Have you?

Now we'll drop a few degrees southward to visit three strikingly different globular clusters. The first is **M80**, which you'll find 2.8° north-northwest of Sigma (σ) Scorpii and

The field surrounding the globular star cluster M4 (below center) is one of the most colorful in the Milky Way. It is awash with emission and reflection nebulosity. Among the latter seen here is the nebulosity around Sigma Scorpii (bright star at upper right), with the familiar blue color, and the unusual orange-tinted glow at upper left due to light from the brilliant star Antares, just outside the field of view, which is 1¾° wide with north up. The small globular cluster (left of center) is NGC 6144. Photo: Marco Lorenzi

just 4' southwest of an 8.5-magnitude star. Through my 105mm scope at 17x, it's visible as a very small but quite bright fuzzball that grows much more brilliant toward its center. At 87x, the 4' halo appears granular, but only a few individual stars are seen — the brightest one at the north-northeastern edge. At 166x in my 10-inch scope, M80 is very pretty, with many stars resolved in the halo and the outer core. The small inner core is intensely bright.

Renowned 18th-century observer William Herschel described M80 as "one of the richest and most compressed clusters of small stars I remember to have seen." He was also struck by the fact that it was perched on the western edge of a large starless void, stating that it "would almost authorise a suspicion that the stars, of which it is composed, were collected from that place, and had left the vacancy."

The outstanding globular cluster **M4** sits halfway between and below a line connecting

Searching the Scorpion

Object	Type	Magnitude	Size/Sep.	Right Ascension	Declination	*MSA*	*U2*
δ Scorpii	Variable star	1.6–2.3	—	16ʰ 00.3ᵐ	−22° 37'	1398	147L
ω Scorpii	Double star	3.9, 4.3	15'	16ʰ 06.8ᵐ	−20° 40'	1398	147L
ν Scorpii	Multiple star	4.2, 5.3, 6.6, 7.2	41", 2.4", 1.3"	16ʰ 12.0ᵐ	−19° 28'	1374	147L
IC 4592	Reflection nebula	—	3.3° × 1°	16ʰ 13.1ᵐ	−19° 24'	1374	147L
B40	Dark nebula	—	15'	16ʰ 14.6ᵐ	−18° 58'	1374	147L
B41	Dark nebula	—	40'	16ʰ 22.3ᵐ	−19° 38'	1373	147L
IC 4601	Reflection nebula	—	20' × 12'	16ʰ 20.2ᵐ	−20° 04'	1374	147L
SHJ 225	Double star	7.4, 8.1	47"	16ʰ 20.1ᵐ	−20° 03'	1374	147L
SHJ 226	Double star	7.6, 8.3	13"	16ʰ 20.5ᵐ	−20° 07'	1374	147L
M80	Globular cluster	7.3	10'	16ʰ 17.0ᵐ	−22° 59'	1398	147L
M4	Globular cluster	5.6	36'	16ʰ 23.6ᵐ	−26° 32'	1397	147L
NGC 6144	Globular cluster	9.0	7'	16ʰ 27.2ᵐ	−26° 01'	1397	147L

Angular sizes or separations are from recent catalogs. The columns headed *MSA* and *U2* give the chart numbers of objects in the *Millennium Star Atlas* and *Uranometria 2000.0*, 2nd edition, respectively.

First published July 2005

Sigma Scorpii and Antares. In my small refractor at 87x, M4 is a gorgeous blizzard of faint stars with partial resolution right to the cluster's heart. A bright and more condensed bar of stars runs north-south through the 7' core, and the halo is about 13' across. Several brighter stars of around 11th magnitude decorate the halo and the outer fringes of the core. M4's stars are loosely scattered for a globular cluster, and it almost looks like a very rich open cluster. In my 10-inch scope at 115x, the raggedy edges of M4 spread this cluster across 25' of sky, and the core shows stars over an unresolved haze. Many of M4's outliers are arranged in southward-drooping curves, like hair hanging down from the fuzzy "head" of the core.

Herschel felt that his suspicions were supported by the fact that M4 sits at the western border of another vacancy. He believed that gravity irresistibly draws stars into groups that grow richer and more compressed over time. He further supposed that this gradual precipitation of the "milky way" could give us a measure of its age and future duration.

Herschel also hypothesized that branches in a forming cluster could condense into separate clusters, and he pointed out a "miniature cluster" near M4. This is the globular **NGC 6144**, and it is best sought at a magnification that keeps the brilliant star Antares out of your field of view. My little refractor at high power shows only a 3' glow, granular around the edges. A lone 12th-magnitude star sparkles at its western edge. In my 10-inch scope at 220x, however, the cluster stretches across 5'. Some stars are resolved right down to the center, all resting on a bed of mist.

Astronomy has certainly progressed since Herschel's day. We now know that the inky patches in the Milky Way are not true vacancies but, rather, vast clouds of dark gas and dust blocking the light of the stars beyond. Open clusters form from nebulae and slowly dissipate, not condense, as they age; and most globular clusters acquired their concentrated form early in the life of our galaxy.

It's the nature of science to evolve and grow. Which of today's cherished assumptions will fall by the wayside?

The Galactic Dark Horse

During July, southern Ophiuchus is a fertile pasture for deep-sky observers.

I was introduced to the Galactic Dark Horse in the 1980s at the Texas Star Party. A remarkable complex of dark nebulae with a striking resemblance to a prancing ebony steed, the Galactic Dark Horse struts across 9° of far southern Ophiuchus, and it can be admired with the unaided eye in the relatively pristine skies of southwest Texas. Astronomy writer Richard Berry gave the feature its name after spotting it in a series of Milky Way mosaics he made with a 35mm camera in the 1970s.

The Galactic Dark Horse takes up two plates in Edward Emerson Barnard's 1927 *Photographic Atlas of Selected Regions of the Milky Way*. It includes no fewer than 18 of Barnard's dark nebulae, opaque clouds of dust and gas that show up in silhouette against the light of stars beyond. The horse's hindquarters are formed by the well-known Pipe Nebula, a prominent naked-eye sight in its own right. Many inky voids blot this outstanding region of the Milky Way. If you get a chance to observe under dark skies, it's a wonderful area to peruse with binoculars or the unaided eye.

Ophiuchus also harbors many telescopic treats, including a wealth of globular clusters. While open clusters are confined to the plane of our galaxy, globular clusters swarm in a nearly spherical distribution centered on the galactic core. Because we live in the galactic boondocks roughly 25,000 light-years from the center of the Milky Way, we see most globular clusters when we direct our gaze toward the hub of our great star city. Indeed, we find the majority in just three constellations: Sagittarius, Scorpius, and Ophiuchus.

The paddock of the Galactic Dark Horse contains three globular clusters from Charles Messier's 1781 catalog. The brightest is **M62**, whose southernmost stars spill over the Ophiuchus border into Scorpius. You can see the cluster as a small fuzzy spot in binoculars or a finder. Through my 105mm refractor at 153x, M62 displays a faint 5' halo with a few elusive points of light in its outer reaches. The much brighter core is off center to the southeast and very mottled. My 10-inch reflector at 219x shows some resolution of stars in the halo and outer core with the brightest ones favoring the western half of the cluster. At 245x in my 14.5-inch reflector, the cluster spans 10' and appears elongated northeast to southwest. I see partial resolution across the entire face with greatly increasing concentration toward the center. If I drop the magnification on my 10-inch scope to 44x, the dark nebula **Barnard 241** (B241) shares the field. This bit of turf kicked up by the Galactic Dark Horse lies off the western side of M62 and appears about ¼° long and one-third as wide.

M62 shares a finder field with **M19**, which hovers 4° to the north. This globular is very pretty in my 105mm scope at 87x. It shows a faint 8' halo enclosing a bright 3½' core.

Dark clouds of dust and gas weave myriad patterns against the distant glow of countless suns populating the center of our Milky Way Galaxy. During the 1970s, astronomy writer Richard Berry noted one such pattern on his wide-field photographs that resembled a rearing steed. Wearing Saturn as a necklace and filling the lower left quadrant of this 1987 image, Berry's Galactic Dark Horse is visible to the unaided eye in a clear, dark sky. Brilliant orange Antares is at right in this 24°-wide field. North is up. Photo: *Sky & Telescope* / Dennis di Cicco

While the core is distinctly oval north-south, the halo is tipped a little east of north. Quite a few stars are sprinkled across the cluster at 127x, and the core clasps a small, brilliant heart.

My little refractor at low power places two smaller and fainter globulars within the field, **NGC 6293** to the east and **NGC 6284** to the north-northeast. At 153x, NGC 6293 reveals a 2' halo and a relatively large, bright core. NGC 6284 is a bit smaller and exhibits slight mottling. Its elongated core is about one-third the diameter of the cluster and has a tiny, intense center. A few faint stars sparkle just beyond the borders of the halo, and a 12th-magnitude star sits off the eastern side.

Our third Messier globular is **M9**, located 2½° north of Xi (ξ) Ophiuchi. It's bright in my 105mm scope at 17x and shares the field with two more globulars and two dark nebulae. **NGC 6356** is obvious but smaller than M9. It has a large bright core, a dimmer halo, and a stellar center. **NGC 6342** is a very small, faint smudge that's a shade brighter in the center. The patchy dark area nuzzling NGC 6342 is **Barnard 259** (B259), the nose of the Galactic Dark Horse. West of M9, I see **Barnard 64** (B64), which Barnard aptly described:

"Cometary in form [with] a very black core or head that sharply abuts against the thick stratum of stars; from this it spreads out into a large dark area with much dark detail, filling quite a space close southwest of M9. It thus resembles a dark comet with a dense and well-defined head and diffused widening tail."

When I boost the power to 153x, M9 stands alone, and its 3' halo and large core appear slightly granular. A few pinpricks of light punctuate the halo, the most conspicuous one on the eastern edge.

The challenging globular cluster **Haute-Provence 1** was first reported in 1954 in *Publications de l'Observatoire de Haute-Provence*. It lies 49' east and 7' south of 45 Ophiuchi. A long line of 11th- and 12th-magnitude stars trending east-southeastward from 45 Ophiuchi points to a short line of three evenly spaced stars. The easternmost of these three is 9th magnitude and the other two are 11th magnitude. Haute-Provence 1 sits 5' north of the westernmost star. With my 10-inch scope at 118x, the globular is a very elusive, barely perceptible haze. Its northern side is cradled by a 1½' curve of three evenly spaced stars, 12th magnitude or so.

Globular clusters and dark nebulae aren't the only deep-

Trotting Through Southern Ophiuchus

Object	Type	Magnitude	Size/Sep.	Right Ascension	Declination	*MSA*	*U2*
M62	Globular cluster	6.5	15'	$17^h 01.2^m$	–30° 07'	1418	164R
B241	Dark nebula	—	18' × 6'	$16^h 59.5^m$	–30° 12'	1418	164R
M19	Globular cluster	6.8	17'	$17^h 02.6^m$	–26° 16'	1395	146R
NGC 6293	Globular cluster	8.2	8.2'	$17^h 10.2^m$	–26° 35'	1395	146R
NGC 6284	Globular cluster	8.8	6.2'	$17^h 04.5^m$	–24° 46'	1395	146R
M9	Globular cluster	7.7	12'	$17^h 19.2^m$	–18° 31'	1370	146R
NGC 6356	Globular cluster	8.3	10'	$17^h 23.6^m$	–17° 49'	1370	146L
NGC 6342	Globular cluster	9.7	4.4'	$17^h 21.2^m$	–19° 35'	1370	146L
B259	Dark nebula	—	30'	$17^h 22.0^m$	–19° 18'	1370	146L
B64	Dark nebula	—	25'	$17^h 17.3^m$	–18° 31'	1371	146R
Haute-Provence 1	Globular cluster	11.6	1.2'	$17^h 31.1^m$	–29° 59'	1416	164L
Trumpler 26	Open cluster	9.5	7.0'	$17^h 28.5^m$	–29° 30'	1416	164L
NGC 6369	Planetary nebula	11.4	30"	$17^h 29.3^m$	–23° 46'	1394	146L
IC 4634	Planetary nebula	10.9	11" × 9"	$17^h 01.6^m$	–21° 50'	1395	146R
ο Ophiuchi	Double star	5.2, 6.6	10"	$17^h 18.0^m$	–24° 17'	1395	146R
36 Ophiuchi	Double star	5.1, 5.1	4.9"	$17^h 15.4^m$	–26° 36'	1395	146R

Angular sizes or separations are from recent catalogs. The visual impression of an object's size is often smaller than the cataloged value and varies according to the aperture and magnification of the viewing instrument. The columns headed *MSA* and *U2* give the chart numbers of objects in the *Millennium Star Atlas* and *Uranometria 2000.0*, 2nd edition, respectively.

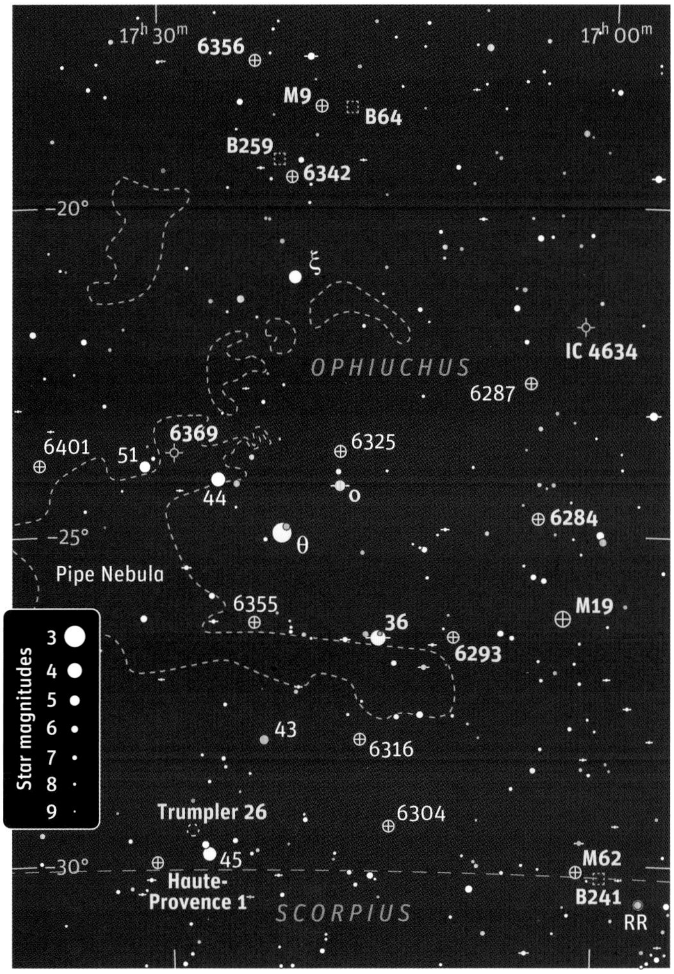

sky denizens of southern Ophiuchus. The open cluster **Trumpler 26** rests ½° north-northeast of 45 Ophiuchi. It's a 7' group of stars, 10th magnitude and fainter, with the brightest forming a Y shape. I count 18 stars in my 105mm scope and 35 in my 10-inch.

The planetary nebula **NGC 6369**, known as the Little Ghost, dwells ½° west-northwest of 51 Ophiuchi. In July 1972, *Sky & Telescope* columnist Walter Scott Houston wrote that he was surprised to find the nebula easily with 7x50 binoculars. Although it appeared stellar, Houston found its telltale greenish tinge obvious. In 6-inch or larger scopes at high power, the nebula appears annular and brighter along its northern rim. The view is improved with an oxygen III filter.

An even smaller planetary lies 4½° west of Xi Ophiuchi. **IC 4634** is tiny but bright in my little refractor at 87x. Almost everything but the planetary disappears from the field when I use an oxygen III filter. At 127x, the planetary is fuzzy around the rim and has a brighter center. My 10-inch reflector at 219x occasionally shows the glint of a faint central star.

The area also hosts some colorful double stars. As seen with my 105mm scope at 68x, two of the finest are **Omicron** (ο) **Ophiuchi**, a golden primary with a pale yellow companion, and **36 Ophiuchi**, a nicely matched pair of golden suns.

First published July 2006

Heavenly Hero

Hercules takes his place of honor among the stars.

So when Alcides mortal mold resign'd,
His better part enlarg'd, and grew refin'd;
August his visage shone; almighty Jove
In his swift carr his honour'd offspring drove;
High o'er the hollow clouds the coursers fly,
And lodge the hero in the starry sky.
— Ovid, *Metamorphoses*

Here, the Latin poet Ovid chronicles the celestial ascent of the mythological hero Hercules (Alcides), son of the god Jupiter (Jove) and a mortal woman. When Hercules' earthly life ended, his godly half was given a place of honor among the stars. As great in death as he was in life, Hercules lays claim to the fifth-largest constellation. His starry head, Rasalgethi, is near the head of Ophiuchus, Rasalhague, as though our hero is offering help with the monstrous serpent that Ophiuchus is wrestling.

Rasalgethi, Alpha (α) Herculis, is a red-giant star 400 light-years away and large enough, if it replaced our Sun, to extend beyond the orbit of Mars. It's also a variable star ranging between magnitude 2.7 and 4.0 during a six-year period that underlies a complex cycle of smaller oscillations over shorter intervals. Through my 105mm refractor at a magnification of 127x, deep golden Rasalgethi closely nuzzles a 5th-magnitude companion sitting east by south. The fainter star is white, but comparison with its yellow-orange partner often seems to lend it a blue or green tint. With components too tight to be split with a telescope, the companion consists of a yellow giant in close proximity to a white star perhaps twice as massive as our Sun.

At low power, Rasalgethi shares the field with **Dolidze-Dzimselejsvili 7** (DoDz 7), a little knot of faint stars 1.3° northwest. Boosting the magnification to 87x, I am able to see a loose collection of several stars in a shape that Finnish observer Jere Kahanpää likens to a sailboat. Six 10th- to 12th-magnitude stars form a curved hull with its open side facing west-southwest. A sail-less mast juts out in that direction, topped by a 10th-magnitude star. Data from proper-motion surveys indicate that these stars are moving in different directions through space and do not form a true, gravitationally bound cluster.

In a dark sky, a naked-eye splash of stars can be seen a few degrees west of Rasalgethi. Astronomy writer Tom Lorenzin calls this asterism **Sudor Ophiuchi** (Sweat of Ophiuchus) and writes, "Hey! If you were wrestling with a giant snake, you wouldn't care where your sweat splashed, either!" In my 8x50 finder, I see eight stars in the shape of an integral sign (∫) about 2½° long with a boxy extension at its northwestern end. Through my little refractor at 17x, the bright star near the center of the integral sign is prominently orange. South of it, the distinctive double star **Σ I 33** displays a nearly matched pair of white and gold suns, which are easily split in binoculars.

Harrington 7, a large asterism 2° west-southwest of Omega (ω) Herculis, is a wonderful group for dot-to-dot games. My 105mm scope at 28x shows a score of 8th- to 10th-magnitude stars in a 1.3° zigzag leaning north-northwest. It's 14' wide in the south and tapers to a point in the north. Writer Philip S. Harrington calls it the Zigzag Cluster.

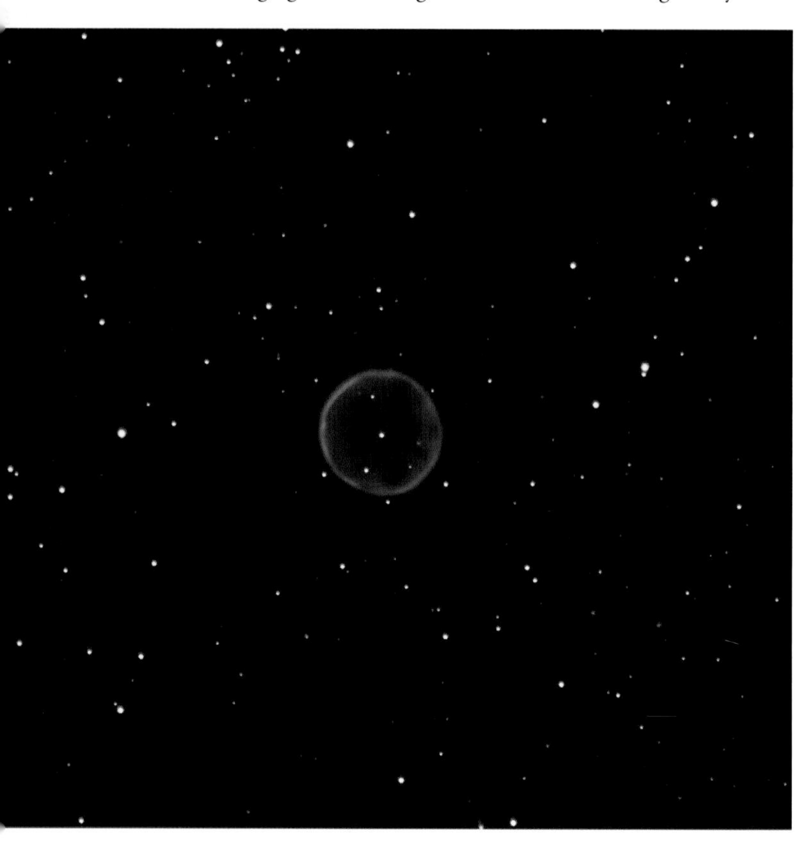

Hercules is home to Abell 39, a nearly textbook-perfect planetary nebula created when the outer atmosphere of the star at the center was ejected as an expanding shell of material thousands of years ago. Estimated to be 7,000 light-years from Earth, the planetary is about 5 light-years in diameter. Even under good sky conditions, Abell 39 can be a challenge to view with a 10-inch telescope. North is to the upper right. Photo: Don Goldman

An asterism that's more apparent to the eye than would be inferred from photographs, the Keystone of Hercules is a jumping-off point for the objects in the southern part of the constellation covered in this section. Of special note is the asterism shaped like an integral sign (∫) west (right) of Rasalgethi, Alpha (α) Herculis. Photo: Akira Fujii

It reminds me of the Chinese dragons carried in parades. The brightest star (third from the tip of the dragon's tail) is gold, so I call this the Golden Dragon. However, the view through my 10-inch reflector at 44x gives me an entirely different impression. A large spray of stars at the southern end seems to outline a flower, perhaps an iris, whose stem winds northward. Another twist of the imagination turns the asterism into a lit fuse with a sparking end.

A white 7th-magnitude star lies 1¼° west of the dragon's head. Drawing a line from it to a golden 8th-magnitude star 13' south-southwest and continuing three times that distance takes you to the double star **Σ2016**. The white 8.5-magnitude primary has a yellow 9.6-magnitude companion 7.4" south-southeast. Planetary nebula **IC 4593**, sometimes called the White-eyed Pea, dwells 11' north-northwest of this pair. The nebula's 10.7-magnitude central star makes it easy to locate. My little refractor at 28x reveals a very small, gray-green glow around the star. The nebula shows up better at 47x with an oxygen III filter. Large scopes render a more striking hue, variously reported as green, blue-green, or blue. At high power, the nebula is oval and seems to grow with averted vision. Look for a brighter patch in the northwest.

Let's move north and pay a visit to the attractive double star **Kappa (κ) Herculis**, whose bright components are easily split at low power. They appear a lovely deep yellow and gold to me, which agrees fairly well with their cataloged spectral classes of G8 and K1.

Hercules plays host to many galaxies, none dazzling. **NGC 6181**, near Beta (β) Herculis, is one of the brightest. Drop 1° south from Beta to a yellow 5th-magnitude star, and then slide 47' south-southeast to the galaxy. Just a few arcminutes east of an 11th-magnitude star, NGC 6181 is faintly visible in my little refractor at 47x. Boosting the magnification to 87x, I see a small north-south oval with a broadly brighter core.

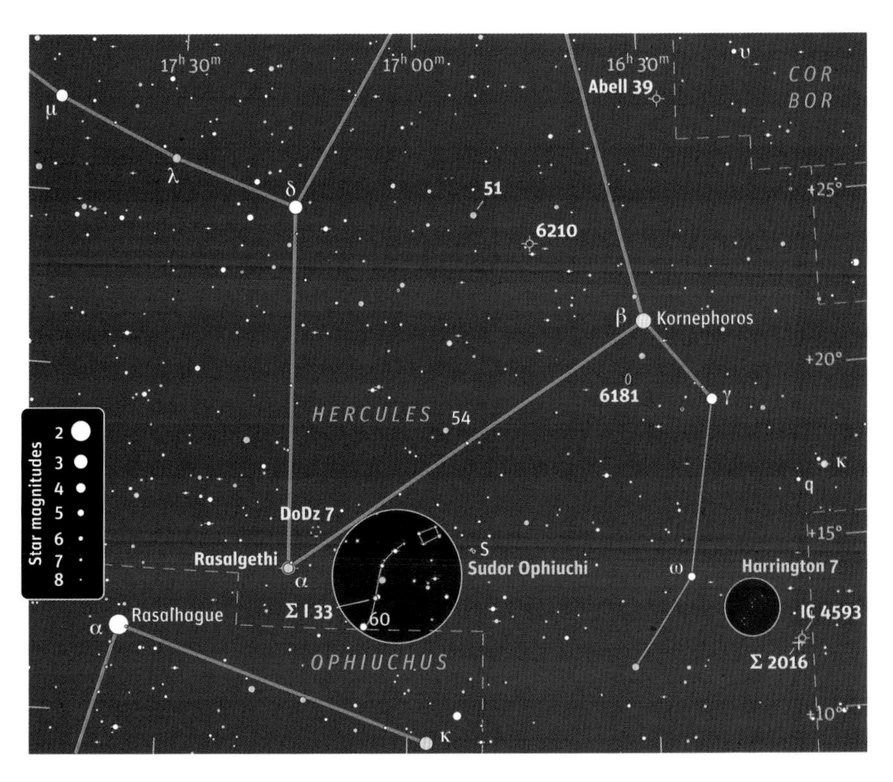

A pair of 7th-magnitude stars spaced 17' apart lies two-thirds of the way from Beta to 51 Herculis. The bright planetary nebula **NGC 6210** sits 9' west-northwest of the duo's northeastern star. I'm able to spot the nebula even at 17x through my 105mm refractor. At 87x, this pretty, blue-gray nebula almost manages to overpower its 12th-magnitude central star. A faint halo rims the planetary. NGC 6210 looks greenish blue through my 10-inch reflector, and at high power, it's elongated east-west. Deep images of NGC 6210 show unusual projections that

First published July 2007

Harbored in the Hero

Object	Type	Magnitude	Size/Sep.	Right Ascension	Declination	*MSA*	*U2*
Rasalgethi	Double star	3.5, 5.4	4.6"	17h 14.6m	+14° 23'	1251	87L
DoDz 7	Asterism	—	10'	17h 11.4m	+15° 29'	1251	87L
Sudor Ophiuchi	Asterism	—	~3½°	17h 01.1m	+14° 13'	1251	87L
Σ I 33	Double star	5.9, 6.2	305"	17h 03.7m	+13° 36'	1251	87L
Harrington 7	Asterism	—	100' × 15'	16h 18.1m	+13° 03'	1254	87R
Σ2016	Double star	8.5, 9.6	7.4"	16h 12.1m	+11° 55'	1254	87R
IC 4593	Planetary nebula	10.7	13" × 10"	16h 11.7m	+12° 04'	1254	87R
κ Herculis	Double star	5.1, 6.2	27"	16h 08.1m	+17° 03'	1230	87R
NGC 6181	Spiral galaxy	11.9	2.5' × 1.1'	16h 32.4m	+19° 50'	1229	69L
NGC 6210	Planetary nebula	8.8	20" × 13"	16h 44.5m	+23° 48'	1204	69L
Abell 39	Planetary nebula	13.0	170"	16h 27.6m	+27° 55'	1181	69L

Angular sizes are from recent catalogs. The visual impression of an object's size is often smaller than the cataloged value and varies according to the aperture and magnification of the viewing instrument. The columns headed *MSA* and *U2* give the appropriate chart numbers in the *Millennium Star Atlas* and *Uranometria 2000.0*, 2nd edition, respectively. All the objects this month are in the area of sky covered by Charts 54 and 55 in *Sky & Telescope's Pocket Sky Atlas*.

earn this planetary its fitting nickname, the Turtle Nebula.

Our final target is another planetary nebula, a very challenging one: **Abell 39** (PK 47+42.1). The nearest star with a Bayer designation is Upsilon (υ) Coronae Borealis. From Upsilon, hop 1.7° east-southeast to a golden 7.5-magnitude star, the brightest in the area, and then drop 39' south-southeast to an 8.6-magnitude star. Next, sweep 26' due east to a 9.8-magnitude star at the northern corner of a 15' trapezoid formed with three slightly brighter stars. A line drawn from the trapezoid's western star through its northern one and continued for the same distance again will take you right to Abell 39.

When I first visited Abell 39 with my 10-inch reflector, I could spot it only using averted vision and an oxygen III filter. Since familiarizing myself with it, I've been able to view the nebula while gazing directly at it. Abell 39 is round, vaguely annular, and moderately large for a planetary — almost 3' across. A magnification of about 70x gives me the best view. Other observers have managed to nab this elusive planetary with 8-inch scopes.

The 12 Labors of Hercules

The celestial strongman travels high across the sky on evenings in July.

With his broad heart to win his way to heaven;
Twelve labours shall he work.
— Theocritus, "Idyll XXIV"

Hercules is probably the best-known mythological character that dwells among the constellations. The 12 Herculean tasks he performed over a period of 12 years have been immortalized in everything from classical poetry to Garrison Keillor's *A Prairie Home Companion*. The hero has won his place among the stars in the sky and the stars of the silver screen. In recognition of Hercules' labors, let's set ourselves the less arduous undertaking of visiting 12 deep-sky wonders adorning his celestial abode.

The star of our 12-step program is the globular cluster **Messier 92** (M92). It lies about two-fifths of the way from Iota (ι) to Eta (η) Herculis, and it's visible in my 12x36 binoculars as a small hazy glow with a brighter center. Although often passed by in favor of its more prominent neighbor M13, M92 is a wonderful cluster in its own right. My 105mm refractor at 127x displays an 8' x 7' halo of loosely strewn stars. M92's bright 2½'-wide core glitters with stars nearly down to the brilliant blaze of its center. In my 10-inch reflector at 115x, the cluster's entire 14' face is heavily freckled with stars. The ones in the core are enmeshed in a gleaming haze of unresolved suns. Many cluster stars seem to form short lines arrayed in haphazard directions.

The dimmer globular cluster **NGC 6229** sits 4.8° east-northeast of Tau (τ) Herculis, where it makes a nice little triangle with two yellow-white 8th-magnitude stars. At 28x through my small refractor, NGC 6229 is a rather bright

Messier 92 looks spectacular through any telescope with at least 8 inches of aperture. But no matter how big your scope, you're unlikely to see as many stars through the eyepiece as are shown in this 40-minute CCD exposure through a 20-inch telescope. Photo: Doug Matthews / Adam Block / NOAO / AURA / NSF

spot with a fainter halo. At 68x, it spans 2' and sports a nearly stellar nucleus. Even with my 10-inch scope at 213x, I can spot only five stars in the 4' halo. The core appears half as large and is mottled with brighter patches. The lack of resolution can be attributed to this globular's exceptional distance of 99,000 light-years. Star-flecked M92, on the other hand, is only 27,000 light-years away.

Now drop down to the close double star **Rho (ρ) Herculis**, which sits off the northeastern corner of Hercules' Keystone asterism. This attractive white pair has a 4.5-magnitude primary with a 5.4-magnitude secondary 4.1" to its northwest. On a night when the atmosphere was steady, I was able to split the duo at 28x through my 105mm refractor. Don't be surprised if this requires a lot more magnification when the seeing isn't as good. I needed 105x (with an 8-inch reflector) to separate the stars on a lesser night. Rho is about 400 light-years distant, and its components take at least 4,600 years to complete an orbit.

Two wider and more colorful doubles reside farther south. As seen through my little refractor, **Delta (δ) Herculis** bears a 3.1-magnitude white primary with an 8.3-magnitude yellow companion 12" west-northwest. This is merely an optical pair — a chance alignment of two stars that are not physically related. Sweeping 4½° west of Delta will take you to the triangle of 5th- and 6th-magnitude stars formed by 51, 56, and 57 Herculis. The northernmost star, **56 Herculis**, is a fine double with a yellow primary and a faint companion 18" east. The pair is easily split in my small scope, but it takes the 10-inch reflector to reveal the yellow-white hue of the secondary. Although these stars share a common motion across the sky, they may be too far apart to form a true pair.

On the opposite side of Delta, we find the asterism **Dolidze-Dzimselejsvili 8** (DoDz 8). From Delta hop 1.4° east-southeast to 5th-magnitude 70 Herculis and then make an equal hop in the same direction to reach DoDz 8. The group is sparse but has an interesting symmetry when viewed through my little refractor at 47x. Four 8th- and 9th-magnitude stars in two pairs make a north-south zigzag with a dim star that lies about halfway between them. Four 10th- to 12th-magnitude stars, one to the east and three to the west, help give this 15' asterism a radial starburst pattern.

Next, we'll visit the intriguing multiple star **Mu (μ) Herculis**. My 105mm scope at 17x shows the bright, yellow primary with a much fainter, orange companion a generous 35" west-southwest. The primary is an elderly version of our Sun. The hydrogen supply in its core has run out, and it's now an expanding subgiant.

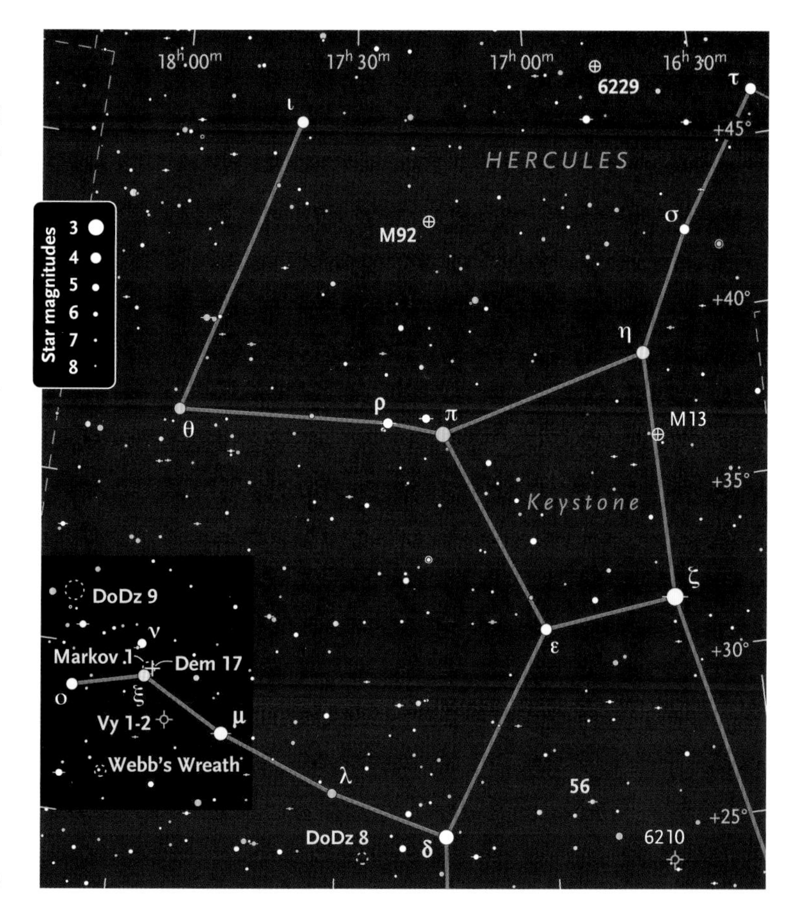

12 Stargazing Prizes in Hercules

Object	Type	Magnitude	Size/Sep.	Right Ascension	Declination
M92	Globular cluster	6.4	14'	$17^h\ 17.1^m$	+43° 08'
NGC 6229	Globular cluster	9.4	4.5'	$16^h\ 47.0^m$	+47° 32'
ρ Herculis	Double star	4.5, 5.4	4.1"	$17^h\ 23.7^m$	+37° 09'
δ Herculis	Double star	3.1, 8.3	12"	$17^h\ 15.0^m$	+24° 50'
56 Herculis	Double star	6.1, 10.8	18"	$16^h\ 55.0^m$	+25° 44'
DoDz 8	Asterism	6.8	15'	$17^h\ 26.4^m$	+24° 12'
μ Herculis	Triple star	3.4, 10.2, 10.7	35", 1.0"	$17^h\ 46.5^m$	+27° 43'
Vy 1-2	Planetary nebula	11.4	5"	$17^h\ 54.4^m$	+28° 00'
Markov 1	Asterism	6.8	15'	$17^h\ 57.2^m$	+29° 29'
Dem 17	Double star	9.9, 10.3	24"	$17^h\ 56.7^m$	+29° 29'
DoDz 9	Asterism	—	34'	$18^h\ 08.8^m$	+31° 32'
Webb's Wreath	Asterism	—	11' × 7'	$18^h\ 02.3^m$	+26° 18'

Angular sizes and separations are from recent catalogs. Visually, an object's size is often smaller than the cataloged value and varies according to the aperture and magnification of the viewing instrument.

Double seeing

Splitting tight double stars into separate components depends largely on the seeing, or steadiness of the air. This is completely distinct from transparency, which measures how clear the air is. In fact, hazy summer nights often offer the best seeing, allowing excellent high-power telescopic viewing.

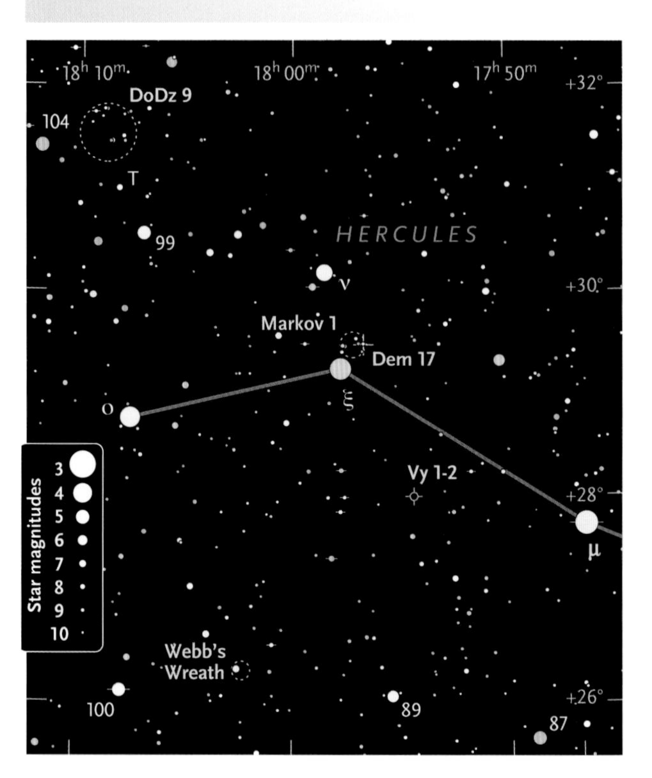

The secondary actually consists of a very close pair of red-dwarf stars, magnitudes 10.2 and 10.7. I was barely able to split this ruddy couple with my 14.5-inch reflector at 245x when they were only 0.6" apart. Late in 2011, these dim stars are 1.0" apart, a test for a 6-inch scope when the seeing is very good. Their separation closes to 0.5" in 2015, and then widens to a maximum of 1.6" in 2030. As you might suspect from the apparent brightness of its red-dwarf pair, the Mu Herculis system is a nearby neighbor — only 27 light-years away from us.

Sitting just 1.7° east and 17' north of Mu is the tiny planetary nebula **Vyssotsky 1-2** (Vy 1-2; PN G55.3+24.0 or PK 53+24.1). It's easily visible as an 11th-magnitude "star" through my little refractor at 87x. Vy 1-2 sits at the center of a 5' bow-shaped arc formed with a slightly dimmer star northwest and a 10th-magnitude star south. A close pair of 12th-magnitude stars lies 3' west-southwest. The planetary is a minuscule disk in my 10-inch scope at 213x. Both narrowband and oxygen III nebula filters make the planetary stand out better, the latter more so.

Now let's move on to the delightful asterism **Markov 1**, located 16' north-northwest of yellow Xi (ξ) Herculis and visible in the same low-power field. When Canadian amateur Paul Markov noticed this group in July 2000, he thought it looked similar to the much larger Teapot asterism formed by the bright stars in southwestern Sagittarius. With my 105mm refractor at 47x, I see nine 9th- and 10th-magnitude stars spanning 17'. Three of them form a skinny isosceles triangle pointing south-southeast, and the rest make a sideways capital T with a slanted top. Several fainter stars are scattered across the group. The double star **Dembowski 17** (Dem 17) resides at the center of the T's long bar. Its 9.9-magnitude primary holds a

First published July 2008

It's a lot harder than it looks to take a good picture of a simple asterism. This photograph of Markov 1 is a collaboration between two Canadian astrophotographers. Paul Mortfield manned the CCD camera, and Stef Cancelli did the image processing.

10.3-magnitude companion 24" southeast.

Another DoDz asterism is found nearby. Climb 2.8° north of Omicron (ο) Herculis to spot **Dolidze-Dzimselejsvili 9** (DoDz 9). At 28x through my little refractor, the group is enshrined off center within a colorful triangle formed by yellow-white 99 Herculis, orange 104 Herculis, and a golden 5.7-magnitude star (HIP 88636) to its northwest. This pretty asterism displays 30 stars, magnitude 8 and fainter, loosely dispersed within 32' and devoid of bright stars near the center. In my 10-inch scope, some of the brighter stars are tinted orange.

Moving back to Omicron and dropping 2.7° south-southwest takes you to a golden 7th-magnitude star. It ornaments the eastern side of **Webb's Wreath**, a little-known asterism first mentioned in the 4th edition (1881) of Thomas William Webb's observing guide *Celestial Objects for Common Telescopes*. My 105mm scope at 68x reveals 13 additional stars, magnitudes 11 and 12, outlining an 11' x 7' oval that leans northeast and is dented inward at the bright star.

This brings us to the end of our 12 star-filled labors in Hercules. I hope you find the quest a labor of love.

Magical Things

Planetary nebulae sprinkled along the Milky Way beckon observers outside on warm summer evenings.

The universe is full of magical things patiently waiting for our wits to grow sharper.
— Eden Phillpotts, *A Shadow Passes*, 1918

Planetary nebulae are fascinating deep-sky beauties. We've learned that aged stars, initially one to eight times the mass of our Sun, form planetary nebulae by shedding material at different speeds and times. But the specific processes leading to their bewildering variety and intricate structures are still a matter of debate. Only 2,500 or so planetary nebulae are known in our galaxy, yet large numbers of them have been identified in galaxies as far-flung as the Virgo Cluster. These extragalactic planetaries aid in determining distances to their host galaxies, while the distances to most of the planetaries in our own backyard remain poorly known.

Let's drop in on some less-visited representatives of these wondrous objects draped along the summer Milky Way, as well as their chance comrades in the sky.

Starting in southern Sagittarius, we'll star-hop our way to NGC 6563 and pause at an interesting star along the way. From Epsilon (ε) Sagittarii, sweep 1.4° west-northwest to a 6th-magnitude star, which is both the brightest in the area and the primary of the quadruple **h5036**. Its three companions are visible at low power and range from magnitude 8.7 to 10.2. Look for two of the attendants a spacious 1.7' northeast of the primary and the third one less than half as far to the east. Making things even more interesting, the primary star is the eclipsing binary **RS Sagittarii**, with a period of 2.4 days. Normally magnitude 6.0, the binary dims once by about 0.3 magnitude and, a half cycle later, by 0.9 magnitude as its stars alternately pass in front of each other as seen from Earth.

Returning to our star-hop, make a second leap of the same length and direction to reach another 6th-magnitude star. **NGC 6563** lies 15' east-southeast of this star, making a trapezoid with it and two 7th-magnitude stars farther south. Through my 105mm refractor at 87x, this planetary nebula is fairly small, faint, and roundish. There is a minor improvement in contrast with a narrowband nebula filter, but the nebula stands out much better with the use of an oxygen III filter. Through my 10-inch reflector at 115x, NGC 6563 is round and ¾' across with a uniform surface brightness. It's enshrined in a little triangle of three 12th- and 13th-magnitude stars, the closest one north and the others northwest and southeast. Boosting the magnification to 213x and adding an oxygen III filter reveals that the nebula is oval and tipped about 60° east of north. There's a slightly brighter arc along the northern edge and a more subtle brightening on the opposite rim.

Farther north in Sagittarius, the planetary **NGC 6445** sits 22' north-northeast of the globular cluster **NGC 6440**. Even at 17x in my little refractor, NGC 6440 is an easily visible, little hazy patch 1.8° northeast of the pale yellow star 58 Ophiuchi. At 47x, the planetary nebula joins the scene as a small, fairly faint spot with the wide double star **h2810** about 5' to its east. The double has a 7.6-magnitude primary with a 10.4-magnitude companion a bit west of south. At 87x, NGC 6445 is about 40" long and roughly oval, and it leans north-northwest. The north-northwestern edge appears brighter and has a faint star a short distance beyond. The globular cluster still shares the field of view. It spans 2.3' and grows much brighter toward the center. NGC 6440 adorns the middle of an 11½'-long line of 11th- and 12th-magnitude stars, two to the cluster's north-northwest and two off the opposite side.

Arizona amateur Frank Kraljic says that through his 10-inch reflector at 112x, NGC 6445 resembles a slightly top-

A Midyear Hunt for Magical Things

Object	Type	Magnitude	Size/Sep.	Right Ascension	Declination
NGC 6563	Planetary nebula	11.0	59" × 43"	18h 12.0m	–33° 52'
h5036	Multiple star	6.0, (8.7, 10.2), 9.5	(1.7'), 40"	18h 17.6m	–34° 06'
RS Sagittarii	Eclipsing binary	6.0–6.9	2.4 days	18h 17.6m	–34° 06'
NGC 6445	Planetary nebula	11.2	45" × 36"	17h 49.2m	–20° 01'
NGC 6440	Globular cluster	9.2	4.4'	17h 48.9m	–20° 22'
h2810	Double star	7.6, 10.4	44"	17h 49.6m	–20° 00'
NGC 6765	Planetary nebula	12.9	40"	19h 11.1m	+30° 33'
M56	Globular cluster	8.3	8.8'	19h 16.6m	+30° 11'
NGC 7027	Planetary nebula	8.5	18" × 11"	21h 07.0m	+42° 14'
NGC 7044	Open cluster	12.0	7'	21h 13.2m	+42° 30'

Angular sizes and separations are from recent catalogs. Visually, an object's size is often smaller than the cataloged value and varies according to the aperture and magnification of the viewing instrument.

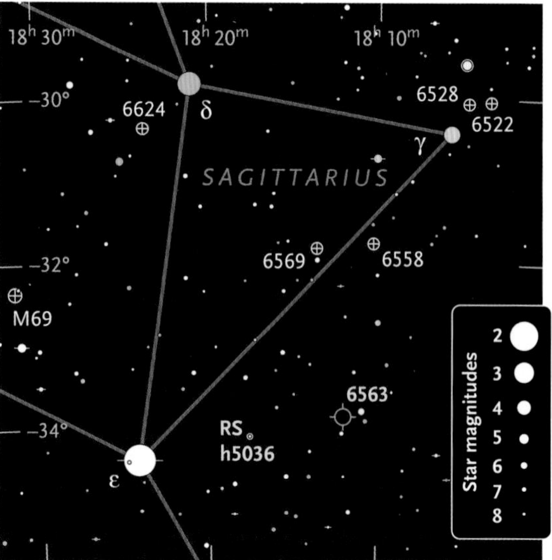

A short star-hop from Kaus Australis, the southernmost star in the Sagittarius Teapot asterism, will lead you to the interesting planetary nebula NGC 6563, floating in a rich star field. North is up in this 8'-wide view. Photo: Adam Block / Mount Lemmon SkyCenter / University of Arizona

Above: Midyear in the Northern Hemisphere means the summer Milky Way is well placed for viewing in the evening sky. Deep-sky observers can spend a lifetime among the rich assortment of star clusters and nebulae scattered across the galactic bulge in Scorpius and Sagittarius. They are prominent in this view by Japanese astrophotographer Akira Fujii.

heavy, hollowed-out rectangular box. The northern and southern walls are distinctly brighter, with the former shorter but more intense. There are hints of faint extensions reaching out from the perimeter of the nebula. Kraljic says that an oxygen III filter worsens the view but a narrowband filter enhances contrast of the planetary's walls. Also, he finds NGC 6445's interior to be darker in its northern half.

Next, we'll move northward to Lyra, where the planetary nebula **NGC 6765** resides with its neighbor **Messier 56** (M56), a globular cluster. Let's start with M56, which is brighter and reasonably straightforward to locate 1.7° west-northwest of the star 2 Cygni. Even with my 15x45 image-stabilized binoculars, M56 is an easily noticeable hazy patch with a broadly brighter center and a faint star off the western side. The cluster is quite pretty through my 105mm refractor at 87x. Its $1\frac{3}{4}$' core is mottled, bright, irregular, and surrounded by a star-spattered halo that fades away near the western star. A magnification of 127x pulls out a few stars in the core. My 10-inch scope at 166x shows an irregular, partly resolved core and a nice mixture of faint to fairly bright stars scattered across the cluster.

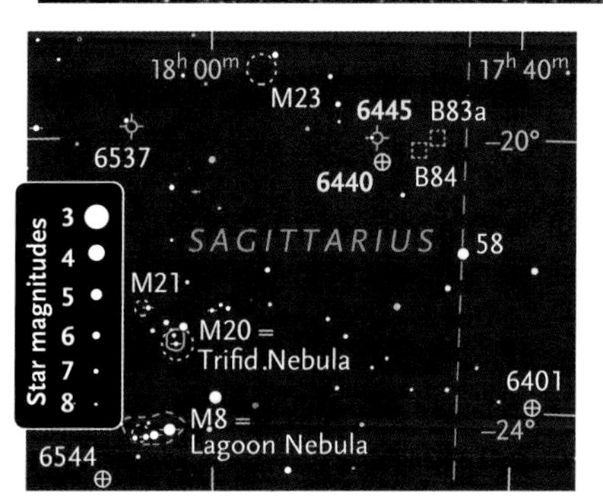

While visible in modest telescopes, the intricate structure present in NGC 6445 invites observers to examine this planetary nebula with large apertures and high magnification. Photo: POSS-II / Caltech / Palomar

A 6th-magnitude orange star lies 26' northwest of M56, and NGC 6765 is 1° west from there. It's a very faint planetary for my little refractor at 87x, and increasing the magnification to 122x helps more than any nebula filter. The

First published July 2009

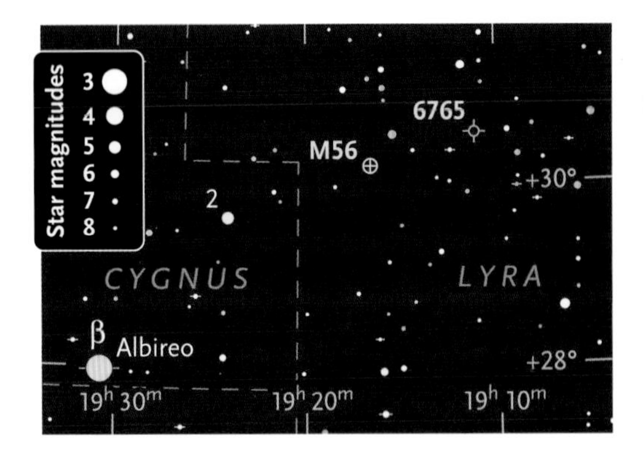

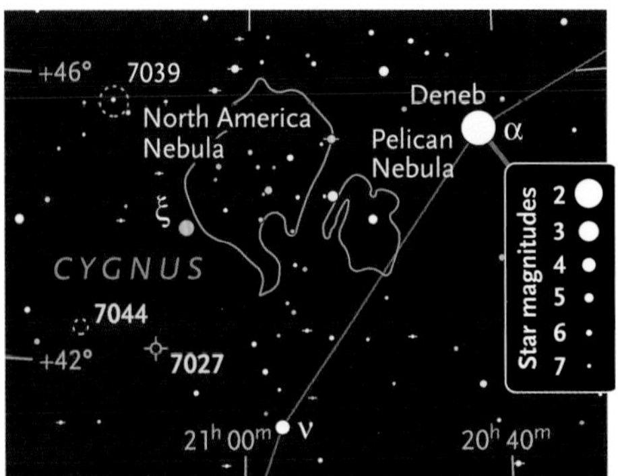

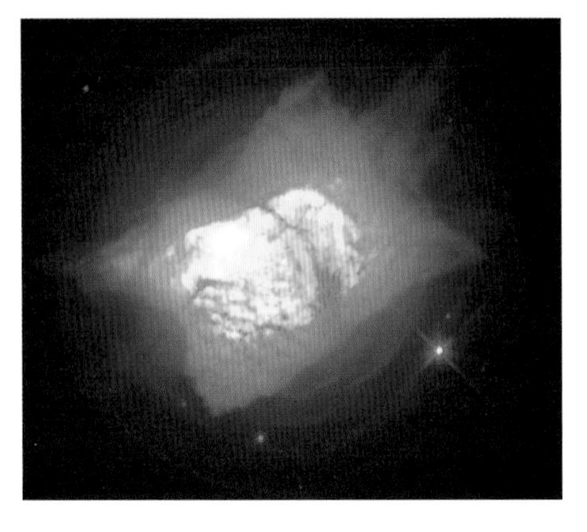
This Hubble Space Telescope view of NGC 7027 captures amazing detail. Only the two lobes and halo (displayed in yellow and orange) can be seen in a backyard scope. Note the bright spot in the northwestern lobe. Northwest is to the lower left in this photo. Photo: H. Bond / STScI / NASA

planetary appears elongated north-northeast to south-southwest and about ½' long. An oxygen III filter is much more helpful with my 10-inch reflector. At 219x, NGC 6765 displays a patchy bar that's brightest at the northern end. There seems to be a faint haze around the bar, especially to the east. The nebula spans about 35" through my 10-inch scope and 40" through my 14.5-inch.

Our final planetary nebula is **NGC 7027** in Cygnus. It's parked at the right angle of a triangle that it makes with Xi (ξ) Cygni and the open cluster **NGC 7044**. Through my little refractor at 87x, the cluster is a hazy ball with two faint stars in the eastern side. My 10-inch scope at 118x reveals a lovely diamond-dust cluster with many very faint stars enmeshed in a misty glow 4½' across.

NGC 7027 is an aqua nebula with a tiny bright center in my 105mm scope at 47x, while at 127x with an oxygen III filter, I see it as an oval tipped northwest. Through my 10-inch scope at 213x, the nebula has a striking bluish green color. It shows two distinct lobes separated by a narrow lane and surrounded by a faint halo. The northwestern lobe is larger with a minute bright spot in its western edge. The southeastern lobe is a bit dimmer and elongated east-northeast to west-southwest. At 299x, the color is not as strong, but the nebula appears wonderfully complex. The halo becomes prominent and the division between the lobes even more distinct. The area surrounding the intense spot is quite bright. The spot itself is nonstellar and remains bright when viewed through an eyepiece with an oxygen III or narrowband filter, attesting to its nebular nature.

The planetaries showcased here give us just a little taste of the varied forms these nebulae take, intriguing forms that place planetary nebulae among the many magical things in our universe.

Night of the Scorpion

Diverse splendors perch on the stinger of Scorpius.

Scorpius is an exceptionally striking constellation, requiring no great effort to imagine the scuttling creature fashioned from its stars. Backdropped by an opulent tapestry of deep-sky wonders, Scorpius is also glorious to peruse with a telescope.

Our trailhead will be the bright star Shaula, or Lambda (λ) Scorpii, in the Scorpion's upraised stinger. Several people have shared their views of this area of the sky, and I'm pleased to feature their observations here.

Sweeping 1.3° eastward from Shaula brings us to the open cluster **NGC 6400**. Three California observers sent me their notes. Steve Waldee sees the cluster as a small ghostly glow through his 9x50 finder, while his 120mm refractor at 100x shows faint stars strewn across a fractured mat of unresolved Milky Way suns. Robert Ayers notes that NGC 6400 makes a double cluster with Ruprecht 127 (Ru 127) in his 8-inch refractor at 65x, the former more prominent than its sparsely populated companion 49' northwest. With a 10-inch reflector, Kevin Ritschel perceives "a nice chairlike pattern of stars radiating out from the center of the cluster" composed of "three bright chains, one with a kink." With my 5.1-inch refractor, I count 16 stars, while my 10-inch reflector

Astronauts installed the Wide Field Camera 3 in the Hubble Space Telescope in May 2009. One of its first images was this incredibly detailed view of NGC 6302. Photo: NASA / ESA / Hubble SM4 ERO Team

reveals 25 stars, 11th and 12th magnitude, in a 10' group.

Farther east, golden G Scorpii burnishes the western edge of the globular cluster **NGC 6441**. The Roman letter designation was assigned by American astronomer Benjamin Apthorp Gould after the constellation Telescopium was cut down to size and the star was reallocated to Scorpius.

Even from an urban site, Ritschel was able to nab NGC 6441 through 50mm binoculars. He finds it very easy to see as a distinct round glow in his 4-inch refractor and reports: "At 90x, it is grainy in appearance. The core is bright and fades gradually to the edge." The cluster spans 2.8' in my 10-inch scope. A 10th-magnitude star tacks down its west-southwestern edge, while a 13th-magnitude star pins its north-northwestern rim.

NGC 6441 is difficult to resolve because most of its members are dimmer than 15th magnitude. Can you spot any of them?

Many people who have sent me their notes seem more fascinated by the symbiotic star near NGC 6441 than they are with the cluster itself. Symbiotic stars are close binary systems in which a cool giant star transfers high-temperature streams of gas to a hot dwarf companion. In the past, they were often confused with planetary nebulae, because both produce strong emission lines. The symbiotic star near NGC 6441 bears several planetary-nebula designations, the earliest being PN H 1-36 or **Haro 2-36**. (It's often incorrectly called Haro 1-36.)

Haro 2-36 nests 1.3' north-northwest of G Scorpii, which helps pinpoint the symbiotic star's position but hinders the view with its glare. Fortunately, Haro 2-36 is the brightest starlike object close to the correct position.

Waldee, Virginia amateur Kent Blackwell, and Finnish amateur Jaakko Saloranta have all caught sight of Haro 2-36 in their 80mm refractors. Waldee comments that he

The 3rd-magnitude star G Scorpii lies just 127 light-years from Earth. The symbiotic star Haro 2-36 is intrinsically brighter than G Scorpii but appears dimmer because it's 100 times farther from us. And the delicate globular cluster NGC 6441 is more than twice as distant as Haro 2-36. Photo: Daniel Verschatse / Observatorio Antilhue / Chile

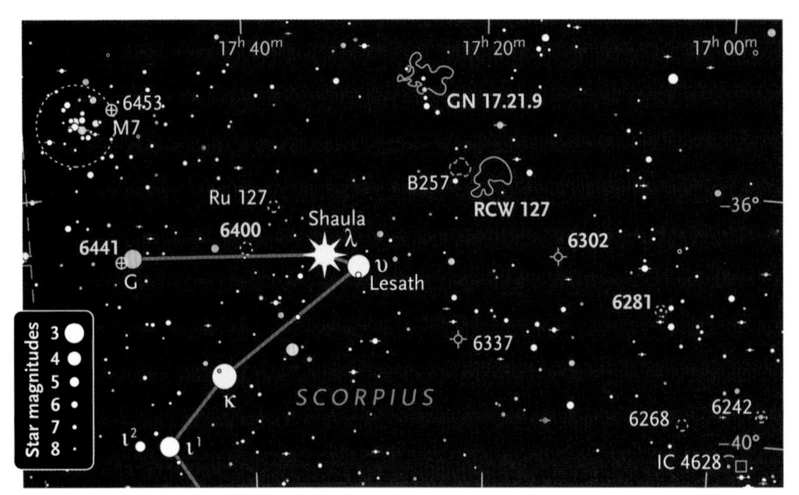

can see it about 25 percent of the time at 114x while using an oxygen III filter. In my 10-inch scope, the symbiotic star is quite easy. Several observers agree that it looks best with an oxygen III filter, which also dims G Scorpii. A narrowband filter provides some improvement as well.

Now we'll move 4° west of Shaula to the planetary nebula **NGC 6302**, nicknamed the Bug Nebula. Through her 4-inch refractor at low power, Virginia amateur Elaine Osborne sees a fuzzy, round nebula that's brighter in the center and a "baby Corvus" star pattern in the same field of view. Turning to her 10-inch reflector, Osborne exclaims, "Ah! A butterfly is born." The butterfly's narrow wings spread east-west from the bright core and span 1.5'.

Waldee enjoys filtered views through his 11-inch Schmidt-Cassegrain at a whopping 466x. With a narrowband filter and averted vision (that is, by looking a bit off to the side of the object), he detects a brighter spot in the western wing. Waldee states that the view through his oxygen III filter is dim, but the core looks "speckled and broken" and the wings have "wispy, cloudy edges." He says the fainter eastern wing extends farther when observed through a broadband filter.

The seldom-visited star cluster **NGC 6281** rests 2° west-southwest of the Bug Nebula. Irish amateur Kevin Berwick pictures the group as a broken arrowhead through his 90mm refractor at 71x. Waldee calls it a wonderful object. Seen in his 120mm scope at 67x, NGC 6281 is a trapezoidal group of a dozen stars, the brightest one at the northeastern corner. My 10-inch scope shows 25 stars in this trapezoid plus some stragglers to the south.

Our next target is **RCW 127**, the Cat's Paw Nebula. The brightest toe of the Cat's Paw, **NGC 6334**, engulfs a

Highlights of the Scorpion's Stinger

Object	Type	Magnitude	Size/Sep.	Distance (l-y)	Right Ascension	Declination
NGC 6400	Open cluster	8.8	12'	3,100	$17^h 40.2^m$	−36° 58'
NGC 6441	Globular cluster	7.2	9.6'	38,000	$17^h 50.2^m$	−37° 03'
Haro 2-36	Symbiotic star	12.1	—	15,000	$17^h 49.8^m$	−37° 01'
NGC 6302	Planetary nebula	9.6	90" × 35"	4,000	$17^h 13.7^m$	−37° 06'
NGC 6281	Open cluster	5.4	12'	1,560	$17^h 04.7^m$	−37° 59'
RCW 127	Emission nebula	—	50' × 25'	5,500	$17^h 20.4^m$	−35° 51'
GN 17.21.9	Emission nebula	—	35'	8,000	$17^h 25.2^m$	−34° 12'

Angular sizes and separations are from recent catalogs. Visually, an object's size is often smaller than the cataloged value and varies according to the aperture and magnification of the viewing instrument.

9th-magnitude star 2.8° west-northwest of Shaula. Ritschel can detect a vague patch of haze even with his 66mm "Refractor Blue" at 16x, with or without a filter. Waldee's 120mm refractor at 19x reveals nebulosity with an oxygen III filter. A filterless view at 100x shows the open cluster **van den Bergh-Hagen 223** (vdB-Ha 223), in the northernmost toe, as a sparse group of at least a half dozen stars. With his 6-inch refractor at 28x and an oxygen III filter, Ayers can make out a patch of mist with two extensions corresponding to the southern toes. All three observers say the nebula is quite challenging through their scopes.

Osborne calls the nebula beautiful and exciting through her 10-inch reflector under a dark, rural sky. She can discern all three toes; our celestial cat seems to be shy a digit. By tapping the scope, she can distinguish some of the foot pad as well. (Imparting a little motion to the view sometimes makes faint, extended objects more obvious.) Osborne's scope is armed with an oxygen III filter fitted to an eyepiece yielding a magnification of 88x and a 67' field of view. The longest diameter of RCW 127 is about 50', and NGC 6334 is approximately 10' across.

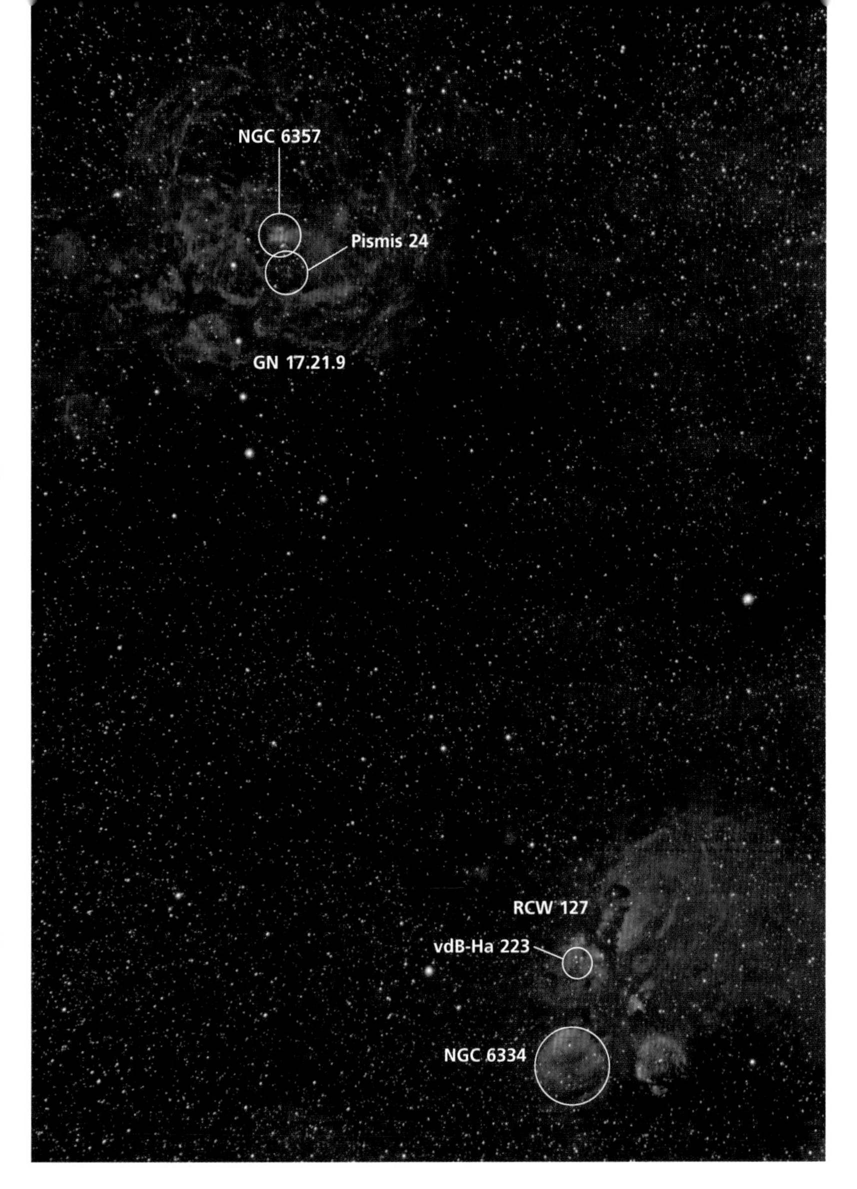

The nebulosity at upper left is variously known as NGC 6357, Sharpless 2-11, RCW 131, and GN 17.21.9. But NGC 6357 properly applies only to the brightest part of the nebula, while the Sharpless and RCW designations include portions that are not visible through the eyepiece of a telescope. Likewise, although RCW 127 (the Cat's Paw Nebula) is sometimes called NGC 6334, that name properly applies to only one of the cat's three toes. Photo: Johannes Schedler

A bright asterism of five 6th- and 7th-magnitude stars lies about 1½° north-northeast of NGC 6334. Perry Vlahos, president of Australia's Astronomical Society of Victoria, calls this group the Golf Club and says, "A right-hander could play a shot toward the greens of Ophiuchus!" The four-starred handle runs north to south for 27' and then takes a bend west-southwest to form the head of the club. Vlahos can easily spot this asterism in his 50mm binoculars.

In Chile, Victor Ramirez calls the four stars of the Golf Club's handle Las Cuatro Juanitas. They lead us into the heart of the faint emission nebula **GN 17.21.9**, also known as the Lobster Nebula. It spans about 35' and includes **NGC 6357**, a 3' bright patch perched 8' west-northwest of the northernmost Juanita.

With a borrowed 4.3-inch refractor at 24x, Ritschel sees the large nebula carved into sections by indistinct dust lanes, but he warns us that even with a nebula filter, it's difficult to distinguish from starclouds in the region. Brighter NGC 6357 permits a direct-vision view.

In his 4.7-inch scope at 171x, Waldee sees a few faint stars just south of NGC 6357, the brightest members of a cluster known as **Pismis 24**. It holds some of the brightest, bluest, and most massive stars known — several approaching 100 times the mass of our Sun. Waldee's 10-inch reflector at 46x with an oxygen III filter displays the Lobster Nebula as a 25' patchy haze with a bright spot at the location of NGC 6357 and Pismis 24.

On your next clear night, see whether you can pot a Lobster, catch a Butterfly, or meet some pretty Juanitas.

First published July 2010

Night of the Trifid

In the nebula's north end, smoke-size dust particles reflect blue light from an embedded star.

Daytime Triffids may be the giant, man-eating plants of science fiction fame, but the Trifid of the night is even more remarkable. The Trifid Nebula is an amazing complex, for it contains nebulae of three different types, an open star cluster, and a multiple star. But the best way to locate the Trifid Nebula is by first visiting its spectacular neighbor, the Lagoon.

Look for a concentrated patch of mist above the spout of the Teapot asterism formed by the bright stars of Sagittarius. From my northerly location in upstate New York, the **Lagoon Nebula** (M8) sails low across the southern sky. Yet I can pick it out with my unaided eye even in suburban skies. When you turn a small telescope toward M8, the first thing you'll notice is the embedded star cluster **NGC 6530**. With my 105mm refractor at 87x, I count 25 stars in a wedge-shaped group that points toward the northeast. The stars are 7th to 11th magnitude, and the longest dimension is about 10' (one-third of the Moon's breadth in the sky). Two smaller gatherings of several stars each lie north and west of the main group. The latter contains the 6th-magnitude star 9 Sagittarii.

The nebula itself is designated **NGC 6523**. At first blush, you may notice nebulosity only near the concentrations of stars. The most obvious patch envelops 9 Sagittarii and includes a particularly bright knot 3' to the star's west-southwest. The knot's pinched shape when viewed at high magnifications inspired its popular name: the Hourglass. The 9.5-mag-

Under optimal conditions, when the Trifid (center) and Lagoon (below) are high in a very dark sky, observers using small telescopes at a variety of magnifications can trace nearly all the nebulosity visible in this image. The main difference is that the human eye is almost completely color-blind in dim light — nebulae, like cats, look mostly gray at night. Photo: Akira Fujii

nitude star alongside the Hourglass is its source of illumination. The second-brightest area of the Lagoon, involving the starry wedge, is best seen where it's smeared to the cluster's southwest. A slash of blacker sky runs between these glowing clouds; it's part of the dark nebula that gives the Lagoon its name.

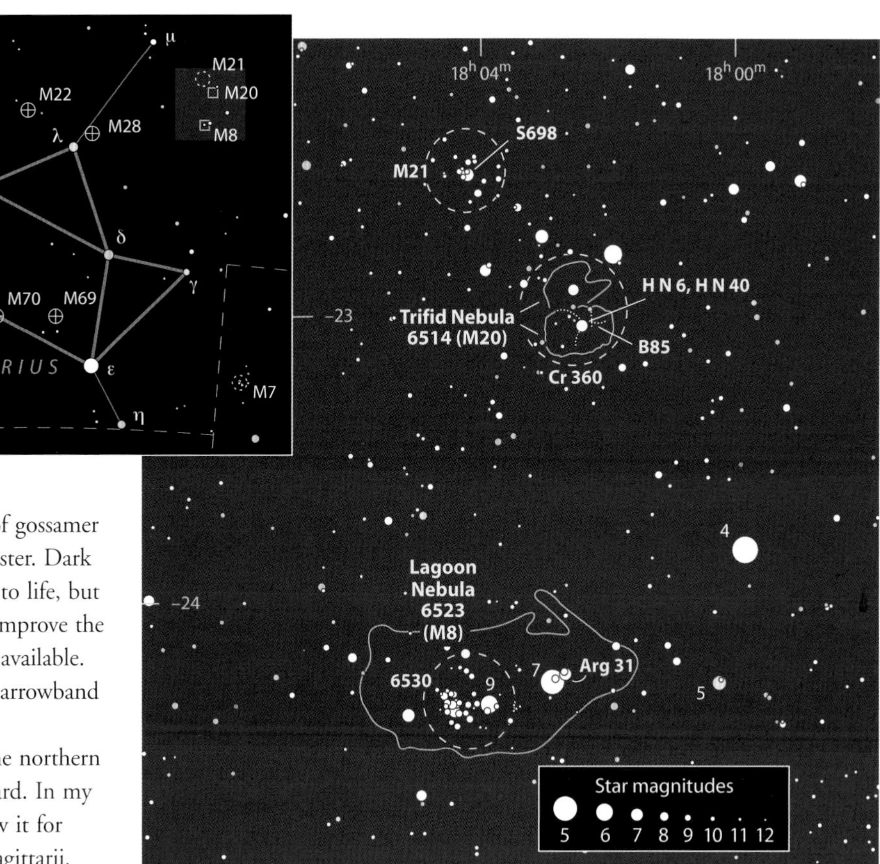

Chart: Adapted from *Millennium Star Atlas* data

A large but more diaphanous region of gossamer light spreads east and south from the cluster. Dark skies will make such faint sections come to life, but if your skies are light-polluted, you can improve the view considerably with the special filters available. The nebula responds very well to both narrowband and oxygen III filters.

Look next for a dim haze engulfing the northern smattering of stars and extending westward. In my 105mm scope at 87x, I am able to follow it for about 26' to the 5th-magnitude star 7 Sagittarii. The nebula fades just before it reaches the double star **Argelander 31** (Arg 31). The pair's 7th-magnitude primary has a 9th-magnitude secondary 34" north-northeast. This wide and gently curving swath of light cradles the brighter nebulosity around 9 Sagittarii, from which it is separated by another leg of the Lagoon's dark channel.

From the Lagoon, finding the **Trifid Nebula** (M20) is simple. Center the star 7 Sagittarii in a low-power field, scan slowly north for 1.3°, and look for a 7th-magnitude star surrounded by a tenuous haze. This is the glow of the nebula itself, which bears the designation **NGC 6514**. But the name Trifid comes from the fact that a dark nebula, **Barnard 85** (B85), divides it into three lobes nearly centered on the 7th-magnitude star. B85 has been seen in scopes as small as 60mm; even without a filter, it is easily visible under semirural skies in my husband's 92mm refractor at 64x. When light pollution interferes, oxygen III and narrowband filters can help improve the view. In photos, B85 divides the nebula into four sections. Noted observer Stephen James O'Meara likens it to a four-leaf clover, since you'd be lucky to glimpse the fourth lobe through a small telescope.

The central star of the Trifid is a multiple system with six known components. My 105mm scope at 153x nicely displays the two brightest stars. Designated **H N 6**, this pair has a 7.6-magnitude primary with an 8.7-magnitude secondary 11" south-southwest. On a steady night, you might spot a third member in a nearly straight line with the other two. Look for a faint 10.4-magnitude star 6" north-northeast of the 7.6-magnitude star. This pair is known as **H N 40**.

It is the small group of hot stars at the heart of the Trifid Nebula that allows us to see the nebulosity surrounding them. The energy from these stars ionizes the hydrogen gas around them. When escaped electrons recombine with the hydrogen atoms, they emit the characteristic red light that is the hallmark of emission nebulae and is so stunning in photographs.

Photographs also show a bluish nebula surrounding the 7th-magnitude star 8' north-northeast of the triple. Here, smoke-size dust particles reflect blue light from the embedded star. The reflection nebula is visible in a small scope, though it seems fainter than its glowing cousin to the south. In dark, haze-free skies, it looks nearly as large as the emission nebula, but in reality, it's much larger. Enhanced images show the reflection nebula completely enveloping its red counterpart, like a blue crab clutching a treasured morsel. Human eyes can't see the beautiful colors so striking in photographs, but if I look closely, I get the impression that the parts have subtly different hues. I also find that

First published August 2002

Trifid Nebula (M20) and Environs

Name	Type	Magnitude	Size/Sep.	Right Ascension	Declination
NGC 6514 (main part)	Emission nebula	—	16'	18h 02.4m	–23° 02'
B85	Dark nebula	—	16'	18h 02.4m	–23° 02'
NGC 6514 (north part)	Reflection nebula	—	20'	18h 02.5m	–22° 54'
H N 6 and H N 40	Multiple star	7.6, 8.7, 10.4	11", 6"	18h 02.4m	–23° 02'
Cr 360	Open cluster	6.3	13'	18h 02.5m	–23° 00'
M21	Open cluster	5.9	12'	18h 04.6m	–22° 30'
S698	Double star	7.2, 8.5	30"	18h 04.2m	–22° 30'

Parts of the Lagoon Nebula (M8)

Name	Type	Magnitude	Size/Sep.	Right Ascension	Declination
NGC 6530	Open cluster	4.6	15'	18h 04.8m	–24° 20'
NGC 6523	Emission nebula	5.8	90' × 40'	18h 03.8m	–24° 23'
Arg 31	Double star	6.9, 8.6	34"	18h 02.6m	–24° 15'

filters do little to bring out the reflection nebula — they may even make it less apparent.

In many deep-sky references, M20 is listed as a "nebula + cluster." The many stars scattered in and around the nebula presumably belong to **Collinder 360** (Cr 360). The cluster is not conspicuous against the rich background of the Milky Way, but a small scope will show the area spangled with one or two dozen stars.

The open cluster **M21** lies 40' northeast of the Trifid. The two bright stars of the Trifid and three additional 7th- and 8th-magnitude stars form a gentle arc that will lead you right to it. About 20 stars are gathered in a small knot with ill-defined borders. High magnifications help bring out some of the fainter stars. The brightest star is the double **S698**. It has a 7.2-magnitude primary with an 8.5-magnitude secondary 30" northwest.

All the objects I've mentioned here fit in a 2½° circle of sky, meaning that a small scope of short focal length could embrace them all in one field of view. Most lie within the amazing complex of M20. When gazing at the Trifid, remember that you're seeing an area where new stars are forming. Will one of them eventually warm a planet with huge, carnivorous plants — or with something still more wondrous and strange?

After Tea

There's more to see in Sagittarius than what lies in and around its celestial Teapot.

The eye-catching Teapot outlined by the brightest stars of Sagittarius sits low in the south on the August all-sky star map on page 309. The Milky Way rises above it, like steam from the Teapot's spout, and captivates observers with its deep-sky largesse. Rapt skygazers seldom turn their attention to the barren-looking region that follows the Teapot in its westward march across the sky. Yet this area holds two Messier objects and other intriguing sights waiting to be added to your list of small-scope captures.

Our targets can be a challenge to find in the empty, star-poor regions of eastern Sagittarius — so we'll begin with the globular cluster **M55**, which has the redeeming quality of being visible through a finderscope as a little fluff ball. Here are a few tactics to get you in the correct area. Visualize a line from Sigma (σ) through Tau (τ) Sagittarii (the end stars in the Teapot's handle) and continue for 2¾ times that distance again. Also note that M55 makes a long isosceles triangle with the handle stars Tau and Zeta (ζ).

You can start a bit closer if your sky is dark enough to spot 52 and 62 Sagittarii, two stars that make a nearly equilateral triangle with M55. Put 52 Sagittarii at the western edge of a low-power field and then sweep 6° southward. With a little luck, any of these methods should get you close enough to recognize M55 through a finder or low-power eyepiece.

With a 60mm telescope at around 50x, M55 appears large and mottled with a broad, slightly brighter core. A 4- to 6-inch scope at 75x to 100x will bring out many faint stars loosely scattered across the cluster. Including the sparse outer halo, the group seems about 10' across. Dark lanes fracture it, and the borders of the cluster are ragged with nibbles taken out of the edge, including a rather large one in the southeast side.

M55 is one of the closest globular clusters, approximately 17,000 light-years away and 100 light-years across. It shines with nearly 90,000 times the luminosity of our Sun.

Our second Messier object is the globular cluster **M75**, which hugs the eastern border of Sagittarius. It's easy to scan past without noticing, so you must precisely pin down its position. Start at 4.4-magnitude, reddish orange 62 Sagittarii, and place it outside the southern edge of a finder field to reveal a pair of 6th-magnitude stars to the north. With a low-power eyepiece in your telescope, position the eastern member of the pair near the southwestern edge of your field of view. This should put M75 near the northern perimeter. If you use a reflex sight (an illuminated 1x targeting device), note that M75 lies about halfway between Psi (ψ) Capricorni and Rho (ρ) Sagittarii. You'll see these stars labeled on the chart on page 182.

Through a small telescope at low power, M75 resembles a fuzzy star. My 105mm refractor at 87x shows a little ball of haze about 2' in diameter with no resolved stars. The cluster brightens sharply toward the center, where a tiny, intense nucleus resides.

At 68,000 light-years, M75 is one of the most remote Messier globulars, making Charles Messier's description quite curious. He said that M75 seemed to be composed only of very faint stars, containing some nebulosity — and yet Messier saw no stars in M55 or other bright globulars much easier to resolve. M75 appears small and faint only because of its great distance. It is about 130 light-years across and over 220,000 times more luminous than our Sun.

A floating ball of stars, M55 would be much better known if it could be seen high in the sky from north temperate latitudes. North is up in this ½° square view. Photo: POSS-II / Caltech / Palomar

Now let's move to Rho Sagittarii at the northern end of an asterism known as the Teaspoon. **NGC 6774** lies just 2° northwest of this star. It is a large and sparse open cluster about 30' across, best viewed through binoculars or a small scope at low power. It hides in a rich star field, and a wide view is needed to help it stand out from its background. My little refractor at 28x shows about 40 stars of magnitudes 8 to 12 arranged in little bunches and chains.

NGC 6774 shares the finder field with 4.5-magnitude Upsilon (υ) Sagittarii. Placing Upsilon a little way outside the western edge of the field will bring a curve of three

Sagittarius Sidelights

Name	Type	Magnitude	Size	Distance (l-y)	Right Ascension	Declination	*MSA*	*U2*
M55	Globular cluster	6.3	19'	17,000	$19^h\ 40.0^m$	−30° 58'	1410	162R
M75	Globular cluster	8.5	6.8'	68,000	$20^h\ 06.1^m$	−21° 55'	1386	144L
NGC 6774	Open cluster	—	30'	820	$19^h\ 16.6^m$	−16° 16'	1365	125L
NGC 6822	Galaxy	8.8	16' × 14'	1.6 million	$19^h\ 44.9^m$	−14° 48'	1363/39	125L
NGC 6818	Planetary nebula	9.3	22" × 15"	5,500	$19^h\ 44.0^m$	−14° 09'	1339	125L

The columns headed *MSA* and *U2* give the chart numbers of objects in the *Millennium Star Atlas* and *Uranometria 2000.0*, 2nd edition, respectively. Distances are from research papers that were published in the early 2000s. Angular sizes are from various catalogs, but NGC 6774 has such indefinite borders that size claims vary widely.

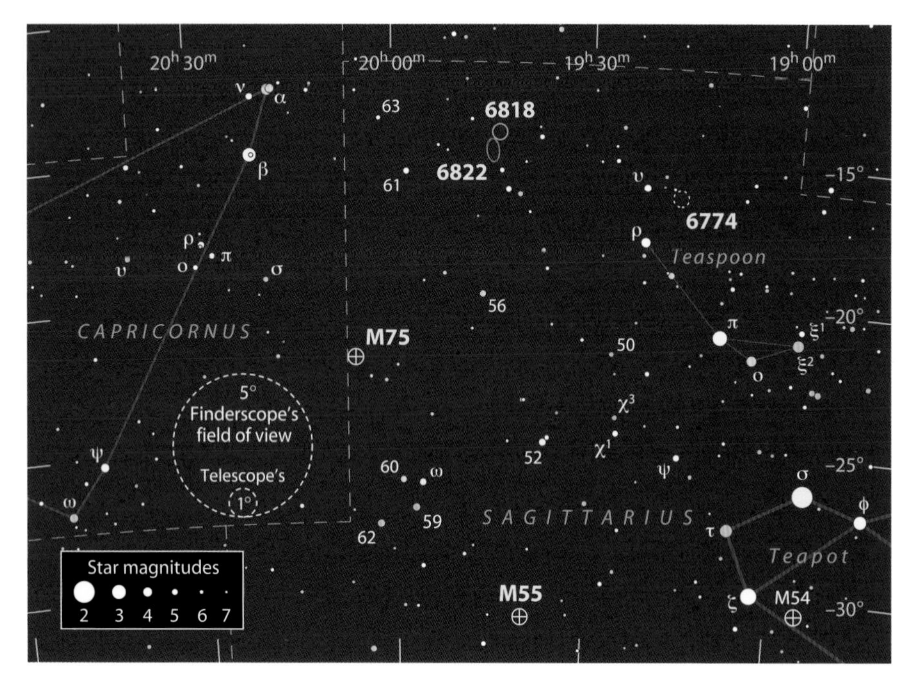

5th-magnitude stars into the eastern side. Moving from the middle star to the northern one and continuing for that distance again will take you to the location of **NGC 6822**, Barnard's Galaxy. It was discovered in 1884 by American astronomer Edward Emerson Barnard while he was still an amateur in Tennessee, but it was not recognized as an extragalactic object until the 1920s. Barnard's Galaxy (also called C57) is a member of the Local Group, a small cluster of 36 known galaxies, including our own.

Barnard described his discovery as "an excessively faint nebula...very diffuse and even in its light. With 6 inch Equatoreal it is very difficult to see, with 5 inch and a power of 30± (field about 1¼°) it is quite distinct. This should be borne in mind in looking for it."

Barnard's advice is worth heeding. It's often easier to catch sight of NGC 6822 in a small scope that allows a wide field than in a large scope with its more restricted view. While Barnard's Galaxy has been seen in 7x35 binoculars, I recommend at least a 60mm telescope under fairly dark skies. With my 105mm refractor at 17x from my semirural home, the galaxy is elusive, appearing very faint and oblong. The long dimension measures about 11' and runs north-south.

Barnard's 1884 discovery, reported in the prestigious journal *Astronomische Nachrichten*, is entitled "New Nebula Near General Catalogue No. 4510." Today, we know GC 4510 as **NGC 6818**, a small, bright planetary nebula nicknamed the Little Gem. It lies 41' north-northwest of Barnard's Galaxy, and the two fit within the same low-power field. If you were unable to find Barnard's Galaxy, return to the curve of 6th-magnitude stars below it and sweep 1.3° north from the northernmost star.

Through my little refractor at 28x, the Little Gem looks like a bloated, bluish star. At 87x, it can be recognized as a tiny, round nebula. At 153x, the planetary hints at being oval with perhaps some slight patchiness. In a sky largely populated with ancient wonders, this aquamarine jewel is comparatively young — a mere 3,500 years old.

Because it is relatively close to our own Milky Way system, NGC 6822 looks more like a telescopic version of a Magellanic Cloud than a typical galaxy. The bright star ¾° to its south-southwest (lower right) is magnitude 5.5. Martin Germano hauled his 8-inch f/5 reflector up Mount Pinos in California to take this photograph on hypered Kodak 2415 emulsion.

NGC 6818, the Little Gem, appears as a tiny bluish oval floating in a black sky. Photo: POSS-II / Caltech / Palomar

First published August 2003

Leagues of Stars

This region of Sagittarius offers good views to observers south of +50° latitude.

Pure leagues of stars from garish light withdrawn
Behind celestial lace-work pale as foam.
— William Hamilton Hayne, "Indian Fancy"

The Milky Way's countless stars paint a misty river of light across the summer sky. As it plunges into Sagittarius, we see one of the finest portions of its vast richness. The area is so populated with deep-sky wonders that you can often capture more than one in a single field of view. Let's turn our attention to the northern reaches of Sagittarius to discover what treasures they hold.

We'll start with easy quarry. The open cluster **M23** lies 4½° west-northwest of Mu (μ) Sagittarii, and both fit in the same field of binoculars or a finderscope. With 10x50 binoculars, you'll see a misty glow with several minute flecks of light sparkling in the mottled haze. Large 14x70 binoculars reveal a beautiful cluster rich in faint stars, and a small telescope shows 50 to 75 stars in a group nearly the size of the full Moon. In *The Messier Album* by John H. Mallas and Evered Kreimer (Sky Publishing, 1978), Mallas describes the view through his 4-inch refractor as a glorious sight and says, "The brightest stars in this irregularly shaped cluster form a pattern resembling a bat in flight."

From M23, you can hop about 2° west-southwest to the planetary nebula **NGC 6445** and the globular star cluster **NGC 6440**. They are close enough together to fit in the same telescopic field, and both have been glimpsed with apertures as small as 60mm. A 4- to 6-inch scope will make the task fairly easy.

First, look for h2810, a wide double star with a 7.6-magnitude primary and a 10.4-magnitude companion 43" to the south. NGC 6445 lies 5' west of the pair. This little planetary is only half the apparent size of the famed Ring Nebula in Lyra, so use a magnification of at least 100x for a good view. NGC 6445 appears oval or perhaps rectangular, elongated south-southeast to north-northwest. It is slightly brighter at the long ends, and a faint star sits off the northwest side. Under less than pristine skies, a narrow-band or oxygen III filter helps improve the view.

NGC 6440 is 22' south-southwest of NGC 6445. It looks about 2' across (more than twice the diameter of its neighbor) and grows brighter toward the center. The globular is framed between a pair of faint stars: one of 12th magnitude off the north-northwest side and the other slightly brighter that is twice as distant to the south-southeast. The cluster remains unresolved even in my

The Small Sagittarius Starcloud, M24, is the white oval glow near the center of this 25°-wide view. Punctuating the cloud's northwestern (upper) edge are two tiny and distinctive dark nebulae, Barnard 93 (left) and Barnard 92. Photo: Akira Fujii

SUMMER
August

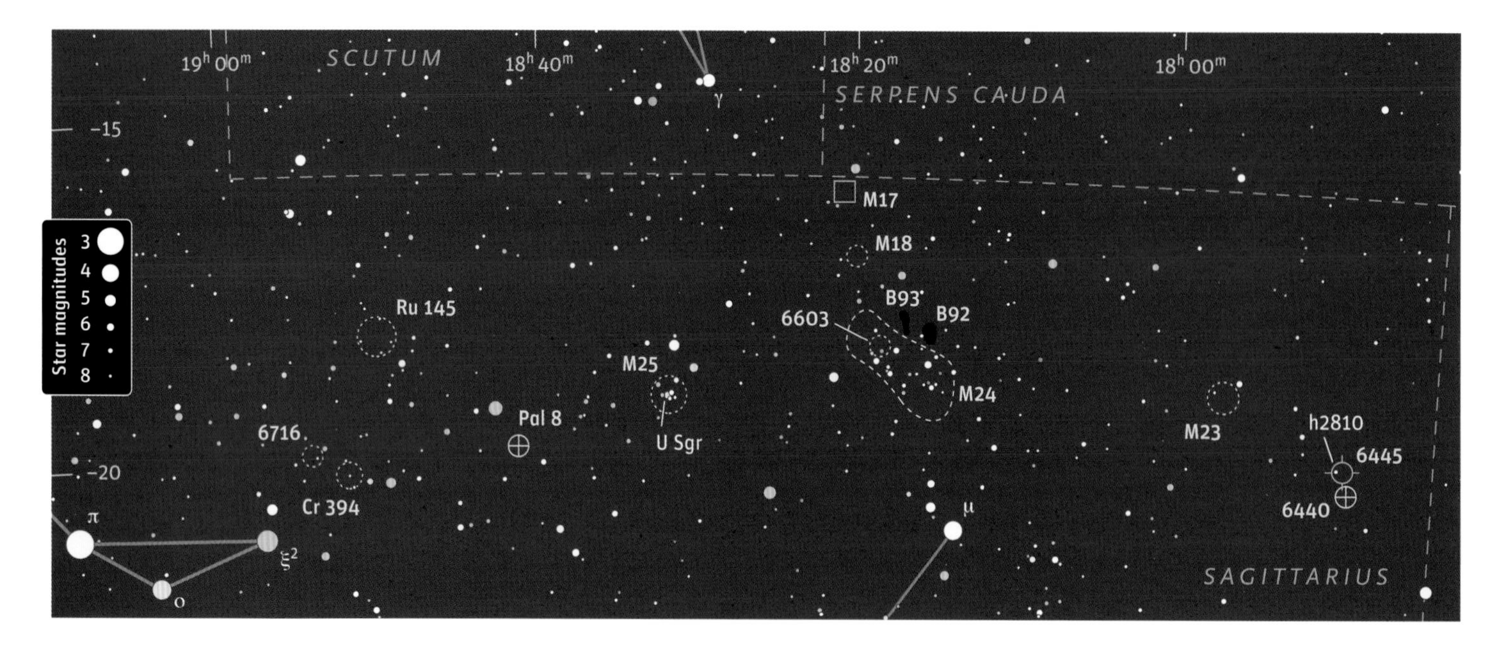

Big binoculars can show more than half of the deep-sky sights plotted here, but it helps to have an exceptionally clear night. Chart: Adapted from *Sky Atlas 2000.0* data

10-inch reflector. NGC 6440 and NGC 6445 are a compelling combo when viewed in the same field.

M24, the Small Sagittarius Starcloud, is visible to the unaided eye as a large, bright patch in the Milky Way just north of Mu Sagittarii. A chance gap in the Milky Way's obscuring dust clouds opens a window on this remote swarm of stars, whose members seem to range from 10,000 to 16,000 light-years away.

Sprawling across 2° of sky, M24 is best appreciated in small, wide-field instruments. This is one of the prettiest star fields for binoculars, and a small telescope will reveal myriad distant suns. M24 enfolds several deep-sky objects, including two dark nebulae nestled on its northwest edge. **Barnard 92** (B92) is oval and quite distinct. Skinny **Barnard 93** (B93) to its east is not as prominent. Northeast of M24's center, the little open cluster **NGC 6603** is a small, misty patch when seen through a small telescope. A 10-inch reveals a rich gathering of faint stars.

A lovely pair of celestial treasures is perched just above M24's northern end. Only 1° apart, the open cluster **M18** and the emission nebula **M17** are nicely framed in binoculars or a small scope at low power. Through 14x70 binoculars, M18 is a small knot of moderately bright stars, while M17 is a fairly bright glow shaped like the numeral 2 with a long bottom bar. Depending on your optics, it may appear upside down and/or backward.

M17 looks very much like a swan in my 105mm scope. The long bar of the 2 is the swan's body, and the curve of the 2 forms its neck and head. We know this must be a heavenly swan, because a small halo hovers over its head. M17 is often called the Swan Nebula because of this striking resemblance. Curiously enough, veteran observer Stephen James O'Meara sees a black swan in M18 with his 4-inch refractor. The bright stars of M18 fashion this ebony swan, and it, like M17, may likewise appear to hang upside down in the sky. But this swan has a partly raised wing —

Several stars in the fine open cluster M25 have an orange hue, notably the variable star U Sagittarii (U Sgr) near its very heart. George R. Viscome of Lake Placid, New York, used a 6-inch f/8 reflector for this 1°-wide view on 3M 1000 slide film.

Sights in a Sagittarius Star Field

Name	Type	Magnitude	Size/Sep.	Right Ascension	Declination	*MSA*	*U2*
M23	Open cluster	5.5	27'	$17^h 56.9^m$	–19° 01'	1369	146L
NGC 6445	Planetary nebula	11.2	44" × 30"	$17^h 49.2^m$	–20° 01'	1369	146L
NGC 6440	Globular cluster	9.3	4.4'	$17^h 48.9^m$	–20° 22'	1369	146L
M24	Starcloud	2.5	1° × 2°	$18^h 17.0^m$	–18° 36'	1368	145R
B92	Dark nebula	—	15' × 9'	$18^h 15.5^m$	–18° 13'	1368	145R
B93	Dark nebula	—	12' × 2'	$18^h 16.9^m$	–18° 04'	1368	145R
NGC 6603	Open cluster	11.1	4.0'	$18^h 18.5^m$	–18° 24'	1368	145R
M18	Open cluster	6.9	9.0'	$18^h 20.0^m$	–17° 06'	1368/67	145R/126L
M17	Emission nebula	6.9	11' × 6'	$18^h 20.8^m$	–16° 10'	1368/67	126L
M25	Open cluster	4.6	32'	$18^h 31.8^m$	–19° 07'	1367	145R
Pal 8	Globular cluster	10.9	5.2'	$18^h 41.5^m$	–19° 50'	1366	145L
NGC 6716	Open cluster	7.5	10'	$18^h 54.6^m$	–19° 54'	1366	145L
Cr 394	Open cluster	6.3	22'	$18^h 52.3^m$	–20° 12'	1366	145L
Ru 145	Open cluster	—	35'	$18^h 50.3^m$	–18° 12'	1366	145L

Angular sizes are from recent catalogs; most objects appear somewhat smaller when a telescope is used visually. The columns headed *MSA* and *U2* give the chart numbers of objects in the *Millennium Star Atlas* and *Uranometria 2000.0*, 2nd edition, respectively.

a starless black wing outlined by a curve of suns on the cluster's eastern side.

Next we'll move to the open cluster **M25**, located 3½° east of the Small Sagittarius Starcloud. They can be seen together in binoculars or a finderscope. My 14x70 binoculars show 30 mixed bright and faint stars in ½°. Through a 4-inch refractor, many pairs decorate the group. A knot of seven 9th- and 10th-magnitude stars at the heart of the group forms a capital letter D. Four cluster stars are bright enough to show a yellow or golden color. The one near the center is U Sagittarii (U Sgr), a Cepheid-type variable star that fades from magnitude 6.3 to 7.2 and then brightens again every 6.7 days.

Our challenge for this sky tour is the globular cluster **Palomar 8** (Pal 8), located 2.4° east-southeast of M25. Most of the Palomar clusters are difficult targets, but Palomar 8 has been spotted in instruments as small as 80mm and described as fairly easy in a 4-inch. I find this globular surprisingly beautiful in my 10-inch telescope. Although faint, it is fairly large at 4' and is clothed in a rich field of faint stars. Some very faint stars are visible over its hazy glow, but these are probably foreground objects, since the cluster's brightest members shine even more feebly at 15th magnitude. At high powers, Palomar 8 appears slightly mottled with no pronounced brightening toward the center.

Moving 3.1° east will bring us to **NGC 6716**. If you didn't find Palomar 8, look for this open cluster 1.4° north-northwest of Xi² (ξ^2) Sagittarii. With my 105mm scope at low power, the cluster is an elongated group of about 25 stars arranged in a wide and rounded M shape about 10' long. A splash of stars more than twice the size of NGC 6716 shares the field to the southwest. This is **Collinder 394** (Cr 394), a rich group of 8th-magnitude and fainter suns. A pretty pair of yellow and blue 7th-magnitude stars sits off its western side. Putting both clusters in the southern part of

The Swan Nebula, M17, shines with the characteristic red emission from hydrogen gas that's been energized by hot young stars. To most human eyes, however, the nebula appears as a colorless (gray) glow. Compare Akira Fujii's photograph to the author's eyepiece impression. Both are oriented north up; they span almost 1° of sky. The author made the pencil sketch with a Nagler eyepiece giving 87x on her 105mm Astro-Physics Traveler refractor.

First published August 2004

my 3.6° field, I can also see **Ruprecht 145** (Ru 145) to the north-northwest. Ru 145 is a loose group of at least 30 faint stars in 35'. A bright pair lies outside the south-western edge — one yellow, the other white. Identifying this cluster trio against the backdrop of the Milky Way may take a little effort, but the three make a nice field when seen together at low power.

While exploring the star-filled southern Milky Way, take the sage words of early American skygazer Garrett P. Serviss to heart: "Do not imagine the thousands of stars that your opera-glass or field-glass reveals comprise all the riches of this Golconda of the heavens. You might ply the powers of the greatest telescope in a vain attempt to exhaust its wealth" (*Astronomy with an Opera-Glass*, 1888). The Indian city that Serviss names was known during the Middle Ages as a plentiful source of cut diamonds.

More Summer Planetaries

Distinctive shapes and prominent colors make planetary nebulae a favorite target for backyard telescopes.

Planetary nebulae are fascinating and beautiful shells of glowing gas shed by dying, low-mass stars. When seen through a telescope eyepiece, they often have distinctive shapes and are the most colorful of the nebulae.

William Herschel was the first astronomer to call these objects "planetary" nebulae, and he classified them on the basis of their appearance. Indeed, many planetaries look like the blue-green disk of Uranus when viewed through a telescope. (A more powerful tool for categorizing nebulae was

Edward Emerson Barnard captured this rich Milky Way field straddling the border between Scutum and Aquila in July 1905 and published it in *A Photographic Atlas of Selected Regions of the Milky Way*. The box shows the area of the finder chart on page 187, top. Photo: Observatories of the Carnegie Institution of Washington

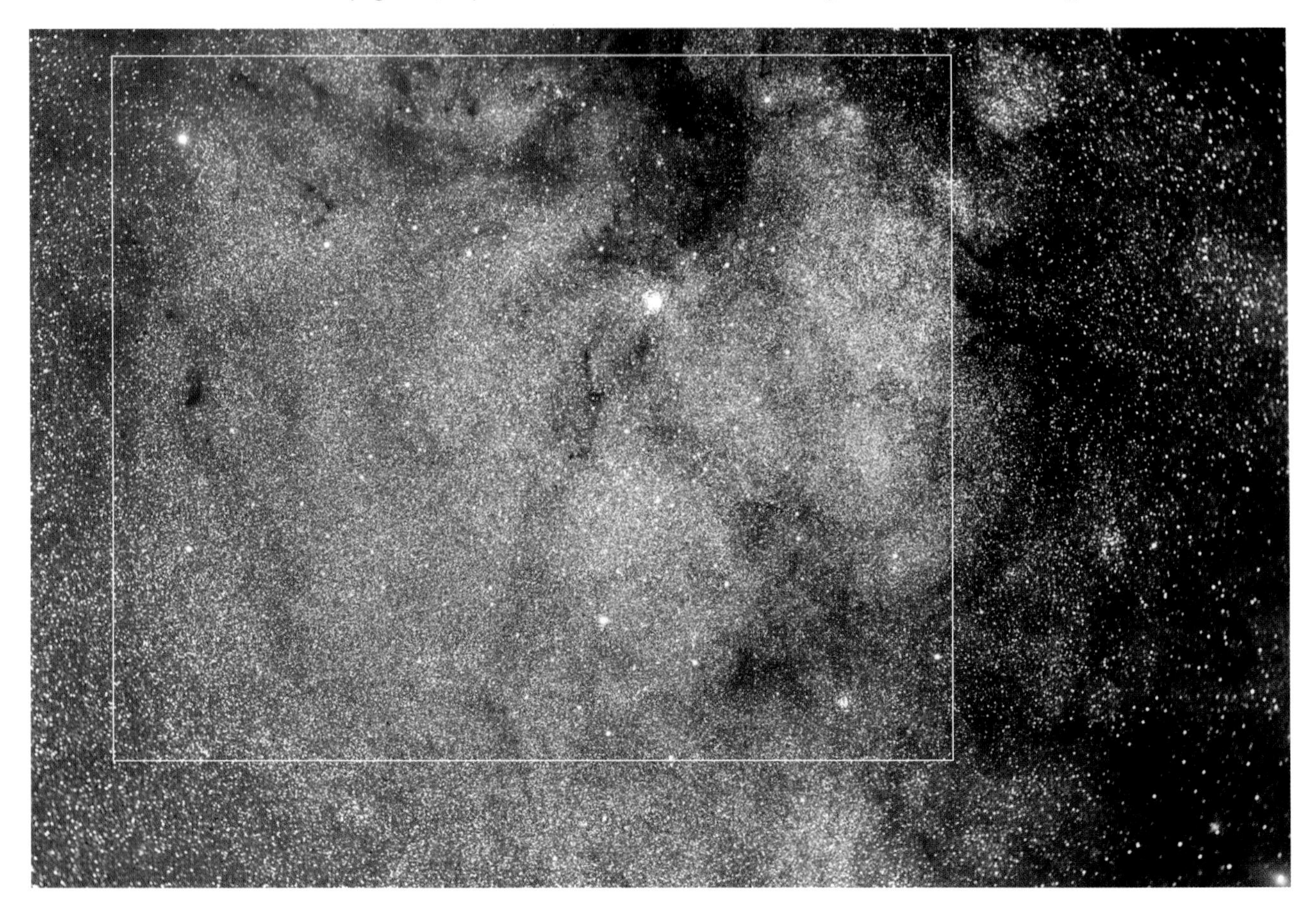

forged in August 1864 when English amateur William Huggins observed the spectrum of the planetary NGC 6543 in Draco and realized that its light originated from a tenuous gas rather than from a mass of unresolved stars.)

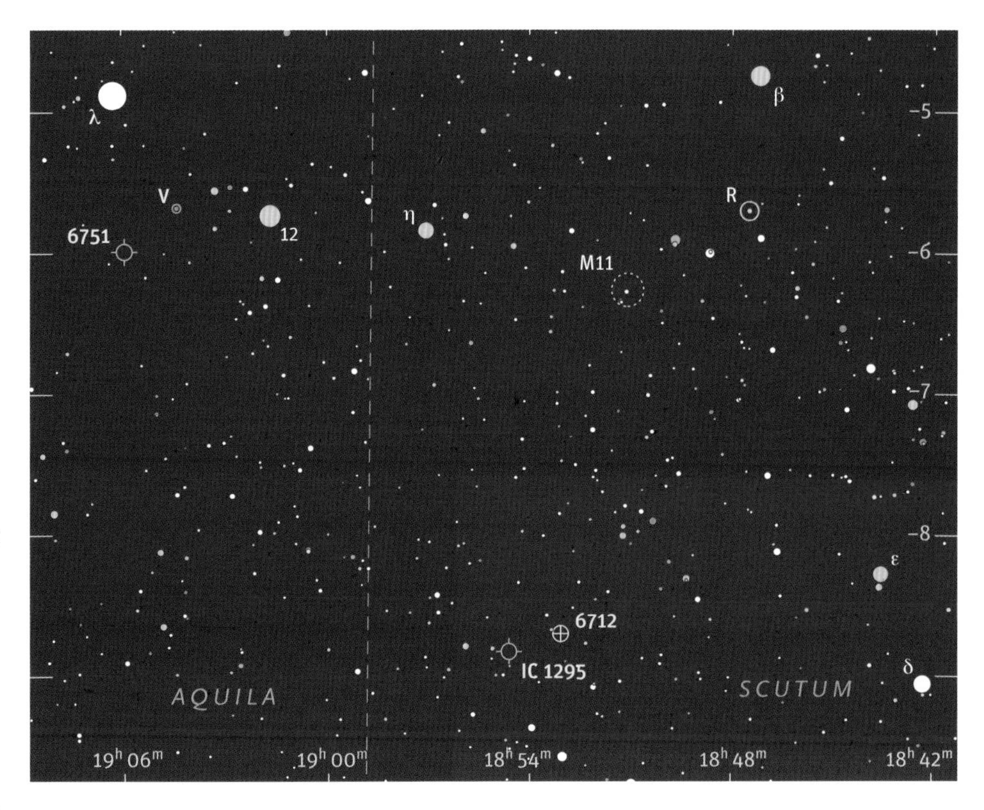

Many of Herschel's planetaries turned out to be other types of objects, mostly galaxies, while some entries in his other classes were later shown to be planetaries. Many years often passed between the discovery of an object and the recognition of its true nature.

We'll begin in Draco with **NGC 6742**, also known as Abell 50. Hunting Abell planetaries is usually a pastime of large-scope aficionados, but this one can be spotted even with modest apertures. NGC 6742 is 1.6° north-northwest of 16 Lyrae and just 3' northeast of an 8.8-magnitude star.

From my semirural home in upstate New York, I need a fairly high magnification to pick out NGC 6742 with my 105mm refractor. I can just spot a faint disk with averted vision (that is, by looking a little to one side of the object) at 153x, but 203x makes it less difficult. A very faint star lies to its northeast. The nebula is considerably easier in my 10-inch reflector. It becomes visible at 70x and shows a round, fairly uniform glow at 219x. About 10 percent of the time, I glimpse an extremely faint star at the western edge. My 14.5-inch reflector at 245x gives hints of a slightly darker region within the disk. A very faint star lies off the north-northeast edge, and another is embedded in the nebula's western edge.

Herschel discovered NGC 6742 in 1788 but did not classify it among his planetary nebulae. Instead, he placed it in his third class: very faint nebulae. The complete list of provisional Abell planetaries first appeared in a 1966 *Astrophysical Journal* article by George O. Abell. The paper described 86 objects identified as possible planetaries by their appearance on plates made with the 48-inch Schmidt telescope at Palomar Observatory. A few have turned out not to be planetaries.

Now let's turn to **NGC 6572**, discovered in 1825 by Friedrich Georg Wilhelm von Struve as he was compiling his famous double-star catalog at Russia's Dorpat Observatory. Huggins first noted its telltale spectrum in 1864. NGC 6572 lies 2.2° south-southeast of 71 Ophiuchi. Various observers have informally called it the Emerald Nebula, the Blue Racquetball, or the Turquoise Orb — names that reflect different perceptions of its color.

I have observed NGC 6572 with a number of telescopes, and I have seen slightly different shades of blue or green

The planetary nebula NGC 6742 is located in Draco near the Lyra border. Photo: William McLaughlin

Summer Planetaries and More

Object	Type	Magnitude	Size/Sep.	Right Ascension	Declination	*MSA*	*U2*
NGC 6742	Planetary nebula	13.4	36" × 30"	18^h 59.3^m	+48° 28'	1111	33R
NGC 6572	Planetary nebula	8.1	11"	18^h 12.1^m	+06° 51'	1272	86L
NGC 6751	Planetary nebula	11.9	24" × 23"	19^h 05.9^m	–06° 00'	1317	125R
V Aquilae	Carbon star	6.6–8.4	—	19^h 04.4^m	–05° 41'	1317	125R
NGC 6712	Globular cluster	8.1	9.8'	18^h 53.1^m	–08° 42'	1318	125R
IC 1295	Planetary nebula	12.5	1.7' × 1.4'	18^h 54.6^m	–08° 50'	1318	125R
NGC 6818	Planetary nebula	9.3	25"	19^h 44.0^m	–14° 09'	1339	125L

Angular sizes or separations are from recent catalogs. The columns headed *MSA* and *U2* give the chart numbers of objects in the *Millennium Star Atlas* and *Uranometria 2000.0*, 2nd edition, respectively.

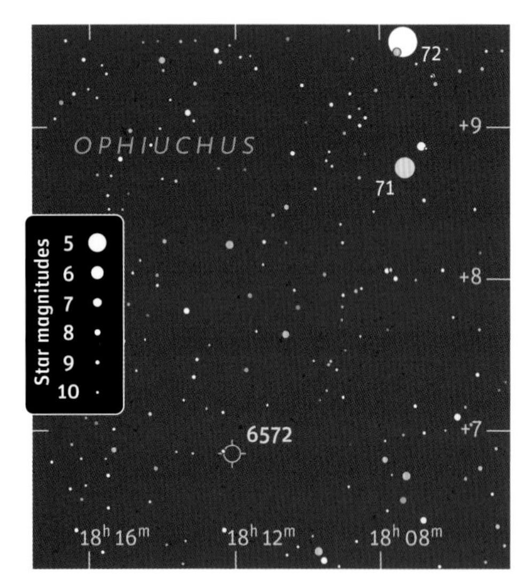

each time. I logged the nebula as tiny, round, fairly bright, and bluish gray with my little refractor. Through an 8-inch refractor, NGC 6572 seemed distinctly robin's-egg blue. It showed a small, bright core and was slightly brighter at opposite edges. In a 10-inch reflector, the planetary looked blue-green and oval with a thin outer fuzz. With a 14.5-inch one night, it seemed greenish, while another night, it appeared strikingly turquoise with a 15-inch. The central star was intermittently visible within the bright core. All observations were made at moderate to high magnification because of the nebula's diminutive size, but its distinctive color gives it away even at low power.

Our next target is **NGC 6751**, one of several hundred deep-sky objects discovered in the 1860s by German-born Albert Marth with William Lassell's 48-inch reflector on the island of Malta. However, it was Scottish-American astronomer Williamina Fleming who first recognized it as a planetary nebula in 1907. NGC 6751 lies 1.1° south of Lambda (λ) Aquilae. Through my 105mm refractor, it's faintly visible at 47x, it's a small, round glow at 87x, and it shows an extremely faint central star (visual magnitude 13.6) at 153x. This star is considerably easier in my 10-inch reflector at high power, and the nebula sports a rim of faint fuzz. Two dim stars bracket the planetary — the closer one a little north of east and the other a little north of west. With my 15-inch scope at 221x, the planetary appears weakly annular.

At low power, NGC 6751 shares the field with impressively reddish **V Aquilae**, 29' northwest. V Aquilae is a carbon-Mira variable star with a magnitude that ranges between about 6.6 and 8.4 and a semiregular period of around 350 days.

We'll use the pretty globular cluster **NGC 6712** to help us find our next planetary. This bright cluster sits 2.4° east and a bit south of Epsilon (ε) Scuti. My 105mm scope at 127x displays a sparse, faint halo surrounding a brighter, more condensed core. Tiny stars sparkle at the limits of vision, and a brighter one stands out at the northeastern edge of the halo. A 9th-magnitude star rests 4' east-northeast of the cluster's center. My 10-inch scope at 219x reveals stars sprinkled over the unresolved mist of the 2½' core, which is flattened along its southern side. The halo's outliers gradually give way to the starry background at a diameter of about 5'.

IC 1295 is 24' east-southeast of NGC 6712 and shares a low-power field of view. Nonetheless, you aren't likely to notice this planetary unless you're looking for it. Although IC 1295 is fairly large, it has a low surface brightness. Search for it 7' west of an 8th-magnitude orange star and just east-northeast of an 11th-magnitude star. The planetary is tough to spot through my little refractor at low power, but I can keep its faint, round, uniform glow steadily in view at 87x with averted vision, and I see it most of the time when looking directly at it. At 127x, I judge it to be about 1½' across. My 10-inch reflector at 219x uncovers a faint star embedded in the southeastern edge of the nebula. Adding an oxygen III filter to the eyepiece, I see hints of structure — darker patches within and some brighter patches along the rim. IC 1295 is a large, lovely, oval annulus in my 15-inch scope at 153x with the oxygen III filter. Without the filter, a few faint stars are superposed.

IC 1295 is one of more than a hundred deep-sky objects discovered in the 1860s by Truman Henry Safford at Dearborn Observatory in Chicago. Heber Doust Curtis of Lick Observatory, near San Jose, California, first recognized it as a planetary nebula in 1919.

First published August 2005

Our final target is **NGC 6818**, the Little Gem, which lies 2° north of 55 Sagittarii in far northeastern Sagittarius. In my 105mm refractor at moderate power, NGC 6818 appears as a small, round, blue-gray disk. The nebula takes magnification well and remains bright at 203x, showing a fat annulus with a slightly darker center. With my 14.5-inch reflector at 245x, I logged the planetary as a nice turquoise-gray with indistinct dark patches in its interior.

William Herschel discovered NGC 6818 in 1787 and classed it among his planetary nebulae. It was another of the objects confirmed as a gaseous nebula by Huggins, who studied it with a spectroscope in 1864. Misclassification of planetary nebulae is not just a thing of the past. Although the statuses of most bright candidates have been sorted out, errors are still found today.

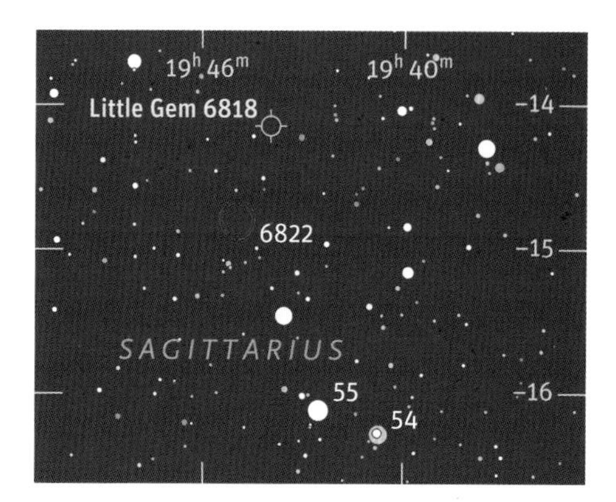

Snake Wrangler

The Milky Way in Ophiuchus and Serpens provides a rich bounty of deep-sky treasures.

Ophiuchus strides the mighty snake,
Untwists his winding folds, and smooths his back,
Extends his bulk, and o'er the slippery scale
His wide-stretch'd hands on either side prevail.
— Marcus Manilius, *Astronomica*

Originally considered one star figure, this scene of celestial herpetology has been divided into the modern-day Ophiuchus, the Serpent Bearer, and Serpens, the Serpent. The division places Serpens in the unique position of being the only constellation split into two disconnected parts. The Serpent's head is west of Ophiuchus, while its tail is east, the halves distinguished as Serpens Caput and Serpens Cauda, respectively.

Here, we'll concentrate on eastern Ophiuchus and the tail of Serpens, where the Milky Way wanders and deep-sky wonders abound. I've received wonderful observations of objects in this area of the sky from other observers and would like to share some of them.

We can begin by slithering down to southern Serpens, where we find the pretty double star **Nu (ν) Serpentis.** Its 4th-magnitude white primary has a 9th-magnitude orange companion to the north-northeast. With a separation of 46", the pair can be split with steadily held binoculars, but you'll need a telescope to detect the secondary's color.

From Nu Serpentis, sweep 1.6° west to **NGC 6309** in Ophiuchus. German amateur Gerhard Niklasch observed this small planetary nebula with his 8-inch reflector and wrote, "At 213x with a UHC [ultrahigh-contrast] filter, it appeared clearly elongated and its two lobes nearly separated." The southern lobe is round and the northern one oval, making the planetary look like a tiny shoe print. The toe of the shoe nearly treads upon a 12th-magnitude star. Many people perceive a greenish hue to this planetary, which is also known as the Box Nebula.

Next, we'll visit **M16** in far eastern Serpens. Philippe Loys de Chéseaux first noted it as a star cluster in an unpublished list he compiled in 1745-46. Charles Messier independently discovered the object in 1764 and described it as a "cluster of small stars, mingled with a faint light." The involved nebulosity is better known today as the Eagle Nebula.

M16 sits 2.6° west-northwest of Gamma (γ) Scuti and is visible through a 50mm finder under moderately dark skies. It's a lovely sight in my 8-inch refractor, and the nebula responds well to a hydrogen-beta filter. At 66x, the nebula covers 25' x 18' and actually looks rather like an eagle with its wings spread northeast and southwest. A large extension northwest speckled with stars is the eagle's tail, and a short bulge southeast forms the head. M16 is distinctly dappled with bright and dark patches. A 97x mirror-reversed view shows a dark L south of a fairly bright star near the heart of the nebula. The short part of the L points north-northeast and the long bar northwest. Other dark areas subtly fleck the eagle's head and northern wing.

Proceeding 2.2° west of M16, we come to a golden 6th-magnitude star. From there, moving 53' west-southwest brings us to the emission nebula **Sharpless 2-46** (Sh 2-46). Steven Coe, an Arizona amateur and author of *Deep Sky Observing: The Astronomical Tourist* (Springer, 2000), sent me a report of the view through his 6-inch Maksutov-Newtonian at 65x. He describes the nebula as a large and extremely faint glow around three stars, and he emphasizes that the object has a very low surface brightness.

Now take a lengthy leap to the north, where we find

another difficult Sharpless object, this one a planetary nebula. **Sharpless 2-68** (Sh 2-68) sits 52' northwest of the star 59 Serpentis. Look for a 7th-magnitude white star with a 9th-magnitude gold star 32' to its west. The southern edge of Sh 2-68 grazes the midpoint of a line connecting the two stars. I'm not certain I've ever really seen this object with my 10-inch reflector. Using low to medium power and an oxygen III filter, I occasionally sense a "presence" when scanning across the area. The planetary is even rather elusive in my 14.5-inch reflector. I've logged it as "possibly annular." Deep photographs show a brighter arc along the rim that may have fostered this impression. I estimate a diameter of about 5', which indicates that I am seeing only the brightest (northeastern) part of this thumbprint-shaped nebula.

A recent study indicates that the interstellar medium is disrupting Sh 2-68 and hindering the expansion of the nebulosity. The 16th-magnitude progenitor star, which is not affected by the slowing, has moved away from the center of the planetary and will eventually leave it behind.

Our next stop is **Theta (θ) Serpentis**, which sits at the bottom of a U-shaped asterism. Theta consists of a widely spaced pair of nearly matched, white suns, while the other four stars of the 15' asterism shine in shades of yellow and orange.

Czernik 38 lies 1.8° west-northwest of Theta. In my 10-inch scope at low power, I see this open cluster as only a 12'

A Midsummer Night's Milky Way

Object	Type	Magnitude	Size/Sep.	Right Ascension	Declination	*MSA*	PSA
ν Serpentis	Double star	4.3, 9.4	46"	$17^h\ 20.8^m$	−12° 51'	1347	56
NGC 6309	Planetary nebula	11.5	16"	$17^h\ 14.1^m$	−12° 55'	1347	56
M16	Cluster & nebula	6.0	28' × 17'	$18^h\ 18.7^m$	−13° 48'	1344	67
Sh 2-46	Bright nebula	—	27' × 17'	$18^h\ 06.2^m$	−14° 11'	1344	67
Sh 2-68	Planetary nebula	13.1	7'	$18^h\ 25.0^m$	+00° 52'	(1295)	(65)
θ Serpentis	Double star	4.6, 4.9	23"	$18^h\ 56.2^m$	+04° 12'	1270	65
Czernik 38	Open cluster	9.7	13'	$18^h\ 49.8^m$	+04° 58'	1270	(65)
IC 4756	Open cluster	4.6	50'	$18^h\ 38.9^m$	+05° 26'	1271	65
NGC 6633	Open cluster	4.6	27'	$18^h\ 27.2^m$	+06° 31'	1271	65
NGC 6572	Planetary nebula	8.1	11"	$18^h\ 12.1^m$	+06° 51'	1272	65

Angular sizes or separations are from recent catalogs. The visual impression of an object's size is often smaller than the cataloged value and varies according to the aperture and magnification of the viewing instrument. The columns headed *MSA* and *PSA* give the appropriate chart numbers in the *Millennium Star Atlas* and *Sky & Telescope's Pocket Sky Atlas*, respectively. Chart numbers in parentheses indicate that the object is not plotted.

Modern photography leads most observers to think of M16 as a spectacular red cloud of glowing hydrogen, but Charles Messier and his 18th-century contemporaries saw it primarily as an open star cluster entangled with a "faint glow," which is an apt description of its appearance in a small telescope. Photo: Russell Croman

First published August 2006

hazy patch with a 10th-magnitude star southeast of center and another near the northeastern edge. At 115x, the haze becomes granular, while 166x draws out many extremely faint stars gleaming like diamond dust.

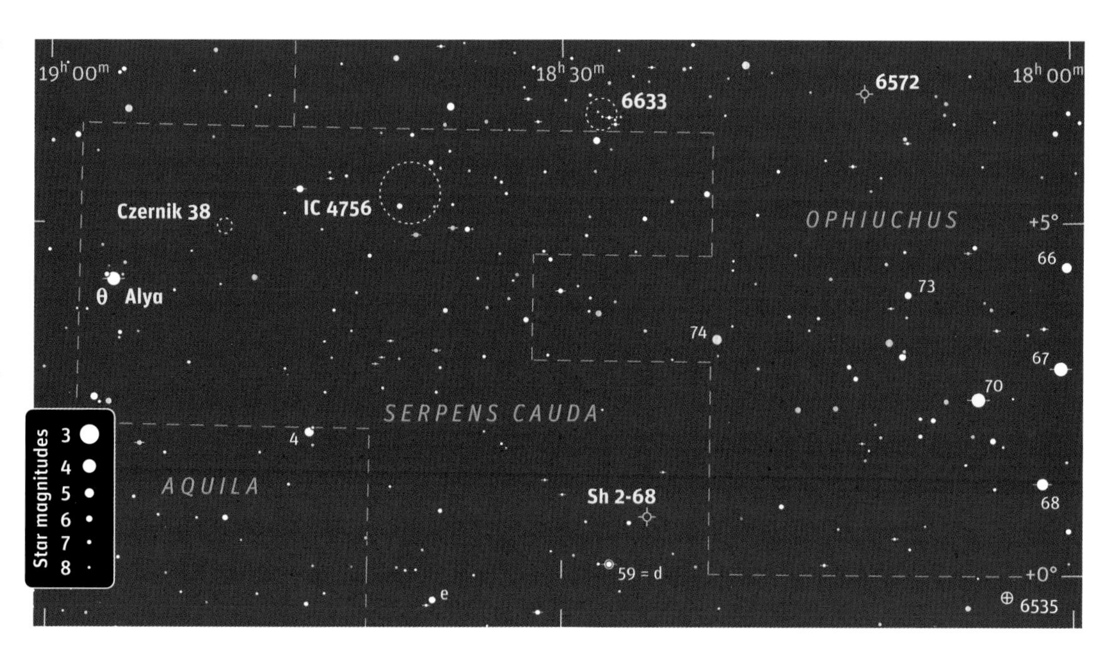

Scanning 2.7° farther west brings us to the very large open cluster **IC 4756**. Coe points out that it's visible to the unaided eye from a dark site and says, "I see it as a glow in the 'off ramp' of the Milky Way that points into Serpens and Ophiuchus. It's an obvious cluster in my 8x42 binoculars; 9 stars are resolved." Through his 6-inch reflector at 26x, Coe counts 55 stars and remarks that a rich-field telescope "provides a great view of this often overlooked, big, bright cluster."

IC 4756 shares the same binocular field of view with another open cluster, **NGC 6633**, which is 3° to the west-northwest. My 15x45 image-stabilized binoculars show about 40 faint stars in IC 4756, while NGC 6633 is a smaller, more elongated group of 15 brighter stars. Niklasch counts 30 stars in NGC 6633 with his 8-inch reflector at 74x. He notes its unusual shape and comments, "With all the summer globulars fresh in one's mind, plain young open clusters sure are an unruly bunch in comparison!" British amateur Richard Westwood calls NGC 6633 a "bright, splashy cluster" and remembers seeing Comet Bradfield against it in 1987.

Finally, we'll call on the bright planetary nebula **NGC 6572**, located 2.2° south-southeast of 71 Ophiuchi. Niklasch saw the nebula as a stellar point in his finder and recounts an interesting effect visible in his 8-inch scope. "I tried both a UHC and oxygen III filter on it," he writes, "revealing somewhat different zones of the nebulosity. Each filter caused a certain blink effect, with a rather small, bright zone showing up in direct vision and a larger one and halo in averted vision."

Westwood always tries to find this planetary at low power by its very green color. He writes, "To me it has the appearance of the green of a traffic light, striking at even a casual glance."

Return of the Hero

From double stars to galaxies, Hercules has lots to offer observers.

We've visited the southwestern reaches of Hercules, but that tour was certainly not enough to do justice to such a large constellation. The Hero abides in the evening sky yet awhile, so let's explore another area, one encompassing his most famous deep-sky wonder: spectacular Messier 13, the Great Star Cluster in Hercules.

M13 is easily found about two-thirds of the way from Zeta (ζ) to Eta (η) Herculis — Zeta, Eta, Epsilon (ε) and Pi (π) Herculis create a trapezoidal asterism known as the Keystone. The cluster is visible as a small misty patch to the unaided eye in a dark sky or through binoculars in a suburban sky. My husband Alan's 92mm refractor at 97x reveals M13 for the beautiful globular cluster it is. A large halo of pinpoint stars surrounds a dense core adorned with glittering flecks tangled in the fleecy glow of unresolved suns.

Through an 8-inch refractor at high power, M13 explodes with stars. Curving arms of outlying stars decorate the globular's 12' halo, and the hazy backdrop of the partially resolved core is riddled with dusky smudges.

Southeast of the cluster's center, three shadowy lanes of

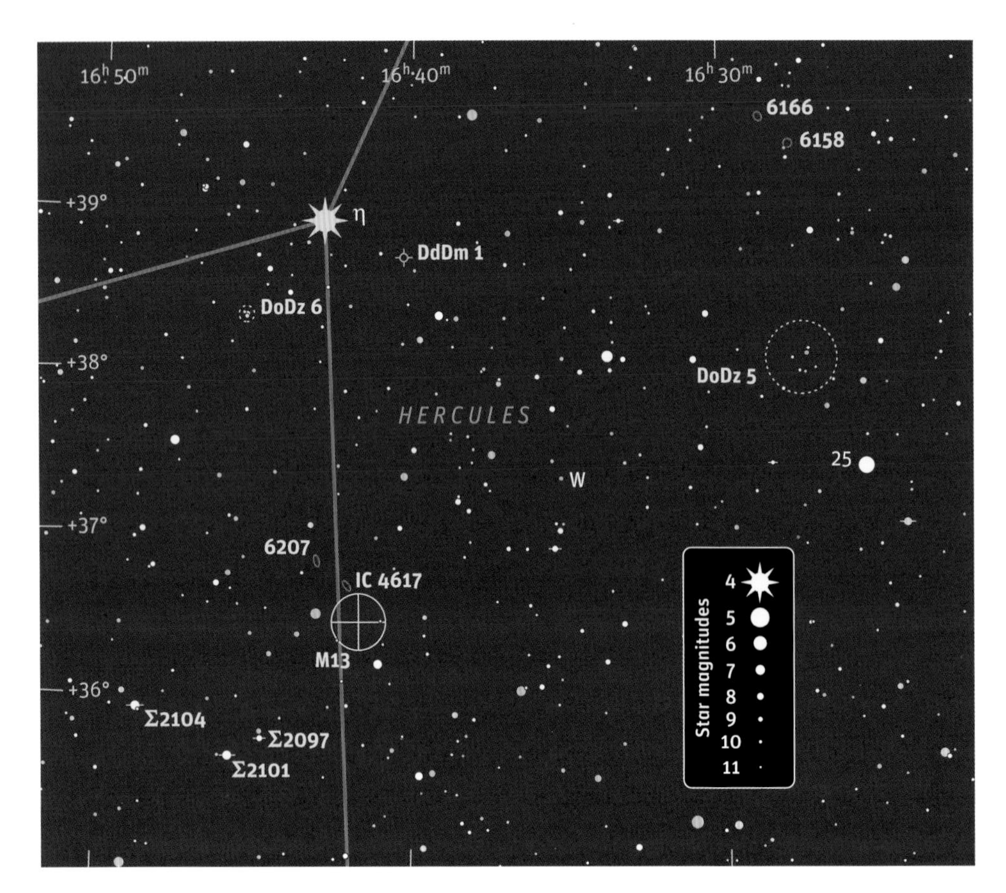

Three tight double stars that are nice tests for a small telescope lie about a degree southeast of M13. The widest pair is **Σ2104**, with a separation of 5.7", split at 47x with my 105mm refractor. Both stars appear white, with the 8.8-magnitude companion north-northeast of its 7.5-magnitude primary. (The extremely faint galaxy CGCG 197-14 sits just 1¼' north of this double. Can you spot it with a large scope?)

Σ2104 shares the field of view with **Σ2101**, whose secondary is barely visible at this power. A magnification of 68x does a better job of parting this 4.1" double and shows the 7.5-magnitude yellow-white primary holding a 9.4-magnitude gold attendant to the northeast.

The toughest of the trio is **Σ2097**; its nearly matched, east-west suns rest a mere 1.9" apart. The yellowish components are kissing at 87x, cozy at 122x, and comfortably split at 153x. Only 14' apart, Σ2097 and Σ2101 make a captivating "double double."

equal length and spacing meet to form a large Y shape. This three-blade propeller was first brought to light in a now famous drawing by Bindon Stoney, who was in charge of Lord Rosse's Birr Castle observatory in the mid-1800s. Spotting the propeller depends heavily on your optics and your sky, but it becomes much easier to discern once you've found it for the first time.

NGC 6207 is 28' northeast of M13 in the same low-power field. This little oval galaxy is easily seen in the 92mm refractor at 97x and improves little with increasing aperture until you get to a 10-inch or larger scope. With my 14.5-inch reflector at high power, NGC 6207 is somewhat patchy and gently brighter toward the center. A 13th-magnitude star is superposed immediately north of the small, fainter core. The galaxy is about 2' long and half as wide, tipped north-northeast.

The 15th-magnitude galaxy **IC 4617** lies midway along and a bit north of a line connecting NGC 6207 and M13. IC 4617 is quite a challenge with my 14.5-inch reflector at high power; I can catch sight of it only on the best nights at my semirural home. Its position is conveniently pointed out by a 1½' parallelogram of stars of about 14th magnitude. Look for the galaxy just west of the parallelogram's southwestern corner. With a large scope under dark skies, IC 4617 is a ½' streak leaning north-northeast. Mediocre conditions may limit the view to a glimpse of the galaxy's tiny core.

Now we'll move northward to the tongue-twisting asterism **Dolidze-Dzimselejsvili 6** (DoDz 6), located 45' southeast of Eta Herculis. With my small refractor at low power, DoDz 6 is just a little knot of four faint stars. At 87x, two very faint stars join the scene and help form a distinctive arrow-shaped group 4' long. Three stars in a little curve create the stem of the arrow. Two more to the southwest make the base of the arrowhead. And farther southwest, the faintest star marks the arrow's point. Since two of these stars are quite dim, the arrow stands out better in larger scopes. Proper-motion surveys indicate that these stars are moving in different directions through space and do not comprise a true cluster.

Our next stop is **Dolidze-Dzimselejsvili 1** (DdDm 1; PN G61.9+ 41.3), one of the few planetary nebulae known to reside in the halo of our galaxy. DdDm 1 is located 52,000 light-years away and 34,000 light-years above the galactic plane. Not surprisingly, this distant planetary looks stellar even at high power with my 10-inch scope. It's recognizable mostly by the fact that it appears brighter relative to the field stars when viewed through an oxygen III filter (which passes the light of doubly ionized oxygen) than it does without the filter. I find it fascinating to spot this

More Highlights of Hercules

Object	Type	Magnitude	Size/Sep.	Right Ascension	Declination	*MSA*	*U2*
M13	Globular cluster	5.8	20'	$16^h 41.7^m$	+36° 28'	1158	50R
NGC 6207	Galaxy	11.6	3.3' × 1.7'	$16^h 43.1^m$	+36° 50'	1158	50R
IC 4617	Galaxy	~15	1.2' × 0.4'	$16^h 42.1^m$	+36° 41'	(1158)	50R
Σ2104	Double star	7.5, 8.8	5.7"	$16^h 48.7^m$	+35° 55'	1158	50R
Σ2101	Double star	7.5, 9.4	4.1"	$16^h 45.8^m$	+35° 38'	1158	50R
Σ2097	Double star	9.4, 9.6	1.9"	$16^h 44.8^m$	+35° 44'	1158	50R
DoDz 6	Asterism	8.3	3.5'	$16^h 45.4^m$	+38° 21'	1158	50R
DdDm 1	Planetary nebula	13.4	1"	$16^h 40.3^m$	+38° 42'	(1159)	50R
NGC 6166	Galaxy	11.8	2.2' × 1.5'	$16^h 28.6^m$	+39° 33'	1159	51L
NGC 6158	Galaxy	13.7	0.5'	$16^h 27.7^m$	+39° 23'	1159	51L
DoDz 5	Asterism	7.8	27'	$16^h 27.4^m$	+38° 04'	1159	51L
NGC 6058	Planetary nebula	12.9	24" × 21"	$16^h 04.4^m$	+40° 41'	1138	51L

Angular sizes are from recent catalogs. The visual impression of an object's size is often smaller than the cataloged value and varies according to the aperture and magnification of the viewing instrument. The columns headed *MSA* and *U2* give the appropriate chart numbers in the *Millennium Star Atlas* and *Uranometria 2000.0*, 2nd edition, respectively. Chart numbers in parentheses indicate that the object is not plotted. All the objects this month are in the area of sky covered by Charts 52 and 53 in *Sky & Telescope's Pocket Sky Atlas*.

First published August 2007

Considered by many to be the showpiece deep-sky object of the northern summer sky, the giant globular star cluster M13 in Hercules can be glimpsed with the naked eye in a clear, dark sky. Large apertures can reveal a subtle dark feature that looks like a three-blade propeller in the cluster's southeastern (lower-left) quadrant. Look carefully, and you'll see it in the image below, which is ¾° wide with north up. Inset: The feature is absent, however, in a sketch of the cluster made by the author, since the dark lanes are too subtle to be seen in her 105mm refractor. Photo: *Sky & Telescope* / Sean Walker

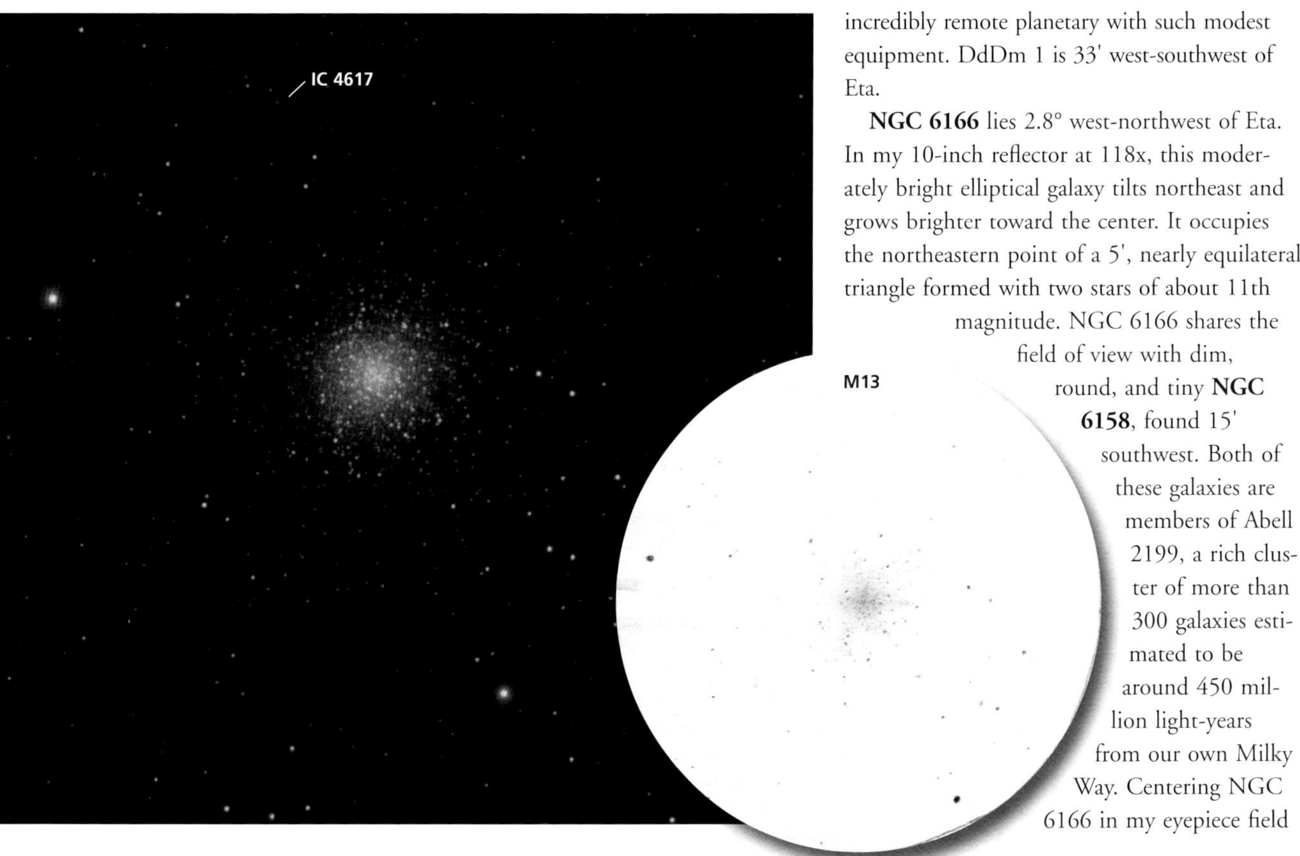

incredibly remote planetary with such modest equipment. DdDm 1 is 33' west-southwest of Eta.

NGC 6166 lies 2.8° west-northwest of Eta. In my 10-inch reflector at 118x, this moderately bright elliptical galaxy tilts northeast and grows brighter toward the center. It occupies the northeastern point of a 5', nearly equilateral triangle formed with two stars of about 11th magnitude. NGC 6166 shares the field of view with dim, round, and tiny **NGC 6158**, found 15' southwest. Both of these galaxies are members of Abell 2199, a rich cluster of more than 300 galaxies estimated to be around 450 million light-years from our own Milky Way. Centering NGC 6166 in my eyepiece field

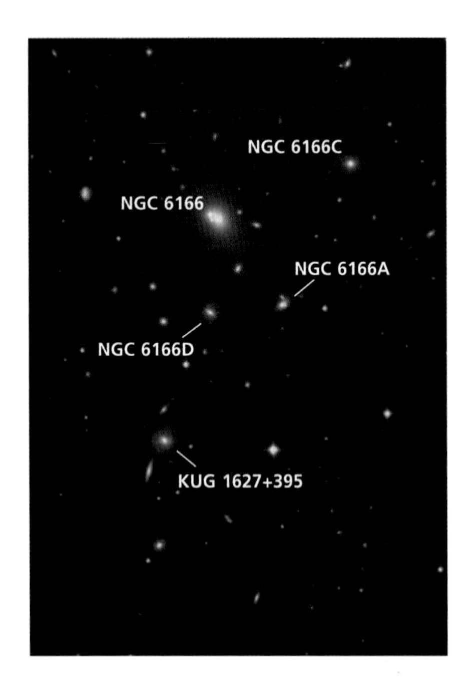

The elliptical galaxy NGC 6166 anchors Abell 2199, a rich cluster of some 300 galaxies estimated to lie 450 million light-years from us. Dark skies, a moderately large backyard telescope, and some patience should allow you to pick out a few members of the cluster. Photo: POSS-II / Caltech / Palomar

and boosting the magnification to 219x lets me tease out a few very faint, very small members of the group: NGC 6166C, 3.2' west-northwest of the middle of NGC 6166; NGC 6166A, 2.3' southwest; NGC 6166D, 2.0' south; and KUG 1627+395, 4.8' south-southeast.

Dropping 1.5° south from NGC 6166 brings us to **Dolidze-Dzimselejsvili 5** (DoDz 5), an asterism best viewed in a small telescope at low power. In my 105mm refractor at 28x, I see a 9' smattering of stars, magnitudes 9 to 12. The two brightest stars are in the north, with a skinny triangle of three to their east. Three stars arc along the southern edge of the group. According to most catalogs, DoDz 5 spans 27', but the outlying stars don't seem to be part of the asterism to me.

Our final target is the planetary nebula **NGC 6058**, located 2.8° southeast of Chi (χ) Herculis and just south of an east-west pair of 9th-magnitude stars. This pair constitutes the upper ends of a Y shape, while three dimmer stars to the south form the Y's slightly curved stem. Little NGC 6058 is nestled in the upper fork of the Y and is easily visible as a round, gray disk in my 105mm scope at 87x. The 10-inch reveals a faint central star. If you have a large telescope, look for brighter arcs along the north-northwest and south-southeast edges. NGC 6058 is only one-fifth as far away as is DdDm 1.

Stars Unnumbered

Explore the clusters and nebulae of eastern Ophiuchus.

Milky Way with stars unnumbered,
Merged in clouds of light;
Beauty, mystery, that arches
Round the summer night,
Yours are times that none can fathom,
Spaces none can span;
Light of ages, see in darkness
Your wee watcher, Man.
— Leland S. Copeland, *Sky & Telescope*, July 1949

This verse captures the grandeur of the summer Milky Way and the awe it continues to inspire. That glorious river of light never fails to evoke a sense of wonder when viewed against a truly dark sky. Our eyes are naturally drawn to the impressive starclouds in Sagittarius, beyond which the heart of our galaxy lies. Yet the muted fringes of the Milky Way harbor many celestial marvels that can be admired with modest optical aid. So let's explore the organdy curtain draped over the eastern reaches of Ophiuchus, the Serpent Bearer, as it's portrayed in *Sky & Telescope's Pocket Sky Atlas.*

The splashy open cluster **IC 4665** is barely within the boundary of the Milky Way on the atlas. It's easy to spot 1.3° north-northeast of Beta (β) Ophiuchi, which shares the field of view through binoculars or a finderscope. In my 15x45 image-stabilized binoculars, the cluster's prominent core spans more sky than the full Moon, and loosely scattered outliers extend to 70'. Half its 40 stars outline a rough circlet with one star at its center and a stem that leads westward to a slightly wavy line tipped north-northeast. To me, it looks like a simplistic drawing of a flower springing up from a gently undulating patch of ground. Finnish amateur Jaakko Saloranta comes away with a different impression. Through his 80mm refractor, he sees this group as the Fish Spear of Poseidon.

Shifting 53' west-northwest from Gamma (γ) Ophiuchi brings us to **NGC 6426**. This ghostly globular cluster is a

Power it up

Many deep-sky objects look great at low power, but planetary nebulae and globular clusters typically improve with higher magnification.

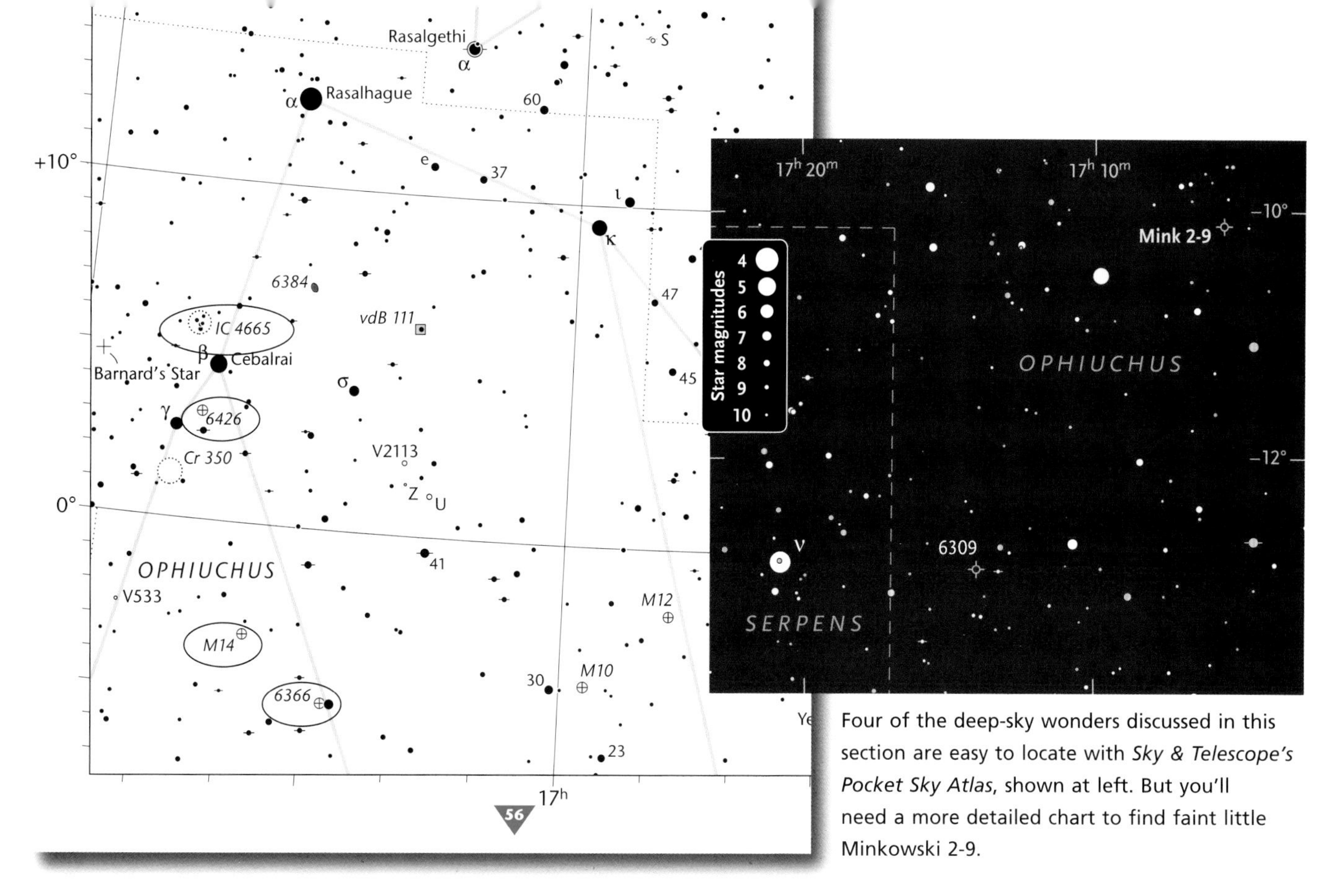

Four of the deep-sky wonders discussed in this section are easy to locate with *Sky & Telescope's Pocket Sky Atlas*, shown at left. But you'll need a more detailed chart to find faint little Minkowski 2-9.

wan 2.3' glow as seen through my 105mm refractor at 87x. A 14' mini-dipper of five 9th- to 11th-magnitude stars sits off the cluster's south-southwestern side. Through my 10-inch reflector at 213x, NGC 6426 is 3' across and unevenly textured, with a slightly brighter core. A few intermittent sparkles indicate stars at the limit of perception.

Since NGC 6426 is much farther from us than IC 4665, it's not surprising that its stars are more difficult to see. The clusters lie at distances of 67,500 and 1,150 light-years, respectively.

Considerably brighter than NGC 6426, the globular cluster **Messier 14** (M14) perches 4.9° due north of Mu (μ) Ophiuchi. Through 15x45 binoculars, it's simply a soft round glow. My refractor at 47x displays a very faint 5½' halo enfolding a bright 4' core that grows more intense toward the center. At 153x, a few extremely faint stars are visible. The core is mottled, and the halo is slightly oval northeast-southwest.

In my 10-inch reflector at 43x, M14 bears a bright 3' inner core, a fairly bright 5' outer core, and a faint halo that fades outward to 8'. At 115x, many stars bead the halo and are netted in haze at the core. The cluster is quite pretty at 213x. It is richly populated with very faint stars, while several brighter ones spangle its countenance.

The globular cluster **NGC 6366** lies 3.1° southwest of M14 and ¼° east of a 4.5-magnitude star. Although small and faint in my 105mm scope at 47x, it shows up well at 122x. A four-starred hockey stick is giving it a nudge. The handle starts west of NGC 6366 and slopes southeast, while the end of the blade prods the cluster's edge. Several very faint stars fleck the globular, the most prominent of them in the east and southeast. The group spans 5' in the refractor but grows to 6' through my 10-inch scope at 213x. This also keeps the distractingly bright star out of the field. At least two dozen faint stars can be distinguished, but the globular shows almost no brightening toward the center.

NGC 6366 is considerably closer to us than M14 (11,700 versus 30,300 light-years), but it appears fainter because it's less than 5 percent as luminous and suffers somewhat greater dimming by intervening space dust.

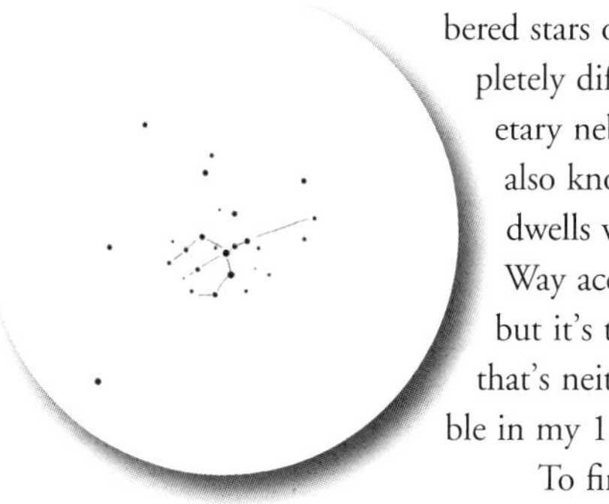

Finnish stargazer Jaakko Saloranta sees IC 4665 as the Fish Spear of Poseidon.

Now let's move from the virtually unnumbered stars of globular clusters to a completely different deep-sky wonder, the planetary nebula **Minkowski 2-9** (Mink 2-9), also known as Minkowski's Butterfly. It dwells within the confines of the Milky Way according to the *Pocket Sky Atlas*, but it's the only object discussed here that's neither plotted on the atlas nor visible in my 105mm refractor.

To find the nebula, hop 3.6° northwest from Nu (ν) Serpentis to a 5.4-magnitude star, the brightest in the area. Then continue for 1° in the same direc-

Above: A large backyard telescope shows the globular star cluster Messier 14 as a swarm of faint stars. Photo: Bernhard Hubl. Right: Minkowski 2-9 is a striking example of a "butterfly" (bipolar) planetary nebula, as evident in this image from the Hubble Space Telescope. Photo: B. Balick / V. Icke / G. Mellema / NASA

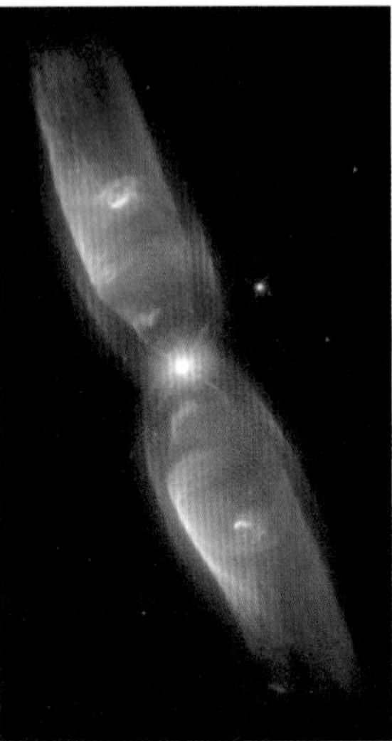

tion to a roughly east-west pair of stars, magnitudes 8½ and 9. West of this duo, in the same low-power field, is a very long, skinny, southward-pointing triangle of 9th- and 10th-magnitude stars. To reach Minkowski's Butterfly, extend the western side of the triangle far enough southward to double its length.

Through my 10-inch reflector at 115x, the planetary is a small faint spot that marks the northeastern corner of a lopsided trapezoid formed with three field stars, magnitudes 11 and 12. At 213x, the nebula is elongated north-south. The contrast is a little better with either an oxygen III or a narrowband filter. Through my 14.5-inch reflector at 170x, Minkowski 2-9 is two times longer than wide and hints at having two lobes. With its narrow wings and bluish tints (as seen in the stunning Hubble Space Telescope photo at left), perhaps this cosmic butterfly is a Sara Longwing.

Summertime Treats Along the Milky Way's Fringe

Object	Type	Magnitude	Size/Sep.	Right Ascension	Declination
IC 4665	Open cluster	4.2	70'	17^h 46.2^m	+5° 43'
NGC 6426	Globular cluster	11.0	4.2'	17^h 44.9^m	+3° 10'
M14	Globular cluster	7.6	11.0'	17^h 37.6^m	−3° 15'
NGC 6366	Globular cluster	9.2	13.0'	17^h 27.7^m	−5° 05'
Mink 2-9	Planetary nebula	14.6	39" × 15"	17^h 05.6^m	−10° 09'

Angular sizes and separations are from recent catalogs. Visually, an object's size is often smaller than the cataloged value and varies according to the aperture and magnification of the viewing instrument.

First published August 2008

The Glorious Eagle

Many lesser-known clusters and nebulae swarm around the mighty Eagle Nebula.

Glorious bird! thy dream has left thee,
Thou hast reached thy heaven —
Lingering slumber hath not reft thee
Of the glory given —
With a bold and fearless pinion,
On thy starry road,
None, to fame's supreme dominion,
Mightier ever trode.
— James Gates Percival, "Genius Waking"

The Eagle Nebula is one of the most celebrated deep-sky wonders, thanks to the stunning 1995 Hubble Space Telescope image showcasing the glorious Pillars of Creation, shown at lower right. These dense columns of gas and dust nestled in the heart of the Eagle Nebula nurture infant stars forming within. Yet these magnificent structures may be mere ghosts of a time long past.

Infrared images from the Spitzer Space Telescope show a hot gas cloud that some astronomers interpret to be a supernova shock wave advancing toward the Pillars. If true, the shock wave may have destroyed the Pillars 6,000 years ago. But because the Eagle Nebula is about 7,000 light-years away, it will be another thousand years or so before the devastation can be seen from the Earth's vicinity. The supernova itself might have graced our sky as a bright star one or two millennia ago.

How strange and rare that we're caught in the astronomically brief moment when we can gaze at a relatively nearby celestial treasure that may no longer exist! And the sight is not reserved for those imaging the sky or armed with immense telescopes. The Eagle can be viewed through small telescopes, and you can even glimpse a bit of the Pillars under dark, transparent skies.

On one particularly fine night, I devoted some time to sketching **Messier 16** (M16; the Eagle Nebula and its embedded star cluster) as seen through my 5.1-inch refractor at 63x. I find sketching to be a wonderful way to garner detail from complex deep-sky objects. As I study an object, I seem to slowly develop an enhanced image — almost as photographic film does over time. My sketch shows a thronelike structure emblazoned on the Eagle's breast. In the three Pillars of Creation, it corresponds to the central shaft and the bar connecting it to the longest pillar. This dark throne is approximately 4 light-years tall.

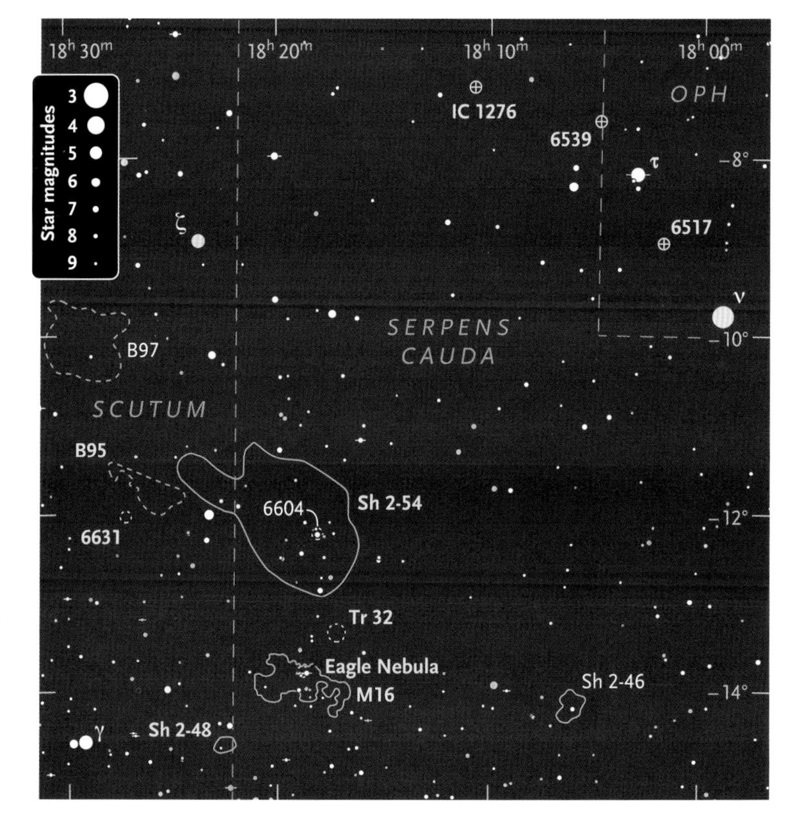

Photo: NASA / ESA / STScI / Jeff Hester / Paul Scowen

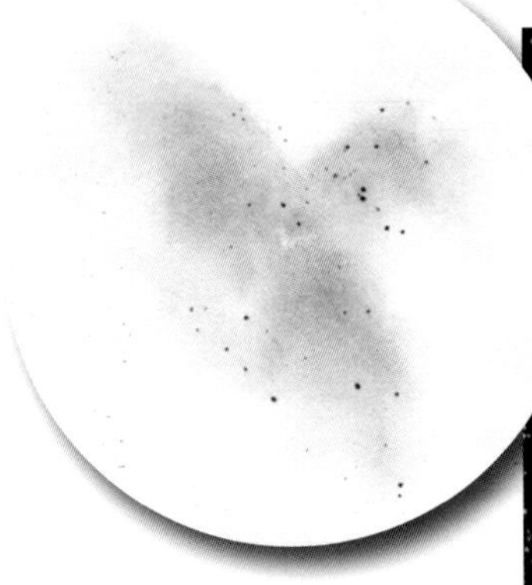

Right: This wide-field shot of the Eagle Nebula shows that the nebulosity extends far beyond what's visible to the human eye. Photo: Bernhard Hubl. Above: The author's sketch of the Eagle demonstrates that many of the features in the photograph are visible through the eyepiece of a 5.1-inch scope.

The Eagle dwells in a realm of the sky rife with nebulae and star clusters. Its unsung neighbor **Trumpler 32** (Tr 32) resides just 38' northwest and shares the field with M16 through my 6-inch reflector at 38x. This nice little cluster is a hazy cobweb with a corona of faint, dewdrop stars glimmering on the sable walls of night. At 154x, about 20 stars overspread 11' of sky. A prominent knot of several stars dominates the southern part of the cluster. At high magnification, my 10-inch reflector reveals a wealth of suns, many so faint that they blink in and out of view.

On the opposite side of the Eagle, we find the emission nebula **Sharpless 2-48** (Sh 2-48). Look 1° southeast of M16 for a 7th-magnitude star, the brightest in the area. Centered 12' south-southeast of that star is a 6' trapezoid of stars 9th magnitude and fainter. Through my 6-inch reflector at 95x, the brightest region of this patchy nebula surrounds the western side of the trapezoid and spreads 4' westward. Dimmer nebulosity extends to the trapezoid's eastern side and becomes more attenuated as it reaches southeast.

The large, faint nebula **Sharpless 2-54** (Sh 2-54) lies 1¾° north of the Eagle. Under a dark sky, binoculars will give you a nice view if you place the Eagle in the southern part of the field. I perused this emission nebula through my 15x45 image-stabilized binoculars while in the northern Adirondack Mountains, one of the darkest areas in my

The Eagle and Its Nestlings

Object	Type	Magnitude	Size/Sep.	Right Ascension	Declination
M16	**Cluster / nebula**	**6.0**	**34' × 27'**	**18^h 18.8^m**	**–13° 50'**
Tr 32	**Open cluster**	**12.2**	**12'**	**18^h 17.2^m**	**–13° 21'**
Sh 2-48	**Emission nebula**	**—**	**10'**	**18^h 22.4^m**	**–14° 36'**
Sh 2-54	**Emission nebula**	**—**	**144' × 78'**	**18^h 19.7^m**	**–12° 04'**
B95	**Dark nebula**	**—**	**30'**	**18^h 25.6^m**	**–11° 45'**
NGC 6631	**Open cluster**	**11.7**	**7.0'**	**18^h 27.2^m**	**–12° 02'**
NGC 6517	**Globular cluster**	**10.2**	**4.0'**	**18^h 01.8^m**	**–8° 58'**
τ Ophiuchi	**Double star**	**5.3, 5.9**	**1.6"**	**18^h 03.1^m**	**–8° 11'**
NGC 6539	**Globular cluster**	**9.3**	**7.9'**	**18^h 04.8^m**	**–7° 35'**
IC 1276	**Globular cluster**	**10.3**	**8.0'**	**18^h 10.7^m**	**–7° 12'**

Angular sizes and separations are from recent catalogs. Visually, an object's size is often smaller than the cataloged value and varies according to the aperture and magnification of the viewing instrument.

home state of New York. They revealed a 1° x 2° star-rich haze that's wide in the west and tapers toward the east-northeast. Much of the haze I see may well be the light of unresolved stars. This section of the Milky Way stands out as a somewhat isolated starcloud bounded on the west by the dusky clouds of the Great Rift and lined with smaller dark nebulae on the east. Nonetheless, it's a pretty sight.

Barnard 95 (B95) is the most obvious dark nebula on the eastern end of Sh 2-54. Through my 6-inch scope at 38x, it seems to grow blacker toward the center. Barnard 95's inkiest area spans about 10', while the entire nebula is perhaps twice that size and quite irregular. The open cluster **NGC 6631** shows as a little misty patch off its southeastern side. The cluster is elongated southeast-northwest with an 11th-magnitude star at the northwestern end. At 95x, NGC 6631 shows 20 faint to very faint stars in a 5' glow, while at 154x, it's crowded with minute stars.

Next, we'll trek northwest to a trio of globular clusters. The first is **NGC 6517**, located 1.1° northeast of Nu (ν) Ophiuchi and 5' north-northeast of a 10th-magnitude star. The cluster is easy to overlook at low power but emerges nicely through my 105mm refractor at 122x. It covers a scant 1½' and grows brighter toward the center. A 9'-long zigzag of six field stars passes west of the cluster. With my 10-inch reflector at 115x, NGC 6517 is 2½' across and clasps a tiny bright nucleus. The outer regions appear granular, while the interior is coarsely mottled. NGC 6517 is elongated northeast-southwest at 213x, and a faint star guards its south-southwestern edge.

Moving 50' north-northeast of NGC 6517 brings us to the lovely double star **Tau (τ) Ophiuchi**. Tau is a visual binary with a period of 257 years. In 2011, the companion is 1.6" west-northwest of its primary. The separation will close to 1.5" in 2015, 1.4" in 2021, and 1.3" in 2027. The pair is comfortably split through my 105mm refractor at 174x. To my eye, the brighter star glows yellow-white and its companion shines yellow. As we'd expect for a double that exhibits a visible shift in position over a relatively short period of time, Tau is a nearby pair only 170 light-years away.

At low power, Tau shares the field of view with the globular cluster **NGC 6539**, which straddles the Ophiuchus-Serpens border 44' northeast of Tau. Through my 105mm scope at 87x, this softly glowing ball of light is 5' across and patchy in brightness. Some faint foreground stars dot the western side of the cluster.

Nudging the scope 1.5° east-northeast takes me to **IC 1276**, nestled in the northern corner of a 13' triangle of 11th- and 12th-magnitude stars. This dim globular cluster presents a 2' halo and a tiny mottled center barely ½' across. At 122x, I see a very faint star punctuating the cluster's western side and an occasional glint in the core. With my 10-inch scope at 213x, several threshold stars form an east-west band across the cluster's face.

NGC 6539 has the same intrinsic brightness as NGC 6517, but it's closer to us and looks correspondingly brighter. These clusters lie at distances of 27,400 and 35,200 light-years, respectively. IC 1276 is comparatively nearby at 17,600 light-years, but it gives off a fifth as much light. This makes it the faintest of the three globulars by a small margin. The natural glory of these clusters is diminished not only by distance but also by interstellar dust between us and them. The dust grains absorb and scatter the light emitted by the globulars, dimming each by about three magnitudes.

The emission nebula Sharpless 2-54 spans more than 2° of sky. When scanning the nebula with binoculars, the author overlooked NGC 6604, the tiny knot of stars in the southwestern sector of the nebula. Can you see this open cluster in your telescope? Photo: Dean Salman

First published August 2009

Gold Dust

Bright star clusters and dark clouds pepper the section of the Milky Way that lies southwest of the galactic center.

A broad and ample rode, whose dust is Gold
And pavement Starrs, as Starrs to thee appeer,
Seen in the Galaxie, that Milkie way
Which nightly as a circling Zone thou seest
Pouderd with Starrs.

— John Milton, *Paradise Lost*

When gazing at the Sagittarius Milky Way, you are seeing the inner realms of our galaxy, where throngs of distant suns that softly powder our summer sky dwell. Our galaxy's true heart is hidden behind rafts of interstellar clouds, yet what a view we have! Let's visit this star-paved celestial boulevard and see what wonders it holds.

We'll begin with the two golden motes that make up the double star **Piazzi 6** (Pz 6), which was initially noted by Giuseppe Piazzi, the Italian astronomer who, in 1801, discovered the first asteroid, Ceres. The pair is often referred to as h5003, its designation in John Herschel's subsequent catalog. Pz 6 is easy to locate 1½° west of Gamma (γ) Sagittarii, and it's visible to the unaided eye in a dark sky.

This colorful duo was pointed out to me by North Carolina amateur David Elosser. Through my 105mm refractor at 76x, the 5th-magnitude primary appears yellow-orange. The 7th-magnitude companion 4.3" east-southeast has a deep yellow tint. The components are red-giant stars about 775 and 85 times the diameter of our Sun.

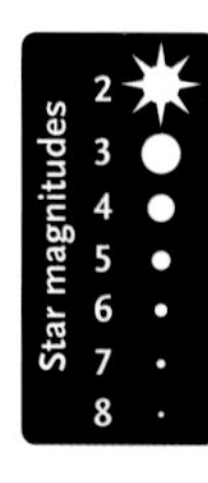

Globular star clusters also have the gleam of gold dust about them. These ancient systems, mostly 10 to 12 billion years old, contain many red giants that affect their overall hue. For example, **NGC 6569** has a yellowish glow similar to a star of spectral type G1. Although some observers note this hue, I'm not particularly sensitive to the colors of globular clusters. Can you detect the pale Sun-yellow glow of NGC 6569?

NGC 6569 sits 2¼° southeast of Gamma and 8' north of a 6.8-magnitude star. Through my 105mm scope at 28x, it's visible as a small hazy globe bordered by a few faint field stars. At 87x, the cluster displays a dim 3' halo enveloping a bright 1' core. In my 10-inch reflector at 43x, NGC 6569 is softly dappled and grows gently brighter toward the center. At 213x, the globular appears granular, but I can't resolve its individual stars, which top out at magnitude 14.7. What size scope will compel NGC 6569 to surrender some of its members to your view?

The dark nebula **Barnard 305** (B305) is centered 13' east of NGC 6569. In his glorious and rare *A Photographic Atlas of Selected Regions of the Milky Way*, Edward Emerson Barnard lists B305 as being irregular and 13' across. He also notes, "Dark streamers radiate from this spot to the north for more than ¾° and broken ones for ½° toward the southwest."

From my disadvantaged latitude of 43° north, the most prominent part of this nebula is a 9½' x 7½' dusky blot from which sooty tendrils branch out to fracture the starry background.

NGC 6569 and Barnard 305 share the field of view through my 105mm scope at 76x, and when I nudge my scope westward until NGC 6569 is at the edge of the field, **NGC 6558** comes into view. Even though I spotted it right away, this globular cluster is really quite ghostly in my 105mm refractor at 76x. It has a 1' core surrounded by an ethereal halo that doubles its size. At 122x, the core appears to be mottled and the halo is very patchy. A number of faint field stars closely fence the cluster.

NGC 6569 is farther away than NGC 6558, about 35,000 light-years compared with 24,000. In fact, they're the most distant and the closest of the six globulars we'll visit in this sky tour. Despite this, NGC 6569 looks brighter than NGC 6558 because it's intrinsically more luminous. It emits more than three times as much visible light as NGC 6558, its apparent neighbor on the sky. On the other hand, NGC 6558 is the least luminous of the globulars we'll explore.

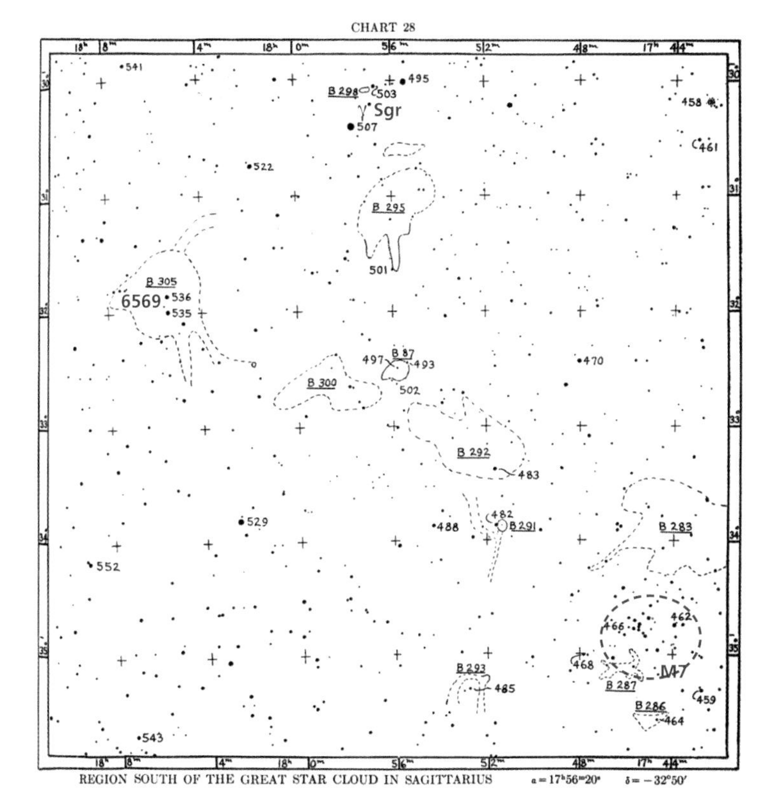

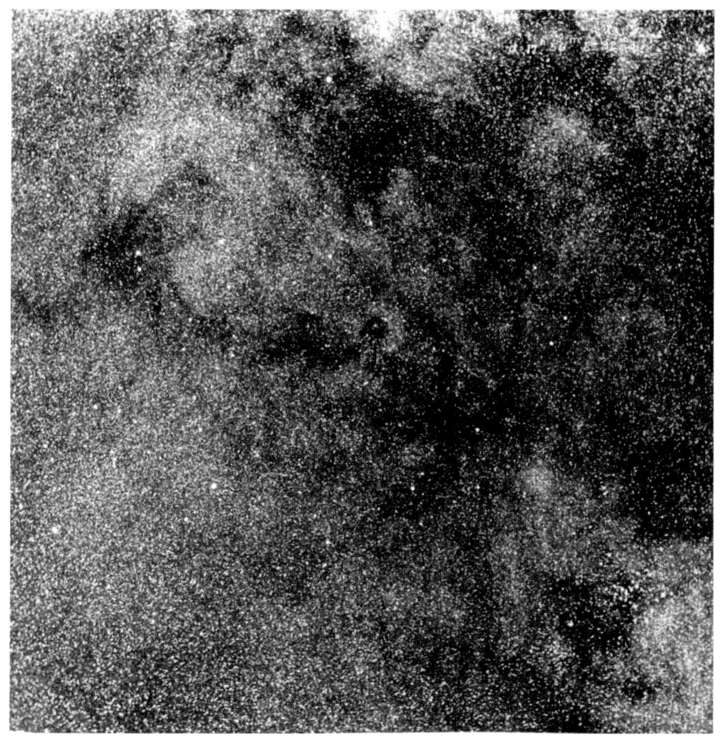

Edward Emerson Barnard's *A Photographic Atlas of Selected Regions of the Milky Way* was the first work to identify many of the dark nebulae observed by amateur astronomers. Plate 28, covering southwestern Sagittarius and a piece of Scorpius, is pictured above. Barnard's chart, top, shows his interpretation of the dark nebulae in the photo. The red outline and labels have been added to clarify the correspondence between this chart and the one on page 200.

To the southwest, an intricate tracery of dark nebulae is espaliered against the Milky Way. The most distinct area is **Barnard 87** (B87), also known as the Parrot's Head, which spans about 13'. The Parrot faces east with three 10th-magnitude stars lining his chin and a fourth marking his eye. **Barnard 300** (B300) is larger and sprawls east-southeast of B87, while southwest of B87, **Barnard 292** (B292) is a broken complex of dark filaments splayed across 1°. All three nebulae can be distinguished through my 105mm scope at 76x.

For a brighter globular cluster, seek out **NGC 6624** ¾° southeast of Delta (δ) Sagittarii. In my 5.1-inch refractor at 23x, it shares the field of view with Delta and the lovely yellow and gold optical double star **Rousseau 31** (RSS 31). At 87x, the cluster reveals a 3½' mottled halo enclosing a bright 2' core that intensifies toward a very small, brilliant nucleus. A star pair pins the west-southwestern edge of NGC 6624, its components shining at magnitudes 11 and 13, while a 12th-magnitude star sits just off the cluster's east-southeastern side. Through my 10-inch scope at 213x, a few stars sparkle in the halo.

NGC 6624 is brighter than NGC 6569, and with an overall spectral type equivalent to a G4 or G5 star, it's also a slightly deeper shade of yellow. Is this color easier for you to see?

Brighter yet is the globular cluster **Messier 69** (M69), which has a G2 to G3 spectral type and rests 2½° northeast of Epsilon (ε) Sagittarii. Viewed in my 9x50 finder, M69 is a hazy spot with an 8th-magnitude star near the northwestern edge. In my 5.1-inch scope at 102x, the cluster shows a faint halo 4' across and a bright core half as large with a vivid center. The cluster is very pretty at 164x, with at least two dozen stars of mixed brightness scattered across the halo and inner core.

At low power, M69 shares the field with dimmer **NGC 6652**, just 1° southeast. In the 5.1-inch refractor at 164x, the halo spans nearly 3'. Feeble stars hugging the globular's bright, granular core transform it into a 1' oval tipped a little north of west.

M69 and NGC 6652 are about the same distance away from us (30,000 and 33,000 light-years), but M69 is twice as luminous.

The globular cluster **Messier 70** (M70)

First published August 2010

These Hubble Space Telescope images resolve far more stars in Messier 69 (left) and Messier 70 (right) than you can see through the eyepiece of any telescope. Both images are roughly 5' square, but Messier 69's exposure is considerably deeper. Photos: NASA / STScI / WikiSky

lies 2½° due east of M69. I can barely cram both into the same field of view through my 5.1-inch scope with a wide-angle eyepiece giving a magnification of 23x. At 164x, the cluster spans 3' with a core that's half as large and grows brighter toward the center. M70 is spangled with about 10 stars, some of them foreground, while the inner core is very mottled and bright with an additional star or two weakly glimmering within.

Our last stop is the pint-size planetary nebula **IC 4776**, found 1.2° south-southeast of M70 and forming a squat isosceles triangle with 9th-magnitude stars approximately 7½' northeast and west-northwest. Although the nebula may appear starlike even at high power, it can be recognized by its telltale hue. In my 5.1-inch refractor at 164x, it's easy to spot, very small, blue-gray, and possibly brighter in the center. I've heard other observers describe it as blue or green.

Stardust in Sagittarius

Object	Type	Magnitude	Size/Sep.	Right Ascension	Declination
Piazzi 6	**Double star**	**5.4, 7.0**	**4.3"**	**17^h 59.1^m**	**–30° 15'**
NGC 6569	**Globular cluster**	**8.6**	**6.4'**	**18^h 13.6^m**	**–31° 50'**
B305	**Dark nebula**	**—**	**13' / 80'**	**18^h 14.7^m**	**–31° 49'**
NGC 6558	**Globular cluster**	**9.3**	**4.2'**	**18^h 10.3^m**	**–31° 46'**
B87	**Dark nebula**	**—**	**13'**	**18^h 04.2^m**	**–32° 32'**
B300	**Dark nebula**	**—**	**45' × 30'**	**18^h 07.0^m**	**–32° 39'**
B292	**Dark nebula**	**—**	**1°**	**18^h 00.6^m**	**–33° 21'**
NGC 6624	**Globular cluster**	**7.9**	**8.8'**	**18^h 23.7^m**	**–30° 22'**
RSS 31	**Double star**	**8.1, 8.7**	**45.5"**	**18^h 24.4^m**	**–29° 32'**
M69	**Globular cluster**	**7.6**	**9.8'**	**18^h 31.4^m**	**–32° 21'**
NGC 6652	**Globular cluster**	**8.6**	**6.0'**	**18^h 35.8^m**	**–32° 59'**
M70	**Globular cluster**	**7.9**	**8.0'**	**18^h 43.2^m**	**–32° 18'**
IC 4776	**Planetary nebula**	**10.8**	**8.8" × 4.7"**	**18^h 45.8^m**	**–33° 21'**

Angular sizes and separations are from recent catalogs. Visually, an object's size is often smaller than the cataloged value and varies according to the aperture and magnification of the viewing instrument.

Lyre Lessons

High overhead, harplike Lyra features a treasury of deep-sky sights for September stargazers.

At the center of the September all-sky star map on page 310, you'll find **Vega** — the fourth-brightest star in the night sky and the brightest star in the constellation Lyra, the Lyre. According to one myth, the harplike Lyre belonged to Orpheus, who charmed every living creature with its music. When his young wife, Eurydice, died, Orpheus went into Hades to seek her return. With the beauty of his music, he convinced Pluto, the king of the underworld, to let her go. Orpheus was warned not to look back at his bride until they had both reached the light of the upper world. When Orpheus stepped into the daylight, he eagerly glanced back, but it was too soon. Eurydice was still within the cavern, and as she vanished into the darkness, he heard only a faint "Farewell." Thereafter, Orpheus wandered heartbroken, and when he died, the gods placed his Lyre among the stars.

Lyra is the site of many deep-sky treasures suitable for a small telescope, among them one of the most renowned multiple stars in the sky: **Epsilon (ε) Lyrae**, the Double-Double. Look for a 4th-magnitude star 1.7° east-northeast of Vega. Binoculars or a finder will reveal that this star is really a double. In fact, under good viewing conditions, I can just split this pair with the unaided eye. Each of these stars is itself a double. Through a 92mm refractor, both star pairs are cleanly split at 94x, and 169x puts lots of space between them. In 2011, the close pairs have nearly the same separation, 2.3" for Epsilon^1 (the northern duo) and 2.4" for Epsilon^2. Over the following two dozen years, the components of Epsilon^1 widen close to 2.0" with the secondary remaining roughly north-northwest of its primary, while those of Epsilon^2 widen to 2.5" with the secondary east-northeast. All four stars appear white to slightly yellowish white. To split these tight doubles, you need good optics and atmospheric steadiness and your scope must be nearly at equilibrium with the outdoor temperature.

Four 3rd- and 4th-magnitude stars southeast of Vega outline a parallelogram that forms the body of Orpheus's Lyre. The nearest is the pair made up of **Zeta^1 (ζ^1) Lyrae**, a 4.3-magnitude white star, and **Zeta^2** (ζ^2), its 5.7-magnitude pale yellow-white companion, which lies 44" away. They can be found 1.9° southeast of Vega and are easily separated at 20x.

Next in the parallelogram, we come to a more colorful double, **Delta^1** (δ^1) and **Delta^2** (δ^2) **Lyrae**, 2° east-southeast of Zeta. At magnitude 4.2, Delta^2 is the brighter star and is distinctly reddish orange. Delta^1 is a prettily contrasting blue-white star at magnitude 5.2. This wide pair can be seen through steadily supported binoculars and is actually part of

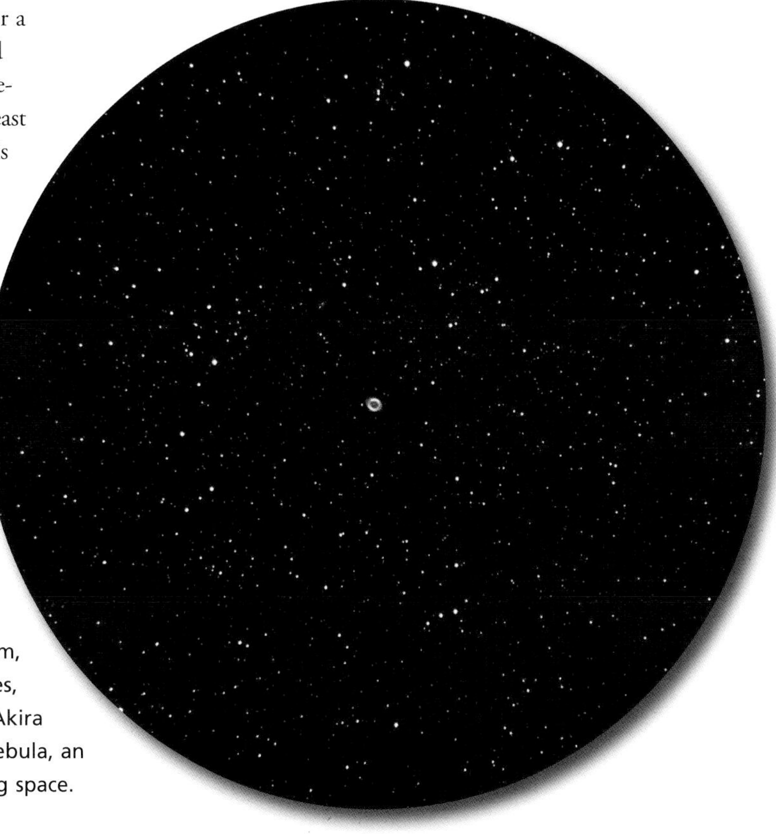

The Ring Nebula (M57) in Lyra appears as a small, dim, gray doughnut of light through most small telescopes, but a time-exposure photograph like this one from Akira Fujii reveals its true colors. The Ring is a planetary nebula, an aging star blowing much of its mass into surrounding space.

the sparse open cluster **Stephenson 1**. A small telescope will show 10 more stars in the group (scattered mostly to the south of Delta[1] and Delta[2]), making this cluster appear 16' across.

Dropping 3.7° south-southwest from Delta, we come to a variable star whose brightness changes can be followed with the naked eye. **Beta (β) Lyrae** is an eclipsing binary whose components are so close that they are distorted into ellipsoids by their mutual gravitation and rapid rotation. The system shows a continuous change in brightness, with the primary minimum coming once every 13 days. At maximum, it is about the same brightness as 3.2-magnitude Gamma (γ) Lyrae, while at minimum, it is roughly the brightness of 4.3-magnitude **Kappa (κ) Lyrae**. Compare Beta with these two stars, and sooner or later, you'll catch it in eclipse.

Our last parallelogram star is **Gamma (γ) Lyrae**, located 2° east-southeast of Beta. This bluish white star makes a nice binocular or low-power double with nearby orange **Lambda (λ) Lyrae**. Look three-fifths of the way from Gamma to Beta to find the showpiece of Lyra: **M57**, the Ring Nebula. This planetary nebula is very small but distinctly nonstellar, even at 20x. Through the 92mm refractor at 94x, M57 is a lovely little oval doughnut of light. The area within the ring appears brighter than the background sky. The star at the center of the Ring Nebula shows well on

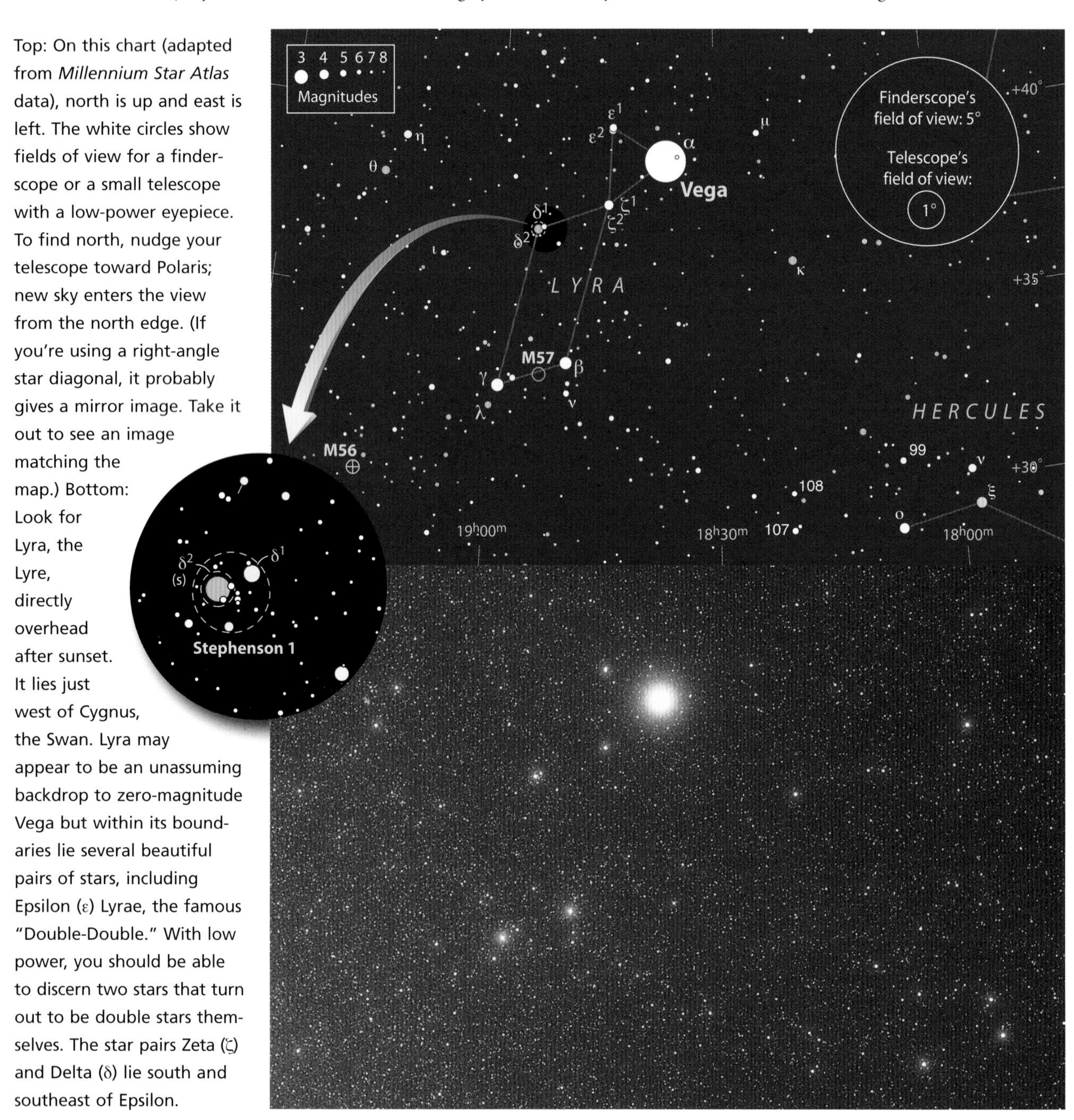

Top: On this chart (adapted from *Millennium Star Atlas* data), north is up and east is left. The white circles show fields of view for a finderscope or a small telescope with a low-power eyepiece. To find north, nudge your telescope toward Polaris; new sky enters the view from the north edge. (If you're using a right-angle star diagonal, it probably gives a mirror image. Take it out to see an image matching the map.) Bottom: Look for Lyra, the Lyre, directly overhead after sunset. It lies just west of Cygnus, the Swan. Lyra may appear to be an unassuming backdrop to zero-magnitude Vega but within its boundaries lie several beautiful pairs of stars, including Epsilon (ε) Lyrae, the famous "Double-Double." With low power, you should be able to discern two stars that turn out to be double stars themselves. The star pairs Zeta (ζ) and Delta (δ) lie south and southeast of Epsilon.

First published September 2000

A Lyre Among the Stars

Object	Type	Magnitude	Distance (l-y)	Right Ascension	Declination
ε^1 Lyrae	Double star	5.1, 6.2	162	18h 44.3m	+39° 40'
ε^2 Lyrae	Double star	5.3, 5.5	160	18h 44.4m	+39° 37'
ζ^1 Lyrae	Star	4.3	153	18h 44.8m	+37° 36'
ζ^2 Lyrae	Star	5.7	150	18h 44.8m	+37° 35'
δ^1 Lyrae	Star	5.6	1,000	18h 53.8m	+36° 56'
δ^2 Lyrae	Star	4.2	900	18h 54.5m	+36° 54'
Stephenson 1	Open cluster	3.8	1,000	18h 54.0m	+36° 52'
β Lyrae	Star	3.3–4.3	900	18h 50.1m	+33° 22'
γ Lyrae	Star	3.2	634	18h 58.9m	+32° 41'
κ Lyrae	Star	4.3	200	18h 19.8m	+36° 04'
λ Lyrae	Star	4.9	1,500	19h 00.0m	+32° 09'
M57 (Ring Nebula)	Planetary nebula	9.7	2,000	18h 53.6m	+33° 02'
M56	Globular cluster	8.2	31,000	19h 16.6m	+30° 11'

most photographs, but it is not visible through a small telescope and is a challenge even in large amateur instruments. The Ring Nebula is a cloud of gas being shed from the dying star. It is actually tunnel-shaped but appears oval because we are seeing it nearly face-on.

Our final target is the globular cluster **M56**, found 4° east-southeast of Lambda with a 6th-magnitude orange star 26' to the northwest. Through a small scope at low power, M56 looks like a small, faint, round blur lying in a beautiful, rich field of stars. It shows considerable brightening toward the center, and a 10th-magnitude star can be seen outside the western edge. When the air is clear, a 6-inch scope at 200x will partly resolve this globular. The designation M56 means that this cluster is the 56th object in a famous catalog compiled by comet hunter Charles Messier. His list was mostly meant to sort out deep-sky objects merely masquerading as comets. Although many of the Messier objects are decidedly not cometary through today's small telescopes, M56 is a prime example of an object that could be mistaken for a distant comet that has not yet grown its tail.

As you wander amid the sights of Lyra on late-summer nights, let your imagination draw you into what John Milton described in "Il Penseroso" as the haunting melody that "drew iron tears down Pluto's cheek and made Hell grant what love did seek."

The Gem of the Milky Way

Star clusters both dense and thin populate the eye-catching Scutum Starcloud.

Behind the tail of Aquila, the Eagle, a bright patch of Milky Way punctuates this otherwise hazy band. In his 1927 *Photographic Atlas of Selected Regions of the Milky Way*, Edward Emerson Barnard devoted a section entitled "The Great Starcloud in Scutum" to this island of distant suns. He began: "This, the gem of the Milky Way, is the finest of the starclouds. It is interesting from many points of view. The main body is apparently made up of extremely minute stars. The great hammer-like head, however, looms up to the west with much coarser stars, as if it were much nearer to us."

Here we see the Sagittarius Arm of our galaxy as it curves toward us. Countless stars along our line of sight create a dense concentration that is a wonder to behold in binoculars. The Scutum Starcloud is bordered by the Great Rift, a dark dust lane that plunges from Cygnus to Scutum before broadening and turning west through Serpens Cauda toward Ophiuchus. This shadowy rift helps make the relatively bright starcloud one of the easiest features to see in the Milky Way.

The constellation where the starcloud resides is Scutum, the Shield. On our September all-sky star map on page 310, this is shown as a long, skinny diamond of 4th-magnitude stars. Scutum's brightest telescopic treasure is **M11**, a beautiful open cluster set against the northern border of the starcloud. In a finder, M11 can be seen as a small,

One of the Milky Way's brightest patches is the Scutum Starcloud, about two fists at arm's length southwest of the 1st-magnitude star Altair. The rectangle at lower right shows the region charted at a much larger scale in the finder chart on the facing page. Photo: Akira Fujii

misty spot sharing the field with yellow Beta (β) Scuti.

Through a telescope, M11 is a magnificent congeries of tiny stars. It is often called the Wild Duck Cluster for a V-shaped structure that I generally find less than evident at the eyepiece. On one occasion, however, when I observed M11 with the light dome of a nearby city washing out the faint stars, the comparison was a bit more apparent. With my 105mm refractor at 153x, the cluster appeared quite rich in faint stars, but a prominent dark lane in the shape of a very wide V sundered the bulk of the cluster from a straggling line of stars that did bring to mind a flock of migrating geese heading southeast. This left the rest of the stars gathered into a wedge-shaped group (another flock?) with the cluster's brightest star near the point. West of the bright star, which is probably in the foreground, the wedge was further broken by complex dark lanes plaited among the stars and reaching to the far edge of the cluster.

M11 was discovered in 1681 by Berlin astronomer Gottfried Kirche, who called it a nebulous star. A quote often attributed to Kirche, but actually from a 1715 paper by Edmond Halley, specifically mentions M11's foreground star: "It is of its self but a small obscure Spot, but has a Star that shines through it, which makes it the more luminous." In 1733, English clergyman William Derham laid claim to being the first to resolve the cluster's fainter stars, but he may have been referring to the Scutum Starcloud as a whole.

Now let's visit **NGC 6664**, a more scattered group 20' east of Alpha (α) Scuti. With 11x80 binoculars, Auke Slotegraaf of South Africa gives this charming description: "Beautiful! A soft glow, round, upon a murky field. A delicate object, surprisingly large, like percolated starlight." My less poetic notes, made while I was observing with a 6-inch reflector at 95x, simply log about 25 stars of 10th magnitude and fainter in an irregular group 16' across. Nearby Alpha is yellow-orange.

NGC 6664's brightest star is usually **EV Scuti** (EV),

The Scutum Starcloud's Secrets

Object	Type	Magnitude	Size/Sep.	Distance (l-y)	Right Ascension	Declination	*MSA*	*U2*
M11	Open cluster	5.8	13'	6,100	$18^h\ 51.1^m$	−6° 16'	1318	125R
NGC 6664	Open cluster	7.8	16'	3,800	$18^h\ 36.5^m$	−8° 11'	1319	125R
EV Scuti	Variable star	9.9–10.3	—	3,800	$18^h\ 36.7^m$	−8° 11'	1319	125R
I68-603	Stellar ring	—	8'	3,800	$18^h\ 36.6^m$	−8° 17'	1319	125R
M26	Open cluster	8.0	14'	5,200	$18^h\ 45.2^m$	−9° 23'	1318	125R
NGC 6712	Globular cluster	8.1	10'	22,500	$18^h\ 53.1^m$	−8° 42'	1318	125R

The columns headed *MSA* and *U2* give the chart numbers of objects in the *Millennium Star Atlas* and *Uranometria 2000.0*, 2nd edition, respectively. Distances (in light-years) are from recent research papers. Angular sizes are from various catalogs; most objects appear somewhat smaller in a telescope used visually.

a Cepheid-type variable that ranges from magnitude 9.9 to 10.3 and back in a cycle of 3.1 days. At its dimmest, EV Scuti has almost the same brightness as the cluster's other 10th-magnitude star 4' to the northwest. Cluster Cepheids are treasured as important calibrators of the cosmic distance scale, and NGC 6664 is one of the few clusters known to possess one.

In the 1960s and 1970s, German astronomer Jörg Isserstedt compiled lists of 1,091 "prolate ellipsoidal stellar aggregates" once thought to be good tracers of our galaxy's spiral arms. **Isserstedt 68-603** (I68-603) is found in NGC 6664. It is visible, though unimposing, in my 6-inch reflector at 137x as a rather incomplete ring of 11th- to 13th-magnitude stars south of the cluster's center. The eastern arc is most obvious, while the northern edge is nearly nonexistent. The ring is oval east-northeast to west-southwest and about 6' long. Most of these stellar rings are now believed to be chance arrangements of stars.

The open cluster **M26** is intermediate in concentration between M11 and NGC 6664. To track it down, scan 2.1° east from Alpha to 5th-magnitude Epsilon (ε) Scuti. Yellow Epsilon has a 7th-magnitude golden star just 6' to the south. Put this pair in the western edge of a low-power field and drop 1.1° south to reach M26. Note also that a line from Alpha Scuti through 4.7-magnitude Delta (δ) Scuti points straight to M26. Delta, Epsilon, and M26 will all fit within a finderscope's field.

Although you may not be able to see M26 through your finder, you can use the triangle it makes with those two stars to pinpoint this impoverished cousin of M11. Indeed, in a casual sweep, M26 might be mistaken for a mere asterism. My 105mm refractor at 87x shows a pretty group of about 10 moderately faint stars within 8' and many fainter stars like diamond dust. The brightest weighs in at magnitude 9.1 and sits at the southwest edge of the bunch.

The globular cluster **NGC 6712** lies 2.1° east-northeast of M26 and makes a nearly equilateral triangle with Alpha and Beta Scuti. My little refractor at 127x reveals a 5' glow set amid a rich star field. A number of exceedingly faint stars at the very limits of vision are scattered over an unresolved blur, and one brighter star stands out in the northeast part of the halo.

Globular clusters have elongated orbits, and over eons, they plunge through our galaxy's disk and then soar far out into its halo. The path of NGC 6712 carries it through the dense central regions of our galaxy, and this cluster is believed to have passed through the disk more times than most others have. One indication is that gravitational disruption has stripped the cluster of its lightest stars, which no doubt continue to move in orbits that take them into the Milky Way's vast halo. Many of the stars now in the halo may have been snatched from globular clusters.

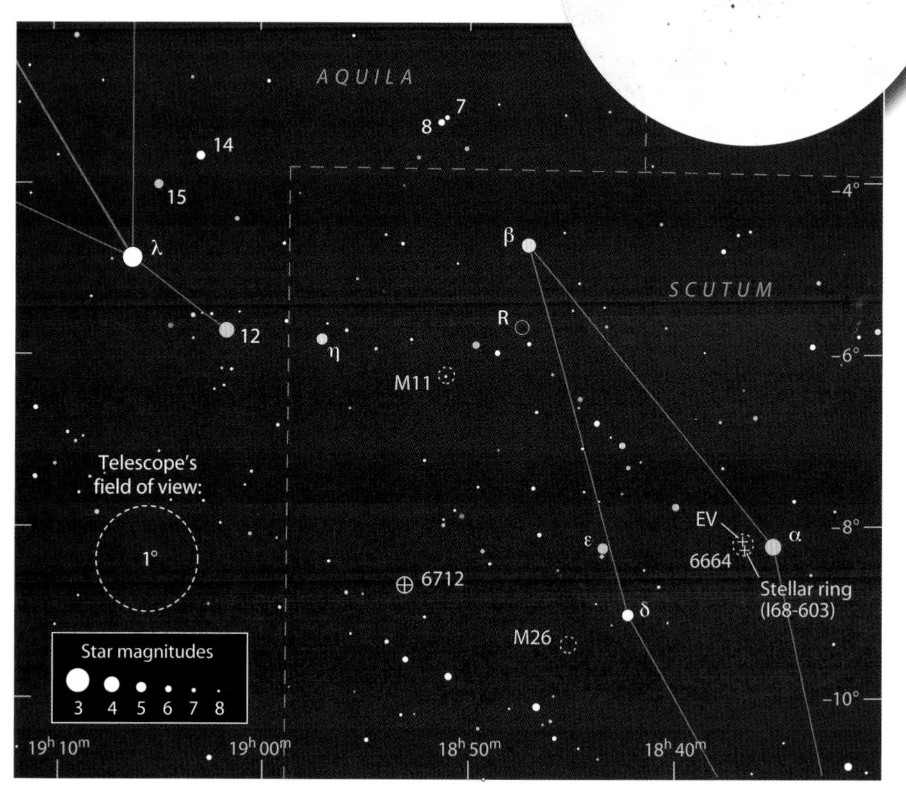

Inset: The author tried to literally plot every star seen in and around M11 in her 105mm refractor at 127x. North is up and the field ⅔° across. Note the cluster's central 8th-magnitude star and the bright pair at the cluster's southeast edge. Visually, these three stars stand out, but they tend to be swamped by their neighbors in photographs. Left: A small finderscope can easily encompass all the objects discussed in this section (those whose names are in yellow). While sampling this region, be sure to look in on R Scuti, the variable star lying only about 1° northwest of M11. Chart: Adapted from *Sky Atlas 2000.0* data

First published September 2003

Summer's Bright Planetaries

Check our pick list of planetary nebulae for a September evening.

Planetary nebulae are remarkable objects. Through a backyard telescope, they are often seen as disks, rings, or butterflies — many painted in distinctive shades of blue and green. A planetary nebula is composed of material shed by a low-mass star as it consumes the last of its nuclear fuel. The collapsing core of the dying star becomes intensely hot, warming the cast-off gases and causing them to glow.

From our vantage point on Earth, many planetaries are very tiny or very faint. But the northern summer sky holds several that can easily be seen in a small telescope. Despite being very well known, they still offer observing challenges through large instruments. Starting high in the north, let's examine a few of these diaphanous bubbles that now grace our evening sky.

NGC 6543, the Cat's Eye Nebula (Caldwell 6), is located in Draco about halfway between the stars Omega (ω) and 36 Draconis. At low power, the Cat's Eye could be mistaken for a star were it not for its pale aqua hue. With my 105mm refractor at 150x to 200x, the planetary is a 20" oval running north-northeast to south-southwest.

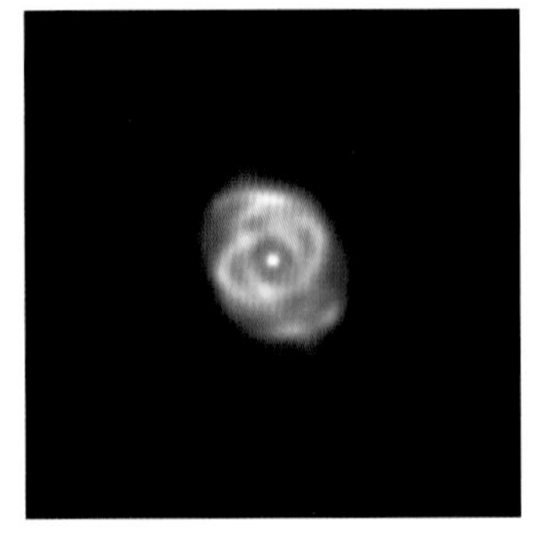

It's a little darker in the middle and shows a faint central star. A thin halo gives the nebula a fuzzy edge. My 10-inch reflector at 220x transforms the oval into a rhombus with well-rounded corners, the southernmost being the pointiest. Subdued areas are visible within the nebula. The small, faint galaxy **NGC 6552** may be seen 9' to the east in the same field of view. It is elongated west-northwest to east-southeast and has a uniform surface brightness.

NGC 6543 has a visually elusive outer halo more than 5' across. This dim corona is difficult to snag even in large amateur telescopes, but its brightest filament can be spotted in my 10-inch. The elongated smudge lies 1.8' west of the Cat's Eye and 1.1' southeast of a 9.8-magnitude star. It shows up with averted vision at magnifications from about 120x to 220x. If you have trouble glimpsing it, an oxygen III filter may help unveil its gossamer glow. This strand of nebulosity bears the designation **IC 4677**. It was discovered and sketched on April 24, 1900, by Edward Emerson Barnard with the 40-inch Yerkes refractor. IC 4677 has been misclassified as a galaxy in several catalogs.

Next, we'll drop down to Cygnus and find **NGC 6826** (Caldwell 15). Look for it 28' east of 16 Cygni, a lovely wide

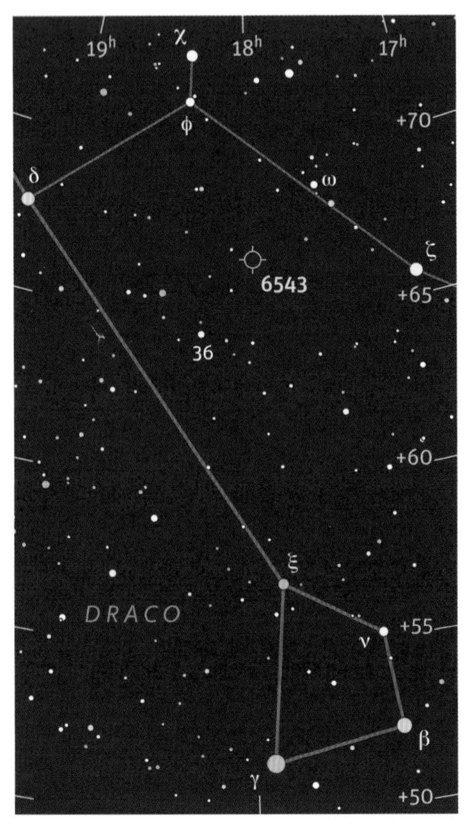

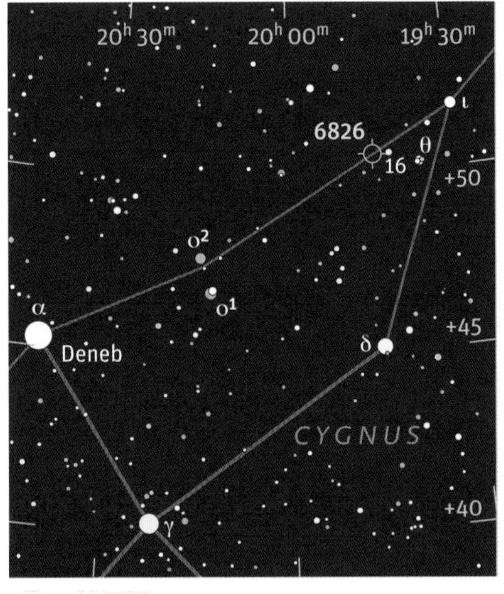

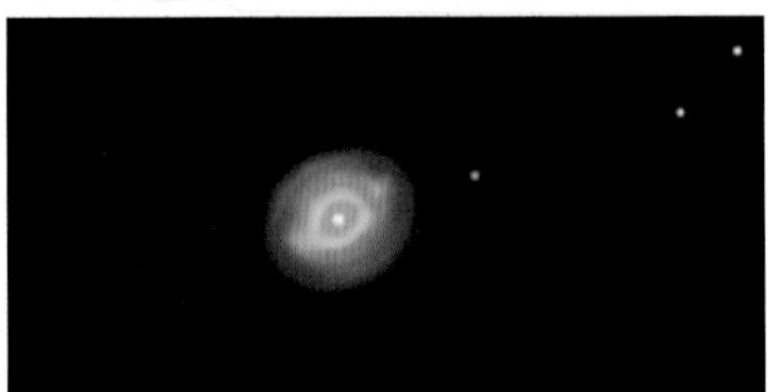

Above: The Cat's Eye Nebula (NGC 6543) in Draco gets its name from the oval shape and glowing pupil-like structure that can be seen in highly magnified images, like this one taken with a Meade 12-inch Schmidt-Cassegrain telescope in Massachusetts. Photo: Sean Walker / John Boudreau. Far right, bottom: It doesn't actually blink, but NGC 6826 in Cygnus gives that impression if you look directly at it, then to one side, and then back at it again. This image by Oregon amateur William McLaughlin shows the Blinking Planetary in a 2.5'-wide field, with north up.

Few celestial objects are imaged as often as is the Ring Nebula (M57) in Lyra, but it is not so common to include the galaxy IC 1296, lying just 4.1' west-northwest. We are seeing light that left the Ring Nebula some 2,000 years ago, but the galaxy is perhaps 100,000 times more remote. After its light began heading our way, 200 million years ago, dinosaurs came and went and our solar system made a complete circuit of the Milky Way Galaxy. Massachusetts amateur Brian Lula used an RC Optical Systems 20-inch reflector.

double consisting of two 6th-magnitude yellow stars. In my 105mm scope, the planetary looks like a turquoise 9th-magnitude star at 17x, while 47x shows a tiny disk. An 11th-magnitude star sits 1.6' south-southwest. This planetary takes on some character at 127x. Its slightly oval shape displays an ashen envelope surrounding a bright ring. A faint central star punctuates its fairly dark heart. Looking directly at the nebula makes the exterior ring vanish, but it pops back into view if you direct your gaze a little to one side (to take advantage of a more sensitive part of your retina). Alternating between direct and averted vision makes the planetary seem to wink, which is the inspiration for its nickname: the Blinking Planetary.

Observers with medium- to large-aperture scopes should look for a tiny bright patch at each end of the planetary's long axis. Each is a FLIER (Fast Low-Ionization Emission Region) moving outward at supersonic speeds. FLIERs have defied explanation, but one study indicates that they may form when a directionally dependent stellar wind interacts with material previously shed by the central star.

Now we'll turn to the famous Ring Nebula in Lyra. **M57** was discovered in January 1779 by Antoine Darquier de Pellepoix and independently by fellow Frenchman Charles Messier later that month. According to Messier, Darquier described the nebula as very dull but perfectly outlined, as large as Jupiter, and resembling a fading planet. This may have been the first time this type of nebula was compared to a planet, but it was William Herschel in England who actually coined the term *planetary nebula*. Ironically, Herschel did not include M57 among the planetary nebulae but referred to it as a perforated nebula. He initially thought that it might be made up of stars.

M57 is conveniently located about two-fifths of the way from Beta (β) to Gamma (γ) Lyrae and is not quite stellar in 10x50 binoculars. My little refractor at 127x shows an east-northeast to west-southwest oval doughnut of light. The long sides are a bit brighter than the ends, and the interior is distinctly brighter than the background sky. A 13th-magnitude star sits just off the nebula's eastern side.

Many observers hotly pursue the central star of the Ring Nebula, which is visually quite difficult. The key to a successful sighting lies neither in excellent transparency nor in exceptionally dark skies. Atmospheric steadiness (good seeing) and high magnification are what you truly need. As far as I know, the smallest scope that has snared this star is a 9-inch. My imagination has allowed me to suspect fleeting glimpses of the star with a similar aperture, but my 14.5-inch reflector is the smallest instrument with which I've seen it with certainty. (While on the subject of fugitive stars, see whether you can spot a second one west-northwest of center within the Ring's dark cavity, along with a third star embedded in the southwestern edge of the annulus itself.) The 14.5-inch scope also gives me the impression that the outer rim of the Ring has a slightly different hue than the rest of the annulus. Other observers have described the edge as red and the rest as green, but for me, the colors remain too subtle to name.

When looking at the Ring Nebula, you also have a galaxy in your field of view. **IC 1296** has been seen in scopes as small as a 10-inch, but at home, I find it barely visible in my 14.5-inch. Look 4.1' west-northwest of M57 for a circlet of 11th- to 14th-magnitude stars that would comfortably hold the nebula. A 14th-magnitude star sits near the center of the

The Dumbbell Nebula (M27) in Vulpecula was once described by Leland S. Copeland as looking "like a comfortable pillow." Chris Schur used a homemade 12.5-inch f/5 reflector and SBIG ST8i camera at Payson, Arizona. This ½°-wide field has north up.

circlet, and the core of IC 1296 lies a mere 28" east-southeast of that star. The tiny core may be all you can see. IC 1296 is a barred spiral galaxy, and its twin arms have very low surface brightness.

"Some of my readers may perhaps feel that I have allotted an undue proportion of space to minute and inconspicuous objects. It may be so. I may have erred in supposing that others might receive as much pleasure as myself from their contemplation... [But this list] will be closed with a nebula which I think will not be found disappointing." These words express my thoughts, but they were written in 1859 by Thomas William Webb as an introduction to the very last entry in his *Celestial Objects for Common Telescopes*, the beautiful Dumbbell Nebula (M27).

M27 was the first planetary nebula ever discovered, swept up by Messier in 1764. It is located in Vulpecula 3.2° north of Gamma (γ) Sagittae, and I can recognize it in my 8x finder. Through 14x70 binoculars, the bright section looks more like an apple core than a dumbbell to me. Faint extensions filling in the sides of the apple can also be seen. At 127x, my 105mm refractor shows brighter areas within the apple core, including its caps, a diagonal bar running east-northeast to west-southwest, and a large patch in the southwestern lobe that hugs the diagonal bar.

At high magnifications, a number of faint stars can be seen against M27's nebulosity. In his 1998 book *The Messier Objects*, Stephen James O'Meara sketches nine as seen in his 4-inch refractor. I count 17 in my 10-inch reflector at 220x. Six, including the planetary's central star, dot the apple core, and the rest are scattered across the faint extensions. The extensions give M27 the shape of a fat football, except that its pointy ends have been erased.

How common are planetary nebulae? About 2,500 are known, but thousands may remain undiscovered or hidden from our view. Planetary nebulae are formed by stars that began their lives with no more than eight solar masses. This includes 95 percent of all stars, and were it the only criterion, our own Sun would eventually die wrapped in one of these celestial shrouds. But new evidence suggests that a dying star might need a binary companion to help create a visible nebula. If so, single stars such as our Sun may end their lives unheralded, slowly fading into obscurity.

Bright Planetary Nebulae and Their 'Companions'

Object	Type	Magnitude	Size	Right Ascension	Declination	*MSA*	*U2*
NGC 6543	Planetary nebula	8.1	22" × 19"	$17^h 58.6^m$	+66° 38'	1065	11L
NGC 6552	Spiral galaxy	13.7	52" × 35"	$18^h 00.1^m$	+66° 37'	1065	11L
IC 4677	Part of planetary	14.5	40" × 20"	$17^h 58.3^m$	+66° 38'	1065	11L
NGC 6826	Planetary nebula	8.8	28" × 25"	$19^h 44.8^m$	+50° 32'	1109	33L
M57	Planetary nebula	8.8	86" × 63"	$18^h 53.6^m$	+33° 02'	1153	49L
IC 1296	Spiral galaxy	14.0	66" × 54"	$18^h 53.3^m$	+33° 04'	1153	49L
M27	Planetary nebula	7.4	8' × 6'	$19^h 59.6^m$	+22° 43'	1195	66R

Angular sizes are from recent catalogs; most objects appear somewhat smaller when a telescope is used visually. The columns headed *MSA* and *U2* give the chart numbers of objects in the *Millennium Star Atlas* and *Uranometria 2000.0*, 2nd edition, respectively.

First published September 2004

The Towering Eagle

Winging its way along the plane of our galaxy, Aquila, the Eagle, is home to many lesser-known deep-sky treats.

The towering Eagle next doth boldly soar,
As if the thunder in his claws he bore;
He's worthy Jove, since he, a bird supplies
The heaven with sacred bolts, and arms the skies.
— Marcus Manilius, *Astronomica*

The eagle is the servant of Jupiter in Roman mythology, and some writers portrayed him carrying the thunderbolts that the king of the gods hurled at hapless victims of his wrath. On late-summer evenings, we see an eagle eternally soaring among the stars as Aquila, no doubt on an errand for his master.

The realm of the Eagle boasts no renowned deep-sky wonders that are easily brought to mind by most stargazers. Yet the Milky Way plunges through Aquila, and thus it would be most surprising if nothing noteworthy lay within. Indeed, Aquila harbors quite a variety of interesting sights, so let's sample a few of these lesser-known denizens of the deep.

An object's obscurity doesn't limit it to the grasp of large telescopes. In fact, our first target is visible in 50mm binoculars under a dark sky. Popularly known as Barnard's E, this inky letter written on the sky is composed of the dark nebulae **Barnard 142** and **Barnard 143**. They lie in the same binocular or finder field with Gamma (γ) Aquilae, or Tarazed, which is 1.4° to the east. The nebulae are intricately patchy in my 105mm refractor at 17x. The most obvious features include a fat, irregular C shape (B143) and a broad oblong patch (B142) just south of it that combine to form Barnard's E. Shadowy lacework reaches eastward from the C to a pair of 7th-magnitude stars near Tarazed. The oblong contains two 9th-magnitude stars and has a discontinuous extension that dips southward before turning toward Tarazed. Barnard's E is about 1° tall and half as broad.

These dark bars also captured the attention of German astronomer Max Wolf, who discovered them in a photograph

Although popularly known as Barnard's E, these dark clouds silhouetted against the rich Milky Way background in Aquila were called B143 (upper section) and B142 when Edward Emerson Barnard made this photograph at Mount Wilson Observatory in August 1905. Brilliant Altair, the star at the southern tip of the Summer Triangle, is at lower left, some 3° southeast of the E feature. Made on a plate sensitive to mainly blue light, this image from *A Photographic Atlas of Selected Regions of the Milky Way* greatly suppressed the apparent brightness of golden-colored Gamma (γ) Aquilae to the east (left) of the E. Photo: Observatories of the Carnegie Institution of Washington

Where the Eagle Flies

Object	Type	Magnitude	Size/Sep.	Right Ascension	Declination	*MSA*	*U2*
Barnard 142	Dark nebula	—	40' × 15'	19h 39.7m	+10° 31'	1243	85L
Barnard 143	Dark nebula	—	30'	19h 41.4m	+11° 00'	1243	85L
NGC 6709	Open cluster	6.7	15'	18h 51.5m	+10° 20'	1246	85R
Burnham 1464	Double star	9.2, 9.7	22"	18h 51.5m	+10° 19'	1246	85R
h870	Double star	9.8, 11.3	12"	18h 51.6m	+10° 19'	1246	85R
NGC 6781	Planetary nebula	11.4	1.8'	19h 18.5m	+06° 32'	1269	85L
NGC 6755	Open cluster	7.5	15'	19h 07.8m	+04° 14'	1269	105R
NGC 6756	Open cluster	10.6	4.0'	19h 08.7m	+04° 42'	1269	105R
θ Serpentis	Double star	4.6, 4.9	22"	18h 56.2m	+04° 12'	1270	105R
NGC 6760	Globular cluster	9.0	9.6'	19h 11.2m	+01° 02'	1293	105R
Palomar 11	Globular cluster	9.8	10'	19h 45.2m	–08° 00'	1315	125L

Angular sizes or separations are from recent catalogs. The visual impression of an object's size is often smaller than the cataloged value and varies according to the aperture and magnification of the viewing instrument. The columns headed *MSA* and *U2* give the chart numbers of objects in the *Millennium Star Atlas* and *Uranometria 2000.0*, 2nd edition, respectively.

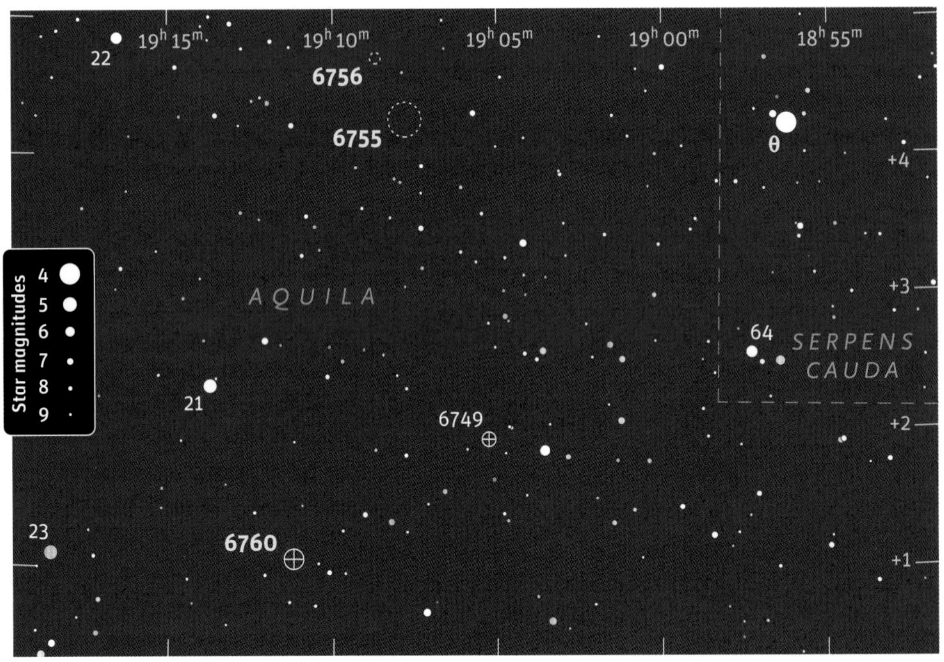

he took in 1891. Wolf called this formation the Triple Cave and wrote: "The broadest arm of the dark structure appears as if it were the nearest, and the smallest arm as if it were the furthest from the observer, so that it would seem to give a perspective view into space of the heavens in the Milky Way. But this is probably mere illusion." See whether you can imagine Barnard's E tipped away from you.

The open cluster **NGC 6709** is another binocular object in Aquila. It lies 4.9° southwest of Zeta (ζ) Aquilae and makes a slender right triangle with Zeta and Epsilon (ε) Aquilae. In 50mm binoculars, the cluster is just a bright misty glow, but 14x70s show me 10 faint stars over haze. It appears about 10' across and is set amid a rich Milky Way star field.

NGC 6709 is very pretty through my little refractor at 87x. Within 15', there are about 60 stars arranged in chains and bunches, the brightest gathered into a roughly triangular group. Starless voids are prominent, especially near the center, including a C-shaped void about one-third the cluster's diameter. The eastern point of the triangle is marked by three of the brightest stars, the westernmost pair forming the double **Burnham 1464** (or β1464). The primary is orange with a white companion 22" north-northeast. The third star is also a double, **h870**. (The lowercase h denotes doubles cataloged by John Herschel, rather than those by his father, William.) Look for its faint companion 12" southwest. Another orange star lying west-southwest of center decorates the cluster.

Moving southeast, we come to a quartet of objects lying along the arc of a circle whose center is near Delta (δ) Aquilae. The first is **NGC 6781**, which makes a short isosceles triangle with Delta and Mu (μ) Aquilae. My 105mm scope at 47x shows an easily spotted, round, fairly large planetary nebula. Its brightness is very slightly uneven at 87x. In my 10-inch reflector at 166x, NGC 6781 spans 1½' and has unevenly brighter segments along the rim, but the annulus is interrupted by a much dimmer area in the north. Adding an oxygen III filter accentuates the planetary's variations in brightness.

While not nearly as well known as other planetary nebulae in the northern summer sky, NGC 6781 in Aquila is easily within reach of 4-inch telescopes. It lies about 8° west-southwest of Altair. Photo: Adam Block / NOAO / AURA / NSF

The next objects in the arc are **NGC 6755** and **NGC 6756**, a pair of open clusters that share a low-power field of view. The brighter one, NGC 6755, lies 2.9° due east of the nicely matched, white double **Theta (θ) Serpentis**. Through my 105mm refractor at 68x, the cluster displays 18 faint to very faint stars sprinkled over a 15' mottled background. Just ½° to the north-northeast, NGC 6756 embraces a few very faint stars in a 3½' haze. In a 10-inch scope at moderate power, NGC 6755 is a rich cluster of several dozen faint stars, its dim northern reaches separated from the main mass by a comparatively starless lane. Clumps and chains of stars embellish the group. NGC 6756 exhibits a score of stars and an unresolved knot just northeast of its center.

NGC 6755 is about 50 million years old and 4,600 light-years away. NGC 6756's similar age and distance — 60 million years and 4,900 light-years — may imply a physical relationship.

The last object in our arc is the small but fairly bright globular cluster **NGC 6760**. With 21 and 23 Aquilae, it forms an isosceles triangle that fits within the field of a finderscope. In my little refractor at 87x, a faint 4' halo surrounds a large, bright core. Several elusive points of light occasionally blink into view. My 10-inch at 170x reveals a sparse scattering of stars across the halo and outer edge of the core. An 11th-magnitude star sits just beyond the northeast edge.

Our final target is **Palomar 11**, located halfway between Kappa (κ) and 56 Aquilae. Look for it 4' south-southeast of an 8.6-magnitude star. This globular cluster is much more challenging than NGC 6760, but that's not to say that it's impossible to see in a small telescope. Finnish amateur Jaakko Saloranta has achieved this with his 80mm refractor at a high elevation in the Canary Islands. This is not an accomplishment I expect to duplicate, but I did tackle Palomar 11 with a somewhat more substantial telescope — my 14.5-inch reflector. At 245x, this low-surface-brightness globular appears about 3' x 2¼' aligned northeast to southwest. It looks mottled, and there are a half dozen superposed stars, at least some of which may be foreground objects. (The brightest stars of Palomar 11 itself are red giants with apparent magnitudes of around 15½.) A skinny triangle of 11th- and 12th-magnitude stars sits off the northeast edge. A 13th-magnitude star hugs the eastern side of the cluster, and a slightly more distant one lies a bit west of south.

Palomar 11 is one of several globulars discovered in the 1950s by American astronomer Albert G. Wilson while he was examining photographs from the *National Geographic Society–Palomar Observatory Sky Survey*. Palomar 11 and NGC 6760 both lie near the plane of our galaxy, where they suffer from obscuration by dust clouds. They are 42,000 and 24,000 light-years away, respectively. Compare NGC 6760 with the relatively unobscured cluster M5 in Serpens, which lies at the same distance. NGC 6760 is intrinsically less than one magnitude fainter than M5, but because of obscuration, it appears over three magnitudes fainter in our sky.

Observers up for a challenge can pursue the globular cluster Palomar 11 in southern Aquila. What's the smallest aperture you use to see it? Photo: STScI / Digitized Sky Survey

First published September 2005

Foxfire Nights

Vulpecula's dearth of bright stars is offset by its wealth of deep-sky objects.

But if you tame me, then we shall need each other.
To me, you will be unique in all the world. To you,
I shall be unique in all the world.

— Fox to the title character in *The Little Prince,*
Antoine de Saint-Exupéry, 1943

Vulpecula, the Little Fox, has no bright star to draw our eyes his way and is overshadowed by the showier constellations around him. Yet this inconspicuous star figure has an amazing wealth of deep-sky wonders. If you take the time to visit them and make them your own, then the Little Fox will hold a special place in your heart and never seem ordinary again.

Our first fiery splendor in the Little Fox is its brightest star, 4.4-magnitude **Alpha (α) Vulpeculae**, a red giant 60 times bigger across and 400 times more luminous than our Sun. Alpha is a lovely optical double designated Σ I 42, meaning the 42nd entry in the first appendix to Friedrich Georg Wilhelm von Struve's double-star catalog. Its unrelated suns are more than 7' apart and are easily seen with steadily held binoculars. The colorful components are widely separated in my 105mm refractor at 17x. The primary appears orange, and its 5.8-magnitude yellow neighbor sits north-northeast. The bright star is about 200 light-years closer to us than its apparent companion.

The open cluster **NGC 6800** is located 36' northwest of Alpha. My little refractor at 17x reveals a very pretty sprinkling of faint stars spanning 15'. At 87x, I count two dozen stars, the brightest strung along the outline of an oval with a large central void. My 10-inch reflector at 118x doubles the star count, but the edges of the cluster become very ill defined. William Herschel discovered this cluster while sweeping the sky with his 18.7-inch speculum-metal reflector in 1784. He aptly described the group as a cluster of coarsely scattered bright stars intermixed with faint stars.

On September evenings the prominent Summer Triangle, comprising the stellar beacons Vega, Deneb, and Altair (right, upper left, and bottom, respectively), sparkles amid the Milky Way's glow. In the accompanying text, the author tours deep-sky objects near the center of this wide-field photograph. Photo: Akira Fujii

NGC 6793 is a confusing open cluster 2.8° south-southwest of Alpha and 1.8° east-northeast of 1 Vulpeculae. It hides in a starry field that renders it tricky to pick out, which is probably why the object isn't plotted on many modern star atlases. Herschel discovered the group in 1789 and called it a scattered cluster of considerably bright stars, pretty rich, of an irregular figure, above 15' in extent. Yet most modern catalogs list the group as being only 6' or 7' in diameter.

My own notes reflect this ambiguity. While observing with my 105mm scope at 17x, I logged a small, granular-looking patch with a few faint stars. At 87x, the cluster showed two triangles of stars — one bright, one faint — plus a few very dim stars in 3½'. Outliers stretch the group to about 6'. The northernmost star in the bright triangle is the double star **h886**, its 10.5-magnitude primary nestled against an 11.5-magnitude secondary to the northeast. With

Little Fox Hunt

Object	Type	Magnitude	Size/Sep.	Right Ascension	Declination	*MSA*	*PSA*
α Vulpeculae	Double star	4.4, 5.8	7.1'	19h 28.7m	+24° 40'	1196	64
NGC 6800	Open cluster	—	15'	19h 27.1m	+25° 08'	(1196)	64
NGC 6793	Open cluster	—	7' core, 18' halo	19h 23.2m	+22° 09'	(1196)	(65)
h886	Double star	10.5, 11.5	8.2"	19h 23.22m	+22° 09.9'	(1196)	(65)
Cr 399	Asterism	3.6	90'	19h 26.2m	+20° 06'	1220	64
NGC 6802	Open cluster	8.8	5.0'	19h 30.6m	+20° 16'	1220	64
Stock 1	Open cluster	5.2	34' core, 80' halo	19h 35.8m	+25° 10'	1196	64
Σ2548	Double star	8.5, 9.9	9.4"	19h 36.5m	+25° 00'	1196	(64)
LDN 810	Dark nebula	—	18' × 9'	19h 45.6m	+27° 57'	(1172)	(64)

Angular sizes or separations are from recent catalogs. The visual impression of an object's size is often smaller than the cataloged value and varies according to the aperture and magnification of the viewing instrument. The columns headed *MSA* and *PSA* give the appropriate chart numbers in the *Millennium Star Atlas* and *Sky & Telescope's Pocket Sky Atlas*, respectively. Chart numbers in parentheses indicate that the object is not plotted.

my 10-inch scope at 43x, I saw these same stars near the center of a larger and brighter collection with an 8th-magnitude star at its western edge. This makes the cluster about 20' across, embracing 50 mixed bright and faint stars.

Herschel obviously cataloged the bigger group. *The Catalogue of Open Cluster Data* lists NGC 6793 as being 18' across with a 7' core, so it seems that references giving a size of 7' limit the cluster to the stars of the more highly condensed core.

Collinder 399 (Cr 399) is a striking asterism commonly known as Brocchi's Cluster or the Coathanger. It has several stars bright enough to be seen with the unaided eye in a very dark sky. From my semirural home in upstate New York, I merely see a nebulous haze 4½° south of Alpha. Binoculars or a finder show six stars in a curved bar plus four more forming a hook to the south. Since the bar is 1.4° across, seeing the Coathanger in a telescope requires a low-power, wide-angle field of view.

Amateur astronomer Stephan Ruchhöft from northern Germany reads *Sky & Telescope*'s German edition, *Astronomie Heute*. He wrote to tell me that he "discovered" Collinder 399 while scanning the sky with binoculars. At the time, he was planning a family trip to the Alps, which inspired him to see the star figure as a ski lift. The bar of the Coathanger is the ski lift's cable, and the hook is its chair. This inventive image has the advantage over the Coathanger of being seen right side up in binoculars by Northern Hemisphere observers.

Just 18' off the eastern end of the Coathanger's bar, we come to the open cluster **NGC 6802**. My little refractor at 28x displays a broad, north-south band of mist about 5' tall. A pair of 9th- and 10th-magnitude stars lies 6½' northwest, and another pair of 10th and 11th magnitude lies the same distance northeast. A smattering of extremely faint stars dots the haze at 87x. With my 10-inch reflector at 170x, NGC 6802 is a beautiful diamond-dust cluster rich in faint and very faint stars. It lies at an estimated distance of about 3,700 light-years, and it's dimmed by intervening dark clouds that seem to split the softly shining band of Milky Way in this area of the sky.

Now let's move 1.7° east-northeast of Alpha to **Stock 1**. Through my 105mm scope at 17x, this is a large, loose cluster of 20 moderately bright and many very faint stars. An 8th-magnitude star sits at the heart of a core group measuring 30' x 20'. This, in turn, is surrounded by a 1.4° halo of stars, mostly encircling the eastern half of the core.

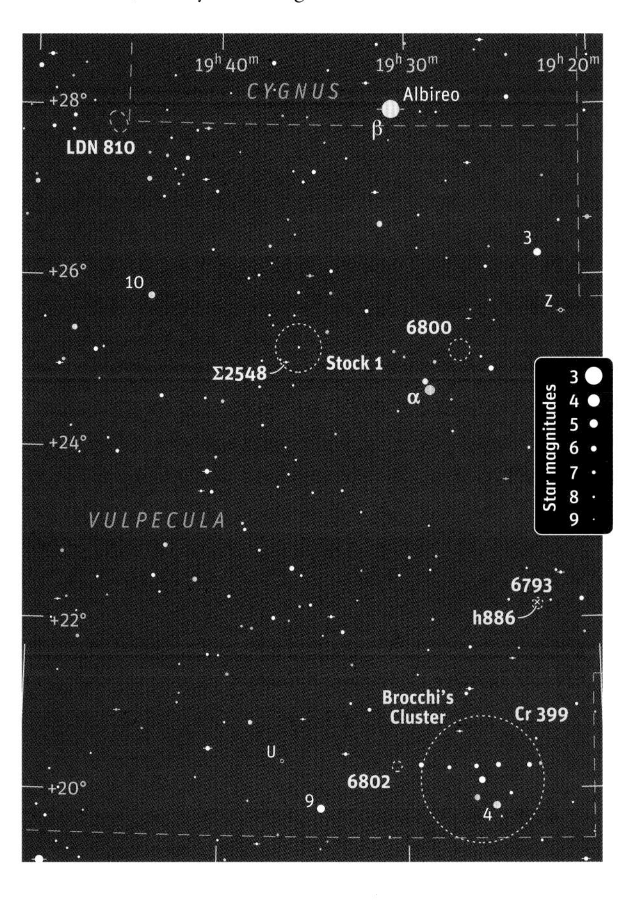

First published September 2006

A reddish orange star adorns the western reaches of the cluster. Stock 1 contains many double stars. The brightest is **Σ2548**, located 14' southeast of the cluster's central star. Its 8.5- and 9.9-magnitude components can be split at about 50x. Studies indicate that most of this cluster's true members are roughly concentrated within the core group's ½° diameter.

Our final stop will be the dark nebula **LDN 810**, which I like to call the Coalman. The simplest way to find it is to scan 3.3° due east from Albireo, the beautiful gold and blue double star that marks the head of Cygnus, the Swan. With my 10-inch scope at 68x, I see a north-south oval of inky darkness containing only a few extremely faint stars. A slightly less conspicuous dark oval perched atop it turns the nebula into a filled-in 8, or the negative image of a snowman. The southern patch is about 9' x 6' and the northern one 6' x 5'.

These are just a few of the foxy residents of Vulpecula. Later, we'll continue our tour with more vulpine delights, including the constellation's brightest planetary nebula and its finest cluster.

The author sees the dark nebula LDN 810 as the negative image of a snowman and thus calls it the Coalman. The field is about 1° wide with north up. Photo: POSS-II / Caltech / Palomar

The Graceful Swan

Cygnus is a wonderland for deep-sky observers.

Down the broad galactic river,
Where the star-beams dance and quiver,
Flies the swan with grace transcendent.
— Anonymous, from *The Stars in Song and Legend*, 1901, Jermain G. Porter

Gliding along the foggy river of the Milky Way, Cygnus, the Swan, is one of the easiest figures to imagine among the stars. In *The Stars in Our Heaven* (1948), Peter Lum writes: "The Swan is by far the most magnificent of the feathered creatures of the star world. There is no other constellation that has such a feeling of outstretched wings, no other which so successfully suggests the movement of flying." Let's soar with the Swan and visit some of the lesser-known wonders scattered along his starry flyway.

Our first port of call is **NGC 6819**, located 2.9° north of 15 Cygni and 8½' west-southwest of a 6th-magnitude star. This small open cluster is chock-full of so many faint stars that its total brightness is magnitude 7.3. Through 14x70 binoculars, it appears as a misty patch with a few pinpoint stars intermittently visible, especially using averted vision.

With a telescope, high magnifications help tease out the dim members of this cluster. My 105mm refractor at 127x reveals a 5' group of 25 sugar-grain stars. About 15 of them, including the brightest ones, form a nearly sinusoidal curve woven into a haze of unresolved suns. My 10-inch reflector at 139x doubles the star tally and apparent diameter of this pretty cluster and gives it a slightly ragged edge.

NGC 6819 is an ancient open cluster estimated to be about 2½ billion years old. It rests on the northern edge of our galaxy's Orion Arm (the one we're in) and precedes the Sun by 7,500 light-years in its orbit around the galactic center.

Our next target is **NGC 6888**, the Crescent Nebula. Look for it one-third of the way from Gamma (γ) to Eta (η) Cygni or 1.2° west-northwest of 34 Cygni. There you'll find a pair of 7th-magnitude stars embedded in the nebula.

Arizona astronomer Brian Skiff and California amateur Kevin Ritschel have seen NGC 6888 both with and without filters through their 70mm and 80mm refractors, respectively. My 105mm scope at 47x shows an 18', northeast-southwest numeral 3 of nebulosity. The northeastern half of the figure is more conspicuous and runs through three bright stars: a deep yellow one at the tip, a golden orange star with a wide, faint companion **(OΣ 401)** decorating the curve, and the third pinned to the interior point of the 3. Narrowband and oxygen III nebula filters slightly improve the view.

The numeral fills in as seen through my 10-inch reflector at 44x. The resulting thumbprint shape is visible without a filter but is easier to see with one. Observers with large scopes report ripples of material adorning the entire nebula.

The Crescent Nebula is the offspring of the 7th-magnitude, intensely hot Wolf-Rayet star seen near its center. As this massive star aged, it expelled its outer layers and denuded its core. Intense radiation from the exposed core drove away more gas at speeds upward of 3 million miles per hour. This fierce stellar wind collided with the material shed earlier, thus creating a dense shell and shock waves that are propagating both inward and outward. In perhaps 100,000 years, this tortured star will exhaust its dwindling reserve of nuclear fuel and end its life in a supernova explosion.

Astronomy writer Richard Berry says that the Crescent Nebula's Wolf-Rayet star appears distinctly pink to him, unlike any other star he's seen. Berry attributes this to the overall blue-white color of the star combining with the crimson hue of its unusually strong hydrogen-alpha emission.

A memorable asterism, dubbed the **Fairy Ring** by Utah amateur Kim Hyatt, lies just 1.6° west of the Crescent Nebula. My little refractor at 47x displays a nifty 22' ring of star pairs. Four bright duos make up the northwestern arc of the circlet. Several dimmer pairs complete the ring, and a scattering of stars inhabits its interior.

The Fairy Ring is even more striking through my 10-inch reflector, which better shows the stars' colors. Working clockwise, the four bright pairs appear blue-white and yellow-white; gold and white; blue-white and white; and white and reddish orange. The two brightest stars within the ring are yellow-orange and orange.

Let's return to 34 Cygni and then slide 28' south-southwest to the open cluster **IC 4996**. My little refractor at low power shows only one fairly bright star flanked by two dimmer ones making a very squat triangle. At 122x, several faint and extremely faint stars join the scene. My 10-inch scope at 68x shows a little C-clamp of eight 8th- to 12th-magnitude stars — five in the clamp's C curve (including the close double **β442 Aa**), two in the handle, and one in the partly closed jaw. At 213x, IC 4996 becomes a rich group of mostly faint stars spanning 3' and elongated north-northeast to south-southwest.

The Crescent Nebula, NGC 6888, is visible in small telescopes under dark, transparent skies. As explained in the accompanying text, the nebula surrounds an extremely hot Wolf-Rayet star that is destined to explode as a supernova. The field is 25 arcminutes wide with north up. Photo: Johannes Schedler

Soaring With the Swan

Object	Type	Magnitude	Size/Sep.	Right Ascension	Declination	*MSA*	*U2*
NGC 6819	Open cluster	7.3	5'	$19^h 41.3^m$	+40° 12'	1129	48R
NGC 6888	Bright nebula	8.8	18' × 8'	$20^h 12.0^m$	+38° 23'	1149	48L
OΣ 401	Double star	7.3, 10.6	13.0"	$20^h 12.2^m$	+38° 27'	1149	48L
Fairy Ring	Asterism	—	22'	$20^h 04.1^m$	+38° 10'	(1149)	(48L)
IC 4996	Open cluster	7.3	5'	$20^h 16.5^m$	+37° 39'	1149	48L
β422 Aa	Double star	9.7, 10.8	4.2"	$20^h 16.50^m$	+37° 38.6'	1149	48L
P Cyg Cluster	Open cluster	—	5'	$20^h 17.7^m$	+38° 02'	(1149)	(48L)
P Cyg	Variable star	4.8 var.	—	$20^h 17.8^m$	+38° 02'	1149	48L
χ Cygni	Variable star	3.3–14.2	—	$19^h 50.6^m$	+32° 55'	1150	48R

Angular sizes and separations are from recent catalogs. The visual impression of an object's size is often smaller than the cataloged value and varies according to the aperture and magnification of the viewing instrument. The columns headed *MSA* and *U2* give the appropriate chart numbers in the *Millennium Star Atlas* and *Uranometria 2000.0*, 2nd edition, respectively. Chart numbers in parentheses indicate that the object is not plotted. All the objects this month are in the area of sky covered by Chart 62 in *Sky & Telescope's Pocket Sky Atlas*.

IC 4996 is thought to form a double cluster with the **P Cygni Cluster**, which has the bright star 34 Cygni at its eastern edge. These groups share the field of view in my small refractor at 87x, but only a few of the stars in the P Cygni Cluster are visible. Through my 10-inch reflector at 249x, the cluster shows 11 stars, most in a curve that spans 3' and runs clockwise around 34 Cygni from north-northeast through south-southwest.

The cluster takes its name from its brightest star, **P Cygni** (P Cyg), which is an alternative moniker for 34 Cygni. In Johann Bayer's 1603 atlas, *Uranometria,* he used Greek letters and then lowercase Roman letters to designate stars. Uppercase Roman letters were used to identify assorted special objects on his charts. Bayer assigned the letter P to the nova of 1600, discovered by the Dutch chart maker Willem Blaeu when it suddenly rose from obscurity to 3rd magnitude. P Cygni gradually faded below 6th magnitude and then experienced a similar meteoric rise in 1655, followed by smaller fluctuations. It stabilized at the end of that century and now shines at about magnitude 4.8, with only small variations at somewhat irregular intervals.

P Cygni and its associated clusters are approximately 7,500 light-years away. This makes P Cygni one of the most distant stars you can see with the unaided eye and one of the intrinsically brightest stars in our galaxy. It's about 30 times as massive and 350,000 times as luminous as our Sun. Luminous blue variables such as P Cygni shed enormous amounts of material during their eruptions. P Cygni may eventually evolve into a Wolf-Rayet star.

An impressive variable star to watch is Chi (χ) Cygni, labeled on the September all-sky map on page 310. Every 204 days, Chi plunges from naked-eye visibility to a feeble 14th magnitude, a range you could pursue with a 6-inch scope. Although the cycle's fall and rise proceed like clockwork, its high points vary from 3rd to 5th magnitude. Chi is a pulsating giant star much like its famous cousin Mira, in the autumn constellation Cetus. You can discover the current brightness of Chi by entering "Chi Cyg" in the star-name box of the American Association of Variable Star Observers Light Curve Generator at www.aavso.org/data/lcg.

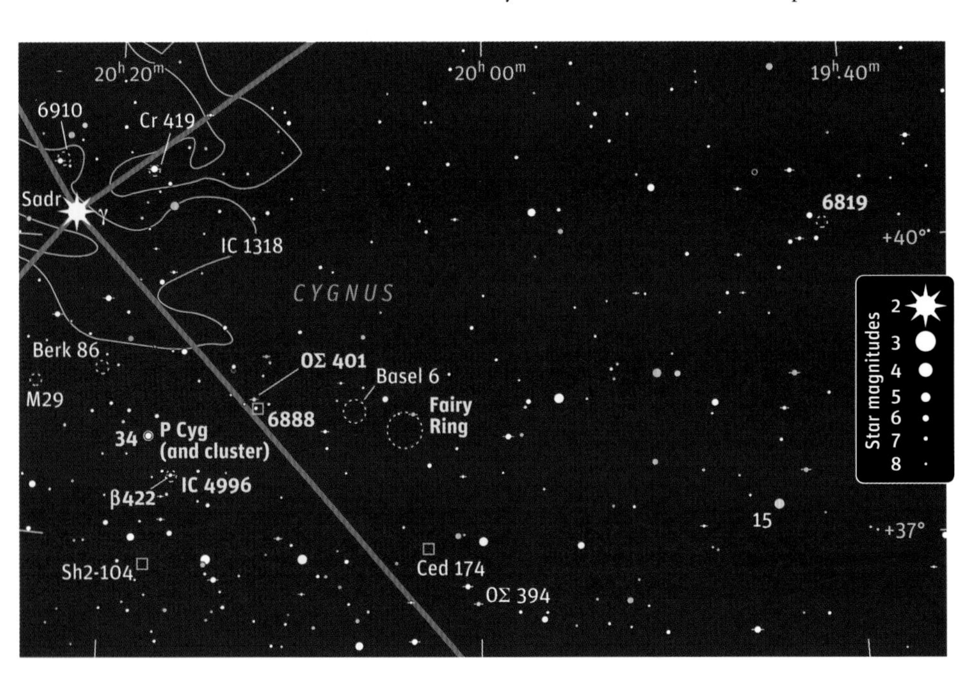

First published September 2007

Magic Nights of Stars

Explore some of the bewitching lesser-known objects in Sagittarius.

Then of a sooth 'twas Timpinen who played to you.... Now, however, for such is the eeriness of that sprite, you will ever love the evenings of summer and the nights of stars, and their magic will cause your heart to ache unquenchably.
— J. R. R. Tolkien, *The Book of Lost Tales*

On enchanting summer evenings, the frothy cataract of the Milky Way cascades into Sagittarius, where it kicks up a bright spume of distant suns. This celestial foam is populated with a captivating array of deep-sky wonders. But with 15 showpiece objects from the Messier catalog wading in this spray pool, observers often bypass the other alluring riches it holds.

One of the prettiest non-Messier sights in Sagittarius is the side-by-side combination of the open cluster **NGC 6520** and the dark nebula **Barnard 86** (B86), also known as the Ink Spot. There are two relatively simple ways to locate the cluster. You can sweep 3.5° east from the star 3 Sagittarii, or you can follow an imaginary line from Gamma (γ) to W Sagittarii and continue for twice again that distance. W is a Cepheid variable star whose pulsations bring it from magnitude 5.1 to 4.3 and back over a period of 7½ days.

In my 14x70 binoculars, NGC 6520 is a very small fuzzy spot with two faint points of light inside, like glittering eyes. Two stars, about the same brightness as the eyes and roughly in line with them, flank the haze. They mark the outer edges of NGC 6520 as seen through my 105mm refractor at 87x. The group totals 23 stars and spans 5½'. B86 abuts its western side and looks as though it's the place from which the cluster's stars were scooped. This black-velvet nebula spans 5' x 3', and its northwestern edge is adorned with a yellow-orange star.

With a wide-angle eyepiece giving 213x, my 10-inch reflector captures the duo nicely. NGC 6520 is bright with 35 to 40 stars, and B86 stands out clearly. Photos show dark tendrils reaching out from B86 to cradle the cluster, but I can't see them well from my northerly and moderately light-polluted observing site. Perhaps you'll have better luck.

The apparent intertwining of nebula and cluster suggests that they're physically related. One study places the cluster about 6,200 light-years away with an age of 150 million years. But the mean lifetime of a dark molecular cloud such as B86 is only a few tens of millions of years, which is puzzling if the objects are associated.

Left: Jeremy Perez of Flagstaff, Arizona, sketched NGC 6520 and Barnard 86 as they appeared at 120x through his 8-inch telescope.

Right: Barnard 86 appears like a dark twin of NGC 6520 in this image from Kitt Peak's Advanced Observing Program. Photo: Fred Calvert / Adam Block / NOAO / AURA / NSF

SUMMER September

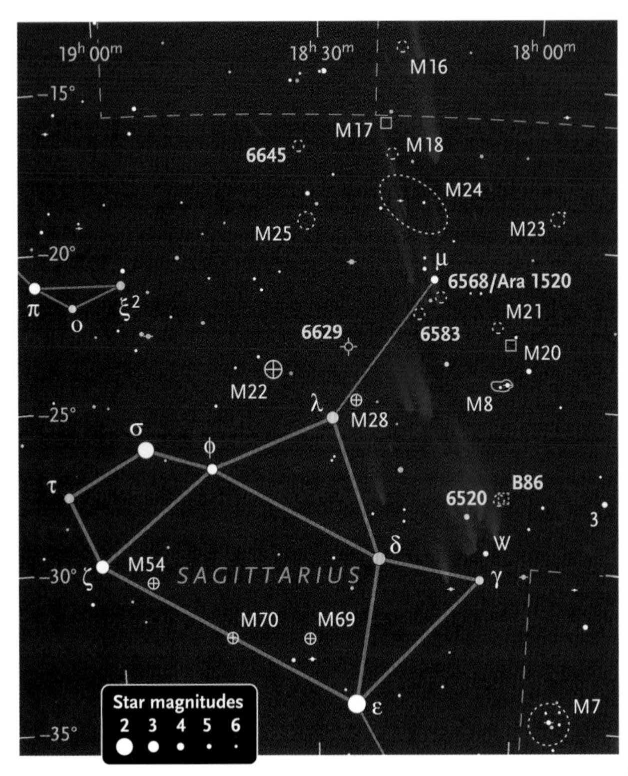

Few parts of the sky are as densely packed with deep-sky wonders as is Sagittarius.

Another appealing open cluster sits 34' south-southwest of Mu (μ) Sagittarii. **NGC 6568** is not showy, but it displays an interesting shape. Observing next to me at a Stellafane convention in Vermont a few years ago, astronomy artist Joe Bergeron studied NGC 6568 with his 6-inch reflector at 62x. He described it as a large, patchy, scattered group of 30 stars and noted arcs of stars that he envisioned as a bird with curved, outspread wings.

In my 10-inch scope at 68x, I count 40 stars spanning 12', many of the brightest in a 6'-long S lying on its side (east-west). The center of this S holds the double star **Ara 1520**, discovered by S. Aravamudan of the Nizamiah Observatory in India. The 11th-magnitude primary holds a slightly dimmer companion 13" to its north-northeast.

Bergeron also turned his scope toward **NGC 6583**, 54' southeast, and called it a small ghostly group, very rich, like NGC 7789 in Cassiopeia but probably obscured by dust. In fact, the cluster is 6,700 light-years away and dimmed 1.6 magnitudes by interstellar extinction. My 10-inch reflector at 213x shows many faint to very faint stars over a $3\frac{1}{2}$' haze with a curve of three brighter stars off its southern side.

Overshadowed by the bright globular clusters nearby, the petite marvel **NGC 6629** often goes unnoticed. Look for this planetary nebula 1.1° west of 23 Sagittarii with an 8th-magnitude star 7' to its north. NGC 6629 is barely nonstellar through my 105mm refractor at 47x, but at 127x, it presents a small, round, blue-gray disk with a faint central star.

The nebula's distinctive color makes it fairly easy to pick

Lesser-Known Treasures in Sagittarius

Object	Type	Magnitude	Size	Right Ascension	Declination
NGC 6520	Open cluster	7.6	6.0'	18^h 03.4^m	−27° 53'
B86	Dark nebula	—	5.0'	18^h 03.0^m	−27° 52'
NGC 6568	Open cluster	8.6	12'	18^h 12.8^m	−21° 35'
Ara 1520	Double star	11.2, 11.7	13"	18^h 12.7^m	−21° 35'
NGC 6583	Open cluster	10.0	4.0'	18^h 15.8^m	−22° 09'
NGC 6629	Planetary nebula	11.3	16"	18^h 25.7^m	−23° 12'
NGC 6645	Open cluster	8.5	10'	18^h 32.6^m	−16° 53'

Angular sizes and separations are from recent catalogs. Visually, an object's size is often smaller than the cataloged value and varies according to the aperture and magnification of the viewing instrument.

First published September 2008

The double star Ara 1520 nestles at the heart of NGC 6568, as shown in this red-filtered photo. Photo: POSS-II / Caltech / Palomar

out at low power through my 10-inch scope. At 213x, NGC 6629 has a relatively large, brighter center elongated approximately north-south. This feature stands out better at a magnification of 300x with a narrowband nebula filter or an oxygen III filter.

William Herschel discovered NGC 6629 in 1784. Observing 53 years later, his son, John, remarked that it might be a "very distant and highly compressed globular" and described it as "one of the smallest if not the very smallest nebulous object I remember to have seen. It is a very remarkable object."

Our final target is much easier: the open cluster **NGC 6645**. The simplest way to locate it is to scan 2° north from the large, bright cluster Messier 25 (M25). NGC 6645 is beautiful diamond dust in my little refractor at 17x. An interesting dot-dash-dash pattern of stars leads up to its eastern side. At 87x, I count 35 stars, many in little bunches and lines, in an 8½' x 10' cluster. My 10-inch scope at 115x reveals a lovely group of 70 irregularly strewn stars with a sable void near its center.

The brightly spangled canopy of a summer night can be deeply compelling. When you find your heart filled with longing for "the magic of the world beneath the stars," perhaps it is Timpinen's flute that calls you forth.

Sagittarius Star Cities

The globular clusters that hover near the galactic core are amazingly diverse.

Sagittarius embraces more globular clusters than any other constellation, each cluster a radiant multitude of tens to hundreds of thousands of stars. Let's pay homage to these magnificent star cities, which range in appearance from a starry blizzard to a touch of frost on the dome of the sky.

Our first stop will be **NGC 6522** and **NGC 6528**, easily located ½° northwest of Gamma (γ) Sagittarii. (See the detailed chart on page 222.) These globulars are only 16' apart and share the field of view through my 105mm refractor at 87x. NGC 6522 appears moderately bright, very granular, and 3' across. It grows much brighter toward the center, and faint foreground stars caress its eastern and south-southwestern edges. NGC 6528 is dimmer and half as large, and it becomes only moderately brighter toward the center.

Through my 10-inch reflector at 219x, NGC 6522 shows several faint stars in its halo and outer core, but NGC 6528 simply appears mottled. This is understandable since the brightest stars in the former cluster are roughly magnitude 14 whereas those in the latter are magnitude 15½.

NGC 6522 is about 25,000 light-years distant and 70 light-years across, while NGC 6528 is a bit farther away

Messier 22 appears extraordinarily big and bright because it's the fourth-closest globular cluster, just 10,500 light-years distant. Photo: Jim Misti

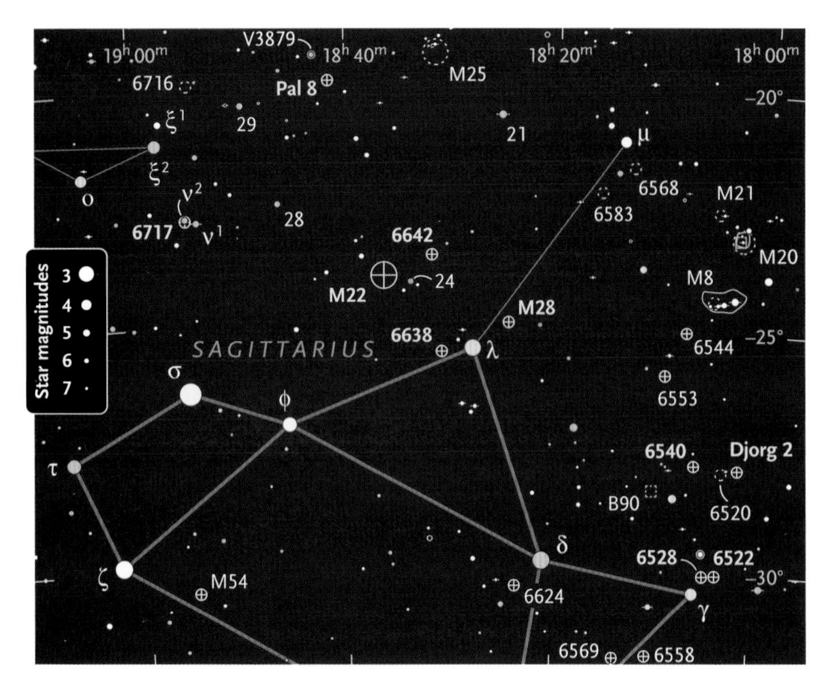

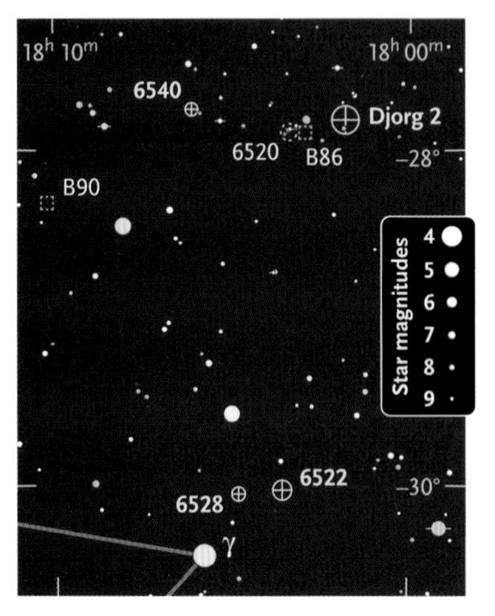

and 37 light-years across. They form a true double cluster less than 500 light-years apart, center to center. Either would be impressive from the other's sky.

Found 2.7° north of Gamma, **NGC 6540** was thought to be an open cluster for many years. Then, in 1987, S. George Djorgovski (Caltech) announced the discovery of three globular clusters in an *Astrophysical Journal* paper. The third entry in his list is NGC 6540, now also known as Djorgovski 3. Djorgovski did not, however, equate the cluster with NGC 6540 at that time.

Visually, it's easy to see why NGC 6540 could be mistaken for a partly resolved open cluster. Through my little refractor at high power, it's just a faint bit of mist harboring a few starlike dots, but in the 10-inch scope, it's a 1.2' haze with a wavy, east-west line of stars splashed across its face. Some of the stars may be foreground objects.

Djorgovski 2 (Djorg 2) was also misclassified as an open cluster when it made its 1978 debut as ESO 456-SC 38 in Part VI of the *ESO/Uppsala Survey of the ESO (B) Atlas of the Southern Sky*. This globular cluster lies 1° west of NGC 6540, on the opposite side of the dark nebula Barnard 86 and the attractive open cluster NGC 6520.

I've never viewed Djorgovski 2 myself, but California amateur Dana Patchick told me that he happened upon the cluster in 1981 while observing with his 16-inch reflector. His notes read: "A very faint, round nebulous object. Perhaps 2' to 3' diameter, showing no hint of resolution. Uniform magnitude across whole object. I suspect it to be a faint, distant open cluster or globular cluster." The following month, Patchick managed to spot Djorgovski 2 in his 8-inch scope.

NGC 6540 and Djorgovski 2 appear faint largely because their light is heavily extinguished and reddened by interstellar dust associated with our galaxy's Great Rift, as shown in the photograph on the facing page. In fact, Djorgovski 2 is 22,000 light-years distant. If there were no dust obscuring the view, it would appear brighter than NGC 6522 and NGC 6528. And NGC 6540 is one of the closest globular clusters, just 12,000 light-years from Earth.

Next, we'll drop in on a quartet of clusters arrayed near Lambda (λ) Sagittarii. The king of the foursome is **Messier 22** (M22), one of the most spectacular globulars in the sky. Its brightest gems glitter near magnitude 10½, so partial resolution is possible with almost any telescope. North Carolina amateur David Elosser tells me that M22 "looks like a ball of rice" in his 4-inch scope.

Maryland amateur Peter Gertson says that he likes to

If you ever thought that all globular clusters look the same, these photos should set you straight. They were all shot through a red filter with identical exposures, and they're reproduced at the same scale. Photos: POSS-II / Caltech / Palomar

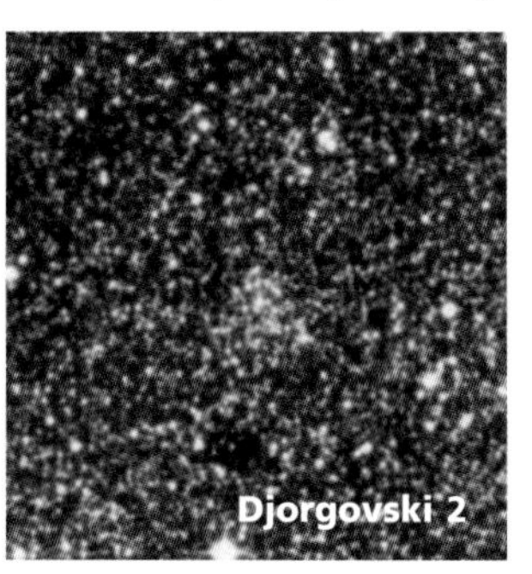

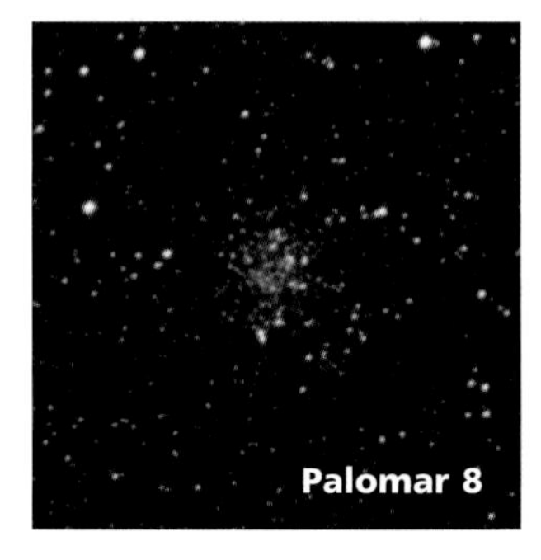

First published September 2009

observe M22 through a "twilight globular filter." Gertson tries to acquire M22 as soon as possible during evening twilight, when it's just a dim haze bounded by a few stars. As the sky darkens, the brightest stars pop into view. Gertson writes: "These begin by flickering in and then become steady under direct vision. I can map the individual stars as they appear fairly reliably. After about 25 stars, the view gets a little crowded and soon becomes a swarm." He feels that this is especially appealing because it makes the globular seem "dynamic and changing before your eyes."

M22 shares the field with **NGC 6642** through my 105mm refractor at low power. It appears very small, with a brighter core and a lone star at its northern edge. At high power through my 10-inch reflector, the 1.5' globular grows sharply brighter toward the center. It presents a granular core and halo, the latter showing a few very faint stars. A small tail of suns trailing north-northwest inspired California amateur Ron Bhanukitsiri to nickname NGC 6642 the Tadpole Cluster.

The second-brightest cluster in the quartet is **Messier 28** (M28). Arizona amateur and astronomy author Steven Coe enthusiastically described the view for me through his 6-inch Maksutov-Newtonian at a magnification of 136x. "Five stars consistently resolved, another eight come and go with the seeing. With averted vision, there is a bizarre effect, as I look at the outer edge of the cluster, different stars come and go as I use averted vision on them 'inadvertently.' Fascinating!"

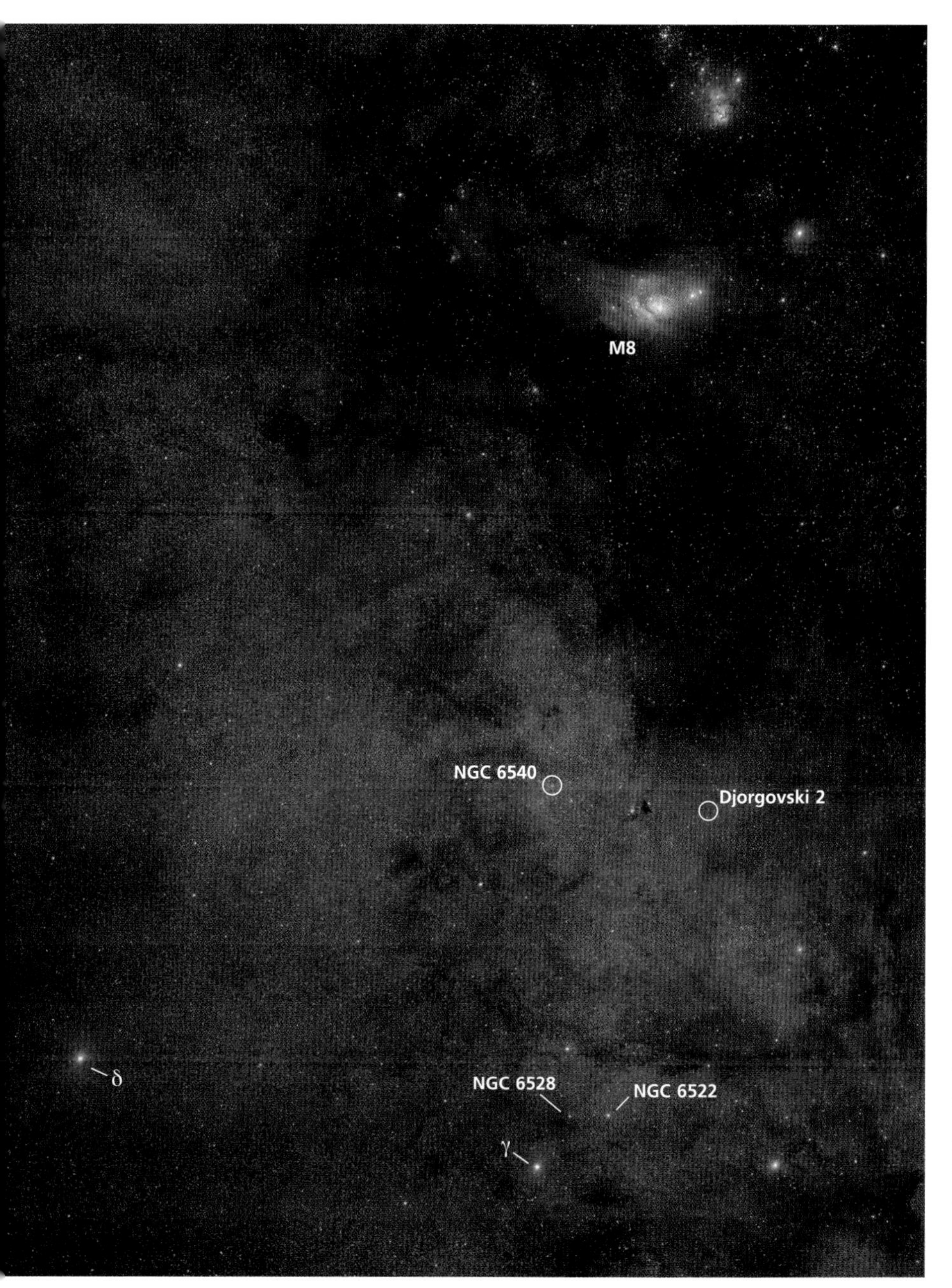

Posing as a little hazy spot with a brighter center, **NGC 6638** shares the field with M28 through my little refractor at 28x. With his 6-inch reflector at 94x, New York amateur Joe Bergeron describes this globular as "an easy, small, condensed, neat little thing containing a hint of some coarse sparkles."

Now we'll sweep farther east to **NGC 6717**, which sits immediately south of golden Nu² (ν^2) Sagittarii. Also known as Palomar 9, it's one of 15 Palomar clusters discovered or rediscovered in the 1950s on photographic plates from the *National Geographic Society–Palomar Observatory Sky Survey*.

Most Palomar globulars are inconspicuous, but NGC 6717 is the brightest of the bunch. Still, I was

This section of Robert Gendler's new Milky Way mosaic matches the lower-right corner of the chart at the top of page 222. The three-dimensional appearance is no illusion. The bright stars are closest to Earth, followed by M8 and, behind that, the dark clouds of the Great Rift. The veil of stars left of center is part of a nearby spiral arm. Between that arm and the Rift, you can see the Great Sagittarius Starcloud, where our galaxy's central bulge shows through. The globular clusters NGC 6540 and Djorgovski 2 are heavily obscured by outlying dust from the Great Rift.

Globular Clusters Bright and Faint

Object	Magnitude	Size/Sep.	Right Ascension	Declination
NGC 6522	8.3	9.4'	18h 03.6m	–30° 02'
NGC 6528	9.6	5.0'	18h 04.8m	–30° 03'
NGC 6540	9.3	1.5'	18h 06.1m	–27° 46'
Djorg 2	9.9	9.9'	18h 01.8m	–27° 50'
M22	5.1	32.0'	18h 36.4m	–23° 54'
NGC 6642	9.1	5.8'	18h 31.9m	–23° 29'
M28	6.8	13.8'	18h 24.5m	–24° 52'
NGC 6638	9.0	7.3'	18h 30.9m	–25° 30'
NGC 6717	9.3	5.4'	18h 55.1m	–22° 42'
Palomar 8	11.0	5.2'	18h 41.5m	–19° 50'

Angular sizes and separations are from recent catalogs. Visually, an object's size is often smaller than the cataloged value and varies according to the aperture and magnification of the viewing instrument.

surprised by how easy it was to see when I aimed my 105mm refractor in its direction. On a mostly cloudy night, I painstakingly star-hopped to the cluster as clear patches drifted by. Using an eyepiece that gave me 87x, I was amazed to find that the cluster was easily visible as a ¾' glow whenever it popped through a hole in the clouds.

Examined at 213x through my 10-inch scope, NGC 6717 contains a very small bright patch at its core that leans north-northwest. An even tinier patch rests a short distance northeast and is tipped northwest. The brightest cluster stars are about 14th magnitude, so you might be able to pick out some individuals with a 10-inch scope under good observing conditions.

The more difficult globular cluster **Palomar 8** lies exactly 2° west-northwest of 29 Sagittarii and 38' south-southwest of the semiregular variable star V3879 Sagittarii, notable for its reddish orange hue. This variable ranges between magnitude 6 and 6½ with a mean period of 50 days.

With my 105mm scope at 122x, I can see Palomar 8 with averted vision. It sits just south of a 4' triangle of three stars, magnitudes 10½ and 11. Its faint glow spans about 1', and there's a tiny detached spot west-northwest.

Palomar 8 triples in size through my 10-inch scope at 170x and is prettily nestled amid many faint field stars. The cluster is slightly dappled but shows little brightening toward the center. A few extremely faint stars are visible over the haze. Even viewing at high power in my 14.5-inch reflector, I can pick out only eight stars, some of which are probably foreground in such a rich field. The brightest members of Palomar 8 feebly glimmer at a mere 15th magnitude.

Cosmic Capture

Southern Sagittarius hosts many fine clusters and nebulae.

No two ways about it, our galaxy is a cannibal. Large galaxies such as our own often grow at the expense of small companions, slowly devouring them during close encounters. The Milky Way is currently munching on the Sagittarius Dwarf Spheroidal Galaxy (Sgr dSph), whose core coincides with the globular cluster **M54**, located 87,000 light-years away from us on the opposite side of our galaxy. Astronomers suggested that M54 could be the dwarf galaxy's nucleus, but new research indicates that while Sgr dSph does have a nucleus, it's distinct from M54. The cluster seems to have formed in the halo of the dwarf galaxy and then settled into the core due to orbital decay. M54 may also be home to a black hole with as much mass as 10,000 Suns, a devourer within the devoured.

In my 9x50 finder, M54 shares the field of view with Zeta (ζ) Sagittarii and is visible as a small fuzz-spot with a stellar center. The cluster is fairly bright through my 5.1-inch refractor at 117x. It displays a very faint 6' halo, a mottled 1¾' core, and a bright ½' inner core.

My 10-inch reflector at 213x shows a few faint foreground stars that are superposed on the halo. M54's brightest stars feebly shine at 15th magnitude because of their great distance from us. I begin to see hints of

The globular cluster Messier 54 has an extraordinarily dense core, as shown in this Hubble Space Telescope image. Photo: NASA / STScI / WikiSky

Roughly two-thirds of the way from M55 to Gamma (γ) Coronae Australis, we find the globular cluster **Terzan** 7. Working at France's Haute-Provence Observatory, Agop Terzan discovered the cluster in 1968 while conducting a near-infrared photographic study of heavily obscured regions near the galactic center. Of the 11 globular clusters that bear his name, it's the easiest one to observe with a backyard telescope.

On a good night, I can spot Terzan 7 in my 105mm refractor as a small, extremely faint hazy patch in an oval ring of stars. The ring stands 10' tall north-south. The two brightest stars, magnitude 8½, are at the northern end of the oval, while Terzan 7 rests within the southern end. Comparing the cluster's apparent size to the distance between nearby field stars, I put its diameter at 0.6'. Terzan 7 is fairly easy through my 10-inch scope. It spans about 1' and slightly brightens toward the center.

Although it looks much different, Terzan 7 shares a little something with M54. It is also thought to be part of the Sagittarius Dwarf Spheroidal Galaxy, yet it's 12,000 light-years closer to us. Eventually, Terzan 7 and M54 will be fossil remnants of the dwarf galaxy, whose more loosely bound stars will spread far and wide.

Our final globular cluster is **NGC 6723**, hovering just above the southern Sagittarius border ½° north-northeast of Epsilon (ε) Coronae Australis.

Observing with his 6-inch reflector at the Stellafane convention in Vermont, Joe Bergeron of New York logged: "Surprisingly bright and easy given it was only about 8° above the horizon. 94x. About 3' diameter. Round, granular, with a few fugitive stars visible. Seen in 50mm finder."

NGC 6723 is quite beautiful through my 10-inch reflector at 213x. It shows many stars of mixed brightness across its 10' face. The cluster's intense core seems to have two brightness steps. The outer core is oval, spans 4', and shows

resolution only with my 14.5-inch reflector at high power.

Nearly 10° east, **M55** is a lovely globular cluster appearing twice as large as M54 in my 5.1-inch scope. At 117x, it's spangled with many stars boasting a large range of brightnesses. The most prominent one is an 11th-magnitude star 5½' southeast of center.

Gorgeous M55 spans about 19' when viewed with my 10-inch scope and is rich in countless stars. They show a patchy distribution, with the brightest forming little chains crossing the core. The stars almost seem to stop at a diameter of 12½', with only a very faint halo haze beyond. M55 is breathtaking in the 14.5-inch reflector at 200x. Hundreds of stars populate the core and form a starburst pattern of arms radiating from it, while unresolved stars softly mist the background.

In their *Observing Handbook and Catalogue of Deep-Sky Objects*, authors Christian Luginbuhl and Brian Skiff note that "a large bite" has been chomped from the southeastern part of M55's halo.

Even though M55 is intrinsically less luminous than M54, its brightest stars shine at magnitude 11. M55 is easy to resolve primarily because it's relatively nearby — only 17,000 light-years away from us, about one-fifth the distance of M54.

Top right: The box near the lower-right corner of this star chart includes most of the objects described. Labels have been omitted for clarity. Bottom right: This photograph is a detailed view of the 2°-wide section of sky shown in the star-chart box. Chilean amateur Stéphane Guisard acquired the data, and Robert Gendler performed the image processing.

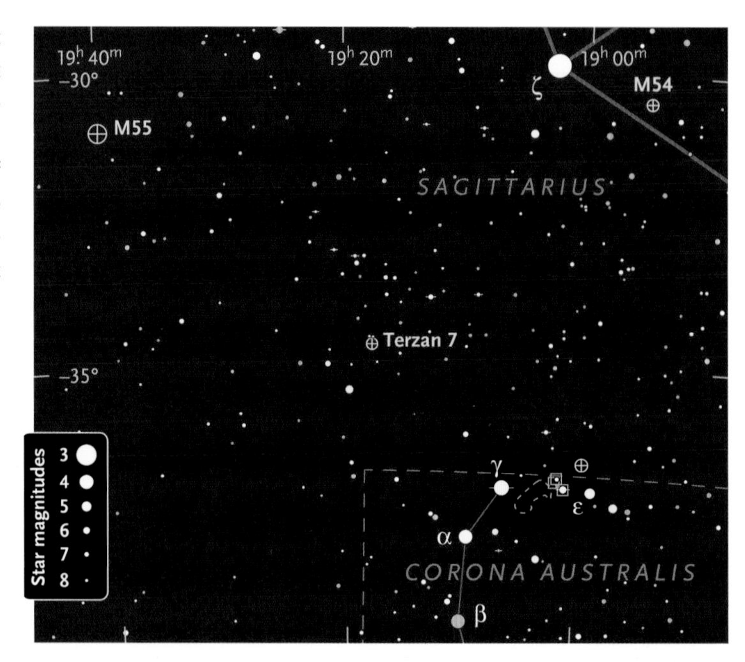

a little unresolved haze, while the inner core covers about 2¼'. A 10th-magnitude star sits in the northeastern part of the sparse halo, and a slightly fainter star lies just off the halo's eastern edge.

Some interesting nebulae lurk beneath the Sagittarius border in the constellation Corona Australis. The most striking surround **B 957**, one of thousands of double stars discovered by Willem Hendrik van den Bos, whose careful work greatly impressed his associates. In 1926, Robert Innes, head astronomer at South Africa's Union Observatory, expressed his fervent wish that when the chief assistant took charge upon his retirement, van den Bos would be given the assistant's position. Innes wrote, "If he is not appointed, then the 26-inch and double stars go to blazes."

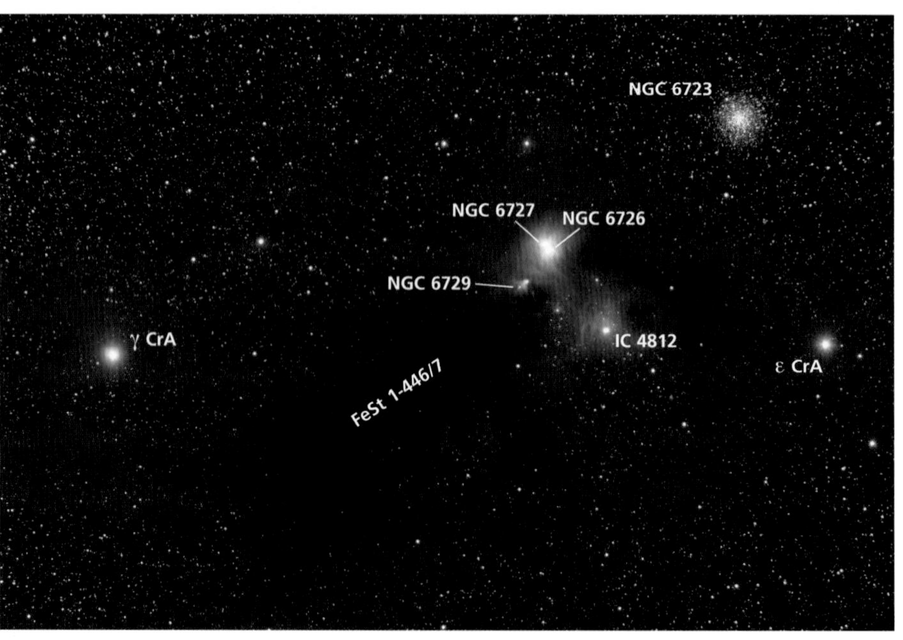

NGC 6726 surrounds the double's primary, a 7th-magnitude blue-white star, while **NGC 6727** envelops a wide companion 57" to the north-northeast. Both reflection nebulae are fairly bright in my 105mm refractor at 76x. They blend together for an overall length of 4½'.

The star illuminating NGC 6727 is the eclipsing binary TY Coronae Australis, whose combined light dips from magnitude 9.4 to 9.8 every 2.9 days. The binary is also subject to intermittent obscuration by dust clouds in the nebula, while its brighter component undergoes irregular fluctuations in brightness. The pair was once seen to dip to 12th magnitude.

Two related reflection nebulae share the field of view. Little **NGC 6729** rests 5' south-southeast of NGC 6726/7. It was about 1' across when I viewed it, but NGC 6729 is a variable nebula much like the celebrated Hubble's Variable Nebula (NGC 2261) in Monoceros. NGC 6729 is powered by the youthful star R Coronae Australis (R CrA), an erratic variable generally lingering between magnitude 10 and 14. Apparent changes in the nebula's form are thought to arise when moving dust clouds very close to the star cast shadows on the reflection nebula. NGC 6729 often displays a cometlike shape, with R CrA embedded in its "head" at the northwestern end. **IC 4812** sits 12' southwest of NGC 6726/7 and is almost as large as their combined glow, but much fainter. It looks vaguely triangular and holds **BrsO 14** (one of the double stars cataloged at Brisbane Observatory, Australia, in the 1800s) in its southern side. The blue-white components are magnitudes

This close-up is less saturated than the photo on page 226, so it shows very clearly both components of the bright, wide double star B 957 inside NGC 6726 and NGC 6727. The doubled diffraction spikes of BrsO 14, in the center of IC 4812, indicate that this star is double too. Photo: Bernd Flach-Wilken / Volker Wendel

6.3 and 6.6, with the dimmer star 13" west of its primary.

Where bright nebulae abound, dark nebulae are also likely to be found. Viewed with my 5.1-inch scope at 37x, the dark nebula combo **FeSt 1-446/7** (Feitzinger-Stüwe 1-446 and Feitzinger-Stüwe 1-447) starts south of NGC 6729 and stretches southeast for about 40'. Its width is somewhat irregular, but I'd say it averages around 12'. The blended nebulae are fairly obvious when the sky grants me good transparency low in the south.

All the nebulae we've visited are collectively known as the R Coronae Australis cloud and are thought to lie approximately 400 light-years away. Contrast this with their neighbor on the sky, NGC 6723, which is impressively distant at 28,000 light-years.

Objects in the Archer and the Southern Crown

Object	Type	Magnitude	Size/Sep.	Right Ascension	Declination
M54	Globular cluster	7.6	12.0'	18^h 55.1^m	−30° 29'
M55	Globular cluster	6.3	19.0'	19^h 40.0^m	−30° 58'
Terzan 7	Globular cluster	12.0	2.6'	19^h 17.7^m	−34° 39'
NGC 6723	Globular cluster	7.0	13.0'	18^h 59.6^m	−36° 38'
NGC 6726/7	Reflection nebula	—	9' × 7'	19^h 01.7^m	−36° 53'
NGC 6729	Variable nebula	—	1'	19^h 01.9^m	−36° 57'
IC 4812	Reflection nebula	—	10' × 7'	19^h 01.1^m	−37° 02'
FeSt 1-446/7	Dark nebula	—	50' × 17'	19^h 03.5^m	−37° 14'

Angular sizes and separations are from recent catalogs. Visually, an object's size is often smaller than the cataloged value and varies according to the aperture and magnification of the viewing instrument.

First published September 2010

Autumn

Hunting Messiers With the Horse and the Water Boy

Globular clusters and planetary nebulae in Pegasus and Aquarius give Messier hunters a chance to bag more prey.

Observing all the Messier objects is a goal pursued by many amateur astronomers and a fine way to hone your observing skills. A good 80mm telescope under moderately dark skies is sufficient to bag them all. Here, we'll ferret out the four Messier objects of Pegasus and Aquarius. While two objects are easy targets for a small telescope, the other two may prove to be a challenge. Let's start with the constellation Pegasus, which contains the brightest and easiest to find of the four.

The 18th-century astronomer Jean-Dominique Maraldi's interest in de Chéseaux's Comet (which appeared in August 1746) led him to discover two of these globular clusters, M15 and M2. On September 7, 1746, Maraldi observed M15 as "a fairly bright, nebulous star which is composed of many stars," a somewhat perplexing description.

Look for **M15** just off the nose of Pegasus, the Winged (or Flying) Horse. This cluster may have the densest core of any globular in our galaxy. It is theorized that M15's stars fell toward its center within the first few million years of the group's 12-billion-year lifetime. Although no globular cluster is unambiguously known to contain a central black hole, M15 is considered a good candidate, possibly containing one with 2,000 solar masses. Alternatively, the core density may simply be due to the mutual

On this chart, north is up and east is left. The circles show fields of view for a typical finderscope and a small telescope with a low-power eyepiece. To find north through your eyepiece, nudge your telescope toward Polaris; new sky enters the view from the north edge. (If you're using a right-angle star diagonal, it probably gives a mirror image. Take it out to see an image matching the map.) Chart: Adapted from *Millennium Star Atlas* data

This month's deep-sky objects lie southeast of the Great Square of Pegasus, beginning with M15 near Enif, the star on the horse's nose. Photo: Akira Fujii

gravitation of M15's 200,000 stars. M15 is also the first globular discovered to contain a planetary nebula, Pease 1. It has been observed in telescopes as small as 10 inches but is considered a tough visual target even in large amateur instruments.

To find M15, draw a line from Theta (θ) Pegasi to Epsilon (ε) Pegasi and continue for a little more than half that distance again. Golden-hued Epsilon is also known as Enif, and it marks the horse's nose. At magnitude 6.4, M15 has been spotted with the unaided eye under very dark skies by some keen-sighted observers. Through a small finder it resembles an out-of-focus star.

A 70mm scope at 20x shows a fairly bright, round, hazy patch surrounding a very small, bright nucleus. Through my 105mm refractor at 47x, the halo of the cluster looks mottled. At around 200x, the cluster appears slightly oval and some of the outer stars pop into view, but the center remains unresolved.

Our next globular cluster, **M2**, lies 13° south of M15 and makes a nearly right triangle with Alpha (α) Aquarii and Beta (β) Aquarii. M2 is a near twin to M15 in size, brightness, and distance, but it is less condensed. At 20x, a 70mm scope shows a bright, round patch of light surrounding an elusive, starlike nucleus. In my 105mm refractor at low power, the bright nucleus is more obvious and nearly stellar. Magnified at 200x, the cluster appears distinctly oval and sparkles with many very faint stars.

Two dim Messier objects, M72 and M73, are located in southwestern Aquarius, and if you'd like to observe the

Look in Pegasus for M15 and about a fist-width at arm's length below it for M2 in Aquarius. These globular clusters rival each other in brightness, though M2 is a little dimmer and farther away from us. Photos, left: Adriano Defreitas; right: Bob and Janice Fera

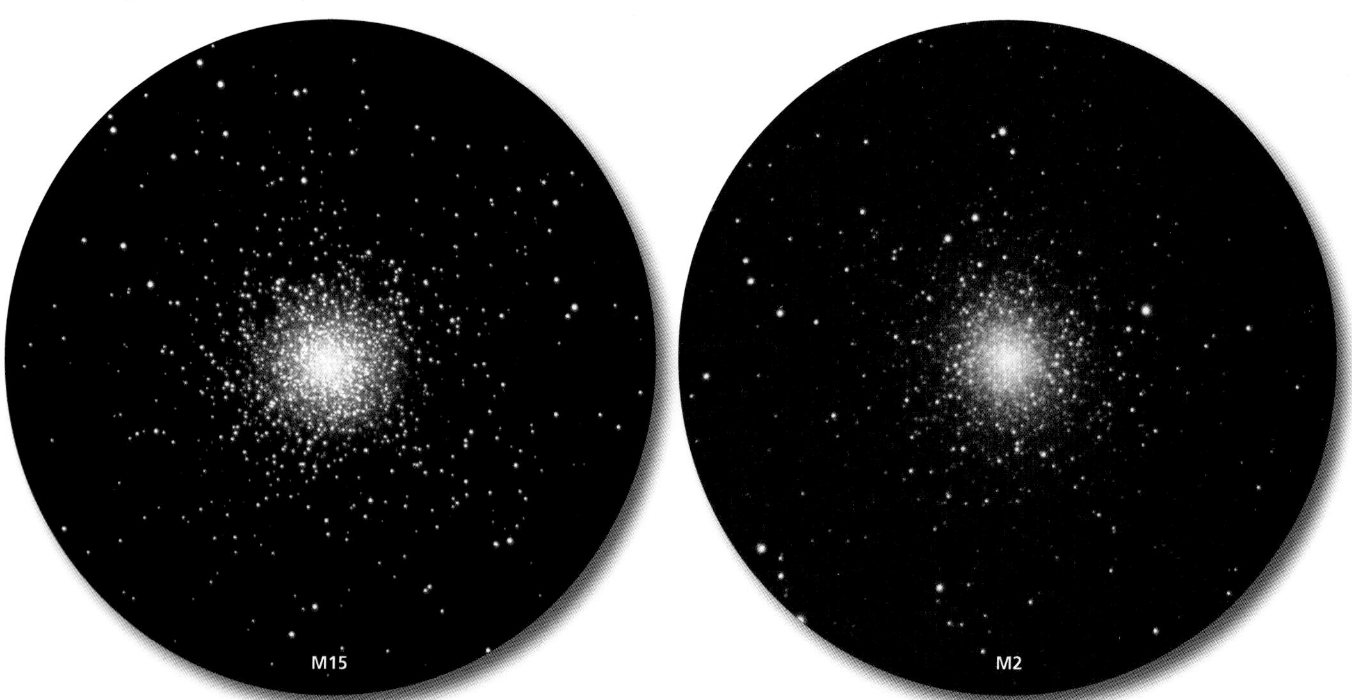

First published October 2000

From Messier 15 to the Saturn Nebula

Object	Type	Magnitude	Distance (l-y)	Right Ascension	Declination
M15	Globular cluster	6.4	34,000	$21^h 30.0^m$	+12° 10'
M2	Globular cluster	6.5	37,000	$21^h 33.5^m$	– 0° 49'
M72	Globular cluster	9.4	55,000	$20^h 53.5^m$	–12° 32'
M73	Asterism	9.0	—	$21^h 00.0^m$	–12° 38'
NGC 7009	Planetary nebula	8.3	3,000	$21^h 04.0^m$	–11° 22'

entire Messier catalog, you'll need to tackle these more difficult targets. **M72** is located 3.3° south-southeast of 3.8-magnitude Epsilon (ε) Aquarii and is the faintest globular cluster on Messier's list, at magnitude 9.4. An impressive 55,000 light-years away, the stars of M72 are very difficult to resolve in a small telescope. A 70mm scope at low power shows a very small and dim fuzzy patch. A 9.4-magnitude star lies 5' to the east-southeast. At 127x in my 105mm refractor, the cluster appears grainy, but there is no true resolution into stars.

Our last Messier object is **M73**, found 1.3° east of M72. Don't blink or you'll miss it! M73 is usually classified as an asterism of unrelated stars, but some believe it might be a multiple-star system or the remains of a dispersed cluster. M73 consists of four stars at magnitudes 10.4, 11.3, 11.7, and 11.9 in a Y-shaped group a mere 1' across. Use magnifications of around 100x to darken the sky background and help bring out the fainter stars.

If you don't find M73 sufficiently challenging, try for the nearby planetary nebula **NGC 7009**, often called the Saturn Nebula. You can star-hop from M73 using an eyepiece that gives about a 1° field of view. Put M73 at the western edge of the field, and you will see a 7.1-magnitude star on the opposite side. Place this star in the southern part of your field, and there will be a 7.0-magnitude star near the northern edge. Position that star near the southwestern edge of the field, and 8.3-magnitude NGC 7009 will be the brightest object in the northeast. At 30x in my 105mm scope, the planetary looks stellar, and at 150x, it appears small, oval, and bluish gray. The faint extensions (or ansae) that give NGC 7009 its Saturn-like appearance are not easily visible in scopes less than 10 inches in aperture.

Hunting Messier objects is a popular pursuit, but they comprise only a small fraction of the sights visible through a small telescope. If you turn your gaze northward to the pretty but dim constellation Lacerta, you'll see there is nary a Messier object to be found.

On the Wings of a Swan

Cygnus offers many treats that are best seen with your lowest-power eyepiece.

Cygnus, the Swan, glides through the zenith on the October all-sky star map on page 311. The leading edge of its outstretched wings is formed by Iota (ι), Delta (δ), Gamma (γ), Epsilon (ε), and Zeta (ζ) Cygni. Myriad deep-sky marvels, many within the grasp of small telescopes, nestle up to this distinctive line of stars.

We'll launch our tour with the planetary nebula **NGC 6826**, also known as C15, near Iota in the Swan's northern wing tip. Iota shares the same finder field with 4th-magnitude Theta (θ) and 6th-magnitude **16 Cygni**. Centering 16 Cygni and viewing it with a low-power eyepiece will reveal a pretty pair of nearly matched yellow suns. Wait three minutes, and the slow westward march of the stars will bring NGC 6826 to the center of the field for you. The nebula will appear tiny, but nonstellar, at 50x.

Boosting the magnification to 127x on my 105mm refractor unveils a slightly oval robin's-egg blue disk with a 10th-magnitude central star. If you stare straight at the star, the nebula will seem to fade, but if you glance a bit to one side, it will brighten again. (The soft light of the nebula is falling upon areas of the eye's retina with different sensitivities to light.) Switching your gaze back and forth creates the impression that has given NGC 6826 its nickname, the Blinking Planetary.

Now let's move to the open cluster **NGC 6811**, located 1.8° northwest of Delta in the same finder field. My small refractor at 87x shows a 15' swarm of about 40 faint to very faint stars in a nearly equilateral triangle. The cluster resembles an arrow with a fat head pointing west-southwest and a

On the Wings of the Swan

Object	Type	Magnitude	Size/Sep.	Dist. (l-y)	Right Ascension	Declination	*MSA*	*U2*
NGC 6826	Planetary nebula	8.8	27" × 24"	5,100	$19^h 44.8^m$	+50° 32'	1091	33L
16 Cygni	Double star	6.0, 6.2	40"	71	$19^h 41.8^m$	+50° 32'	1091	33L
NGC 6811	Open cluster	6.8	12'	4,000	$19^h 37.2^m$	+46° 22'	1109	33L
NGC 6866	Open cluster	7.6	10'	4,700	$20^h 03.9^m$	+44° 10'	1128	33L
NGC 6910	Open cluster	7.4	7'	3,700	$20^h 23.1^m$	+40° 47'	1128	32R
IC 1318	Emission nebula	—	4.0°	3,700	$20^h 22^m$	+40.3°	1127/28	32R/48L
M29	Open cluster	6.6	6'	3,700	$20^h 24.0^m$	+38° 30'	1127	48L
Ru 173	Open cluster	—	50'	4,000	$20^h 41.8^m$	+35° 33'	1148	47R
X Cygni	Variable star	5.9–6.9	—	4,000	$20^h 43.4^m$	+35° 35'	1148	47R
Veil Nebula	Supernova remnant	—	2.9°	1,400	$20^h 51^m$	+30.8°	1169	47R

The columns headed *MSA* and *U2* give the chart numbers of objects in the *Millennium Star Atlas* and *Uranometria 2000.0*, 2nd edition, respectively. Distances (in light-years) are from recent research papers. Approximate angular sizes are from catalogs or photographs; most objects appear somewhat smaller in a telescope used visually.

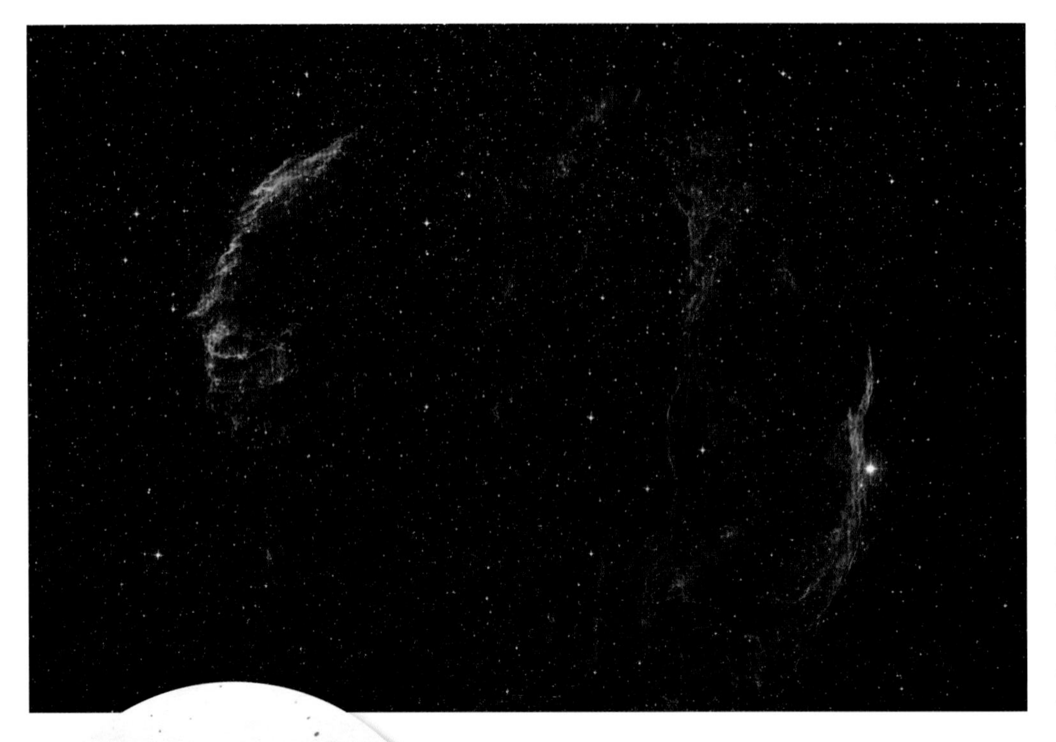

Encompassing the entire Veil Nebula, the author's drawing (below left) was made October 5, 1997, with the use of a 105mm Astro-Physics Traveler with a 35mm eyepiece (17x) and an oxygen III filter. Compare her drawing with Robert Gendler's image, left, composited from exposures with a Takahashi 106mm f/5 astrograph. Small scopes are good for tracing the overall shape and extent of large nebulae, but the human eye has a problem detecting their fine structure.

slender, slanted stalk. There's a paucity of stars in the center of the triangle.

Noted deep-sky observer and writer Walter Scott Houston called for observations of NGC 6811 after a reader from Denmark likened it to a smoke ring of stars with a dark band running through its middle. The ensuing reports included descriptions that compared it to the Liberty Bell, a butterfly, a frog, a clover (three-leaf and four-leaf), and even "Nefertiti's headpiece." Those using small scopes were the most apt to notice a dark interior enshrined by stars. What shape does this cluster suggest to you?

NGC 6866 also shares a finder field with Delta, which lies 3.5° west-northwest. The brightest star near this cluster is a 7th-magnitude, reddish orange star 24' to the west. With my little scope at 87x, I see about 30 stars, faint to very faint, arrayed in a 10' pattern that reminds me of a stunt kite. The kite is soaring to the northwest, and its southerly wing tip is bent. A 6' knot of 17 stars forms the main body of the kite, the brighter ones gathered into a north-south bar.

The cute little cluster **NGC 6910** sits 33' north-northeast of Gamma, the midpoint of the Swan's wings. At 87x, two

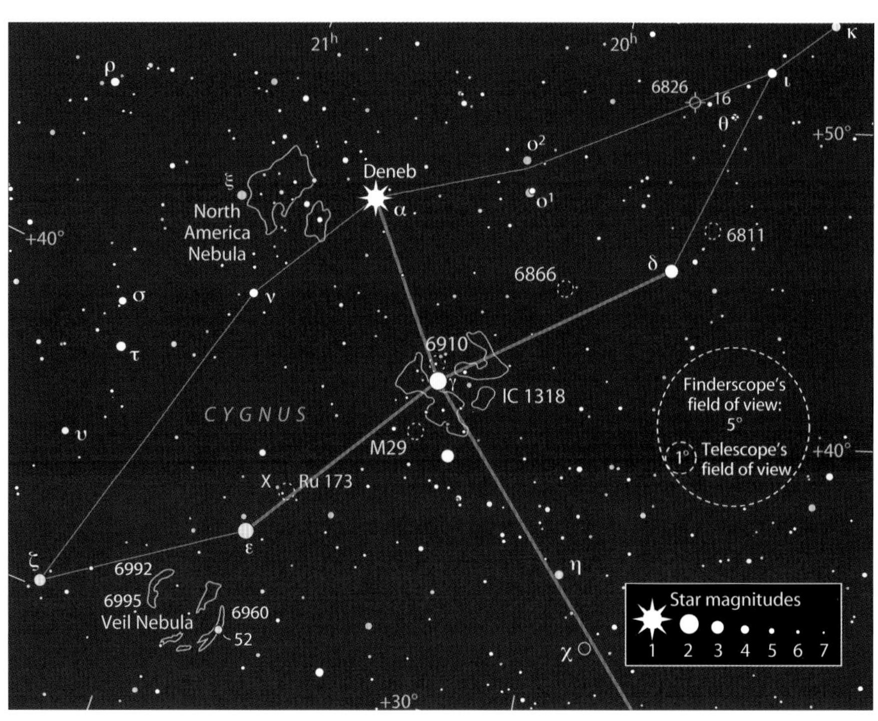

yellowish stars of 7th magnitude and a pearly, split chain of eight 10th-magnitude stars unite in a Y-shaped pattern about 5' long. A half dozen much dimmer stars join the scene.

NGC 6910 is embedded in an obscure section of **IC 1318**, the elaborate expanse of broken nebulosity surrounding Gamma Cygni. With a wide-field eyepiece giving 17x and a field of 3.6°, my little refractor reveals this to be an astounding complex. The three brightest patches lie 1.9° northwest, 0.8° east-northeast, and 1.1° east-southeast of Gamma. Each is aligned roughly northeast to southwest and appears about 30' to 40' long. To me, the field stars seem to grow fainter toward Gamma, almost as if it were sitting at the bottom of a funnel.

Putting Gamma Cygni in the western part of a low-power field and scanning 1.8° south will bring us to **M29**. At 87x, I see a small, pretty gathering of six 9th-magnitude stars arranged in back-to-back parentheses of three stars each. Ten dimmer stars are sprinkled across the cluster. M29 has been described in some observing guides as a miniature dipper or a tiny Pleiades. Arizona observer Bill Ferris finds the resemblance so remarkable that he calls M29 the "Little Sisters." Color images of the group display a nice mix of blue and yellow stars.

Just north of Gamma Cygni (off the frame's bottom edge) sits the open cluster NGC 6910, whose several stellar strings are dominated by two yellow 7th-magnitude stars in this ½°-square photograph by George R. Viscome. Chart: Adapted from *Sky Atlas 2000.0* data

We'll find our next target, **Ruprecht 173** (Ru 173), about three-quarters of the way from Gamma to Epsilon Cygni. Only low powers show this very large, coarse open cluster well. At 17x, I see 60 stars of 6th magnitude and fainter in 50'. Many of the brighter stars are arranged in a nearly cluster-spanning figure 8 with a fatter and dimmer southern half. A rich Milky Way star field skirts along and partly into the eastern edge of the group, where the variable star **X Cygni** resides. This pulsating yellow supergiant goes from around magnitude 5.9 to 6.9 and back in a period of 16.4 days. At its peak, X Cygni is the cluster's brightest star.

Our final stop is the beautiful **Veil Nebula**. This supernova remnant is a lovely sight in a small telescope, but unless you observe under very dark skies, you'll need an oxygen III filter (one blocking most light except green) for a good view.

The brightest section of the Veil bears the designations NGC 6992 and NGC 6995, and together they are C33. It is found a little southwest of a spot halfway between Epsilon and Zeta Cygni. This gently curving arc of gossamer light is more than 1° long and runs approximately north-south. The southern end widens and feathers out into ghostly tendrils reaching toward the west.

The Veil's other major part harbors the naked-eye star 52 Cygni, and for this reason, many people prefer to start their visit to the Veil here. This section, called NGC 6960, or C34, is a bit fainter than NGC 6992/5 but quite charming. It widens and forks to the south of 52 Cygni, while to the star's north it tapers to a point. If you center this northern tip in a low-power field, you can scan eastward to find NGC 6992/5.

A wide-angle eyepiece taking in both of these large arcs offers perhaps the most engaging view. My sketch at 17x (facing page) shows a 3.6° field that frames the entire Veil. The very faint wedge of nebulosity suspended between the northern ends of the brighter arcs is known as Simeis 229, or Pickering's Triangular Wisp. Harvard astronomer Edward C. Pickering mentioned this feature in 1906 after a long-exposure Veil photograph was examined by the observatory's first curator of astronomical plates, Williamina Fleming.

First published October 2003

Navigating North America

Knowing your geography puts you one step ahead in finding your way around this great nebula.

The North America Nebula is one of the most impressive nebulae glowing in our sky. This nebula's remarkable resemblance to the North American continent makes it better known by its common name — bestowed not by a resident of North America but by German astronomer Max Wolf. In 1890, Wolf became the first person to photograph the North America Nebula, and for many years, this remained the only way to fully appreciate its distinctive shape. With today's abundance of short-focal-length telescopes and wide-field eyepieces, we can more readily enjoy this large nebula visually.

The North America Nebula, **NGC 7000**, or Caldwell 20, is certainly easy to locate. Just point your telescope to a spot in the sky about one-quarter of the way from 3.7-magnitude Xi (ξ) Cygni to Deneb. This will put you in the region of the celestial Gulf of Mexico. Be sure to use your lowest-power eyepiece. The nebula spans more than 2°, giving small telescopes a decided advantage. Even with modern eyepieces, large telescopes never show a wide enough field of view and must display the nebula a piece at a time.

The pencil sketch on the facing page was made from the view through my 105mm refractor at 17x. It shows the entire North America Nebula as well as the soft glow of **IC 5070**, the Pelican Nebula, off its East Coast (which lies westward on the sky). I'm often asked whether the nebulae really look like this drawing. The answer is yes — if you view them the way they were sketched, in the dark with a dim red flashlight. Yet these nebulae aren't difficult to see. I've shared this view with many others at public star parties. Few have trouble seeing NGC 7000, and most can see IC 5070. A third nebula, shy by nature, occupies much of the field. Do you see it? The Gulf of Mexico and the space between the East Coast and the offshore Pelican are filled by the dark nebula **LDN 935**.

I use a greenish oxygen III filter to enhance the view at my moderately light-polluted observing site, but a narrowband filter works well too. Those blessed with darker skies may find a filter unnecessary. Although I didn't try to sketch the wealth of stars that crowd the field, the view is impressive and even contains a few star clusters.

NGC 6997 is the most obvious cluster within the confines of the North America Nebula. To me, it looks as though it's been plunked down on the border between Ohio and West Virginia. Putting 4.8-magnitude 57 Cygni at the western edge of a low-power eyepiece field should bring NGC 6997 into view. My 105mm scope at 17x displays a dusting of very faint stars. At 47x, it's a pretty cluster, rich in faint stars, spanning 10'. Through my 10-inch reflector, I count 40 stars, mostly of magnitudes 11 and 12. Many are arranged in two incomplete circles, one inside the other.

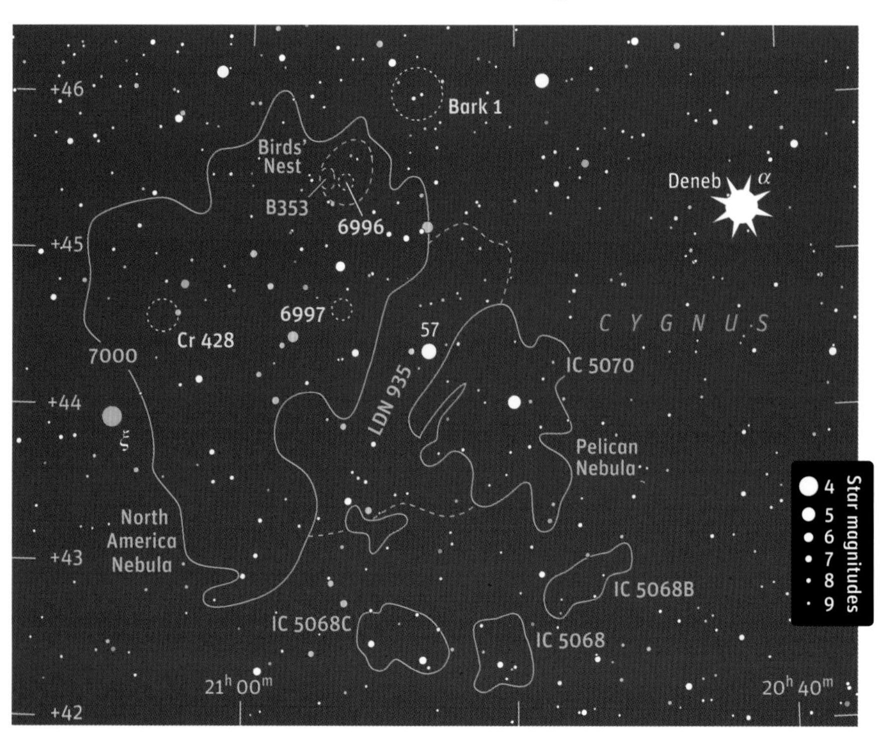

Is NGC 6997 actually involved in the North America Nebula? It's difficult to tell because the distances are poorly known. A 2004 article in the journal *Astronomy and Astrophysics* put NGC 6997 at around 2,500 light-years and adopted a value of approximately 3,300 light-years for the nebula. These figures were higher than those stated in many previous references. If valid, they identify the cluster as a foreground object.

Dave Riddle, an avid deep-sky enthusiast from Georgia, brought the **Birds' Nest** to my attention. It has remained one of my favorite sights in this area ever since. The name comes from a 1927 article in *Popular Astronomy* magazine by Daniel Walter Morehouse. Enti-

Left: Connecticut amateur Robert Gendler made this mosaic of the North America Nebula and its environs with a Takahashi FSQ-106 f/5 refractor and SBIG STL-11000M CCD camera. North is up, and the field is 5° across. Note the dark oval of the Birds' Nest, 1 to 1½ inches from the top and just left of center. The image combines conventional color frames with others taken in hydrogen-alpha light to bring out fine structure in the nebula's predominantly red emission. For more about the technique, visit www.robgendler-astropics.com. Below: For this sketch of a 3.6° field, including the North America Nebula, the author used a 17x eyepiece on her 105mm Astro-Physics Traveler — essentially the same aperture employed for Gendler's digital image.

tled "A Ring Nebula (Dark) in Cygnus," it discusses an interesting feature visible in photographs of the North America Nebula. Morehouse commented that he had "been referring to this object for a number of years as 'The Birds' Nest' in the 'Hudson Bay' region." With my 105mm scope at 47x, the dark rim of the nest is a 23' oval ring running north-northwest to south-southeast.

The dark nebula **Barnard 353** (B353) forms the eastern border of the Birds' Nest, its inkiest section. I count 27 stellar "eggs" filling the interior of the nest. In my 15-inch Newtonian reflector, the center of the nest is crowded with stars. This area, or at least the southern part of it, makes up **NGC 6996**, a concentrated portion of the Milky Way enisled by the dark nebulae surrounding it.

NGC 6997 was discovered by William Herschel and NGC 6996 by his son John, both observing from England two centuries ago. Even though the positions given by father and son were fairly good, the two clusters have frequently been mixed up in atlases and professional journals. Over the years, such notables as German astronomer Karl Reinmuth, French astronomer Guillaume Bigourdan, and American astronomers Harold Corwin and Brent Archinal have untangled the identifications.

If the Birds' Nest is in Hudson Bay, then **Barkhatova 1** (Bark 1) must be somewhere on Baffin Island. A pair of 7th-magnitude stars conveniently points straight to it, the more distant one golden and the closer one white. At 47x, my 105mm scope shows a pretty dusting of 30 stars, moderately faint to very faint, with the brightest two in the southern part of the cluster. A large oval gap in the eastern side of the group harbors a lone, very faint star. A reddish star sits at the eastern side of Barkhatova 1, and a golden star rests beyond its western border. My 10-inch scope reveals about 60 stars within 20'.

First published October 2004

Field of Cygnus's North America Nebula

Object	Type	Magnitude	Size	Right Ascension	Declination	*MSA*	*U2*
NGC 7000	Emission nebula	4	120' × 100'	$20^h 58.8^m$	+44° 20'	1126	32L
IC 5070	Emission nebula	8	60' × 50'	$20^h 51.0^m$	+44° 00'	1126	32L
LDN 935	Dark nebula	—	90' × 20'	$20^h 56.8^m$	+43° 52'	1126	32L
NGC 6997	Open cluster	10	8'	$20^h 56.5^m$	+44° 39'	1126	32L
Birds' Nest	Dark nebula, starcloud	—	23' × 18'	$20^h 56.3^m$	+45° 32'	1126	32L
B353	Dark nebula	—	12' × 6'	$20^h 57.4^m$	+45° 29'	1126	32L
NGC 6996	Starcloud	10	5'	$20^h 56.4^m$	+45° 28'	1126	32L
Bark 1	Open cluster	—	20'	$20^h 53.7^m$	+46° 02'	1106	32L
Cr 428	Open cluster	8.7	13'	$21^h 03.2^m$	+44° 35'	1126	32L
IC 5068	Emission nebula	—	25'	$20^h 50.3^m$	+42° 31'	1126	32L
IC 5068B	Emission nebula	—	42' × 14'	$20^h 47.3^m$	+43° 00'	1126	32L
IC 5068C	Emission nebula	—	25' × 18'	$20^h 54.2^m$	+42° 36'	1126	32L

Angular sizes are from recent catalogs; most objects appear somewhat smaller when a telescope is used visually. The columns headed *MSA* and *U2* give the chart numbers of objects in the *Millennium Star Atlas* and *Uranometria 2000.0*, 2nd edition, respectively.

We've visited the East Coast and Canada; now let's move over to northern Idaho, where we find **Collinder 428** (Cr 428). Putting 3.7-magnitude Xi Cygni at the southern edge of a low-power eyepiece field should bring this cluster into view. My little refractor displays a dozen faint stars in 12' with a 7th-magnitude star on the western edge. In my 10-inch reflector, the bright star appears orange and the star count doubles. The cluster looks somewhat like a fragment of the Milky Way isolated by a trapezoid of dark nebulae.

Three challenging patches of nebulosity lie south of the Pelican Nebula, roughly where you'd expect to find the northern coast of South America. The central patch is **IC 5068**, faint but definitely visible through my 105mm scope with an oxygen III filter. It looks blocky, with two 9th-magnitude stars in its eastern side: one near the northern corner, the other near the southern.

Through my 15-inch scope, the dimensions are about ½° north-south and ⅓° east-west. The nebula's brightest star is south of center and shines at 7th magnitude.

Just to the northwest is a swath of nebulosity labeled **IC 5068B** on the software atlas *MegaStar 5.0*. Harold Corwin of the NGC/IC Project (www.ngcicproject.org) has tentatively identified this as IC 5067. It's just a vague presence in my small scope but fairly bright in the large one. As seen with an oxygen III filter at 57x, it runs southeast to northwest for ¾° and is one-third as wide. A line of three 7th- through 9th-magnitude stars nearly parallels its northern edge. From east to west, they look blue-white, orange, and yellow when the filter is removed.

A third nebulous mass lies just east of IC 5068, and it is called **IC 5068C** in *MegaStar*. I haven't managed to see this with my small scope, but it's visible in the 15-inch. What size telescope do you need to spot it? IC 5068C is about 25' across and looks patchy, with a dimmer north-south band west of center. Two 7th-magnitude stars are widely spaced in its southern edge.

When next it's clear and your telescope beckons, why not go out and explore a celestial continent?

A Cygnus Sampler

There's a rich bounty of deep-sky wonders trailing the celestial Swan east of brilliant Deneb.

Thee, silver Swan, who silent can o'erpass?
An hundred with seven radiant Stars compose
Thy graceful form; amid the lucid stream
Of the fair Milky Way distinguish'd.
— Capel Llofft, "Eudosia," 1781

Gliding over the misty river of the Milky Way, Cygnus, the Swan, glitters with stars great and small. To count more than 100 stars within the modern boundaries of Cygnus, you'd need to see down to only about magnitude 5.8 — surely possible under dark skies. Likewise rich in deep-sky wonders, the

Wall-to-wall stars greet the observer sweeping the sky in northeastern Cygnus with any optical instrument, from the smallest binoculars to the largest backyard telescope. This 7½°-wide image includes the brilliant star Deneb at right and encompasses most of the region depicted in the chart on the following page. Photo: *Sky & Telescope* / Dennis di Cicco / Sean Walker

Swan boasts more delights than I could survey in all the clear and moonless nights of a year. In his little volume *Hours with a Three-Inch Telescope*, William Noble calls the region graced by Cygnus "glorious" and writes: "The merest vague sweeping cannot fail to reveal innumerable objects of beauty and interest."

With so much from which to choose, I'll confine this tour to a diverse sample of my favorite targets in northeastern Cygnus. We'll begin with the open cluster **NGC 6991**, which is easy to find. Look behind the Swan's tail (northeast of Deneb) for a 1° V of stars between magnitudes 5 and 7. The V points right at the cluster, which is another degree away and clasps a 6th-magnitude star in its eastern edge.

With my 105mm refractor at 17x, NGC 6991 is a scattering of faint stars alongside the solitary bright one. At 87x, there are about 35 stars arranged in straight and loopy chains that look very deliberate, as though a message in some alien alphabet were writ large across the cluster. A hazy patch is visible east-southeast of the bright star, where one might expect to find the nebula **IC 5076**, but I suspect I'm simply seeing unresolved stars. A nebula filter did nothing to clarify the issue.

My 10-inch reflector at 70x turns NGC 6991 into a right-handed cyclops. The bright star in the eastern side of the cluster marks his eye. Other stars outline his arms and legs, but his right-hand side (south) appears stronger. Altogether, there are around 100 mixed bright and faint stars gathered within 28'. A hazy patch lies immediately west and southwest of the "eye" with faint sparkles inside. Boosting the magnification to 118x brings out the very faint stars responsible for those phantom glints of light. But wondering whether the haze is just a mist of unresolved stars, I added a narrowband filter to the eyepiece. This improved the view, indicating that at least some of the haze must be nebulosity. IC 5076 is variously listed as a reflection nebula, an emission nebula, or a combination of both.

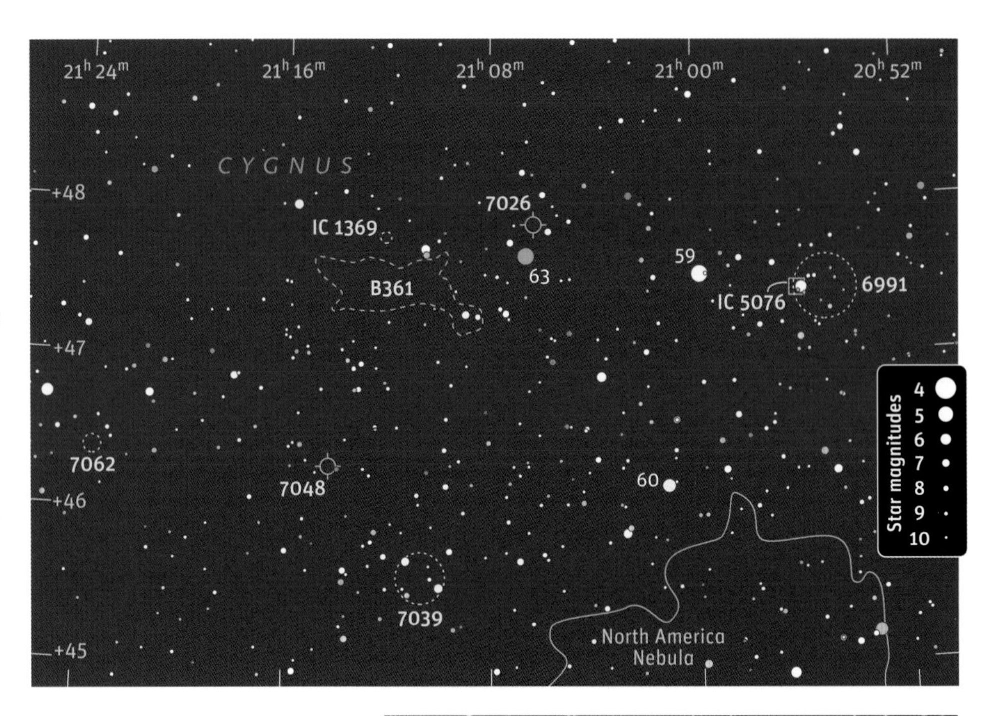

It's not surprising to learn that there's been some confusion in such a rich area of the sky. The group I'm calling NGC 6991 fits the description of the cluster William Herschel cataloged as VIII 76 in the late 18th century. Early in the 19th century, William's son John observed an object that he cataloged as 2091 and equated with his father's cluster, yet his descriptions clearly refer to a different group of stars. John Herschel's 2091 appears to be a 6' concentration of stars in the southern part of the larger group. One could dispute which aggregation rightly owns the title NGC 6991, but the elder Herschel's cluster is the more obvious of the two.

Our next stop is the planetary nebula **NGC 7026**, dubbed the "Cheeseburger Nebula" by Kentucky amateur Jay McNeil. Sweeping 1.8° eastward from the bright star in NGC 6991, you'll pass two 5th-magnitude stars. The second star is golden, and NGC 7026 sits 12½' to its north-northwest. With my little refractor at 87x, this planetary is small and round with a brighter center and a 9.6-magnitude star just off its east-northeastern edge. Through an oxygen III filter, the nebula looks brighter than the star, and it appears slightly oval at 127x.

NGC 7026 takes on a turquoise hue through my 10-inch reflector at 70x, and it looks very unusual at 219x. The east and west sides have large, brighter patches (the cheeseburger's buns) cloven by a thin, north-south line of darkness (the burger). Much fainter extensions elongate the nebula's north-south dimensions.

While scanning the area at 47x with my small refractor, I noticed a small fuzzy spot 1° east of NGC 7026. This is the open cluster **IC 1369**. Increasing the magnification to 87x, I plucked several very faint stars out of a 4' haze. Just south of the cluster, the dark nebula **Barnard 361** (B361) is a fairly large, inky, triangular patch with rounded points. Fingers of darkness reach out from the northeastern corner, and irregular dark areas spread west from the northwestern corner. With my 10-inch reflector at 70x, IC 1369 becomes a lovely rich dusting of faint stars over mist.

Located at the eastern edge of the large, loose star cluster NGC 6991, discovered by William Herschel in the 18th century, this faint blue nebulosity carries its own designation: IC 5076. The small gathering of stars at lower right is likely part of the object mentioned in the accompanying text that was seen by Herschel's son John. The field is ¼° wide. Photo: Chris Deforeit

Under dark skies, IC 1369 appears to lie near the western edge of a great dark nebula that cuts crosswise into the Milky Way. **Le Gentil 3** is probably the first dark nebula ever cataloged. French astronomer Guillaume Le Gentil described

An "easy" object for the author's 10-inch reflector, the planetary nebula NGC 7048 can also be seen under rural skies with a 105mm telescope fitted with an oxygen III filter. Its disk appears roughly the size of Jupiter. Photo: Richard Robinson / Beverly Erdman / Adam Block / NOAO / AURA / NSF

this object in a 1749 paper that was published in 1755. On a nice transparent night at my semirural home, this dark gash in the wake of the Swan is readily visible to the unaided eye.

A degree south-southeast of Barnard 361 there is another interesting planetary nebula, **NGC 7048**. It's larger than NGC 7026, but it has a much lower surface brightness. When I trained my 105mm refractor on the 10th-magnitude star at its south-southeastern edge, I could see no trace of the nebula, and at first I thought it must be too faint for my scope. Not so! Adding an oxygen III filter made all the difference. Through the filter, the planetary was fairly bright and easy at 87x. It appeared round, about 1' across, and had a uniform surface brightness.

Even without a filter, NGC 7048 is an easy target in my 10-inch scope, and its brightness seems more uneven. My 14.5-inch reflector at 245x reveals the planetary with a slightly brighter rim and a vaguely patchy interior. A very faint star is superposed on the planetary, northwest of center. The *Millennium Star Atlas* shows NGC 7048's symbol plotted 2' south-southwest of the planetary's true location.

A low-power sweep 50' southwest of NGC 7048 brings me to a 40' asterism of moderately bright stars that resembles a Christmas tree with its top pointing west in my little refractor. When I increase the magnification to 87x, I see the open cluster **NGC 7039**, which engulfs part of the tree's northern side. It consists of a rich group of faint stars with stragglers reaching south to a distance of 20'. A yellow-white 7th-magnitude star sits at the northeastern edge. In my 10-inch scope, the cluster is a triangle with a large ball on top, the bright star nestled in the middle of the ball. Another 7th-magnitude star lies along the base of the triangle, and an 8th-magnitude star rests on its eastern side. The rest of the stars are mostly 13th and 14th magnitude. Viewed at 115x, the densest band of stars (along the northern side of the cluster) is about 10' x 5'. The position of this cluster is often designated by the coordinates of the yellow-white star. The coordinates in the table below, on the other hand, give the approximate center of the group.

Our final target is the open cluster **NGC 7062**, which lies 1.6° east of NGC 7048. In my little refractor at 87x, NGC 7062 is a nice tight group of 20 moderately faint to extremely faint stars spanning 4½'. I can pick out 30 stars with my 10-inch reflector at 166x. The group is a 5' oblong running east-southeast to west-northwest, with its three brightest stars cradling the southern side of the cluster. On the *Millennium Star Atlas*, the circle for NGC 7062 should be shifted 3' to the east.

There's so much more that I'd like to share in this magnificent region of the sky. I'll continue my Swan tale in "A Cygnus Sampler, Part II," which you'll find on page 259.

In the Swan's Wake

Object	Type	Magnitude	Size/Sep.	Right Ascension	Declination	*SA*	*U2*
NGC 6991	Open cluster	~5	25'	20^h 54.9^m	+47° 25'	9	32L
IC 5076	Diffuse nebula	—	7'	20^h 55.6^m	+47° 24'	9	32L
NGC 7026	Planetary nebula	10.9	29" × 13"	21^h 06.3^m	+47° 51'	9	32L
IC 1369	Open cluster	8.8	5'	21^h 12.1^m	+47° 46'	9	32L
B361	Dark nebula	—	20'	21^h 12.4^m	+47° 24'	9	32L
Le Gentil 3	Dark nebula	—	7° × 2.5°	21^h 08^m	+51° 40'	9	32L
NGC 7048	Planetary nebula	12.1	62" × 60"	21^h 14.3^m	+46° 17'	9	32L
NGC 7039	Open cluster	7.6	20'	21^h 10.7^m	+45° 34'	9	32L
NGC 7062	Open cluster	8.3	5'	21^h 23.5^m	+46° 23'	9	32L

Angular sizes or separations are from recent catalogs. The visual impression of an object's size is often smaller than the cataloged value and varies according to the aperture and magnification of the viewing instrument. The columns headed *SA* and *U2* give the chart numbers of objects in *Sky Atlas 2000.0* and *Uranometria 2000.0*, 2nd edition, respectively.

First published October 2005

Return of the Little Fox

The famous Dumbbell Nebula is only one of Vulpecula's deep-sky treasures.

In "Foxfire Nights" (page 214), I toured some of the deep-sky delights in the den of Vulpecula, the Little Fox. Now I'd like to continue that expedition starting with M27, commonly known as the **Dumbbell Nebula**. Look for it 3.2° due north of Gamma (γ) Sagittae, where it's visible as a small but nonstellar spot in an 8x50 finder.

The Dumbbell is a big, bright, and highly detailed planetary nebula — a wonderful sight in almost any telescope. It's quite compelling through my 105mm refractor at 127x. The most prominent part of the nebula resembles an hourglass or an apple core. Brighter rinds highlight the top and bottom of the apple core. A diagonal stripe connects them, and enhanced glows huddle in the angles where they meet. Faint extensions filling and reaching out from the munched sides of the apple convert it into a football.

With a 6-inch or larger scope, you can crank up the magnification to about 200x and look for foreground stars set against the misty backdrop of the Dumbbell. Several are easier to pick out than the nebula's 14th-magnitude central star. Although it's possible to see the central star with a 6-inch telescope under very good skies, it takes a 10-inch scope to make this a reasonably easy task.

You never know when you'll find out something new about a famous and well-observed object. In 1991, Czech amateur Leos Ondra discovered a variable star while comparing images of the Dumbbell Nebula. With keen attention to detail, he noticed that a star plainly visible on one image was completely missing on another. Further investigation indicates that this may be a pulsating Mira-type variable with a peak visual magnitude around 14.3. Ondra nicknamed it the **Goldilocks Variable**. At maximum light, it should be within reach of a mid-size scope. So far, I haven't managed to catch it. Can you?

The 5th-magnitude star 12 Vulpeculae is exactly 2° west of the Dumbbell. Placing it in the southern part of a low-power field will bring you to the open cluster **NGC 6830**. My 105mm scope at 17x shows only a small hazy patch with a few faint stars, but at 87x, I see 20 stars within 6'. Most are gathered into a cute mushroom shape with the cap west and the stem east-northeast. By dropping to 28x and scanning 1.8° west, I encounter **NGC 6823** — a pretty collection of 30 faint gems including a nice triple at its heart. A shallow S of 9th- and 10th-magnitude stars clings to the cluster's northern edge and wends its way westward for ¾°. Traces of the elusive nebula **Sharpless 2-86** (Sh 2-86) feather the eastern side of the cluster. With my 10-inch scope at 68x, the triple star becomes a quadruple made up of two fairly bright stars with two dimmer sparks perpendicular between them.

Extending the Fox Hunt

Object	Type	Magnitude	Size/Period	Right Ascension	Declination	*MSA*	*PSA*
Dumbbell Nebula	Planetary nebula	7.4	8.0' × 5.7'	$19^h 59.6^m$	+22° 43'	1195	64
Goldilocks Variable	Variable star	~14–18	~213 days	$19^h 59.5^m$	+22° 45'	(1195)	(64)
NGC 6830	Open cluster	7.9	6'	$19^h 51.0^m$	+23° 06'	1195	64
NGC 6823	Open cluster	7.1	12'	$19^h 43.2^m$	+23° 18'	1195	64
Sh 2-86	Bright nebula	—	40' × 30'	$19^h 43.1^m$	+23° 17'	1195	64
Czernik 40	Open cluster	—	4'	$19^h 42.6^m$	+21° 09'	(1195)	(64)
Mini-Dragonfly	Asterism	—	5'	$19^h 43.1^m$	+21° 11'	(1195)	(64)
NGC 6813	Bright nebula	—	1'	$19^h 40.4^m$	+27° 19'	1195	64
Roslund 4	Open cluster	10.0	6'	$20^h 04.8^m$	+29° 13'	(1171)	(64)
IC 4954/5	Bright nebula	—	3' × 2'	$20^h 04.8^m$	+29° 13'	1171	64
NGC 6842	Planetary nebula	13.1	57"	$19^h 55.0^m$	+29° 17'	1172	(64)
NGC 6885	Open cluster	8.1	20'	$20^h 12.0^m$	+26° 29'	1171	64
Cr 416	Open cluster	—	8'	$20^h 11.6^m$	+26° 32'	(1171)	(64)
NGC 6940	Open cluster	6.3	30'	$20^h 34.6^m$	+28° 18'	1170	64
FG	Variable star	9.0–9.5	86 days	$20^h 34.6^m$	+28° 17'	1170	(64)

Angular sizes are from recent catalogs. The visual impression of an object's size is often smaller than the cataloged value and varies according to the aperture and magnification of the viewing instrument. The columns headed *MSA* and *PSA* give the appropriate chart numbers in the *Millennium Star Atlas* and *Sky & Telescope's Pocket Sky Atlas*, respectively. Chart numbers in parentheses indicate that the object is not plotted.

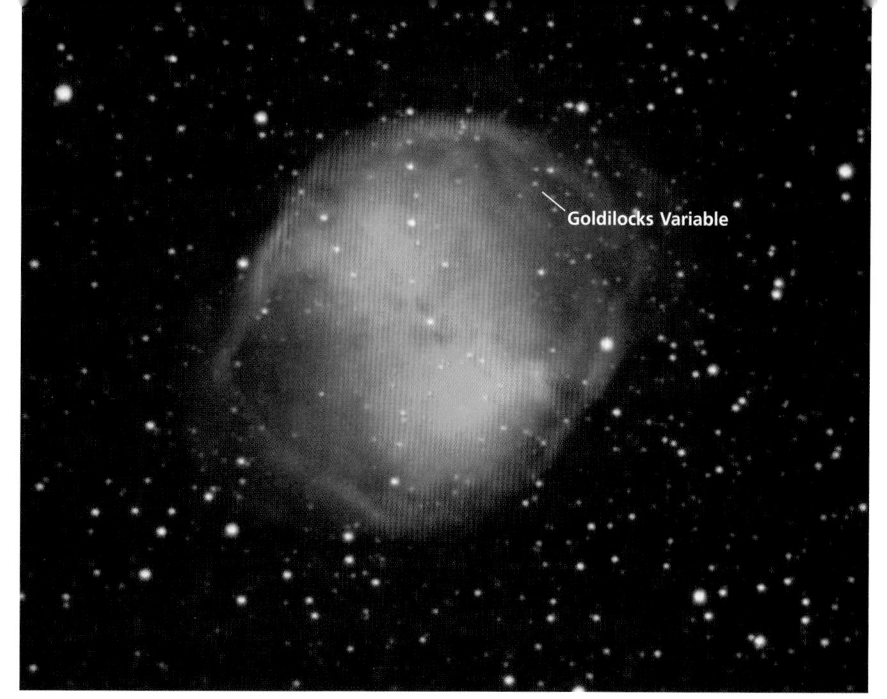

Discovered by Czech amateur astronomer Leos Ondra in 1991 and dubbed the Goldilocks Variable, this distinctly reddish star is believed to be a long-term Mira variable that ranges between 14th and 18th magnitudes. North is up, and the field is 12' wide. Photo: *Sky & Telescope* / Dennis di Cicco / Sean Walker

In my quest to observe the entire list of King clusters, I visited King 27, located 2.1° south of NGC 6823. This very faint group is more commonly known as **Czernik 40**. My 10-inch reflector at 118x shows a 3' patch of haze nestled up to the western side of a 7' trapezoid composed of four stars, magnitudes 8 to 9½. A more conspicuous group of stars nearby reminds me of the star cluster NGC 457 in Cassiopeia, so I call it the **Mini-Dragonfly**. Its tail is formed by six stars just beyond the east-northeastern edge of Czernik 40. Two stars to the east mark the dragonfly's mismatched eyes. The wings run north-south, each wing tip adorned with comparatively bright stars. A sprinkling of extremely faint stars speckles the body and wings. The asterism spans 5', and its brightest star is 11th magnitude. The stars of Czernik 40 are more reluctant to reveal themselves and surrendered to the prying eye of my 14.5-inch reflector only at 170x.

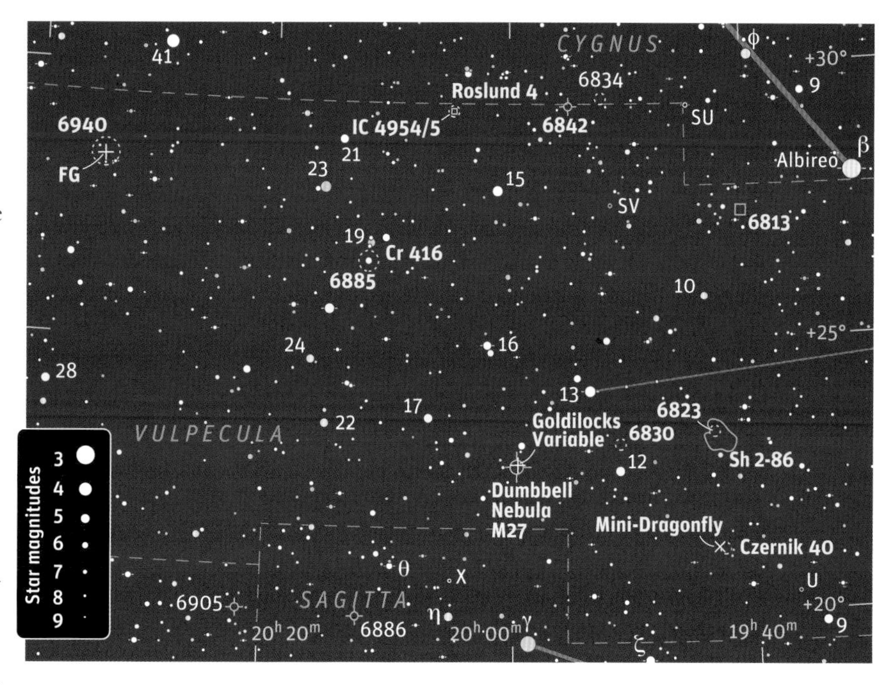

Now let's jump up to the northern border of Vulpecula, where we'll chase down three nebulae. The first is **NGC 6813**, found 1.7° north-northwest of 10 Vulpeculae. My little refractor at 87x displays a patchy nebula only 1' across. A 9th-magnitude orange star rests 2.3' north-northwest, and a faint star is embedded. My 10-inch scope at 311x exposes this star as a double. The pair sits southwest of center in a bright area of the nebula, while a third star is visible in the north.

NGC 6813 is a region of hot interstellar gas that shines by its own light. The brightest parts of our next nebula simply reflect the light of its wedded star cluster, **Roslund 4**. Look for the couple 1.7° north-northeast of 15 Vulpeculae. My 105mm scope displays nine faint stars in a roughly north-south arc enmeshed in dim nebulosity. With my 10-inch reflector at 118x, I count 15 stars in 6'. The nebulosity shows best in the southern part of the cluster, where it bears the designation **IC 4955**. The fragments of nebulosity in the northern part are designated **IC 4954**.

We meet our last nebula, **NGC 6842**, by sweeping 2.1° west. To spot this planetary with my little refractor, I need to use a narrowband nebula filter or an oxygen III filter. Averted vision also helps. At 87x, it's a round disk about 1' across. In my 10-inch scope at 213x, the nebula is vaguely annular with a break in the southwestern border. At low power, NGC 6842 shares the field with an almost perfect 9'-diameter semicircle of stars centered 11' to its east-northeast.

Next, we have an object with a serious identity crisis. It has sometimes been treated as one cluster and sometimes as two. The size, position, and designation of each varies from source to source. I'll simply adopt the data presented in the fascinating book *Star Clusters* by Brent A. Archinal and Steven J. Hynes (Willmann-Bell, 2003). The authors treat **NGC 6885**

First published October 2006

as a large cluster centered on 20 Vulpeculae and **Collinder 416** (Cr 416) as a small superposed concentration of stars. Studying the area with my 105mm scope at 28x, I see a pretty group of 30 stars, magnitudes 9 and fainter, enclosing bluish 20 Vulpeculae. Three bright stars in a curve just north of the cluster shine in shades of blue and gold. Examining the group at 153x brings out a score of faint to extremely faint stars crowded into 8' in the northwestern part of the large group. My 10-inch scope reveals many dim stars that shift NGC 6885's apparent center to the west. Visit the area with your telescope and see what you think.

NGC 6940 provides a foxy finale for our sky tour. This glittering cluster is nearly centered on the reddish variable star **FG Vulpeculae** (FG). My 105mm refractor at 28x uncovers a beautiful group of 70 stars within 35'. The brightest is an 8th-magnitude, yellow-white sun reigning over a relatively star-poor region in the northeastern part of the cluster. An orange-hued consort lingers nearby. At the very limits of vision, many exceedingly faint stars sparkle like diamond dust scattered on the velvet sky.

Unsung Marvels in Cygnus

A road less traveled passes some fascinating scenery.

We strolled among the little-known wonders of the celestial Swan in "The Graceful Swan" (page 216). Let's take the road less traveled again and tread starry pathways few have followed.

I locate deep-sky delights by beginning at a star I recognize. Then I refer to a detailed sky map and follow star patterns to my quarry. Even in this era of GoTo telescopes, old-fashioned star-hopping has its advantages, including chance encounters with intriguing sights along the way. That's how I stumbled upon the delightful quadruple star **Webb 9**.

Scanning the sky with my 10-inch reflector at a magnification of 44x, I spotted this quartet 29' south-southeast of the star 25 Cygni. The 6.7-magnitude blue-white primary is accompanied by a little curve of three companions to the south-southwest, the brightest shining with a deep red-orange hue. The *Washington Double Star Catalog* (*WDS*) lists the westernmost and dimmest companion at magnitude 10.6, but I estimate that it's at least a half magnitude fainter.

This quadruple is not limited to large-scope views. The colors and companions are visible with my 105mm refractor at the same power. This beauty was first noticed in 1878 by Thomas William Webb, a renowned British amateur astronomer and author of the classic observing guide *Celestial Objects for Common Telescopes*.

If four components make a multiple star, how many does it take before we have a star cluster? With 15 components listed in the *WDS*, the **ADS 13292** group seems to blur the boundary. To find this system, start at Eta (η) Cygni and move 1° east-northeast to a long, skinny triangle of stars, magnitude 7½ or so. The lucida, or brightest component, of ADS 13292 shines at magnitude 9.2 and sits just 4½' east-

Winging its way southward along the Milky Way, Cygnus, the Swan, is one of the few constellations that doesn't require a vivid imagination to see among the stars. But finding many of its deep-sky treasures requires a bit more effort. Here, the author continues a tour of some of Cygnus's lesser-known treats, several of which are in this 5°-wide field with bright Eta (η) Cygni right of center and north up. Photo: Davide de Martin / POSS-II / Caltech / Palomar

DECIDE FOR YOURSELF Is it a rich multiple star or a sparse open cluster? This fascinating stellar collection carries the multiple-star designation ADS 13292, and recent studies conclude that six stars (labeled A through F in the author's sketch below) are physically related, but line-of-sight companions give it the appearance of a meager open cluster. The field is ¼° wide with north up.
OBSERVING CHALLENGE While the relatively bright nebulous patch NGC 6857 (left of center) is visible in small telescopes, larger apertures and very good observing conditions are necessary for most observers to catch sight of the fainter nebulosity extending from its edge toward the southwest. The field is ½° wide with north up.
Photos: POSS-II / Caltech / Palomar

Look away

Very faint stars, nebulae, and galaxies can often be seen better if you direct your gaze to the side, placing the object off center in your eye's field of view. Called averted vision, this technique uses the photoreceptors in your retina called rods, which are more sensitive to light than the cones found at the center of the retina's field.

southeast of the triangle's eastern point. My little refractor at 127x shows the four brightest members in a shape that mimics the famous Trapezium embedded in the Orion Nebula. At 311x, my 10-inch scope reveals 14 components. The one that escapes me is cataloged as a 14.3-magnitude star nestled 4.1" south-southeast of the brightest star. The entire group spans less than 2'.

In October 2000, Helmut Abt and Christopher Corbally published the results of their study of 285 possible "Trapezium systems" in *The Astrophysical Journal*. The paper includes Trapezium 748, illustrated with the same 14 stars I saw. The authors conclude that six stars (labeled A through F in my sketch above) are physical members of the group, while the rest just happen to lie along the same line of sight. Unless dimmer stars not included in the study flesh out the group, this makes for a *very* sparse cluster, but it's visually appealing nonetheless.

Our next object is another mimic: the emission nebula **NGC 6857**, found 1.8° south of ADS 13292. NGC 6857 is very small but fairly bright, and I can see it as a little fuzzy spot with a faint star at one edge in my 105mm scope at 28x. Boosting the magnification to 87x helps the nebula stand out and shows two more faint stars off its edge plus a very faint one near its center.

But it was my first view of NGC 6857 through a 10-inch scope that captured my imagination. At 166x, I saw the little nebula surrounded by a kite-shaped asterism of four stars. The star within the nebula then became the center of a miniature cross reminiscent of the Northern Cross outlined by the chief stars of Cygnus. Just as the Northern Cross holds the Gamma Cygni Nebula (IC 1318) at its heart, so, too, does this mini-cross embrace NGC 6857.

Although I learned that NGC 6857 lies within a more challenging nebula, it was two years before I managed to spot it. Through my 10-inch at 311x with the help of an oxygen III nebula filter, the faint haze of **Sharpless 2-100** (Sh 2-100) spreads mostly west and southwest from NGC 6857. The nebula appears roundish, about 2½' across, and is rimmed with faint stars barely visible through the filter.

Now drop southeastward to the planetary nebula **NGC 6894**, which makes an isosceles triangle with the naked-eye stars 39 and 41 Cygni. At 47x through my little refractor, the nebula is visible only with averted vision. Boosting the magnification to 87x and adding an oxygen III or a narrowband nebula filter makes this round 1' nebula considerably easier to view, but I see no detail. At 153x with a narrowband filter, I perceive hints of a slightly darker area in the center. I can confirm the annularity with a filtered view through my 10-inch scope at 213x. The ring is wide and very patchy.

Continuing Flight With the Swan

Object	Type	Magnitude	Size/Sep.	Right Ascension	Declination	*MSA*	*U2*
Webb 9	**Multiple star**	**6.7, 9.0, 9.8, ~11**	**1.2', 1.4', 1.3'**	**$20^h 00.7^m$**	**+36° 35'**	**1149**	**48L**
ADS 13292	**Cluster?**	**8.7**	**19'**	**$20^h 02.4^m$**	**+35° 19'**	**1149**	**48L**
NGC 6857	**Bright nebula**	**11.4**	**38"**	**$20^h 01.8^m$**	**+33° 32'**	**(1149)**	**48L**
Sh 2-100	**Bright nebula**	**—**	**3.0'**	**$20^h 01.7^m$**	**+33° 31'**	**(1149)**	**(48L)**
NGC 6894	**Planetary nebula**	**12.3**	**60"**	**$20^h 16.4^m$**	**+30° 34'**	**1171**	**48L**
NGC 6834	**Open cluster**	**7.8**	**6.0'**	**$19^h 52.2^m$**	**+29° 25'**	**1172**	**48R**
Meerschaum Pipe	**Asterism**	**—**	**22'**	**$19^h 51.2^m$**	**+30° 07'**	**(1172)**	**(48R)**
Minkowski 1-92	**Protoplanetary nebula**	**11.7**	**20" × 4"**	**$19^h 36.3^m$**	**+29° 33'**	**1173**	**48R**

Angular sizes and separations are from recent catalogs. The visual impression of an object's size is often smaller than the cataloged value and varies according to the aperture and magnification of the viewing instrument. The columns headed *MSA* and *U2* give the appropriate chart numbers in the *Millennium Star Atlas* and *Uranometria 2000.0*, 2nd edition, respectively. Chart numbers in parentheses indicate that the object is not plotted. All the objects this month are in the area of sky covered by Chart 62 in *Sky & Telescope's Pocket Sky Atlas*.

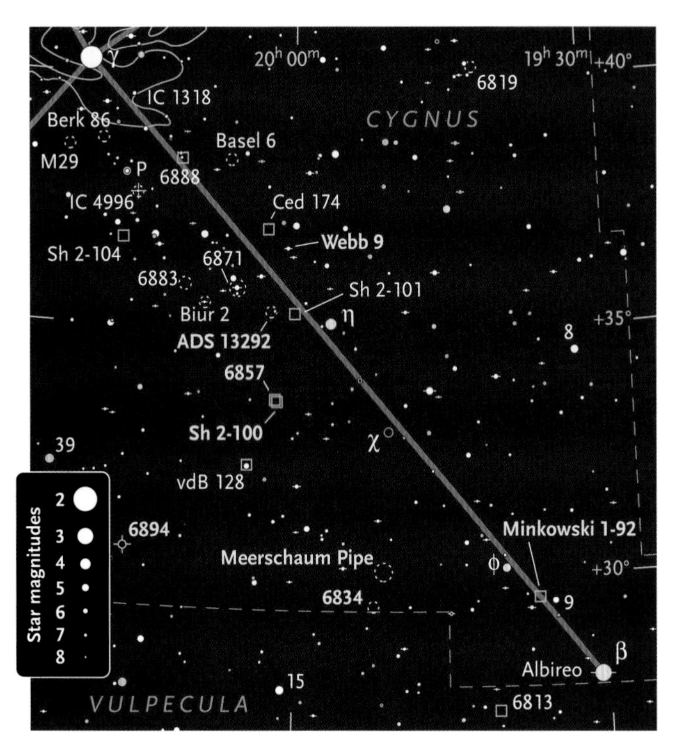

Without the filter, I catch rare glimpses of an elusive star embedded in the rim.

Our next target, the open cluster **NGC 6834**, hugs the Cygnus-Vulpecula border 2.6° northwest of 15 Vulpeculae. Swept up in my 105mm refractor at 17x, it's a conspicuous hazy patch holding one dim star. At 127x, I see an east-west line of five relatively bright stars against a score of faint suns entangled in a net of mist. The cluster is very pretty in my 10-inch reflector at 115x. The bright line blazes across a rich mass of 50 faint stars sparkling like diamond dust.

Now we'll move ¾° north-northwest to an asterism that California amateur Robert Douglas calls the **Meerschaum Pipe**. The pipe is 22' long and upright in the sky. The most prominent part is its long stem, which starts at the asterism's brightest star (9.6-magnitude HIP 97624) and gently curves southeastward to a dimmer bowl shaped like a deep and slightly rounded V. The pipe's 10th- to 12th-magnitude stars are visible in my little refractor, but a larger scope shows off this fumatory asterism better.

We'll finish our sky tour with a very unusual object in southwestern Cygnus. Discovered by German-American astronomer Rudolph Minkowski and reported in 1946, **Minkowski's Footprint**, or Minkowski 1-92, is a protoplanetary nebula. These objects are rare in the sky because they live for an astronomically brief time. When an aging star about the mass of our Sun exhausts its hydrogen fuel, it sheds its outer layers. Meanwhile, its core contracts to become an intensely hot white-dwarf star. Ultraviolet radiation from the white dwarf eventually excites the surrounding gas enough to shine with its own light as a planetary nebula. But there's a short interval, lasting perhaps a few thousand years, when the shrinking star isn't yet hot enough to make its cocoon glow. During this period, we see a protoplanetary nebula, mainly by reflected starlight.

Minkowski's 1-92 is a tiny thing, but fortunately, it's easy to locate. The star 9 Cygni makes a lopsided 38' kite with three stars to the east, magnitudes 6½ to 8. A pretty scattering of faint stars lies along the southern side of the kite with a pair of 10th-magnitude stars to the north. The nebula sits just east of the southern star.

At magnitude 11.7, Minkowski 1-92 isn't hard to see in my 105mm scope. It begins to look nonstellar at 87x and is elongated southeast to northwest at 220x. My 10-inch reflector at 394x shows how the Footprint gets its name. A little oval makes the sole of a shoe print, while a smaller and fainter round patch just to its southwest marks the heel.

Many planetary and protoplanetary nebulae exhibit two polar lobes separated by an equatorial ring. The northwestern lobe of Minkowski's Footprint looks brighter because it's the one tipped toward us, while the lobe at the opposite pole is dimmed by dust in the intervening torus.

First published October 2007

Arion's Dolphin

Tiny but shapely, Delphinus harbors many celestial delights.

But past belief, a dolphin's arched back
Preserved Arion from his destined wrack.
Secure he sits and with harmonious strains
Requites his bearer for his friendly pains.
The gods approve; the dolphin heaven adorns,
And with nine stars a constellation forms.
— Ovid, *Fasti*

Arion was a Greek poet-musician of the 7th century BC who spent much of his time at the court of Periander, ruler of Corinth. Legend has it that Arion was returning to Corinth by ship after a lucrative tour of Sicily and Italy, when the crew decided to relieve him of his riches and his life. Granted his own swan song, Arion took up the lyre with which he charmed both beasts and gods. He finished his plaintive melody and leapt into the sea, where he was rescued by a dolphin and borne to safety. The gods, well pleased, gave the dolphin its own constellation, Delphinus. Some suggest that nine stars were chosen to represent the nine Muses, inspirers of poetry and music. Although early star catalogs listed 10 stars in Delphinus, you can see it represented by nine even today on the mission patch for STS-78, the twentieth flight of the Space Shuttle *Columbia*.

Let's tour the starry realm of the helpful celestial Dolphin and see what deep-sky wonders it has to offer. We'll start with the small, bright planetary nebula **NGC 6891** in far western Delphinus. To locate it, sweep 3.5° due west from Eta (η) Delphini, where we'll pause to admire the pretty double star **Struve 2664** (Σ2664). Its nearly matched, golden suns are easily noticed at low power. From here, the planetary is just 1.1° west-southwest.

With my 105mm refractor at 47x, NGC 6891 is the southernmost "star" in a ½° north-south line of stars, magnitudes 8 to 10. At 153x, I see a very small disk with a tiny brighter center. Through my 10-inch reflector at low power, the nebula is distinguished from the surrounding stars by its telltale color. At 299x, the bright, sky-blue disk appears oval northwest-southeast. It encloses an easily visible central star with a small darker area around it. A very faint, round halo envelops the disk, and there's a faint star just off its eastern side.

NGC 6891 is a triple-shell planetary nebula whose structure reflects different episodes of mass loss by the aging progenitor star. Its dimensions have been measured in the light of hydrogen-alpha emission. The bright inner shell is 9" x 6"; the faint shell surrounding it is 18" across; and its detached halo, which I didn't see, spans 80". At a distance of 12,400 light-years, a figure known with rare accuracy for a planetary, this would make NGC 6891 almost 5 light-years across.

NGC 6905, the Blue Flash Nebula, is the only other planetary from the *New General Catalogue* that resides in Delphinus. It tacks down the northwestern corner of the constellation 4° east of Eta (η) Sagittae. NGC 6905 offers about the same total brightness as NGC 6891, yet it has a lower surface brightness because its light is spread over a larger area. In my 105mm at 153x, the nebula shares the field with a yellow 7th-magnitude star 16' to the south-southeast. Its round and nearly uniform glow occupies most of one side of a tiny triangle of stars, magnitudes 10 to 12, lying approximately north, south, and east of the nebula.

What's in a name?

The Latin *delphinus* comes from the Greek word *delphis*, meaning "womb." The Greeks gave that name to the dolphin because it's the "fish" that gives birth to its young just as humans do.

The Theta Delphini Group (circled) stands out fairly well in this Akira Fujii photograph. Can you make out the bucking bronco and the line of stars representing its skinny rider?

Delphinus Delights

Object	Type	Magnitude	Size/Sep.	Right Ascension	Declination
NGC 6891	Planetary nebula	10.4	18"	$20^h 15.2^m$	+12° 42'
Σ2664	Double star	8.1, 8.3	28"	$20^h 19.6^m$	+13° 00'
NGC 6905	Planetary nebula	10.9	47" × 37"	$20^h 22.4^m$	+20° 06'
θ Delphini Group	Asterism	—	1°	$20^h 38.3^m$	+13° 15'
NGC 6950	Open cluster?	—	14'	$20^h 41.2^m$	+16° 39'
Poskus 1	Asterism	9.6	6.5'	$20^h 46.0^m$	+16° 20'
γ Delphini	Double star	4.4, 5.0	9.1"	$20^h 46.7^m$	+16° 07'

Angular sizes and separations are from recent catalogs. Visually, an object's size is often smaller than the cataloged value and varies according to the aperture and magnification of the viewing instrument.

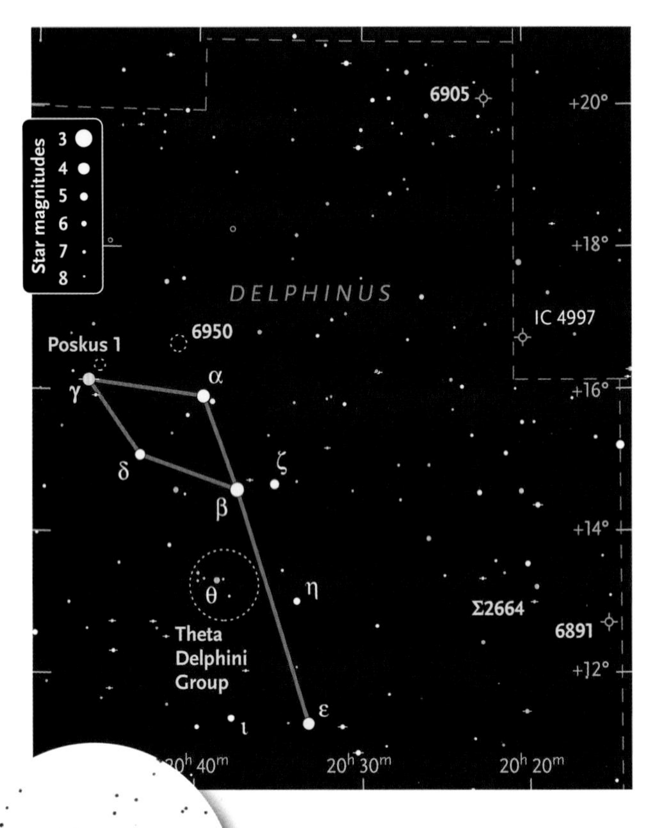

Finnish stargazer Jaakko Saloranta finds NGC 6950 to be "a beautiful, scattered cluster" when viewed at 71x through an 8-inch telescope. His sketch shows only the brightest field stars.

In my 10-inch scope at low power, I can recognize NGC 6905 as a small fuzzy patch. At 311x, the nebula is somewhat patchy, and its faint central star makes an appearance. The nebula is slightly longer north-south but brighter east-west. The shape reminds me of the Dumbbell Nebula (M27) but with a much more subtle brightness difference between the two parts. Despite its nickname, I see no obvious color in the Blue Flash, but other observers have picked up hints of color with 12-inch and larger scopes.

No ordinary white-dwarf star lurks at the heart of NGC 6905 but, rather, a Wolf-Rayet-type planetary nucleus. The petite stars of this class have spectra similar to those of the much heftier classic Wolf-Rayet stars. All Wolf-Rayet stars, regardless of size and origin, are hydrogen-deficient and very hot. The central star of NGC 6905 is a rare type WO planetary nucleus, which indicates strong lines of oxygen emission. It's also an exceptionally torrid 150,000 kelvin, whereas the surface temperature of our Sun is a comparatively tepid 5,780 kelvin.

In the first edition of his *Celestial Objects for Common Telescopes* (1859), Thomas William Webb notes that Theta (θ) Delphini "is in a beautiful field." Massachusetts amateur John Davis pictures the **Theta Delphini Group** as a bucking bronco. After hearing Davis describe this 1° asterism at an annual New England astronomy convention known as The Conjunction, I visited it with my little refractor at 17x. Golden Theta marks the spot where the cowboy meets the horse. A curve of stars trending northward and concave northeast forms the eternal, intrepid rider. East of Theta, a triangle of three bright stars makes the reared-up front legs of the horse, while a few stars to their north are its head. Several stars sprawled southwest of Theta fashion the bronco's hindquarters and tail.

Now let's move 49' north-northeast from Alpha (α) Delphini to **NGC 6950**. My 105mm scope at 47x shows a sparse cluster of 18 stars gathered into the shape of an open umbrella with a canelike handle. The umbrella is tipped north-northeast to ward off a few raindrop stars falling from that direction.

NGC 6950 is better seen visually than photographically. In fact, when Jack W. Sulentic inspected photographic plates while compiling *The Revised New General Catalogue* (1973), he concluded that the group was nonexistent. NGC 6950's stars may not form a true cluster, but they give the impression of being one when seen telescopically. Visual appearance was the only criterion that discoverer William Herschel could use when choosing to log it over two centuries ago.

In 2006, Colorado amateur Bernie Poskus wrote to tell me about an asterism he had noticed 15' northwest of the Dolphin's nose, **Gamma (γ) Delphini**. Now known as **Poskus 1** in the Deep-Sky Hunters database, this group has been referred

First published October 2008

to as a lute, a mandolin, or a flyswatter. Although the least charming of these descriptions, I find the last one the most accurate. The stars, magnitudes 11.5 to 12.8, show up nicely in my 10-inch scope at 68x. The asterism is 6.5' long. Its four-star handle cascades southeast to the rectangular, business end, which seems ready to whack Gamma. Perhaps what we have here is a star-swatter. Gamma Delphini shares the field of view and is seen as a beautiful double with a golden primary accompanied by a yellow-white companion to the west.

Poskus 1, the celestial flyswatter, shows nicely in this photo from the Digitized Sky Survey. The components of Gamma Delphini appear as a single bright blur, but the twin vertical diffraction lines reveal that this star is actually a double. Photo: POSS-II / Caltech / Palomar

Splashing Around the Dolphin

Tiny Delphinus holds a surprising variety of deep-sky objects.

The autumn sky plays host to a flood of aquatic constellations that inundate the southeastern expanse of October's all-sky star map on page 311. Starting in the east and flowing toward the south, you'll see Cetus, the Sea Monster; Pisces, the Fishes; Aquarius, the Water Bearer; Piscis Austrinus, the Southern Fish; Capricornus, the Sea Goat; and Delphinus, the Dolphin. The smallest of these constellations, Delphinus most resembles its namesake — a dolphin leaping playfully from the starry waters.

Let's first take to the water in far southeastern Delphinus, where we'll find the tight double star **Σ2735** sitting 53' west-northwest of the star 1 Equulei in the neighboring constellation Equuleus, the Little Horse. The components of Σ2735 are only 2" apart, but they're cleanly split through my 5.1-inch refractor at 117x. The 6.5-magnitude primary nuzzles a 7.5-magnitude companion a bit to the north of west. To me, the brighter star appears deep yellow and the secondary white.

Swimming 6.1° west-northwest, we come to **NGC 6934**. This globular cluster is readily found by diving 3.9° south from Epsilon (ε) Delphini, the end of the Dolphin's tail. It's visible in 12x36 image-stabilized binoculars as a softly glowing ball nestled against a 9th-magnitude star that makes it look elongated at a casual glance. The cluster's slightly oval form leans north-northeast as seen through my 105mm scope at 153x. A few extremely faint stars sparkle in the halo, which encircles a broadly brighter core. In my 10-inch reflector at 219x, NGC 6934 appears 3' across and partially resolved over a dappled haze. It grows to 4' through my 15-inch reflector and glitters with many stars, incuding some at its very heart.

Next, we'll visit **NGC 6928**, the brightest of a small group

NGC 6934 is the brightest globular cluster north of the celestial equator that's not in the Messier catalog. It's a fine spectacle through medium to large telescopes, though you'll never see it as well resolved as in this Hubble Space Telescope image. Photo: NASA / STScI / WikiSky

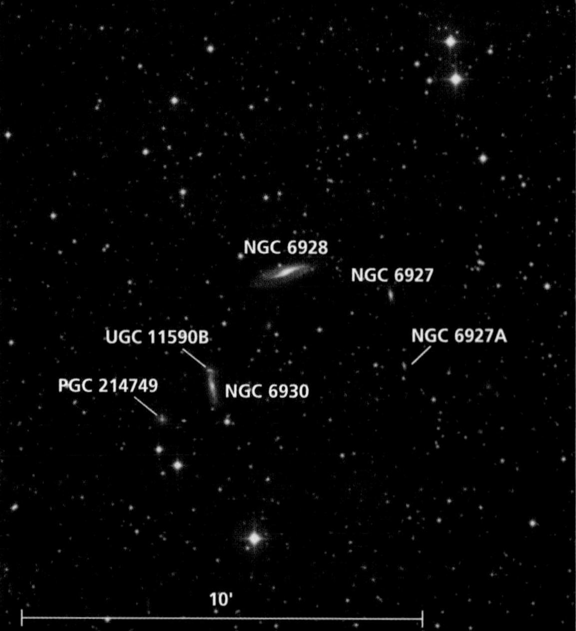

NGC 6928, NGC 6930, and NGC 6927 are the brightest galaxies in their group, at magnitudes 12.2, 12.8, and 14.5, respectively. Faint little NGC 6927A, at magnitude 15.2, is sometimes labeled as PGC 64924 or MCG 2-52-15. Photo: POSS-II / Caltech / Palomar

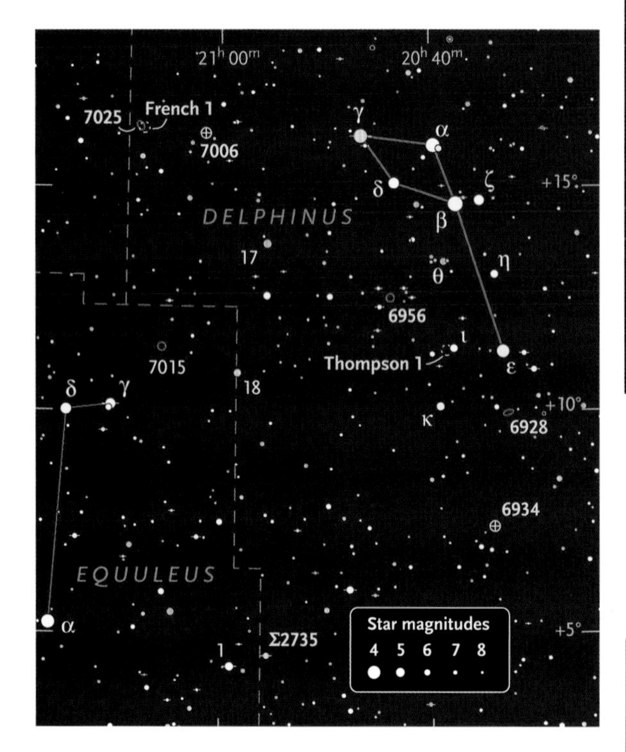

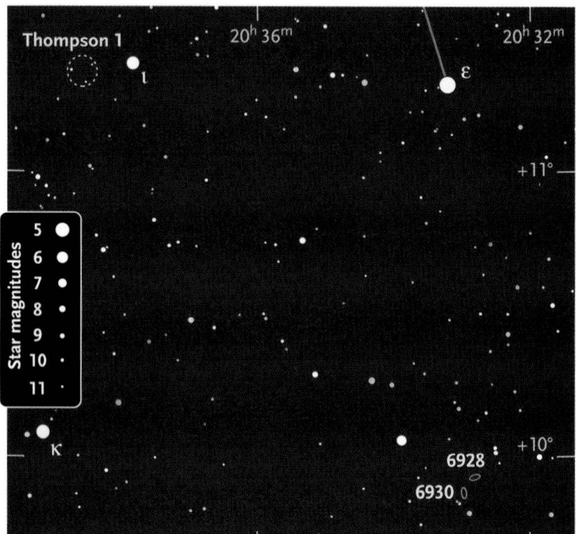

of galaxies treading water 1.4° beneath the Dolphin's tail. Only NGC 6928 is visible with my 105mm refractor at 127x, its diaphanous spindle tipped east-southeast. A 6' chevron of five similarly spaced, 12th-magnitude stars reaches northeast from the galaxy's eastern tip and then bends north-northwest.

Through my 5.1-inch scope at 117x, I can easily distinguish a brighter, elongated core in NGC 6928 when I use averted vision. A faint star adorns the galaxy's northern flank. Boosting the magnification to 164x, I can catch, but not hold steadily, the roughly north-south glow of **NGC 6930**. It lies 3.8' south-southeast of its companion and just north of the pointy end of a 2' triangle of 10.2- to 12.6-magnitude stars.

Large-scope users should keep their eyes open for "extra" stars. NGC 6928 hosted a supernova that peaked at magnitude 15.3 in 2004. The previous year, NGC 6930 bore one perhaps a half magnitude fainter. Both galaxies are about 200 million light-years distant.

I can spot NGC 6928 at low power in my 10-inch scope, but the best view comes at 299x, which shows the mottled core growing brighter toward the center. The core of **NGC 6927** is visible as a very small, very faint spot 3.1' west-southwest of its brighter cousin. While viewing at moderate powers, I noticed something 5.6' west of NGC 6928 that looked tantalizingly like a nebula with a star involved. At 299x, the star became a double accompanied by two more stars southeast and one north. Images of the area reveal a cute upside-down question mark of several stars that might look interesting when viewed through a large telescope.

In my 15-inch reflector, NGC 6927 becomes a small north-south oval, and I can glimpse three additional galaxies. At 247x, tiny **NGC 6927A** is dimly visible 2' south of NGC 6927. A 15th-magnitude star lies off its southern tip, and a fainter one is visible to the north. **PGC 214749** is a very faint round patch 51" north of the easternmost star in the triangle near NGC 6930. I occasionally see the glint of an extremely faint star on its northern edge. Identification of the galaxy is made simple by the fact that it makes a nice pie wedge with the triangle's stars.

The third galaxy, **UGC 11590B**, teeters smack on the northern tip of NGC 6930, but I don't see it as a separate object. The slender profile of NGC 6930 comprises a faint halo, a brighter long axis, and a small oval core, while UGC 11590B merely looks like a brightening at its apex.

If you're looking for still tougher game, download an image

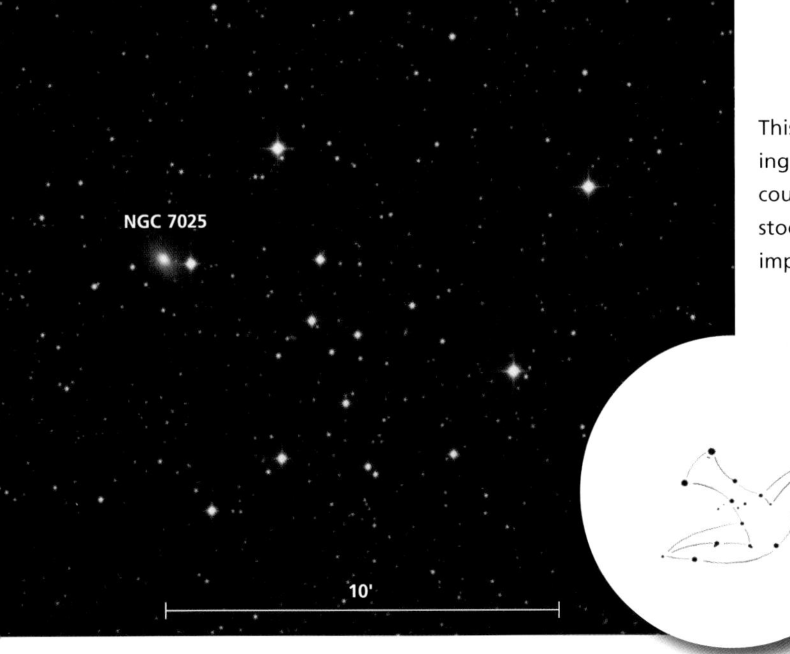

This photo from Caltech's POSS-II sky survey shows NGC 7025 hovering near the easternmost star of French 1. Turn this page 135° counterclockwise, and you'll see why this asterism is called the Toadstool. The sketch, inset, shows Finnish stargazer Jaakko Saloranta's impression of the Toadstool at 96x through his 8-inch telescope.

of the surrounding area and use it to help you hunt down the plethora of faint galaxies that wade in these celestial waters.

Now we'll leap up to **Thompson 1**, whose three brightest stars are shown at upper left in the bottom chart on the facing page. This asterism was brought to my attention by Canadian amateur John Thompson. Through my 5.1-inch refractor at 63x, it shares the field with Iota (ι) Delphini, which glistens 10.4' west-northwest. I count 13 stars, magnitudes 10 to 13, in a distinctive, 5.7'-long triangle whose pointy end aims south-southwest.

Our next port of call is a galaxy that exhibits an interesting structure. **NGC 6956** is a 1.5° voyage east-southeast from golden Theta (θ) Delphini. In my 105mm refractor at 47x, it's a small hazy spot hiding in the glow of the 12th-magnitude star on its eastern edge. At 127x, the galaxy appears inclined north-northwest. My 10-inch scope at high power adds little, other than a faint double star just east of the superposed one. But my 15-inch reflector at 247x uncovers much more. NGC 6956 displays a faint oval halo, tipped a bit south of east, enveloping the elongated form seen in the smaller instruments. NGC 6956's center harbors an oval core aligned north-south.

Another globular cluster, **NGC 7006**, floats 3.6° east of the Dolphin's nose. It's smaller and fainter than NGC 6934, but it's easily noticed at 37x through my 5.1-inch refractor, and it grows brighter toward the center. The cluster is about 2' across and slightly elongated east-west. Two 14th-magnitude stars lie off the edge, a shade west of south, but no stars are resolved in the cluster. Through the 15-inch reflector at 216x, NGC 7006 spans 3' with a mottled core about half as large. Faint foreground stars graze the edge of the cluster approximately north, northwest, south, and east. Several extremely faint stars in the halo glimmer as well as some in the outer core.

NGC 7006 is dimmer and more difficult to resolve than NGC 6934 largely because it's much farther away — 135,000 light-years compared with 51,000 light-years.

Sailing 1.4° east of NGC 7006, we come to **French 1**, a 13-star asterism of 9th- to 12th-magnitude stars. I call it the Toadstool, because it looks like a mushroom with its stem northeast and the 12.5'-wide cap southwest. Washed up against the foot of the Toadstool, we find a bit of celestial flotsam in the form of **NGC 7025**. Through my 105mm refractor at 153x, this ashen galaxy shows a broadly brighter, slightly oval core encircled by a thin, faint, 45"-long halo tipped northeast. In the 15-inch reflector at 216x, it covers about 1½' x 1', its eastern flank guarded by a 13th-magnitude star and its western flank by the golden star at the base of the Toadstool.

For such a small constellation, Delphinus certainly holds a wealth of deep-sky objects.

Delights of the Celestial Dolphin

Object	Type	Magnitude	Size/Sep.	Right Ascension	Declination
Σ2735	Double star	6.5, 7.5	2.0"	20ʰ 55.7ᵐ	+4° 32'
NGC 6934	Globular cluster	8.8	7.1'	20ʰ 34.2ᵐ	+7° 24'
NGC 6928 (group)	Galaxy	12.2	2.2' × 0.6'	20ʰ 32.8ᵐ	+9° 56'
Thompson 1	Asterism	9.0	5.7'	20ʰ 38.5ᵐ	+11° 20'
NGC 6956	Galaxy	12.3	1.9' × 1.3'	20ʰ 43.9ᵐ	+12° 31'
NGC 7006	Globular cluster	10.6	3.6'	21ʰ 01.5ᵐ	+16° 11'
French 1	Asterism	7.1	12.5'	21ʰ 07.4ᵐ	+16° 18'
NGC 7025	Galaxy	12.8	1.9' × 1.2'	21ʰ 07.8ᵐ	+16° 20'

Angular sizes and separations are from recent catalogs. Visually, an object's size is often smaller than the cataloged value and varies according to the aperture and magnification of the viewing instrument.

First published October 2009

Hunting Down the Helix

Despite its dodgy reputation, this planetary nebula is easy to find if you go about it the right way.

Without a doubt, the Helix Nebula is one of the most beautiful objects in the heavens. It earns its name from a double-ringed appearance on photographs, like looking down at two coils of a spring. Long-exposure photos show nebulous fingers pointing inward from the ring toward the central star. The Hubble Space Telescope captured stunning close-ups of these radial filaments, showing each with a cometlike head and a gossamer tail.

Also known as NGC 7293 and Caldwell 63, the **Helix Nebula** is one of the nearest and brightest of the class of objects known as planetary nebulae. A planetary nebula is born of an aging star that exhausts its nuclear fuel and sheds its outer layers into space. As the nebula expands, the core of the star is exposed. This hot, dense cinder will evolve into a white-dwarf star and then slowly cool over billions of years.

Observationally, the Helix is one of the easiest of bright planetaries and one of the most elusive. It can be seen in binoculars and yet remain invisible in large telescopes. These claims may seem contradictory, but they can be explained by the nebula's low surface brightness. If the light of the Helix were gathered into a single point, it would shine with the light of a 7.3-magnitude star. But this light is spread over a rather large area of the sky with about half the angular diameter of the Moon! Low powers help concentrate the nebula's light, and wide fields show plenty of surrounding dark sky for contrast. That's why the Helix is a good small-scope target.

With such elusive quarry, you need to hunt the Helix carefully, being sure that your scope is aimed at the correct spot in the sky. The Helix is never very high in the sky from my semirural site in upstate New York, but I can sometimes see an adjacent star, Upsilon (υ) Aquarii, with the unaided eye. Spotting Upsilon will make your nebula search simple. But if your sky is too bright to show that star, try starting at 3.3-magnitude Delta (δ) Aquarii. From Delta, look 4° southwest for 4.7-magnitude 66 Aquarii. It should fit in the same finder field with Delta and shine with an orangish light through the telescope. Continue along that line another 2.8°, and you will come to yellow-white Upsilon, the brightest star in the area, at magnitude 5.2. The Helix sits 1.2° west of Upsilon and is best sought with a low-power eyepiece. Two 10th-magnitude stars lie halfway between, helping pinpoint the Helix in two easy jumps.

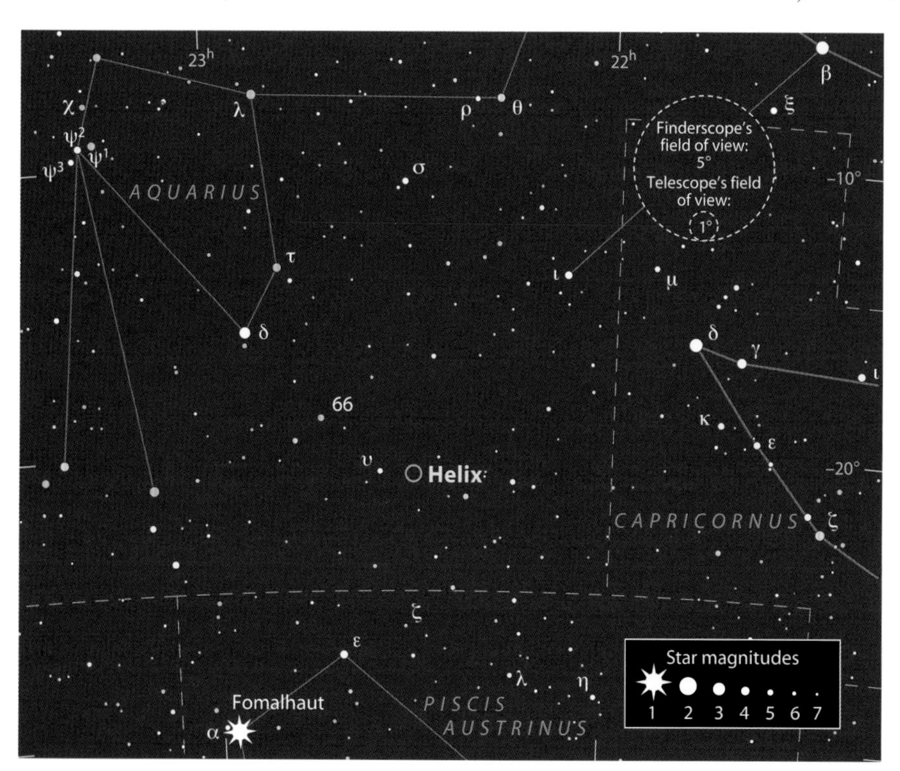

I can pick out the Helix with 50mm binoculars from home. At first glance, a small telescope shows only a featureless oval disk, but keep looking. The Helix Nebula yields up its details

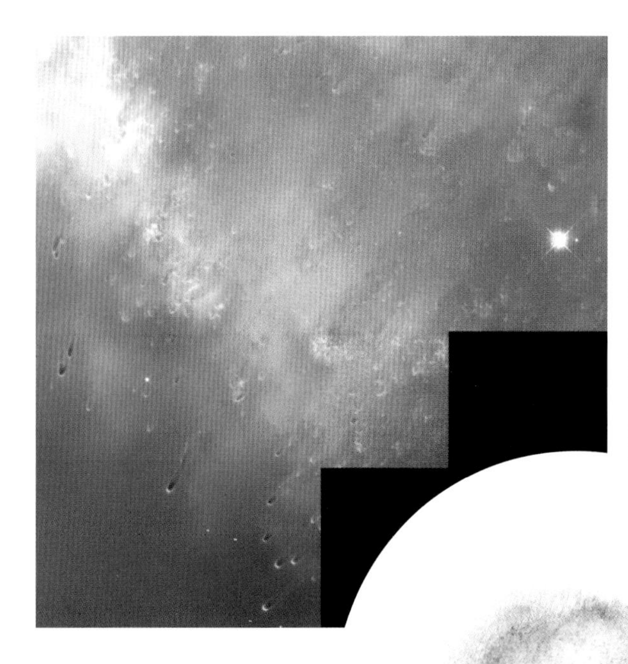

only to the patient observer.

On nights of good transparency, my 105mm scope at low to medium power shows a 14' x 11' oval glow elongated northwest to southeast. The center is slightly darker than the rim. Averted vision (directing your gaze to one side) can help you see the nebula better, and oxygen III and narrow-band light-pollution filters often work well. At high power, without a filter, some faint stars are embedded in the nebulosity.

But what about all the beautiful features visible in photographs? While you can't observe the wealth of colorful detail that appears in deep images, there is more to see than you might think. I've read many observing reports and noted the smallest instruments through which observers have been able to detect various features.

Amazingly, Michael Bakich reports: "From our dark-sky site 50 miles east of El Paso, Texas, three of us saw the Helix with the naked eye on August 26, 2000." And in Western Australia, where the Helix passes almost overhead, Maurice Clark and his friends have used its naked-eye visibility as a guide to sky conditions. Others have found it through 6x30 finders and described it as easy with small binoculars. The annularity of the Helix has been noted in 60mm telescopes. The central star has been spotted in a 6-inch.

After this, we are getting beyond the realm of small scopes, but I know that many observers actually use fairly sizable apertures. Here are a few challenges for larger instruments.

Through 8-inch scopes, the Helix has been seen to shine with a blue-green glow. The human eye is more sensitive to this color than to the reddish hue that dominates photographs. This is why an oxygen III filter improves the view so much; it passes the blue-green light given off by doubly ionized oxygen (designated O III) while blocking the colors common to many sources of light pollution. Irregularities in the brightness of the ring also start to show up with an 8-inch scope.

The two "coils of the spring" have been clearly seen by Bill Ferris of Flagstaff, Arizona, using a 10-inch reflector and an oxygen III filter. Like many planetaries, the Helix has a faint outer halo, but it doesn't show on most photographs. The brightest section emanates from the southeast edge, spirals out clockwise, and fades in the north. While this halo has been suspected in a 10-inch, a 20-inch is the smallest scope in which it has been described as apparent.

Another tough target is a 16th-magnitude galaxy embedded in the northwest edge of the annulus. It is located 1.2' south of the prominent 9.9-magnitude star at that end and

Above: The Helix is shown here in a field that's ½° across; the tiny frame outline shows the region of the Hubble view to the left. Photo: Robert Gendler. Top left: This remarkable close-up, obtained in 1996 with the Hubble Space Telescope, reveals "cometary knots" whose heads face the nebula's central white-dwarf star. Photo: C. Robert O'Dell / STScI. Inset: While observing at the Madison Astronomical Society's dark-sky site near Brooklyn, Wisconsin, Bill Ferris made this sketch of the Helix on September 27, 1995, with a 10-inch f/4.5 Newtonian reflector, a Lumicon oxygen III filter, and a Meade super-wide-angle eyepiece at 63x. "The transparency was exceptional that night," he writes. North is up.

First published November 2002

has been seen in a 17.5-inch scope. And what about those radial spokes? The streamers have been observed in a 16-inch scope and hints of the cometary globules at their ends with a 22-inch.

Many of these observations were made by very experienced deep-sky enthusiasts who frequent the web-based group Amastro (http://tech.groups.yahoo.com/group/amastro/). Some of them observe from darker skies and more southerly latitudes than most of us enjoy, but their accomplishments give us something to strive for. When trying to nab any of these features, choose nights of good transparency and the darkest sky you can find.

The Helix Nebula

Designations	**NGC 7293, Caldwell 63**
Right ascension	**$22^h\ 29.6^m$**
Declination	**–20° 48'**
Angular size	**16' × 12'**
Total visual magnitude	**7.3**
Magnitude of central star	**13.5**
Distance in light-years	**300**

Coordinates are for equinox 2000.0. The catalog size given here includes some of the outlying halo.

A Regal Treasure Trove

High in the November sky, a sparkling array of kingly ornaments awaits.

On fall evenings, the constellation Cepheus, the King, lies temptingly high overhead. Its southern reaches are bathed in the riches of the Milky Way, where deep-sky wonders abound.

Let's begin in the southwest corner of Cepheus with one of the sky's most captivating sights. Here we see a galaxy and an open star cluster just ⅔° apart — close enough to be embraced by a single low-power view. Their nearness is an illusion, of course. The galaxy is nearly 5,000 times more distant than the cluster, giving us a sense of the profound depths of space.

Open cluster **NGC 6939** is the brighter of the two. To pinpoint it, place Theta (θ) Cephei in a low-power field and then drop 2.3° south. Once you succeed in spotting the cluster, seek out **NGC 6946** to the southeast.

With my 105mm refractor at 47x, these two make a beautiful pair. NGC 6939 shows many very faint pinpoints over a hazy background. NGC 6946 is a bit larger and slightly oval, appearing as a nearly uniform hazy glow with only a slight increase in brightness toward the center. A small, squat triangle of 10th- to 12th-magnitude foreground stars is partly embedded in the southern part of the galaxy.

When I boost the magnification to 87x with a wide-angle eyepiece, the duo just squeezes into the same field. The cluster is very rich in faint stars and appears 8' across. Several exceedingly faint stars can be seen scattered across the galaxy, which looks about 10' x 8' and elongated northeast to southwest. Under dark skies with his 4-inch

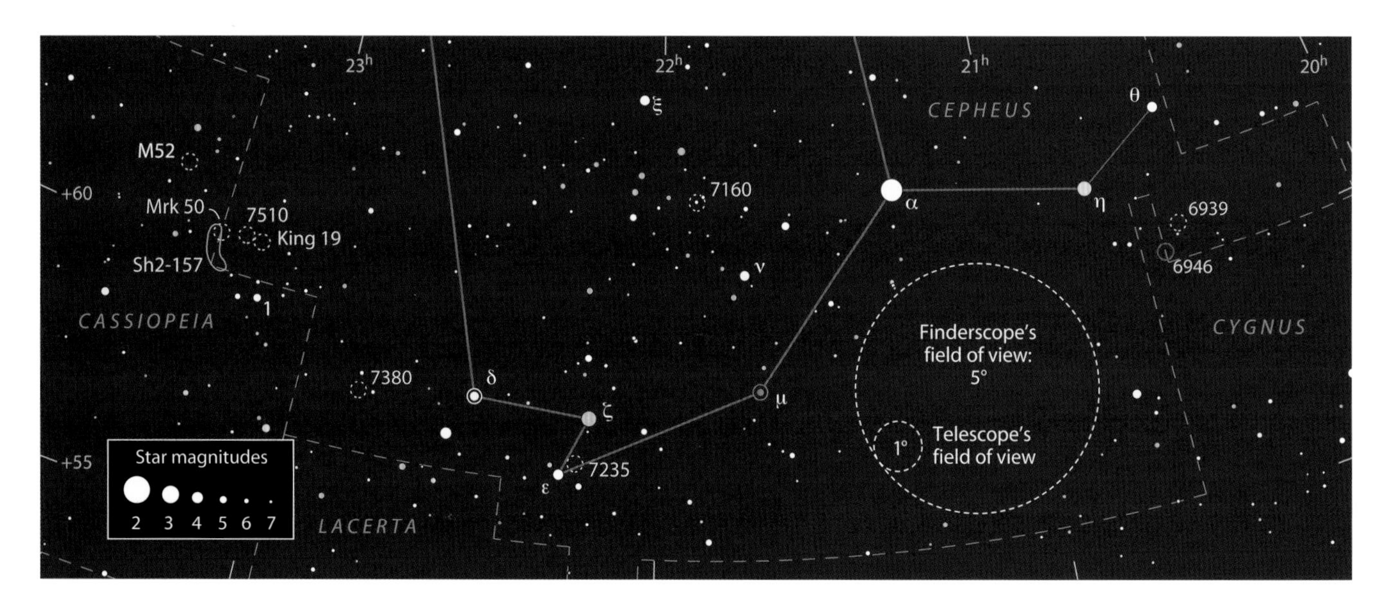

Two colorful star clusters in this month's selection were photographed by George R. Viscome of Lake Placid, New York. NGC 7235 (far left) has the bright star Epsilon (ε) Cephei in the same field. NGC 7510 (left) appears curiously elongated. These are 12- and 10-minute exposures, respectively, with a 14.5-inch f/6 Newtonian reflector on 3M 1000 slide film. Each view is ⅔° tall, with north toward top and top left, respectively.

Some of Cepheus's Speckled Wealth

Object	Type	Magnitude	Size/Sep.	Distance (l-y)	Right Ascension	Declination	*MSA*	*U2*
NGC 6939	Open cluster	7.8	7'	3,900	$20^h\ 31.5^m$	+60° 39'	1074	20L
NGC 6946	Spiral galaxy	8.8	11' × 10'	19 million	$20^h\ 34.9^m$	+60° 09'	1074	20L
NGC 7160	Open cluster	6.1	7'	2,600	$21^h\ 53.8^m$	+62° 36'	1072	19R
NGC 7235	Open cluster	7.7	4'	9,200	$22^h\ 12.5^m$	+57° 17'	1072	19L
NGC 7380	Open cluster	7.2	12'	7,200	$22^h\ 47.4^m$	+58° 08'	1071	19L
NGC 7510	Open cluster	7.9	4'	6,800	$23^h\ 11.1^m$	+60° 34'	1070	18R
King 19	Open cluster	9.2	6'	6,400	$23^h\ 08.3^m$	+60° 31'	1070	18R
Mrk 50	Open cluster	8.5	2'	6,900	$23^h\ 15.2^m$	+60° 27'	1070	18R
Sh 2-157	Diffuse nebula	—	60' × 10'	6,900	$23^h\ 16^m$	+60.3°	1070	18R

The columns headed *MSA* and *U2* give the chart numbers of objects in the *Millennium Star Atlas* and *Uranometria 2000.0*, 2nd edition, respectively. Distances in light-years are from recent research papers. Approximate angular sizes are from catalogs or photographs; most objects appear somewhat smaller when a telescope is used visually.

scope and much patience, noted observer Stephen James O'Meara has been able to spot traces of spiral structure in this galaxy.

Nine supernovae have been observed in NGC 6946, more than in any other galaxy. The first appeared in 1917 and the last in 2008, within a human lifespan. Such a prolific galaxy is worth keeping an eye on! Its neighbor is also unusual. Most open clusters are loosely bound and lose hold of their stars within a few hundred million years, but NGC 6939 is relatively ancient, at 2 billion years old.

Now we'll march eastward across Cepheus through a succession of small open clusters. The first is **NGC 7160**, located about two-fifths of the way from Nu (ν) to Xi (ξ) Cephei. The cluster is easy to identify because its two brightest stars, 7th and 8th magnitude, form the wide pair known as South 800. In 14x70 binoculars, the double and a few fainter stars look like a stubby caterpillar with glowing eyes. My 105mm refractor at 87x shows six fairly bright stars in a distinctive shape. South 800 and a star to the northeast make an arrowhead, while a little curve of stars to the west-southwest forms a bent shaft. A dozen fainter stars are strewn about the area.

Our next cluster is **NGC 7235**, situated 25' northwest of Epsilon (ε) Cephei. It is visible in 14x70 binoculars as a single star offset in a small, dim patch of light. My refractor at 87x shows five moderately bright to faint stars plus seven very faint stars in a little clump. The three brightest mark the corners of a skinny east-west triangle that almost spans the group.

Now center a low-power eyepiece on the striking yellow and blue double Delta (δ) Cephei. Move 2.4° east to a blue-white star of 6th magnitude. Notice that the star is parked at the pointy end of a slender triangle. One corner of its short base is marked by a star of similar brightness, the other by the double star OΣ 480. **NGC 7380** sits just east of this pair. At 87x, I see a pretty group of 30 moderately bright to faint stars in a shape like a witch's hat. The hat's large floppy brim forms the eastern side, and OΣ 480 marks its pointed top. The cluster has a hazy background, but am I seeing unresolved stars or nebulosity? NGC 7380 is, in fact, embedded in a very faint nebula, Sharpless 2-142. At 28x, a green oxygen III filter seems to slightly enhance and

First published November 2003

extend the haziness compared with an unfiltered view. When you use a light-pollution filter, can you detect the nebula with a small scope?

Our final stop will be in far eastern Cepheus, where several interesting targets lie close together. The brightest is **NGC 7510**. To find it, look for the naked-eye star 1 Cassiopeiae. A low-power view will show a 6th-magnitude star to its east. Centering that and sweeping 1.2° north will take you to a conspicuous knot of stars. My little scope at 153x shows a small group of a dozen faint to very faint stars. The brightest of these form a bar running east-northeast to west-southwest; the rest lie north of the bar and turn the cluster into a triangle.

While we're visiting Cepheus, the King, it's only fitting that we look in on one of the clusters reported by Ivan R. King in the Harvard College Observatory Bulletin in February 1949. Four lie in the constellation Cepheus, none very striking in a small scope. **King 19**, however, has the redeeming quality of being easy to locate. It is a faint knot of stars ⅓° west of NGC 7510 in the same low-power field. At 153x, I count 10 faint to very faint stars gathered in a loose, irregular group. The brightest make a triangle in the cluster's eastern side, a three-star line in the western side, and a solitary point between the trios.

Markarian 50 (Mrk 50) is a smaller but brighter knot ½° east and a little south of NGC 7510. At 127x, I see five stars in a tiny curve reminiscent of a miniature Corona Borealis. Avid deep-sky enthusiast and author Tom Lorenzin calls this group the Tiny Tiara. Mrk 50 sits off the northwestern edge of the brightest arc of the large nebula **Sharpless 2-157** (Sh 2-157). Although barely detectable without a filter, the nebulosity is fairly obvious with an oxygen III filter at 17x. It is curved concave westward, wider in the south, and about 1° long north-south. Lorenzin nicknamed this the Californietto Nebula for its resemblance to the much larger California Nebula in Perseus.

We've spanned Cepheus from west to east, sampling the kingly treasures that bejewel this swath of sky. These riches belong to any observer with a clear, dark night and the will to behold them.

The Starred Lizard

Lacerta's deep-sky sights are modest but intriguing, like the constellation itself.

Lacerta, the Lizard, inhabits the softly glowing patch of the Milky Way flanked by the head of Cepheus, the King, and the forelegs of Pegasus, the Winged Horse. The Lizard first leapt into the sky at the bidding of Johannes Hevelius, the great astronomer of Danzig (now Gdańsk, Poland). Hevelius introduced Lacerta and several other inconspicuous constellations in his impressive star atlas *Firmamentum Sobiescianum*, completed in 1687, the year of his death. The atlas saw only limited distribution until 1690, when his widow and fellow observer, Elisabetha, published it as a combined work with his star catalog and supplementary text. Today the works of Hevelius are quite rare.

With its rattish tail and upright legs, the Lacerta in Hevelius's atlas doesn't look very much like a lizard (see pages 257 and 258). The odd creature is actually labeled *Lacerta sive Stellio*. Today the term *stellio* belongs to a lizard species with starlike spots, but the ancients sometimes

Lurking in Lacerta

Object	Type	Magnitude	Size	Right Ascension	Declination	*MSA*	*U2*
IC 1434	Open cluster	9.0	7'	22ʰ 10.5ᵐ	+52° 50'	1086	19L
NGC 7245	Open cluster	9.2	5'	22ʰ 15.3ᵐ	+54° 20'	1086	19L
King 9	Open cluster	9½?	2.5'	22ʰ 15.5ᵐ	+54° 25'	1086	19L
IC 1442	Open cluster	9.1	5'	22ʰ 16.0ᵐ	+53° 59'	1086	19L
IC 5217	Planetary nebula	11.3	8" × 6"	22ʰ 23.9ᵐ	+50° 58'	1086	31L
Merrill 2-2	Planetary nebula	11.5	5"	22ʰ 31.7ᵐ	+47° 48'	1102	31L
NGC 7243	Open cluster	6.4	21'	22ʰ 15.3ᵐ	+49° 53'	1103	31R
NGC 7209	Open cluster	6.7	24'	22ʰ 05.2ᵐ	+46° 30'	1103	31R

Angular sizes are from recent catalogs; most objects appear somewhat smaller when a telescope is used visually. The columns headed *MSA* and *U2* give the chart numbers of objects in the *Millennium Star Atlas* and *Uranometria 2000.0*, 2nd edition, respectively.

equated *stellio* with a newt. It's difficult to picture such a creature among the faint stars of Lacerta, but in a dark sky, you can see his spots zigzag across the Milky Way.

We'll start this month's deep-sky foray with a few open star clusters near Lacerta's nose. The most obvious one is **IC 1434**, located 2.1° west-northwest of yellow-orange 4th-magnitude Beta (β) Lacertae. Looking through a finderscope, you can start at Beta and follow a curve of four 6th- and 7th-magnitude stars that arcs over the cluster. Each star is a little brighter than the last, and they are evenly spaced at just less than 1° apart. IC 1434 is between and south of the two middle stars.

My 105mm refractor at 17x shows four 10th- and 11th-magnitude stars arranged in the shape of a kite that's diving toward the south. The kite overlies a haze of unresolved stars with a 10th-magnitude star at the southeast edge. Pushing the magnification to 153x pulls many faint stars out of the glow. With my 10-inch Newtonian at 115x, this pretty cluster is very rich in faint stars and about 7' across. Several long rays of stars seem to emanate from the cluster.

A close group of three more clusters lies 1½° north-northeast of IC 1434. You can locate them by drawing a line from Alpha (α) through Beta Lacertae and continuing for that distance again. Centered in the trio is **NGC 7245**, but it is easy to overlook at low power through my little refractor. At 87x, I see a 3' patch of mist within a triangle: 11th-magnitude stars at its west and south edges and a 9th-magnitude star outside the northeast edge. NGC 7245 is an attractive group in the 10-inch reflector at 170x. It's about 4' across and rich in faint stars with a bit of unresolved haze in the center. The cluster **King 9** can be seen to the north-northeast in the same field of view. It's a granular-looking smudge with one faint star north of center. King 9 is also visible in the refractor at magnifications of 87x to 153x.

In his atlas frontispiece, Hevelius offers a parade of new constellations (lower left) to the Muse of Astronomy. Leading the pack is Lacerta, the Lizard.

The third member of the trio, **IC 1442**, is just 22' south-southeast of NGC 7245. I failed to find it with my small refractor until I examined the area with the 10-inch. At that point, I could identify a cluster centered about 5½' southwest of the position plotted on my atlases. At 118x, I see about 30 stars of 11th magnitude in 5'. There's a 9th-magnitude reddish orange star just beyond the northeast edge and a slightly brighter orange star at the southeast edge. Inspecting the area later with the 105mm scope revealed 20 faint stars in a group with a large void in the center. The 2003 book *Star Clusters* by Brent A. Archinal and Steven J. Hynes confirms my position for IC 1442.

First published November 2004

Next, we'll turn to the tiny planetary nebula **IC 5217**, located 1.3° south of Beta. A finder will show two 7½-magnitude stars south of and making a north-south line with Beta and IC 5217. The planetary is a scant 14' south-southeast of the southern one. There you will see the planetary hiding among a little patch of stars of similar brightness. Even at 127x in my small refractor, IC 5217 looks stellar.

This is a good place to practice "blinking" with a nebula filter. Holding an oxygen-III filter in front of the eyepiece makes this planetary the brightest object in the field. Moving the filter alternately over and then away from the eyepiece produces a blinking effect that is useful when you're trying to single out very small planetaries. (To avoid seeing stray-light reflections off the filter, try draping a dark cloth over your head and the eyepiece.) Identification of this

The stellar concentration known as NGC 7243 features stars of various colors in George R. Viscome's photograph, below. The Lake Placid, New York, amateur used his 14.5-inch f/6 Newtonian reflector and 3M 1000 slide film for the 10-minute exposure. North is up, and the field is 0.8° wide. Bottom: Hevelius's chart of Lacerta, like others in his atlas, shows the star patterns reversed, as if plotted on a globe. Photo: US Naval Observatory Library

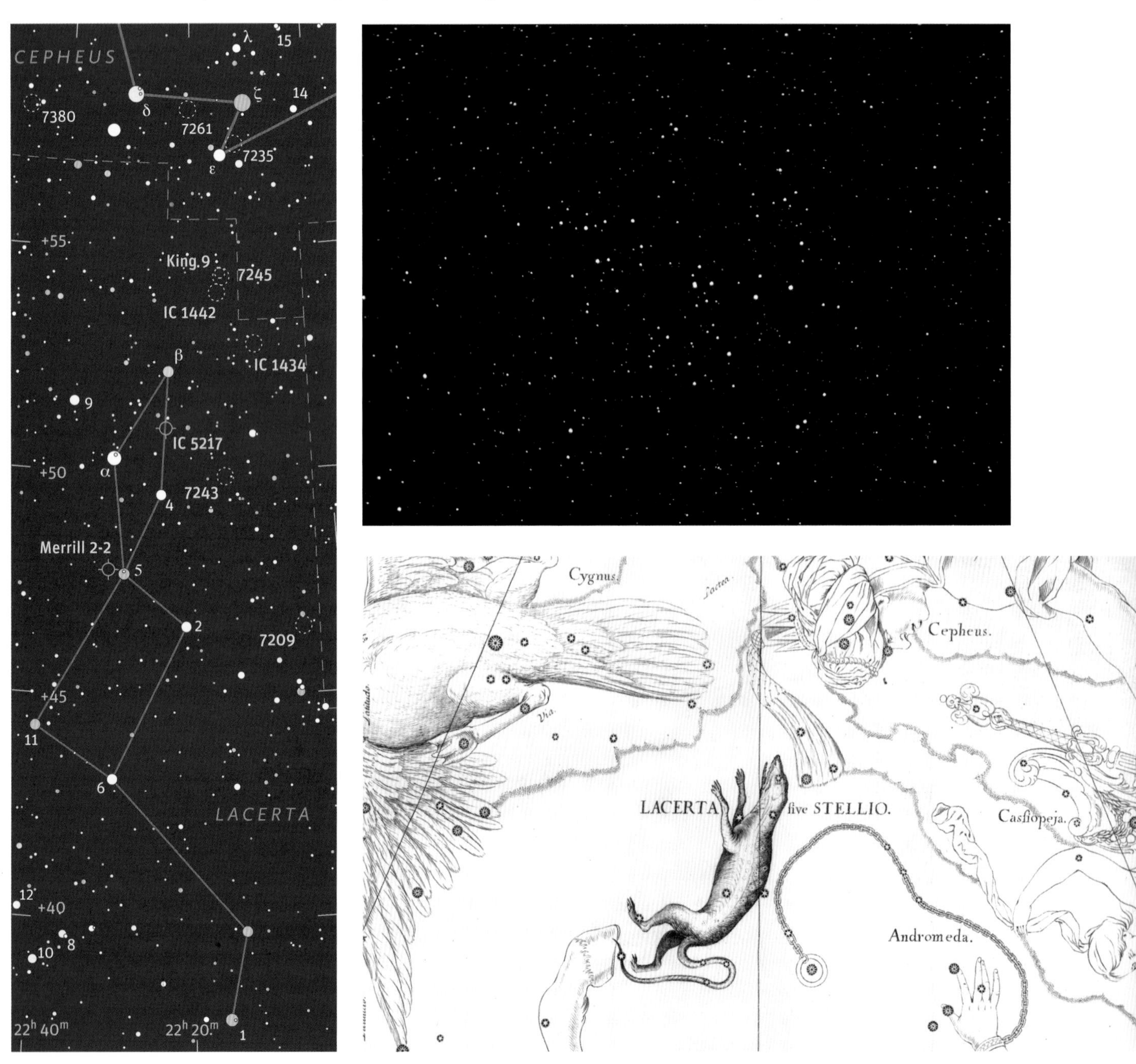

object is easier in the 10-inch scope at 170x, which shows it as a tiny, slightly oval disk of robin's-egg blue.

Merrill 2-2 (PN G100.0–8.7) is another diminutive planetary that's good for blinking. Look for a small, unevenly spaced, and very shallow curve of three 12th-magnitude stars 23' east-northeast of 5 Lacertae. The planetary is the middle "star" in this curve. Although minute, it has a high surface brightness and responds dramatically to an oxygen III filter. I have never tried for this planetary with anything smaller than an 8-inch scope, but it should be visible with less aperture. At 314x on the 8-inch, it is small, round, and blue-gray. Both this planetary and IC 5217 are around 10,000 light-years away.

The cluster **NGC 7243** is a much easier target, visible even in my 8x50 finder. It's located 1½° west-northwest of 4 Lacertae, and the arrowhead formed by Alpha, 4, and 5 Lacertae points toward it. My 105mm scope at 47x shows 45 stars of 9th magnitude and fainter within 20'. A small triangle of moderately bright stars near the center sports a widely spaced double star, Σ2890, at its southeast corner. The 10-inch at 43x reveals 75 stars in a very irregular group with patches, gaps, and ragged edges. The cluster reminds me of a crab with claws reaching north-northwest. The northernmost star in the central triangle shines with an orange hue.

NGC 7209 is one of the most pleasing sights in Lacerta. This open cluster is located 2.7° due west of 2 Lacertae and immediately south of a yellow-orange 6th-magnitude star. You can approximate its position by noting that it forms one end of a half circle with 4, 5, and 2 Lacertae. My small refractor at 68x shows 50 stars of 9th magnitude and fainter in 23'. The brightest of these snake north-south across the cluster. Three stars of around magnitude 8½ cradle the southwestern third of the group's edge, and the easternmost one appears orange.

At 115x, my 10-inch reflector reveals about 100 stars, many arrayed in meandering chains. A 7½-magnitude star just east of center appears golden, and a 9th-magnitude star in the eastern edge looks orange. The 8½-magnitude star at the southwestern edge has a nice concentration of faint stars to its east that looks like a little subcluster. We gaze across approximately 3,000 light-years to view NGC 7209 and NGC 7243.

Overshadowed by the bright constellations surrounding it, Lacerta patiently lingers high in the northern sky, waiting for the curious observer to visit its host of modest treasures.

A Cygnus Sampler, Part II

Autumn evenings beckon deep-sky observers with a wonderland of sights spread along the northern Milky Way.

In "A Cygnus Sampler" (page 238), I plowed the fertile star fields of northeastern Cygnus, but much was left unearthed. This delightful region of the sky is noteworthy not only for the sheer number of celestial treasures it holds but also for their impressive variety.

Now, let's pick up where we left off, starting with the showy star cluster **M39**. On a good night under my semi-rural skies, M39 is visible to the unaided eye as a hazy patch. It's easy to spot in binoculars or a finderscope, where you'll see it 2½° west-southwest of 4th-magnitude Pi² (π^2) Cygni. My 15x45 image-stabilized binoculars reveal the cluster as a very pretty group of 25 stars, mostly rather bright.

Since M39 spans ½°, it requires a low-power view to be fully appreciated. My 10-inch reflector with a 35mm wide-angle eyepiece gives me 44x and a 1½° field. With this setup, M39 is an astonishingly triangular group of bright stars, mostly white or blue-white but with a few touches of color: a yellow star on the western side, a golden one east of the triangle's northern point, and an orange one south-southeast of the cluster's brightest star. A 9th-magnitude star along the triangle's eastern side forms the double star **h1657** with its 12th-magnitude companion 23" north-northeast. A few stars jut out from the southern side of the triangle, making M39 a wide and brightly lit Christmas tree with a very short trunk. I count 25 stars of magnitudes 6½ to 10½ over a rich Milky Way backdrop of three times as many faint stars.

Although the discovery of M39 is generally credited to Charles Messier, it was probably noticed by some early skygazers because of its naked-eye status. In the February 1925 *Journal of the British Astronomical Association*, Peter Doig wrote that according to Irish astronomer John Ellard Gore, Aristotle referred to M39 as a "star with a tail."

The planetary nebula **Minkowski 1-79** (Mink 1-79) lies near M39. To find it, you need only hop ½° east-northeast from the brightest star near the triangle's northern tip to an 8th-magnitude star, the brightest in the area. Then move the same distance and direction again to reach the planetary. My 10-inch scope at 115x shows the planetary as an oval glow a bit less than 1' long running east-west. A 13th-magnitude star rests just off its western end, and when I boost

Above: The Cocoon Nebula, IC 5146, in northeastern Cygnus has an appearance very reminiscent of the emission portion of the Trifid Nebula, M20, in Sagittarius. The Cocoon sits at the eastern end of the 2°-long meandering dark nebula Barnard 168, which does not show in this ½°-wide close-up. Photo: Sean Walker and Sheldon Faworski

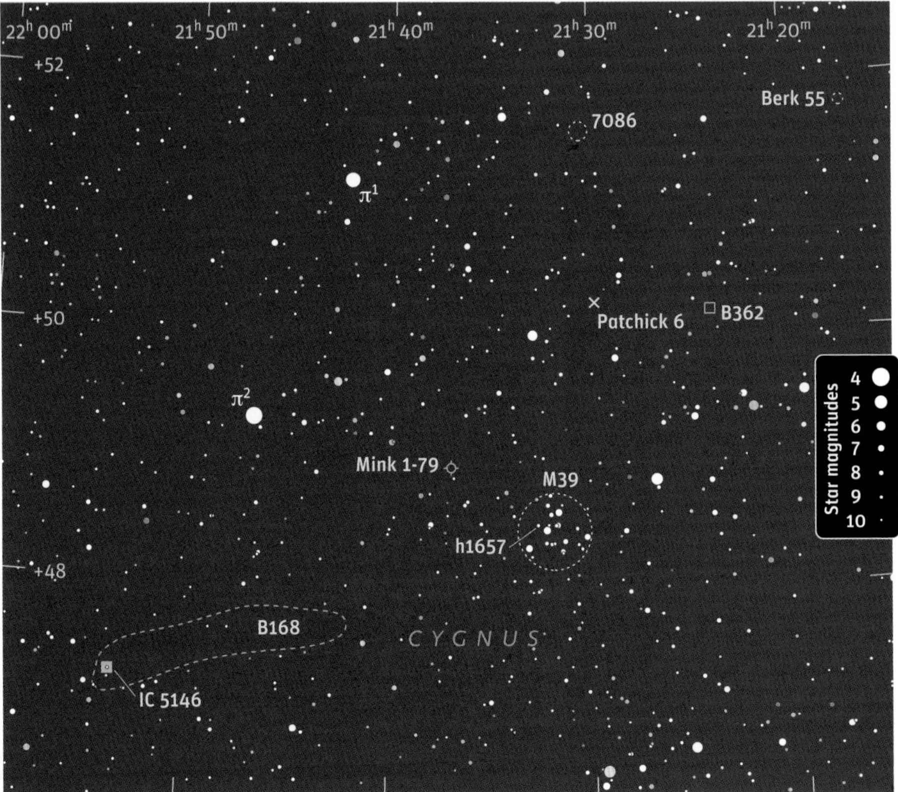

In the Swan's Wake, Part II

Object	Type	Magnitude	Size/Sep.	Right Ascension	Declination	*SA*	*U2*
M39	Open cluster	4.6	31'	21h 32.2m	+48° 27'	9	32L
h1657	Double star	9.0, 12.1	23"	21h 32.7m	+48° 29'	9	32L
Mink 1-79	Planetary nebula	13.2	60" × 42"	21h 37.0m	+48° 56'	9	31R
B168	Dark nebula	—	1.7° × 0.2°	21h 47.8m	+47° 31'	9	31R
IC 5146	Bright nebula	9	11' × 10'	21h 53.5m	+47° 16'	9	31R
B362	Dark nebula	—	12' × 8'	21h 24.0m	+50° 10'	9	32L
Patchick 6	Asterism	10.5	1.6'	21h 29.8m	+50° 14'	9	32L
NGC 7086	Open cluster	8.4	9'	21h 30.5m	+51° 36'	9	32L
Berk 55	Open cluster	11.4	5'	21h 17.0m	+51° 46'	9	32L
NGC 7008	Planetary nebula	10.7	86"	21h 00.6m	+54° 33'	9	19R
h1606	Double star	9.6, 11.7	19"	21h 00.6m	+54° 32'	9	19R

Angular sizes or separations are from recent catalogs. The visual impression of an object's size is often smaller than the cataloged value and varies according to the aperture and magnification of the viewing instrument. The columns headed *SA* and *U2* give the chart numbers of objects in *Sky Atlas 2000.0* and *Uranometria 2000.0*, 2nd edition, respectively.

the magnification to 213x, I see a fainter star on the eastern tip. The nebula responds well to both oxygen III and narrowband nebula filters. Mink 1-79 enfolds a torus of material seen nearly edge-on and tipped a little east of north. The outer edges of the torus produce areas of enhanced brightness on the nebula's rim. This planetary is approximately 9,000 light-years away.

Sweeping 1½° east-southeast of M39 with my 105mm refractor at 17x gives me a wonderful view of the remarkable dark nebula **Barnard 168** (B168). The end nearest M39 is wide and irregular, but the dark cloud's most striking feature is a black velvet ribbon that trends east-southeast for nearly 2°. It's a cinch to follow B168 to its end, which engulfs **IC 5146**, the Cocoon Nebula. At first, I couldn't see the nebula, but when I added a hydrogen-beta filter, it was immediately apparent, with a 10th-magnitude star at its heart and another at its southern edge. An oxygen III filter also helps, but not quite as much. I no longer need the filter at 47x, and a filterless view improves the visibility of the embedded stars. There are two faint stars between the 10th-magnitude pair, two more east of center, and one on the nebula's western edge. Pushing the magnification a bit further to 68x, I see IC 5146 as quite patchy.

In my 10-inch reflector at 118x, the brightest sections of the nebula are arranged in two or three northward-reaching fingers. The middle one stretches up through the central star, the eastern one goes through an arc of three 12th-magnitude stars, and the western one is not as well defined. These branches meet between the two 10th-magnitude stars and combine to form a broad expanse dominating the nebula's southeastern quadrant. The overall effect reminds me of a catcher's mitt.

The stars involved in the Cocoon Nebula compose the nascent star cluster Collinder 470. As with many stellar nurseries, IC 5146 shines by both emitted and reflected light.

A much smaller dark nebula, **Barnard 362** (B362), is located 2° northwest of M39. It shows up in my little refractor at 47x as a 12' inky oblong patch leaning northeast. A pair of stars, magnitudes 7.5 and 7.9, sit ¼° to its north — the brighter one yellow and the dimmer one a slightly deeper yellow.

The little-known asterism **Patchick 6** is parked just 1° east of Barnard 362. In my 105mm scope at 127x, I see a half dozen stars in a tiny 1.1' knot with the brightest at the north-northeastern edge. California amateur Dana Patchick discovered this compact group with his 13-inch reflector. At 200x, he counts nine stars filling a curved V shape.

Moving 1.4° north of Patchick 6, we find the open cluster **NGC 7086**. It's very pretty in my little refractor at 47x, like diamond dust with a dozen brighter gems scattered over it. At 87x, I count 30 stars atop a dappled haze 8' across. My 10-inch reflector at 118x shows 50 stars, including a 3' central concentration with the cluster's brightest star, a 10th-magnitude gold nugget, at its northwest edge.

Berkeley 55 (Berk 55), 2° west of NGC 7086, should be an interesting target for large telescopes. My 105mm at 87x exposes only a couple of dim stars amid a faint 3½' haze. In a 10-inch at 166x, I see a dozen very faint stars over an unresolved glow just begging for inspection with a bigger scope. In the 1960s, Berkeley astronomers Harold Weaver and Arthur Setteducati found 104 "new" open clusters while systematically examining photographs from the *National Geographic Society–Palomar Observatory Sky Survey*. Berkeley 55 is among the 85 Berkeley clusters that appear to be original discoveries.

It's a longer jump to our next deep-sky wonder, but it's well worth the effort. First, scan 2½° northwest of Berkeley 55 to a 6th-magnitude golden star. From there, sweep 1⅓°

First published November 2005

One of just two Messier objects in Cygnus, the open cluster M39 looks better in binoculars and low-power telescopes than it does in this ¾°-wide photograph, which enhances faint stars in the Milky Way background, downplaying the prominence of the cluster's moderately bright stars. Photo: Robert Gendler

north-northwest to a 10th-magnitude star nestled against the unusual planetary nebula **NGC 7008**. My 105mm refractor at 47x displays a fat oval bent into a slight curve that's concave eastward. At 87x, the planetary appears about 1½' long north-south and is distinctly blotchy. The 10th-magnitude star has a 12th-magnitude companion to the south, forming the double **h1606**. An extremely faint star sparkles at the western edge of the nebula.

With my 10-inch scope at 68x, the brighter star of the double is golden, while the companion looks bluish. The nebula is a fat, tightly curved arc opening toward the southeast. It has brighter patches north-northeast and south-southwest, the northern one being more intense. At 166x, NGC 7008 is very pretty and quite intricate. The central star is visible, and another is embedded in the arc east-northeast of center. Oxygen III and narrowband filters faintly fill out the oval, but they also hinder the view of the stars that lend added character.

Many photographs of NGC 7008 show a starlike object 22" west-northwest of the central star. This object was once classified as a separate planetary nebula, Kohoutek 4-44, but is now believed to be a knot in NGC 7008. A recent study indicates that NGC 7008's central star is a close binary with components of roughly equal brightness. NGC 7008 would appear considerably smaller than Minkowski 1-79 were it not one-third as far away.

Sovereign Splendors

A royal tapestry of celestial sights awaits you in the northern autumn sky.

Cepheus illumes
The neighboring heaven; still faithful to his queen

— Capel Lloft, "Eudosia," 1781

King Cepheus and Queen Cassiopeia reign as the celestial royal couple of the boreal sky. Their side-by-side constellations are circumpolar from my latitude of 43° north, but Cassiopeia seems to win most of our attention — perhaps pleasing to a queen often portrayed as quite vain. Nevertheless, Cepheus holds his share of regal treasures, and I'd like to whisk you away on a tour of a king's riches.

Our trailhead is a 4th-magnitude orange star perched one-quarter of the way from Polaris to Gamma (γ) Cephei and a little outside a line between them. It's the brightest star in the area, 2 Ursae Minoris, labeled 2 UMi on some star charts. Although once considered part of Ursa Minor, this star found itself within the borders of Cepheus when the International Astronomical Union published its official constellation boundaries in 1930.

The open cluster **NGC 188** is located just 1.1° south-southwest of 2 UMi. My 105mm refractor at 87x reveals 30 faint to extremely faint stars in ¼°. A dappled background hints at unresolved stars. In my 10-inch reflector at 68x, this is a lovely medallion of 40 diamond chips enmeshed in a gauzy net and set amid a pretty field of bright stars.

With an estimated age of around 7 billion years, NGC 188 is one of the Milky Way's most ancient open clusters. It's about 5,400 light-years from Earth and more than 2,000 light-years above the plane of our galaxy. The cluster's orbit seldom takes it through the inner regions of the galactic disk, thus curtailing disruptive encounters with giant molecular clouds. NGC 188's large mass also helps hold it together. The concentration of its 1,000-plus stars makes the group structurally comparable to a loose globular cluster.

Now drop farther south to the planetary nebula **NGC 40**, located one-third of the way from Gamma Cephei to Kappa

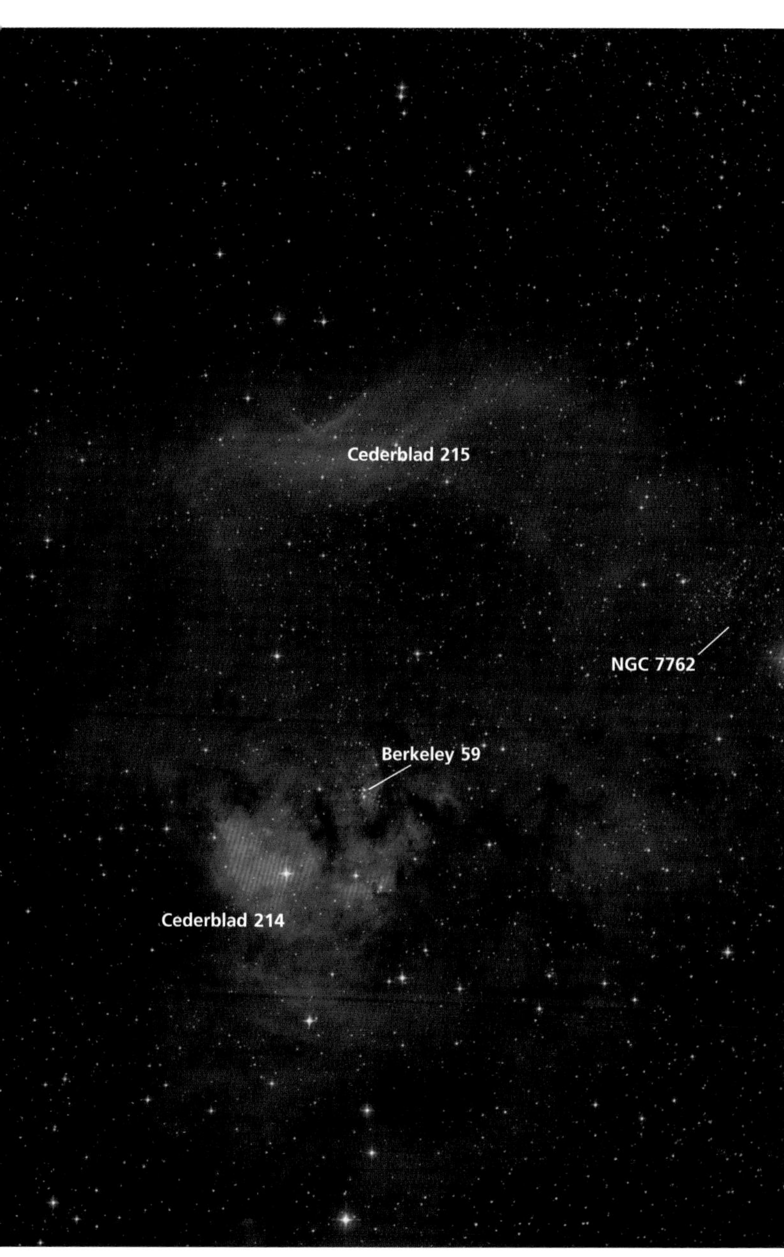

Cederblad 214 and 215 are among the many large clouds of emission nebulosity dotting the Milky Way in Cepheus and Cassiopeia. You can use this image to follow part of the celestial deep-sky tour narrated in the accompanying text. The field is 2.6° wide with north up. Photo: Robert Gendler

(κ) Cassiopeiae in an area devoid of bright stars. You can star-hop from Gamma to a white 6th-magnitude star 2.3° due south and continue for 1.3° to a yellow 6th-magnitude star. A 1° V of stars, fairly distinctive in a finder, sits 1.5° east. The 7th-magnitude star off its southern point is just 1° northwest of NGC 40. Through a low-power eyepiece, look for a 9'-long zigzag of four stars, magnitudes 9 to 11. The dimmest one is the planetary's central star.

My little refractor at 127x shows a fairly small glow surrounding this star and a faint star just off the nebula's southwest edge. In my 10-inch scope at 43x, NGC 40 looks bluish, and at 115x, it appears slightly oval and annular. The most detailed view comes at 231x. The oval leans a bit east of north, and its long sides are brighter than its ends. The planetary responds well to a narrowband nebula filter, but in my semirural location in upstate New York, I prefer a filterless view.

NGC 40's central star is a late-type WC Wolf-Rayet star. Although such stars have spectra similar to classic Wolf-Rayet stars, the likeness is superficial. A classic Wolf-Rayet is a short-lived massive star that has its outer hydrogen envelope blown away by fierce stellar winds. It's doomed to end its life in a supernova explosion. The heart of NGC 40, on the other hand, holds the contracting core of a smaller star that aged into a red giant and then puffed off its outer layers. It's gradually evolving into a white-dwarf star.

Our next target is **Omicron (ο) Cephei**, a colorful double star only 210 light-years distant. To me, the 5th-magnitude primary is a polished topaz gemstone, while the 7th-magnitude companion cozied up to its southwestern side shines golden. Many observers, however, perceive the dimmer star as blue. What color do you see? Omicron is a true binary with an orbital period of 1,505 years. Its components are 3.3" apart in 2011 and widen to 3.5" in 2046, retaining a northeast-southwest alignment.

Sweeping 2.9° east of Omicron, we come to the open cluster **NGC 7762**. In my 105mm scope at 87x, I glimpse many faint to extremely faint stars in a patchy group 13' across. A 5th-magnitude field star rests 17' southwest. The cluster is quite pretty in my 10-inch reflector, where a prominent north-south bar of stars dominates its center. NGC 7762 is about 2,400 light-years away and 270 million years old.

A widely spaced pair of 6th-magnitude stars lies 1.7° southeast of NGC 7762, the northern one gold and the southern one orange. At a dark-sky site in the northern Adirondack Mountains with my little refractor, I can see the large, faint nebula in which they're embedded. **Cederblad 214** (Ced 214) stretches about 45' north-south. It's widest in the north, where it spans about 35', and tapers to a point tacked to the sky by the orange star. A nebula filter accents variations in the nebula's brightness. The most pronounced patch surrounds the golden star and spreads farther east than west. The second-brightest area folds itself around a yellow 8th-magnitude star to the west. Two 9th-magnitude stars in the northwestern part of Cederblad 214 mark the site of the open cluster **Berkeley 59** (Berk 59), but I need my 10-inch scope at 171x to make out much of a cluster here. The star pair sits in a knot at the center of a roughly east-west, 8' x 4' oblong of

21 stars. Several more form a detached clump to the north, and a few others are scattered southward.

Cederblad 214 is equated with NGC 7822 in some astronomical databases, yet the latter is plotted 1½° to the north on most amateur atlases. The confusion arises because its discoverer, John Herschel, gave the position of the northern nebula but a description that seems to fit the southern nebula. I'll avoid the issue by referring to the northern nebula by its alternate designation **Cederblad 215** (Ced 215). Look for a pair of 8th-magnitude stars 1.1° north of Berkeley 59. My 105mm scope at 28x reveals ghostly nebulosity surrounding these stars and stretching eastward through a starry field for about ½°. In my 14.5-inch scope with a nebula filter, Cederblad 215 grows much longer and must be viewed a piece at a time, but it becomes an intriguingly convoluted tapestry.

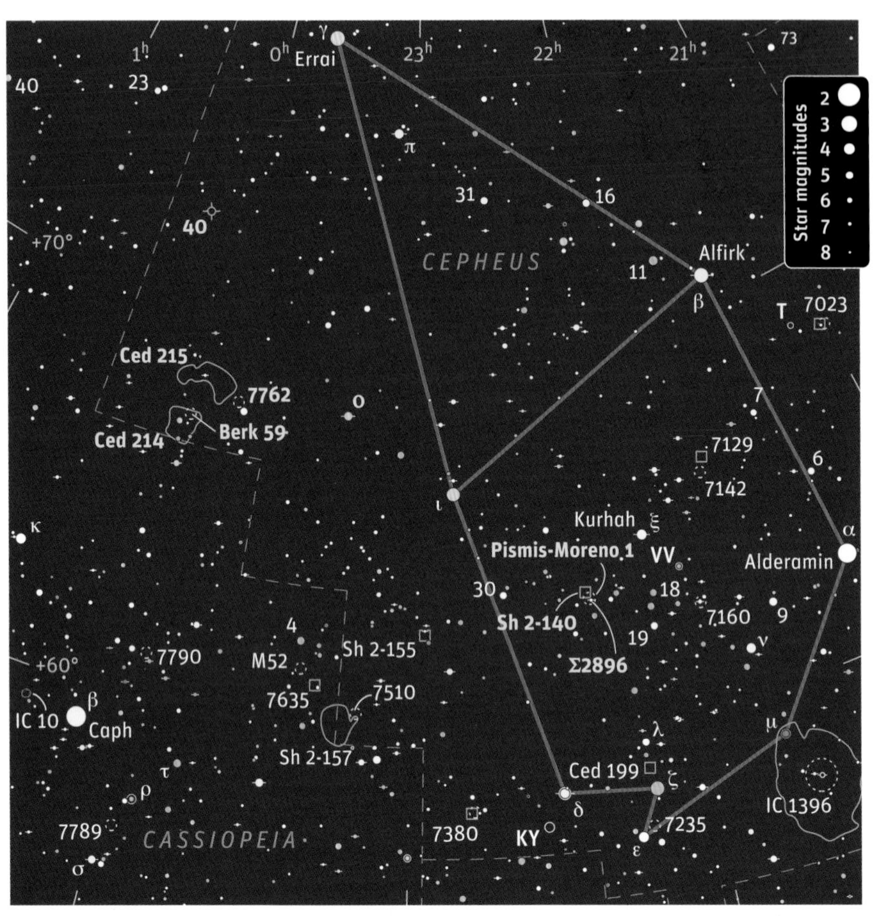

Below: Moderately easy for small telescopes, the planetary nebula NGC 40 lies in the relatively star-poor region between Cassiopeia and the celestial pole. North is up and the field 2' wide. Photo: Steve and Paul Mandel / Adam Block / NOAO / AURA / NSF

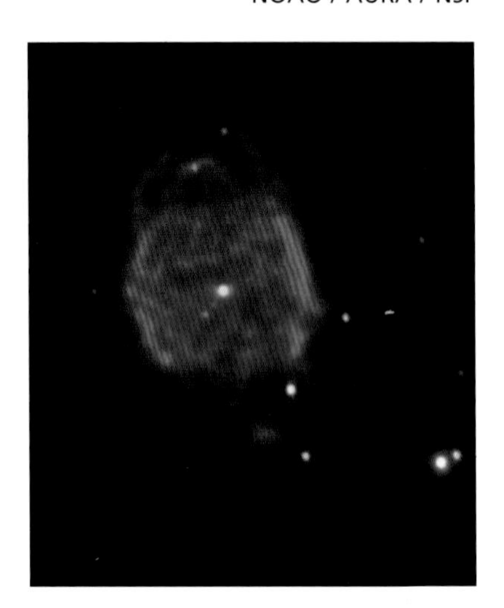

Regal Riches

Object	Type	Magnitude	Size/Sep.	Right Ascension	Declination	*MSA*	*PSA*
NGC 188	Open cluster	8.1	15'	00^h 47.5^m	+85° 15'	6	71
NGC 40	Planetary nebula	12.3	48"	00^h 13.0^m	+72° 31'	24	71
ο Cephei	Double star	5.0, 7.3	3.3"	23^h 18.6^m	+68° 07'	1057	71
NGC 7762	Open cluster	10.3	15'	23^h 49.9^m	+68° 01'	1057	(71)
Ced 214	Bright nebula	—	50'	00^h 03.5^m	+67° 13'	1057	71
Berk 59	Open cluster	—	10'	00^h 02.2^m	+67° 25'	1057	(71)
Ced 215	Bright nebula	—	72' × 20'	00^h 01.2^m	+68° 34'	1057	71
Pismis-Moreno 1	Open cluster	—	19'	22^h 18.8^m	+63° 16'	(1059)	(71)
Σ2896	Double star	7.8, 8.6	21"	22^h 18.5^m	+63° 13'	1059	71
Sh 2-140	Bright nebula	—	11' × 4'	22^h 19.0^m	+63° 18'	(1059)	(71)

Angular sizes and separations are from recent catalogs. The visual impression of an object's size is often smaller than the cataloged value and varies according to the aperture and magnification of the viewing instrument. The columns headed *MSA* and *PSA* give the appropriate chart numbers in the *Millennium Star Atlas* and *Sky & Telescope's Pocket Sky Atlas*, respectively. Chart numbers in parentheses indicate that the object is not plotted.

First published November 2006

Our final stop will be the open cluster **Pismis-Moreno 1**, located 28' north of 25 Cephei. California amateur Dana Patchick introduced me to this charming group of stars some years ago. Patchick describes it as a pleasant cluster that reminds him of a sailboat sitting on a calm sea. In my refractor at 87x, I see 11 stars in an east-west boat about 16' long. The two brightest form the double star **Σ2896**, whose 7.8- and 8.6-magnitude components are 21" apart. Eight faint stars to the north fashion a triangular sail, and a lone star dangling beneath the boat marks its anchor.

My 10-inch reflector reveals **Sharpless 2-140** (Sh 2-140), a gossamer swath of nebulosity running up the eastern side of the sail — perhaps a furled jib awaiting the weighing of our sailboat's anchor.

The Winged Horse

Some fascinating deep-sky objects in Pegasus will delight and challenge you at the same time.

Now heav'n his further wand'ring flight confines,
Where, splendid with his num'rous stars, he shines.
— Ovid, *Fasti*

According to Greek myth, Pegasus is the winged steed that sprang from the blood of the hideous Gorgon Medusa when she was slain by Perseus. This beauty born from ugliness was involved in many tales, including the great feats of the hero Bellerophon.

Bellerophon grew overly proud of his accomplishments and tried to fly Pegasus to the abode of the gods. Zeus, angered by this display of audacity, sent a gadfly to sting Pegasus. Bellerophon was thrown from his steed and fell back to Earth, but Pegasus was accepted into the heavens, where we see him today.

Our flying horse seems intent on the celestial sugar ball suspended before him, **Messier 15** (M15), one of the most beautiful globular clusters in the northern sky. An imaginary line from Theta (θ) Pegasi, the horse's ear, through Epsilon (ε) Pegasi, his nose, points right to it. From my semirural home, M15 is easily visible in 12 x 36 image-stabilized binoculars as a hazy orb. An 8th-magnitude star sits just off one edge, and a ½° box with two or three stars at each corner rests east of the cluster.

My 105mm refractor at a magnification of 17x shows M15 with an intense nucleus, a small bright core, and a large halo that weakens

BRIGHT SPIRAL Moderately large and bright enough to show in binoculars, the spiral galaxy NGC 7331 (Caldwell 30) in Pegasus is a popular destination for deep-sky observers of all experience levels. Four nearby galaxies, collectively known as the Fleas, are far more challenging but still within reach of modest-aperture telescopes. The field is ¼° wide with north up. Photo: Russell Croman

Flying With the Horse

Object	Type	Magnitude	Size	Right Ascension	Declination	*MSA*	*U2*
M15	Globular cluster	6.2	18.0'	21^h 30.0^m	+12° 10'	1238	83R
Pease 1	Planetary nebula	14.7	1"	21^h 30.0^m	+12° 10'	1238	83R
NGC 7094	Planetary nebula	13.4	94"	21^h 36.9^m	+12° 47'	1238	83L
NGC 7332	Galaxy	11.1	4.1' × 1.1'	22^h 37.4^m	+23° 48'	1187	64R
NGC 7339	Galaxy	12.2	3.0' × 0.7'	22^h 37.8^m	+23° 47'	1187	64R
NGC 7331	Galaxy	9.5	10.5' × 3.5'	22^h 37.1^m	+34° 25'	1142	46R
NGC 7320	Galaxy	12.6	2.2' × 1.1'	22^h 36.1^m	+33° 57'	1142	46R

Angular sizes are from recent catalogs. The visual impression of an object's size is often smaller than the cataloged value and varies according to the aperture and magnification of the viewing instrument. The columns headed *MSA* and *U2* give the appropriate chart numbers in the *Millennium Star Atlas* and *Uranometria 2000.0*, 2nd edition, respectively. All the objects this month are in the area of sky covered by Charts 74 and 75 in *Sky & Telescope's Pocket Sky Atlas.*

CONTROVERSIAL CLUSTER Stephan's Quintet is one of the sky's best-known galaxy groups, thanks to a controversy involving discordant distance estimates of the members. The field is 8' wide with north up. Photo: Johannes Schedler

outward. At 87x, the core looks mottled, and many stars glimmer in the halo. Several sugar grains sweeten the core at 153x, and the nucleus reveals a brighter center but remains unresolved. The halo's stragglers stretch the globular to a diameter of 9'.

M15 is positively gorgeous in my 10-inch reflector. At 166x, it boasts a brilliantly blazing center from which it fades sharply outward to a diameter of 12'. The most obvious stars of the sparsely populated halo seem to fill four or five starfishlike arms.

Inward, where the cluster's brightness starts to soar, M15 is very rich in stars that grow more and more crowded toward its luminous heart. M15 is one of the most densely packed clusters in our galaxy. It lies 33,600 light-years from Earth and 15,400 light-years below the midplane of our Milky Way Galaxy.

Another aspect of M15 that sets it apart from the ordinary is a planetary nebula that is visible in amateur telescopes. **Pease 1** is buried deep within M15 and tough to locate among the globular's myriad stars. Fortunately, detailed charts are available on the Planetary Nebulae Observer's website (www.blackskies.org/peasefc.htm). With these charts and an hour of invested time, I identified the little clump of stars that includes Pease 1.

At 284x through my 15-inch reflector, I couldn't isolate the planetary until I added an oxygen III filter. This left just one object standing out well, no doubt the planetary.

Don't worry — you don't need a 15-inch scope to try for Pease 1. Experienced observers under better skies than mine have reported success with telescopes as small as 8 inches.

If the planetary is too much of a tease, pass the Pease and try for **NGC 7094**. Located 1.8° east-northeast of M15, this planetary nebula is rather faint through my 10-inch reflector at 115x. But its 13.6-magnitude central star is easily spotted, and I occasionally glimpse an extremely dim star within its edge. Using a narrowband or an oxygen III filter greatly improves the view. The 1½' nebula appears roundish, somewhat uneven in brightness, and vaguely annular (ring-shaped).

Now, for a bit of extragalactic excitement, we'll zoom up to the pretty galaxy pair **NGC 7332** and **NGC 7339** near

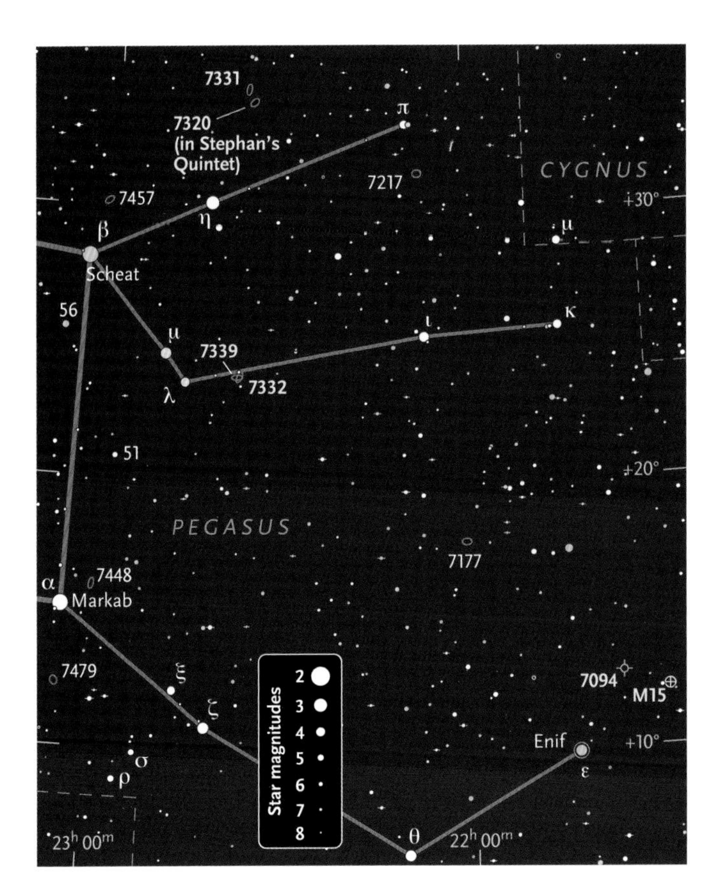

Lambda (λ) Pegasi in the horse's leg. The duo dwells 2.1° west of Lambda, where it's sandwiched between a north-south pair of 7th-magnitude stars.

In my little refractor at 87x, NGC 7332 is a thin spindle tilted north-northwest with an elongated core and stellar nucleus. Also slender, but requiring averted vision, NGC 7339 lounges 5' east of its companion's southern tip and runs east-west. A faint star lies between the galaxies. At 122x, NGC 7339 is visible with direct vision, has a uniform surface brightness, and spans about 2¼'. NGC 7332's halo extends farther south than north of its nucleus and bridges 2¾'.

A dozen degrees northward, a gold and orange pair of 6th-magnitude stars decorates the Pegasus-Lacerta border. Magnificent **NGC 7331** rests 1.2° south of these stars. The galaxy is easily visible in my 105mm scope at 47x. Its elongated core and halo lean slightly west of north and envelop an intense, starlike nucleus.

The galaxy is quite pretty at 87x and measures about 6' x 1½'. The light along the eastern side diminishes gradually outward, whereas an abrupt falloff on the opposite side indicates the presence of a dust lane.

The dark lane is lovely through my 10-inch scope at 166x, and the galaxy covers about 9' x 2½'. The 3' core brightens markedly toward its tiny nucleus. Four dim galaxies, nicknamed the Fleas, harry the eastern flank of NGC 7331. The most prominent is NGC 7335, which measures 1¼' x ½' and tips north-northwest. This galaxy presents a fairly uniform surface brightness with a scumbled edge. Look for it 3.5' east of the northern end of NGC 7331's core.

A pair of 10th- and 11th-magnitude stars 4' farther east points southward to **NGC 7340**, a smaller and rounder version of NGC 7335. A north-south line of four 12th- and 13th-magnitude stars runs between NGC 7340 and NGC 7331. **NGC 7337** sits 1' west-southwest of the southernmost star. Only the galaxy's very small but relatively bright core is visible, and a faint star bedecks its southeastern edge.

NGC 7336 is also petite and is seen only with averted vision. It lies in the four-star line two-thirds of the way from the northernmost star to the next one down.

While preparing this section of the deep-sky objects in Pegasus, I wondered whether any of the Fleas could be seen in my little refractor. Studying the region carefully on the next clear night, I was delighted to discover that, though challenging, all except NGC 7336 were visible with averted vision.

The Fleas aren't true neighbors of NGC 7331 but, rather, reside deep in the background. NGC 7331 is about 50 million light-years from Earth, while three of the Fleas belong to a group nearly six times as distant. NGC 7336 is yet another 100 million light-years farther away.

NGC 7331 is, however, generally thought to be associated with **NGC 7320**, one of the galaxies in nearby **Stephan's Quintet**. Look for an east-west pair of 10th-magnitude stars 24' south-southwest of NGC 7331. Stephan's Quintet huddles 6' south of the western star.

In my 10-inch reflector at 311x, the Quintet's **NGC 7318A** and **NGC 7318B** are so snug that they seem to share an east-west halo, but close inspection reveals separate cores. To the southeast, NGC 7320 is about the same size and brightness as the combined glow of the conjoined twins. This galaxy is elongated southeast-northwest and appears somewhat brighter toward its core.

Target tip

While observers often find that large faint nebulae and galaxies are seen to advantage in a low-power, wide-field eyepiece, the faintest stars and small diffuse objects are usually easier to see with high magnifications. Try using a high-power eyepiece if you can't find your target in a low-power view.

First published November 2007

Due north, elusive **NGC 7319** shows a similar slant, which indicates that I'm seeing only this spiral galaxy's core and bar. West of NGC 7320, tiny **NGC 7317** is nestled against a 13th-magnitude star north-northwest.

Encouraged by my success with the Fleas, I turned my 105mm scope toward Stephan's Quintet. At 203x, I was able to spot NGC 7318A/B as a single patch of mist. NGC 7320 was also discernible, but more subtle. I'm a great fan of small telescopes for deep-sky observing, but their capabilities still manage to surprise me from time to time.

Most astronomers place NGC 7320 at roughly the same distance as NGC 7331, whereas the remaining galaxies of Stephan's Quintet are thought to inhabit the same region as the more distant Flea trio. A vocal minority of researchers, though, maintains that all five galaxies in Stephan's Quintet are related, and they cite this as one piece of evidence indicating that redshifts (recessional velocities) may not be reliable indicators of extragalactic distances. There's a nice up-to-date history of this astronomical controversy in Jeff Kanipe and Dennis Webb's *Arp Altas of Peculiar Galaxies* (Willmann-Bell, 2006).

Frederick's Glory

A defunct constellation houses a diverse set of celestial treats.

Johann Elert Bode devised the constellation Frederick's Glory and introduced it in a paper read at the special assembly of the Academy of Science in Berlin on January 25, 1787. It was intended to honor Frederick II, King of Prussia, who had died the previous year. Bode's published paper included a map on which his creation is labeled Friedrichs-Ehre. The constellation subsequently appeared in the 1795 edition of Jean Fortin's *Atlas Céleste de Flamstéed*, where it bears the name Trophée, while in Bode's own atlas *Uranographia*, published in 1801, the name was Latinized as Honores Friderici.

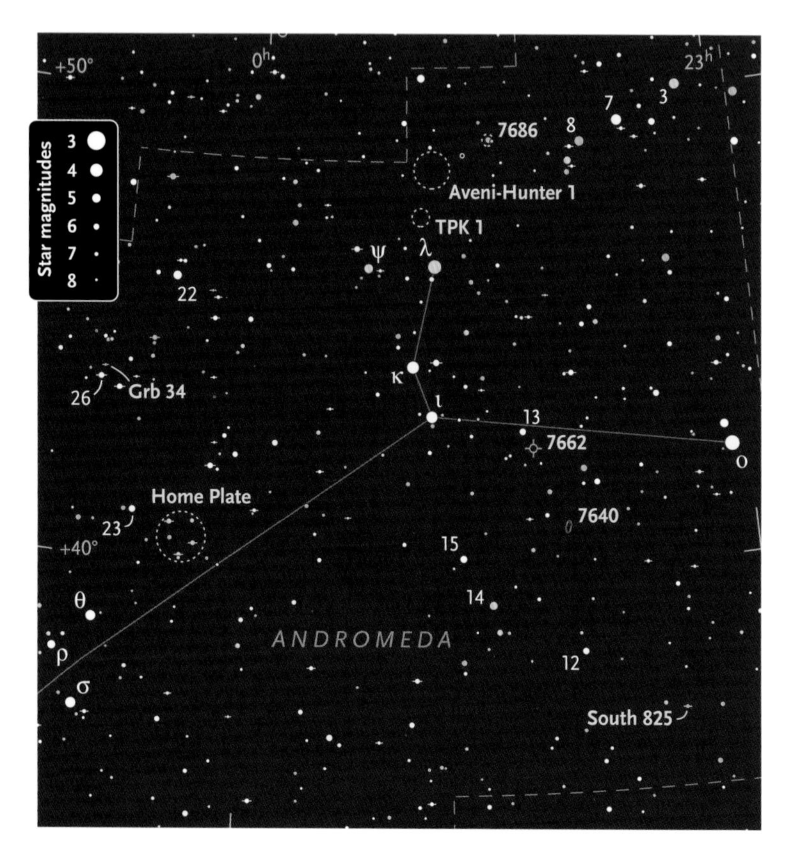

Bode depicted this now obsolete constellation with a crown, sword, quill pen, and olive branch to represent Frederick the Great as ruler, hero, intellectual, and peacemaker. It was composed of stars that now officially reside in the constellations Andromeda, Cassiopeia, and Lacerta. Its brightest star is Omicron (ο) Andromedae in the sword's scabbard. To the east, the misshapen Y formed by Iota (ι), Kappa (κ), Lambda (λ), and Psi (ψ) Andromedae mark the hilt of the sword.

The entire Y-shaped group fits within the field of my 15x45 image-stabilized binoculars, which show Psi and Lambda as deep yellow stars. The interesting asterism **TPK 1** (cataloged by deep-sky hunters Phillip Teutsch, Dana Patchick, and Matthias Kronberger) sits 1.1° north-northeast of Lambda. Through my 105mm refractor at 17x, I see a roughly trapezoidal collection of a dozen stars covering about ¼°. Increasing the magnification to 87x brings out many faint stars that make TPK 1 look more clusterlike, with 45 stars spanning ⅓°. With my 10-inch reflector at 70x, I count 50 stars filling a parallelogram. The asterism measures 23' across its long diagonal and 15' across the short one. A few stars in the eastern half are distinctly orange.

Moving 1° north-northwest takes us to the open cluster **Aveni-Hunter 1**. Although its cataloged diameter is 47', the only part that stands out clearly through my little refractor is a 15' clump near the center. At 47x, I count 14 stars, magnitudes 8 to 12, but several of these are unrelated field stars. Only 18 of 94 stars in the whole 47' area have a greater than 50 percent probability of being cluster members.

Frederick's Deep-Sky Glories

Object	Type	Magnitude	Size/Sep.	Right Ascension	Declination
TPK 1	Asterism	—	23' × 11'	23h 39.3m	+47° 31'
Aveni-Hunter 1	Open cluster	—	47'	23h 37.8m	+48° 31'
NGC 7686	Open cluster	5.6	15'	23h 30.1m	+49° 08'
NGC 7662	Planetary nebula	8.3	29" × 26"	23h 25.9m	+42° 32'
NGC 7640	Spiral galaxy	11.3	11.6' × 1.9'	23h 22.1m	+40° 51'
South 825	Double star	7.8, 8.3	67"	23h 10.0m	+36° 51'
Home Plate	Asterism	—	44' × 31'	0h 07.5m	+40° 35'
Grb 34	Double star	8.1, 11.0	35"	0h 18.4m	+44° 01'

Angular sizes and separations are from recent catalogs. Visually, an object's size is often smaller than the cataloged value and varies according to the aperture and magnification of the viewing instrument.

The emission nebula LBN 534 slashes through the open star cluster Aveni-Hunter 1. Photo: Thomas V. Davis

Aveni-Hunter 1 is an interesting target for astrophotographers. Its core contains a colorful triangle of one deep gold and two blue-white stars seemingly enmeshed in the head of the comet-shaped nebula LBN 534, which trends northeast for 1½°. The brightest part is the small reflection nebula van den Berg 158 surrounding the triangle's southeastern star, the source of its illumination.

In a paper published in 2007 in *The Astrophysical Journal*, Hsu-Tai Lee and Wen-Ping Chen presented two possible scenarios for shaping LBN 534 and triggering the birth of the youthful members of Aveni-Hunter 1. One hypothesis features the action of shock waves from a supernova explosion in the nearby association of young, massive stars known as Lacerta OB 1. A less violent possibility involves compression of the nebular material by ionization fronts from the intensely hot, spectral-type O9V star 10 Lacertae, 410 light-years from LBN 534.

NGC 7686, a better-known cluster, lies 1.4° west-northwest of Aveni-Hunter 1. Its center is ornamented with a 6th-magnitude, fiery yellow-orange star that makes the group easy to find. Through my 105mm scope at 28x, I see eight lesser lights surrounding this gem with a goldenrod-yellow star at the cluster's west-southwestern edge. At 67x, about 25 fainter stars embellish the group, including several bundled between the two bright suns.

Now we'll plunge southward in Frederick's Glory to the planetary nebula **NGC 7662**, which is only 26' south-southwest of 13 Andromedae. Half a century ago, in February 1960, Leland S. Copeland wrote an article in *Sky & Telescope* in which he described NGC 7662 as "looking like a light blue snowball." Now nicknamed Copeland's Blue Snowball, this diminutive nebula does appear faintly bluish in my 105mm refractor at 153x. It's slightly oval and shows traces of a darker center.

Examining the Blue Snowball with my 10-inch reflector at 44x reveals a small, very bright core surrounded by a turquoise halo. At 220x, I see a bright annulus superposed on a fainter oval, both tipped more or less northeast. The

Binary beauty

Double stars often look most impressive at the lowest magnification that splits them cleanly. Extra power can pull the components so far apart that they look like unrelated stars. Experiment to find out what magnification makes the colors most vivid to you.

fainter regions interior and exterior to the ring have about the same brightness and are enhanced when I use a narrowband nebula filter. Just 1' or so northeast of the nebula is a small arc of three stars, magnitudes 13.4, 14.7, and 14.9. At 308x, the annulus appears brightest at its ends and dimmest on its northwestern side. At this magnification, the ring looks markedly turquoise and the fainter parts more subtly so.

An orange 6th-magnitude star lies 1° south-southwest of NGC 7662, and the galaxy **NGC 7640** is a slightly shorter hop south-southwest from there. It is located in a starry field and passes right through a 5' triangle of 11th-magnitude stars (see the sketch below). This makes the galaxy difficult to spot through my little refractor at low power, but at 68x, I see a 6' x 1¼' streak that leans a bit west of north and has a somewhat brighter, elongated core. A faint star rests just off the galaxy's southern tip.

In my 10-inch scope at 118x, NGC 7640 exhibits a subtle curve. The halo is 7½' long, and the 2' core has a faint star at its southeastern edge. At 220x, the galaxy is very pretty, and its gentle S curve becomes more apparent.

Next, we'll visit the attractive double star **South 825**, which shows up in a low-power sweep 2.5° west-southwest of 12 Andromedae. At 17x through my little refractor, this is a nearly matched, deep golden pair with the 8.3-magnitude companion northwest of its 7.8-magnitude primary. South 825 was discovered by English astronomer James South in 1825. It is just one of 152 multiple-star systems that bear his name alone. Many others, labeled South & Herschel, or SHJ, were discovered in collaboration with his great contemporary John Herschel.

Now let's move just outside the western border of Frederick's Glory to visit a catchy binocular asterism introduced to me by Minnesota amateur Pat Thibault. Its five 6.7- to 6.9-magnitude stars mark the corners of the pentagonal home plate used in a game of baseball. Thibault's **Home Plate** is centered 1.2° west-southwest of 23 Andromedae, which easily fits in the same binocular field of view. The asterism is 44' tall, with its pointy end to the south. Through my 15x45 image-stabilized binoculars, the star at the point and the one at the northeast corner each show a fainter companion.

Thibault slid into Home Plate while sprinting for **Groombridge 34** (Grb 34). Only 11.6 light-years distant, this is one of our nearest stellar neighbors. Look for an

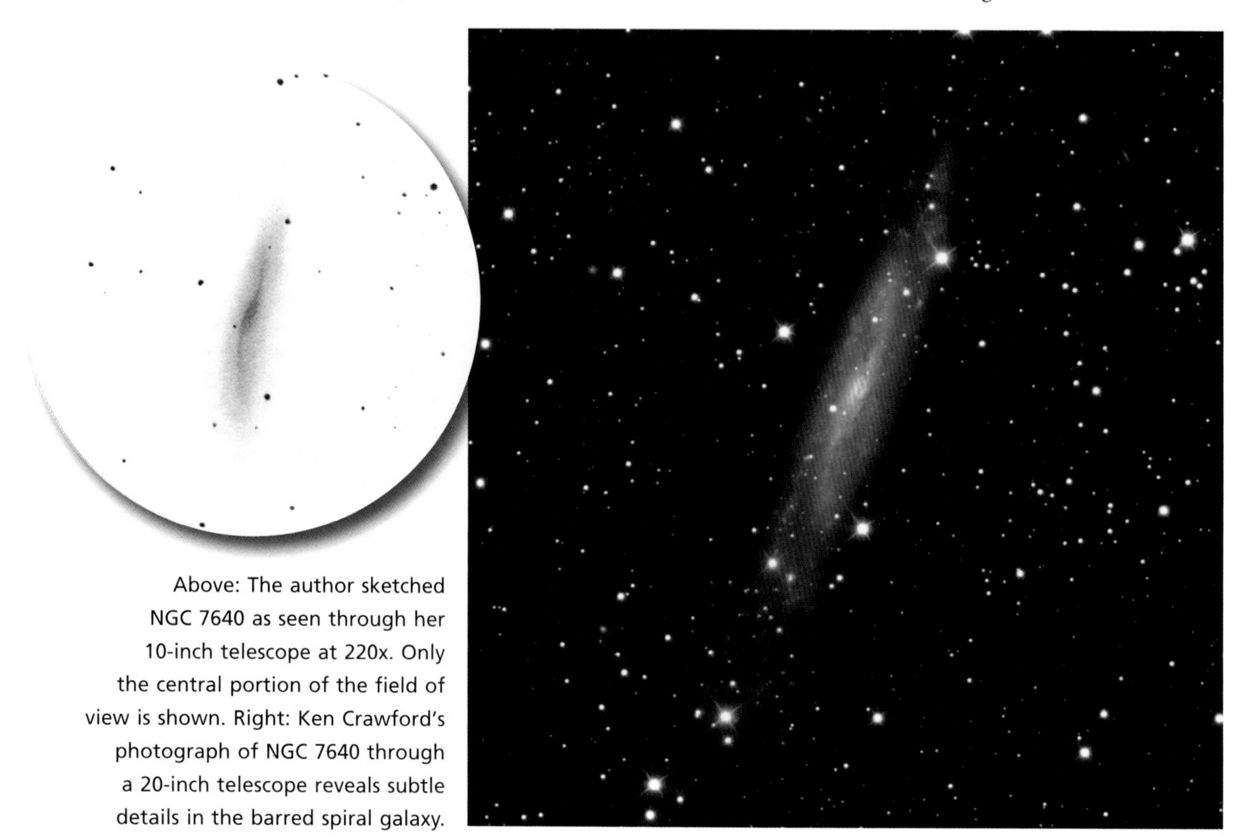

Above: The author sketched NGC 7640 as seen through her 10-inch telescope at 220x. Only the central portion of the field of view is shown. Right: Ken Crawford's photograph of NGC 7640 through a 20-inch telescope reveals subtle details in the barred spiral galaxy.

First published November 2008

orange 8th-magnitude star 14' north-northwest of 26 Andromedae. Through my 105mm scope at 47x, Groombridge 34 is a nice double with an 8th-magnitude orange primary holding an 11th-magnitude companion 35" to the east-northeast. My 10-inch scope brings out some color in the dimmer star, which looks reddish orange to me. In reality, the stars are both red dwarfs with similar spectral types. The brighter star is type M1.5V and the other is M3.5V, only a little redder. Both stars are slightly variable.

If you use a software sky atlas with image overlays, don't try to identify Groombridge 34 by its position with respect to nearby field stars. Groombridge 34 has a very large proper motion; that is to say, it appears to drift across the backdrop of distant stars. Its position changes by 29" per decade, so the pair looks distinctly out of place even on a 10-year-old sky survey. An 11.8-magnitude star is often listed as a third component, but it is merely an optical companion — an unrelated star. In 1904, it was measured at just 35" from the primary, but now it's about 4½' away and no longer bears even a superficial resemblance to being a member of the system.

The open cluster NGC 7686 occupies the central third of this 34' x 25' photograph. Photo: Anthony Ayiomamitis

Flying Horse Fish

The border between Pegasus and Pisces is home to a galaxy cluster and many noteworthy stars.

As Pegasus wings his way across our evening sky, a strange companion escorts him in his flight. The western of the two Fishes that make up Pisces glides just south of the Flying Horse's pinions. But perhaps the fish sailing along with Pegasus is not so out of place. I've seen more flying fish than flying horses.

The boundary between Pegasus and Pisces is home to the densely packed core of the **Pegasus I** galaxy cluster. The cluster is centered about 170 million light-years away and holds challenging targets for both large and small telescopes.

I can spot five of the Pegasus I galaxies with my 105mm refractor. Sitting 6' south-southwest of a 10th-magnitude star, 11.1-magnitude **NGC 7619** is small, round, and fairly easy at 47x. Slightly fainter and about the same distance southeast of the star, **NGC 7626** is similar in appearance. The galaxies show some brightening toward the center, NGC 7619 more so. Both are elliptical galaxies, the only ones we'll visit in this spiral-laden cluster.

The pair is much improved at 87x. NGC 7619 reveals a stellar nucleus, while NGC 7626 embraces a more subtle one. NGC 7626 is attended by a 12th-magnitude star 2½' east-northeast and a 13th-magnitude star symmetrically placed west-northwest. The two galaxies share the field of view with 12.5-magnitude **NGC 7611**, which rests just within the western side of a pentagon of stars, the northernmost and southernmost having wide, faint companions. Little NGC 7611 has a tiny bright center cloaked in an ashen mantle spreading northwest-southeast.

In the northern reaches of the cluster's core group, I now see 12.8-magnitude **NGC 7612**. It's very small and fairly faint, but visible with direct vision (that is, by looking directly at it). A bit dimmer, **NGC 7623** requires

Celestial Treats in Western Pisces and Pegasus

Object	Type	Magnitude	Size/Sep.	Right Ascension	Declination
Pegasus I	Galaxy cluster	—	—	23ʰ 20.5ᵐ	+8° 11'
Σ3009	Double star	6.9, 8.8	7.1"	23ʰ 24.3ᵐ	+3° 43'
Σ2995	Double star	8.2, 8.6	5.3"	23ʰ 16.6ᵐ	–1° 35'
Σ3036	Double star	8.2, 9.6	2.8"	23ʰ 46.0ᵐ	+0° 16'
Σ3045	Double star	8.0, 9.3	1.7"	23ʰ 54.4ᵐ	+2° 28'
TX Piscium	Carbon star	4.8–5.2	—	23ʰ 46.4ᵐ	+3° 29'
Barnard 19	Double star	9.2, 9.5	1.1"	23ʰ 47.0ᵐ	+5° 15'
HD 222454 Group	Asterism	6.8	14' × 8'	23ʰ 40.7ᵐ	+7° 57'

Angular sizes and separations are from recent catalogs. Visually, an object's size is often smaller than the cataloged value and varies according to the aperture and magnification of the viewing instrument.

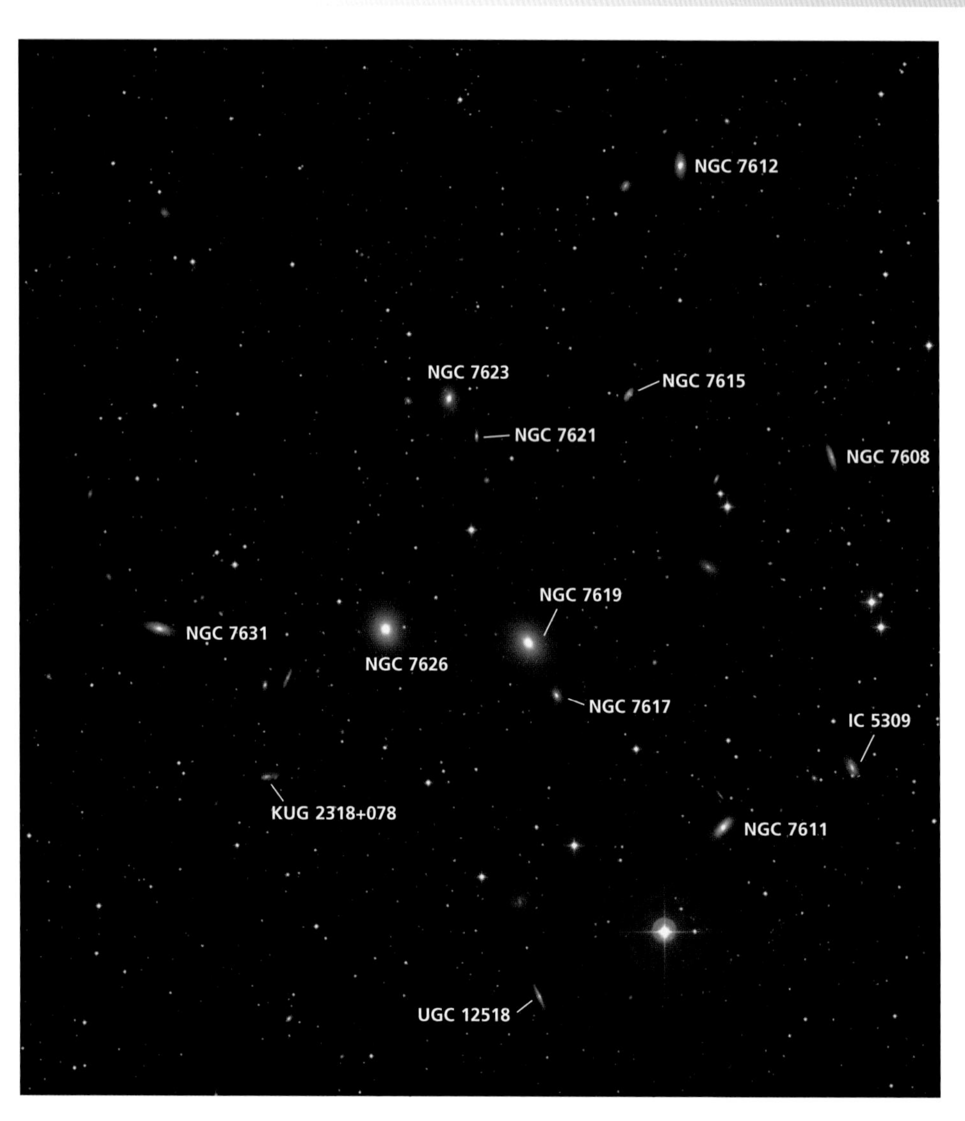

The brightest galaxies in the Pegasus I galaxy cluster are labeled in this color picture synthesized from red and blue plates of the Second Palomar Observatory Sky Survey. Some of the fainter galaxies may also be visible through an 18-inch or larger scope.
Photo: POSS-II / Caltech / Palomar

averted vision until I boost the magnification to 122x, which better shows its north-south oval.

Not surprisingly, in my 10-inch reflector, these galaxies appear brighter and generally greater in extent. A few more features are visible as well. At 220x, NGC 7619 is slightly oval, and NGC 7623 sports a faint star off its western side. NGC 7612 hosts a somewhat brighter core and an intermittently visible, starlike nucleus.

Four additional galaxies grace the scene with the 10-inch scope. In a straight line with the two ellipticals, 13.1-magnitude **NGC 7631** is a fairly faint, elongated glow tipped east-northeast. Dangling beneath the western side of NGC 7619, 13.8-magnitude **NGC 7617** appears small and diaphanous, while in the northwestern part of the group, 14.2-magnitude **NGC 7608** is a mere breath of haze. A 14.6-magnitude galaxy that I can see only with averted vision forms the southern corner of an equilateral triangle with NGC 7626 and NGC 7631. Its primary designation in the NASA/IPAC Extragalactic Database is KUG 2318+078, where KUG stands for the *Kiso Ultraviolet Galaxy Catalogue*. It often bears alternate names in atlases and field guides intended for amateur astronomers, most commonly

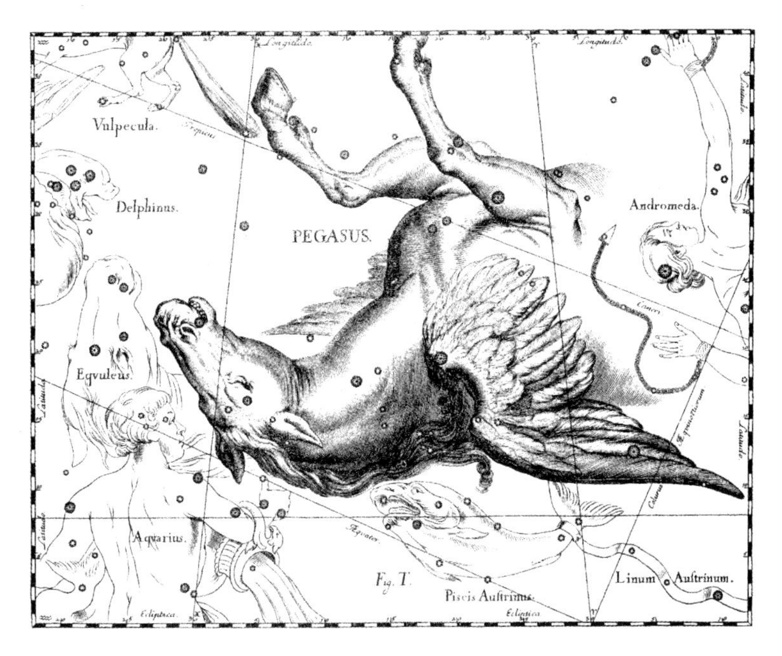

Pegasus flies upside down across the autumn sky, with the western of the two Fishes beneath him. This plate from Johannes Hevelius's *Uranographia* shows the sky mirror-reversed, as though you are looking at the celestial sphere from the outside.

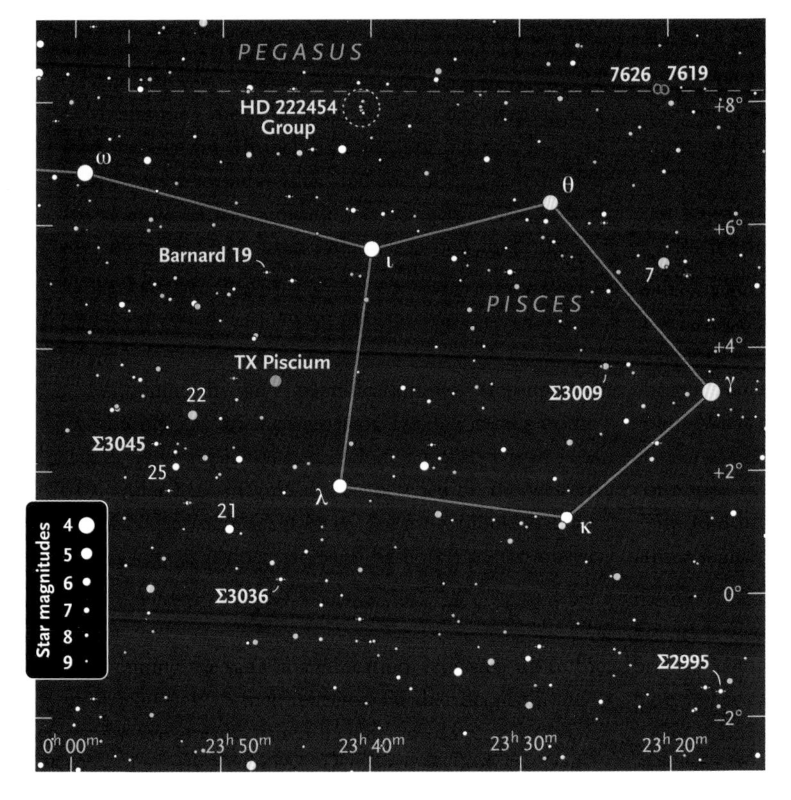

CGCG 406-79 (*Catalogue of Galaxies and of Clusters of Galaxies*) or MCG +01-59-58 (*Morphological Catalogue of Galaxies*).

Maintaining the same magnification, let's step up to the view through my 15-inch reflector. Further details in the galaxies we've already visited include an oval core in NGC 7619, and a 14th-magnitude star perched on the western edge of NGC 7626. Oval NGC 7623 now looks half again as long as it does wide and harbors a stellar nucleus, while NGC 7608 becomes a very faint slash canted north-northeast.

The 15-inch conjures up four new galaxies. **NGC 7615** is a faint smear just west of a 14th-magnitude star, while the remaining three look positively ghostly. **IC 5309** stretches northeast-southwest with a faint star caressing its southeastern flank, and **UGC 12518** (*Uppsala General Catalogue of Galaxies*) is a delicate mirage that vanishes as I shift my gaze. I overlooked **NGC 7621** when first surveying Pegasus I with this telescope, because the galaxy wasn't plotted on the chart I was using. On a later try, I could hold it steadily in view with averted vision, despite the aurora softly brightening my sky.

Many more galaxies populate Pegasus I and the surrounding field, but for a change of pace, we'll visit some deep-sky delights within our own galaxy. Let's make the acquaintance of some double stars around the Circlet of Pisces, starting with the widest and working our way through progressively tighter pairs. All were viewed with my 5.1-inch refractor.

Our first stop is **Σ3009** within the finny Circlet, 1.8° east-northeast of Gamma (γ) Piscium. This 7.1" pair is comfortably split at 63x. The 7th-magnitude yellow-orange primary guards a 9th-magnitude companion southwest. The field is beautified by a reddish orange field star 4.5' east. The Σ (capital Greek Sigma) indicates that this double was discovered by the renowned 19th-century German-Russian astronomer Friedrich Georg Wilhelm von Struve, who described the brighter component as very yellow and the dimmer one as blue. What colors do you see?

For a 5.3" pair, drop down to **Σ2995**, 3.8° southwest of Kappa (κ) Piscium. It's enshrined in a ⅓° triangle of 7th- to 8th-magnitude stars, which in turn has a line of three slightly dimmer, evenly spaced stars pointing toward

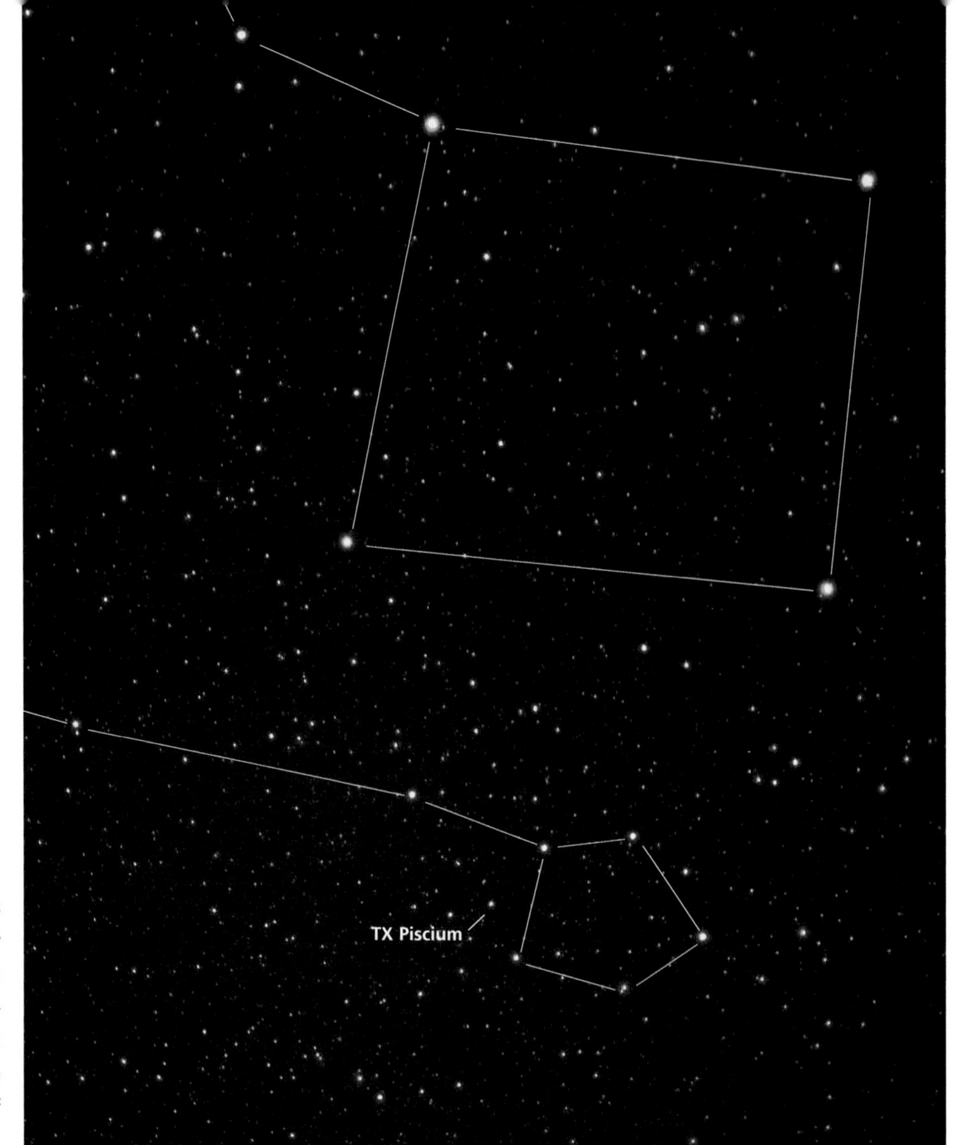

The Great Square of Pegasus and the Circlet of Pisces show well in this excerpt from P. K. Chen's book *A Constellation Album*, available from ShopatSky.com. Note the startling copper color of the carbon star TX Piscium.

it from the southeast. Viewing at 102x, I see a lovely yellow primary with a slightly fainter, gold companion north-northeast. Struve recorded both components as white.

Star-hopping 1.8° south-southeast from Lambda (λ) Piscium brings us to the 2.8" double **Σ3036**. A magnification of 164x reveals a moderately unequal pair, the 9.6-magnitude attendant southwest of its 8.2-magnitude yellow primary. Struve saw the brighter star as yellowish but didn't list a color for the companion.

The components of **Σ3045** are about the same magnitude as those of Σ3036 and split, but very close, at the same power. The brighter star is white, and the dimmer one lies 1.7" to its west. Look for this couple 30' northeast of 25 Piscium. As with the previous double, Struve saw the primary as yellowish but didn't give a color for the fainter star.

While in the neighborhood, let's seek out vividly colored 19 Piscium, also known by its variable-star designation **TX Piscium**. This carbon star is an irregular variable that slowly fluctuates between magnitude 4.8 and 5.2. The cool giant star has a deeply reddish hue because most of the shorter wavelengths of light are filtered out by the carbon molecules and compounds (C2 and CN) in its atmosphere.

Our final double star is **Barnard 19**, discovered in 1889 with the 12-inch refractor at Lick Observatory, near San Jose, California, by the gifted American astronomer Edward Emerson Barnard. This nearly matched 9th-magnitude pair has a separation of only 1.1" and shows a hairline split at 164x. To me, the primary appears yellow-white, while its companion to the north is simply white.

We'll bring our tour to a close with the **HD 222454 Group**, an asterism named for its brightest star. My 5.1-inch refractor at 63x displays four stars, magnitudes 8.2 to 8.8, in an arc concave toward the west. The arc is snugged against a crooked L (backward in my mirror-imaged view) of five dimmer stars to the east. The asterism was first noted by Bruno Sampaio Alessi, founder of the Deep Sky Hunters. This Yahoo group maintains a database of deep-sky wonders previously unpublished in astronomical journals and discovered by amateur astronomers around the world.

First published November 2009

Sculpting at the South Galactic Pole

Here, we have a clear window into the depths of the universe, unobscured by the Milky Way.

You can see Sculptor low in the south on the December all-sky map on page 313. It is a relatively recent constellation, one of 13 introduced by French astronomer Nicolas-Louis de Lacaille to commemorate the instruments of art and science. Sculptor first appeared on a chart in the *Mémoires* of the Royal Academy of Sciences for 1752 (published in 1756) as *L'Atelier du Sculpteur*, the Sculptor's Workshop. The chart depicted a carved bust on a platform and three sculpting tools on a table.

Skygazers will be hard-pressed to see the sculptor's studio among the stars. Sculptor contains only six stars of magnitude 5.0 and brighter, none bearing common names. Its brightest star is 4.3-magnitude Alpha (α) Sculptoris. You can pick it out in the night sky by noting that the brighter stars Iota (ι) and Beta (β) Ceti point toward it.

We'll begin our small-scope tour at blue-white Alpha, first moving 1.7° north-northwest to a 6th-magnitude reddish orange star, then continuing along the same line another 1.4° to the globular cluster **NGC 288**. It's easy to spot with my 105mm refractor at low power as a small, round glow. Boosting the magnification

Photographs reveal stars in the globular cluster NGC 288 more readily than a telescope used visually. The field shown is 0.4° wide, and the brightest star (left of center) is magnitude 10.3. All illustrations in this essay have north up. Photo: POSS-II / Caltech / Palomar

The huge size of the galaxy NGC 253 is apparent from the fact that the 9th-magnitude star off the oval's top edge is 1/3° from the pair of similar stars below center. Photo: Robert Gendler

Right: For deep-sky viewing this month, check out the constellation Sculptor just east of the 1st-magnitude star Fomalhaut. Appearing rather vacant to the naked eye, it has much to offer the owners of small telescopes. Below: The mottled glow of the galaxy NGC 7793 has no easy stars around it to serve as guideposts. Photo: European Southern Observatory

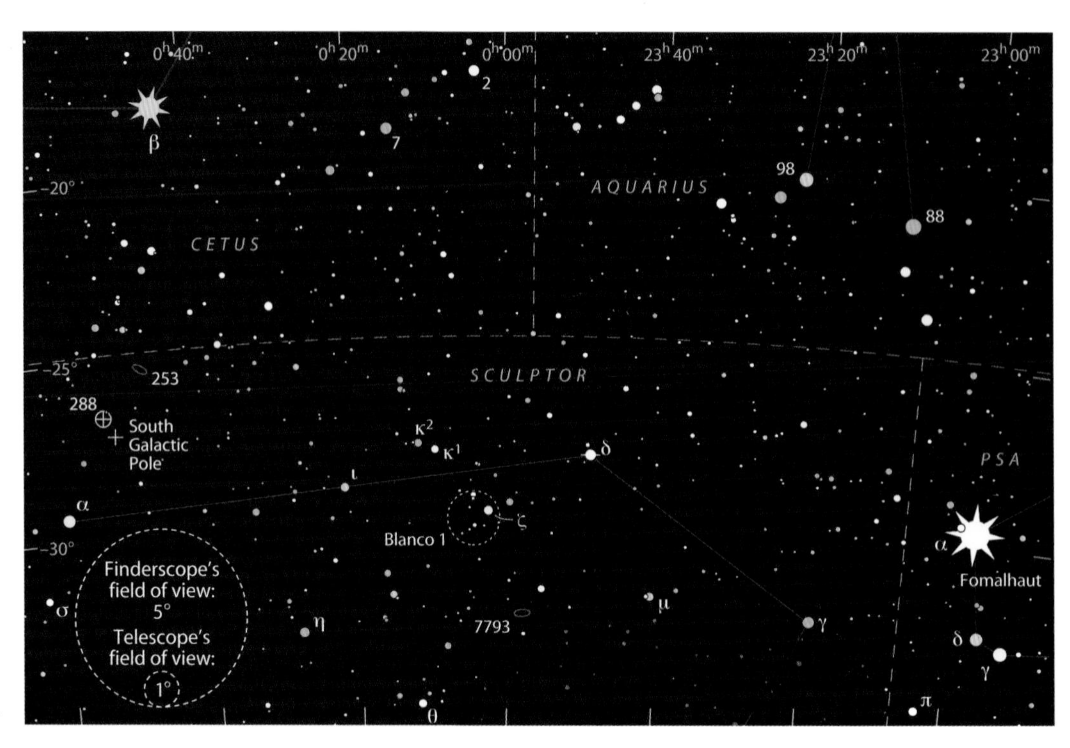

to 87x makes NGC 288 more obvious. The cluster appears a little brighter toward the center, slightly mottled, and about 8' across. Noted observer Walter Scott Houston was able to resolve some of the group's halo stars in giant 5-inch binoculars at 20x.

Just 37' south-southwest of NGC 288 lies the south galactic pole, one of the two spots in the sky farthest from the starry, dust-laden plane of our galaxy's disk. Here, we have a clear window into the depths of the universe, where we can view distant galaxies unobscured by the Milky Way. One of the most impressive is **NGC 253**, sometimes called the Silver Coin Galaxy.

You can find the Silver Coin by scanning 1.8° northwest of NGC 288. It is visible through binoculars or a large finder in a dark sky. This galaxy is fairly bright in my 105mm scope, and at 87x, I can trace it out to 20' x 4' running northeast to southwest. It is broadly brighter along its long axis and subtly dappled with light and dark patches. A pair of 9th-magnitude stars hugs the southeast side.

Caroline Herschel discovered NGC 253 in 1783 while sweeping the sky for comets, a noteworthy accomplishment since this galaxy never climbed higher than 12° above the horizon at her observing site in England. NGC 253 is a barred spiral galaxy, highly inclined to our line of sight. It is the brightest member of a small cluster of galaxies known as the Sculptor Group. At a distance of 10 million light-years, this is the closest galaxy cluster to the Local Group (the one in which our galaxy resides). The Silver Coin is also the closest starburst galaxy, a title it has earned by having an exceptionally high rate of star formation in its core.

Now let's turn our attention to the 4.6-magnitude star **Delta (δ) Sculptoris**, near the center of the chart above. Notice that Delta lies about two-fifths of the way from the bright star Fomalhaut to Beta Ceti and a little below the line between them. It is the brightest star in the area and a low-power double for a small telescope. The white primary has a 9th-magnitude secondary 74" to the west-northwest, but it's too dim for me to make out any color. This companion's spectral class is generally given as G (yellow), but it is described in the Webb Society's *Visual Atlas of Double Stars* as "intensely blue." Do you see any color here?

After visiting Delta, star-hop 3.3° east-southeast to the 5.0-magnitude star Zeta (ζ) Sculptoris. Zeta lies in the foreground of the large open cluster **Blanco 1**. With my 105mm refractor at 17x, I see a dozen fairly bright stars, mostly arranged in a V shape with one curved arm. Another dozen faint stars are scattered in and around

First published December 2001

Highlights of Sculptor

Object	Type	Magnitude	Size/Sep.	Distance (l-y)	Right Ascension	Declination
NGC 288	Globular cluster	8.1	12'	27,000	0^h 52.8^m	–26° 35'
NGC 253	Galaxy	7.6	26' × 6'	10 million	0^h 47.6^m	–25° 17'
δ Sculptoris	Double star	4.6, 9.3	74"	144	23^h 48.9^m	–28° 08'
Blanco 1	Open cluster	4.5	89'	880	0^h 04.2^m	–29° 56'
NGC 7793	Galaxy	9.3	9' × 7'	9 million	23^h 57.8^m	–32° 35'

the degree-wide V, swelling its apparent size to about 1.4°.

Although Blanco 1 does not stand out well from surrounding stars, it is a true cluster. The group was discovered in 1949 by Victor M. Blanco, who noticed that the area contains five times the usual concentration (at similar galactic latitudes) of stars brighter than 9th magnitude of spectral type A0. Studies have shown that this is a young star cluster perhaps 90 million years old, similar in age to the Pleiades. It lies about 880 light-years away and may have as many as 200 members. Its unusual position high above the galactic plane suggests that Blanco 1 may have been formed by a different mechanism than other nearby clusters.

From Zeta in Blanco 1, we'll move exactly 3° south-southwest to our final deep-sky treasure, **NGC 7793**. This galaxy is never more than 14° above my horizon. Even though it's embedded in the skyglow from a nearby city, I find NGC 7793 to be reasonably obvious through my 105mm scope. At 87x, it looks fairly faint overall with a slightly brighter, small center. The 6' x 4' oval is aligned nearly east-west.

NGC 7793 is another spiral galaxy of the Sculptor Group. It displays short, asymmetric arms instead of the long, graceful arms seen in classic spirals. Such arms are probably created by ephemeral star-forming regions that are stretched into spiral-like shreds by galactic rotation. These shreds are short-lived in astronomical terms — about 100 million years.

Sculptor contains several other galaxies suitable for small-scope viewing. Given my viewing location in upstate New York, I have picked out only the brightest of those in the northern reaches of the constellation. More southerly observers might want to take out a good atlas and look for some of the other treasures held by the constellation at our galaxy's south pole.

File Under W

If the rich Milky Way isn't too distracting, Cassiopeia offers quite a selection of nice open clusters.

On the December all-sky star map on page 313, the prominent W traced by Cassiopeia's brightest stars sits slightly above center. The bright and easily recognized pattern will serve as our guidepost to some of the deep-sky delights lying beneath or, rather, south of this famous asterism.

The beautiful double star **Eta (η) Cassiopeiae** will be our first stop. It is visible to the unaided eye one-third of the way from Alpha (α) to Gamma (γ) Cassiopeiae and south of a line connecting them. Eta's components weigh in at magnitudes 3.5 and 7.4 and have been variously described by well-known double-star observers of the 19th century. English clergyman Thomas W. Webb called the pair yellow and pale garnet, Friedrich Georg Wilhelm von Struve saw them as yellow and purple, and John Herschel and James South recorded them as red and green! With my 105mm refractor, I see the brighter star as yellow and the dimmer one as reddish orange, in good agreement with their spectral types of G0V and M0V. The pair is split at 29x, and widely so at 36x. (The suffix "V" in each star's spectral type indicates a star that fuses hydrogen in its core.)

The secondary is one of the very few observable red dwarfs that are paired with a naked-eye star. Shining at about 6 percent of our Sun's luminosity, this ruddy star is visible through small scopes only because it is relatively nearby, at 19 light-years. If you're wondering how our Sun would appear at that distance, simply cast your gaze upon the primary — a very Sunlike star.

Just 1.3° south-southeast of Eta, we find the nebula **NGC 281** and its associated star cluster, IC 1590. The nebula makes a nice isosceles triangle with Eta and Alpha Cassiopeiae. Its faint, irregular glow is easy to spot with my refractor at 68x and occupies about ⅓°. A narrowband light-pollution filter helps to improve the

The soft glow of NGC 281 can't be perceived as truly colorful by human eyes, but its vivid red hue dominates this photographic image. Sean Walker of Methuen, Massachusetts, used an 8-inch telescope at f/6.3 for this 55-minute exposure on Kodak E200 film. North is up and the field 1° wide.

view a bit, while a green oxygen III filter adds a little more.

The nebula was discovered in the late 19th century by American astronomer Edward Emerson Barnard. Only later did Guillaume Bigourdan of France spot the open cluster within it — and little wonder! IC 1590 is not at all obvious through the scope. It is redeemed, however, by Burnham 1, the pretty multiple star at its core. Burnham 1 is the brightest star enshrined in the nebula. At 87x, I see three close components, all bluish white, with magnitudes of 8.6, 8.9, and 9.7. If the air is very steady, try examining the brightest of this trio at a magnification of 200x or more. It has a 9.3-magnitude attendant a scant 1.4" to the east, a severe test for small telescopes. This fourth component turns Burnham 1 into a tiny trapezium ensconced in its own miniature Orion Nebula. Like its larger and more famous counterpart, the trapezium of Burnham 1 supplies much of the radiation that compels its parent nebula to shine.

A line from Epsilon (ε) through Delta (δ) Cassiopeiae points toward Phi (φ) Cassiopeiae, which is visible to the naked eye under moderately dark skies. Phi is a lovely double star that can be split even with binoculars. Its components appear to be the two brightest stars of the open cluster **NGC 457**, although they may actually be foreground objects. In my little scope at 87x, the 5th-magnitude primary appears yellow and the 7th-magnitude secondary looks blue-white. About 45 fainter stars are seen in a cluster that inspires imaginative flights of fancy.

In *1000+: The Amateur Astronomer's Field Guide to Deep Sky Observing*, author Tom Lorenzin calls it the ET Cluster, after the cute little alien in the movie *E.T.: The Extra-Terrestrial.* Lorenzin writes, "ET waves his arms at you and winks!" The two bright stars form ET's eyes, sprinklings of stars northeast and southwest are his outstretched arms, and his feet extend northwest.

Those big eyes lead quite naturally to the nickname of the Owl Cluster, which was created by David J. Eicher. In a similar vein, I always picture NGC 457 as the Dragonfly. California amateur astronomer Robert Leyland is somewhat more creative. "Two wings of stars spread from either side of the cluster," he says, and they "give it the look of an F/A-18 on afterburners, with Phi Cas being the engines."

Its alternate designation is C13. In his book *The Caldwell Objects*, Stephen James O'Meara gives us this haunting impression of the view in his 4-inch scope: "The cluster's bright 'eyes' seem to pierce the night with the fiery gaze

Attractions on High

Object	Type	Magnitude	Size/Sep.	Distance (l-y)	Right Ascension	Declination	*MSA*	*U2*
η Cassiopeiae	Double star	3.5, 7.4	13"	19	0h 49.1m	+57° 49'	49	18L
NGC 281	Diffuse nebula	8.0	28' × 21'	9,600	0h 53.0m	+56° 38'	49	18L
NGC 457	Open cluster	6.4	13'	7,900	1h 19.6m	+58° 17'	48	29R
NGC 436	Open cluster	8.8	5'	9,800	1h 16.0m	+58° 49'	48	29R
Stock 4	Open cluster	—	20'	—	1h 52.7m	+57° 04'	47	29R
NGC 744	Open cluster	7.9	11'	3,900	1h 58.5m	+55° 29'	63	29R

Angular sizes are from catalogs or photographs; most objects appear somewhat smaller when a telescope is used visually. Approximate distances are given in light-years. The columns headed *MSA* and *U2* give the chart numbers of objects in the *Millennium Star Atlas* and *Uranometria 2000.0*, 2nd edition, respectively.

Above: The two colorful little clusters NGC 436 (left) and NGC 457, were recorded by George R. Viscome with his 14-inch Newtonian reflector at Lake Placid, New York. Each view is ⅔° wide. The author sees NGC 457 as the Dragonfly (note the "eyes").

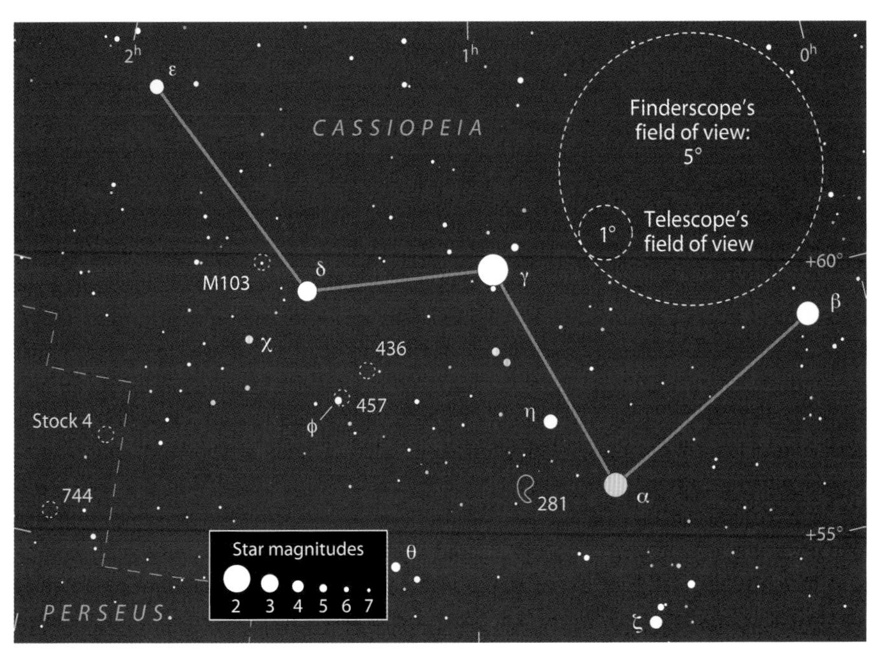

of a specter emerging from the dusty cobwebbed corridors of space. The ghost's clothes hang in tatters from skeletal limbs."

Behind my Dragonfly, ½° northwest of his tail, the open cluster **NGC 436** is a little ball of fluff at low power. At 28x in my 105mm refractor, I can make out only a few faint pinpricks of light in the fuzz. Upping the magnification to 153x teases out a dozen faint to very faint stars embedded in haze. Most of the brighter stars are arranged in diverging arms about 4' long. Leyland speculates that NGC 436 might be the target of his jet fighter, which is aimed in its direction.

Now draw an imaginary line from Delta through Chi (χ) Cassiopeiae; continue for almost 2½ times that distance again to reach the pretty open cluster **Stock 4**. At 68x, I see a group of about 50 faint stars in a more or less rectangular bunch, about 20' x 12', that runs west-northwest to east-southeast. A bar of fairly bright to faint stars lies just north-east of the cluster and runs northwest to southeast. A widely spaced pair of 8th-magnitude stars can be seen to the east.

Stock 4 is one of the clusters discovered by Jurgen Stock and initially published in the first edition of the *Catalogue of Star Clusters and Associations* (Prague, 1958). So far, it has received very little attention from either amateur or professional astronomers. Since contemporary printed and software atlases often include some of the sky's more obscure objects, many pleasant discoveries such as this await the notice of curious observers.

Now drop 1.8° south-southeast to another attractive group, **NGC 744**. At 87x, my little scope shows 15 faint to very faint stars over haze. Disconnected star bunches lie southwest, south, east-southeast, and east-northeast.

It's not unusual to see Earth satellites pass through the field of view when you're observing. During one observation of NGC 744, I saw three in a triangle. Many other observers have noticed satellite triangles with telescopes, binoculars, and even the unaided eye. They are apparently Naval Ocean Surveillance System reconnaissance triads intended to locate and track ships at sea.

First published December 2003

Unsung Star Clusters

Intriguing clumps of stars lie all along the Milky Way.

When I became enamored with the night sky, my only detailed celestial charts were those in Antonín Bečvář's *Atlas of the Heavens*. Prepared at the Skalnaté Pleso Observatory, this classic work came out in 1948 and remained in print for 33 years. Leafing through its pages again, I see that quite a few plotted nonstellar objects are left tantalizingly anonymous, while those that are labeled carry mainly long-standing Messier, NGC, or IC designations. A 1983 article in the popular Slovak science journal *Kozmos* described this atlas as a necessity for both professional observatories and ambitious amateur astronomers.

Today's amateurs have access to a wealth of comprehensive atlases, some displaying more than a hundred different types of designations for nonstellar objects. While many of those with obscure names are fair game only for large telescopes, a surprising number are within the reach of small instruments. This is especially true for numerous star clusters scattered along the Milky Way. Who are these newcomers, and how did they come to be recognized?

Let's begin our celestial star search at 4.8-magnitude 1 Cassiopeiae, a blue-white star located about two-thirds of the way from Beta (β) Cassiopeiae to Delta (δ) Cephei. If you can't spot it with your unaided eye, sweep 8° westward from Beta with your finder. This star shares a low-power eyepiece field with 5.7-magnitude **2 Cassiopeiae**, a nice multiple star with three bright components. The yellow-white primary has a white 8th-magnitude companion to the south-southeast and a reddish 11th-magnitude companion to the west. They are very widely spaced and form a squat isosceles triangle.

While using the second edition of the *Uranometria 2000.0* star atlas, I noticed a little cluster labeled **Bergeron 1** just 42' north-northwest of 1 Cassiopeiae. Examined with

The tiny knot of stars known as either Bergeron 1 or Reiland 1 and the embedded nebulosity BFS 15 show up well in the extreme close-up view, inset, taken with Palomar Observatory's 48-inch Schmidt telescope in California. The field is just 10' wide, with north up, and the two brightest stars (at far right) are only about magnitude 10. The wide view's 1° circle locates the field. Photo, below: Akira Fujii. Inset: POSS-II / Caltech / Palomar

Star Groupings Along the Cassiopeia-Cepheus Border

Object	Type	Magnitude	Size/Sep.	Right Ascension	Declination	*MSA*	*U2*
2 Cassiopeiae	**Multiple star**	**5.7, 8.2, 10.9**	**168", 163"**	**$23^h 09.7^m$**	**+59° 20'**	**1070**	**18R**
Bergeron 1	**Knot of stars**	—	**1'**	**$23^h 04.8^m$**	**+60° 05'**	**1070**	**18R**
BFS 15	**Nebulosity**	—	—	**$23^h 04.8^m$**	**+60° 05'**	**1070**	**18R**
King 21	**Open cluster**	**9.6**	**4'**	**$23^h 49.9^m$**	**+62° 42'**	**1069**	**18R**
King 12	**Open cluster**	**9.0**	**3'**	**$23^h 53.0^m$**	**+61° 57'**	**1069**	**18R**
Harvard 21	**Open cluster**	**9.0**	**3'**	**$23^h 54.3^m$**	**+61° 44'**	**1069**	**18R**
Frolov 1	**Open cluster**	**9.2**	**2'**	**$23^h 57.4^m$**	**+61° 37'**	**1069**	**18R**
Stock 12	**Open cluster**	—	**35'**	**$23^h 36.6^m$**	**+52° 33'**	**1083**	**18R**
Espin 2729	**Double star**	**8.1, 9.5**	**19.8"**	**$23^h 38.0^m$**	**+52° 49'**	**1083**	**18R**

Angular sizes or separations are from recent catalogs. The columns headed *MSA* and *U2* give the chart numbers of objects in the *Millennium Star Atlas* and *Uranometria 2000.0*, 2nd edition, respectively.

my 105mm refractor at 153x, Bergeron 1 is a very small hazy patch with a few extremely faint stars. My 10-inch scope at 213x shows four or five faint stars in a tight group less than 1' across. The cluster is slightly elongated east-west and straddles a faint nebula.

Bergeron 1 made its way into *Uranometria* after New York amateur Joe Bergeron chanced upon it in 1997 with his 6-inch refractor. He reported his find to some of the people who later assisted in the production of that atlas. But an earlier account of it appeared in *Sky & Telescope* in November 1988. Walter Scott Houston wrote that Tom Reiland had turned up the cluster while observing with his 8-inch Newtonian reflector in Pennsylvania. Reiland saw a half dozen stars crowded into 30" with some nebulosity. Houston used a narrowband filter to help him spot the nebula. As a result of that article, the group is also known as Reiland's Object, Reiland's Nebulous Cluster, or simply Reiland 1. The involved nebula had first been described in a professional journal in 1982, where it earned the name **BFS 15** — the 15th new object listed by astronomers Leo Blitz, Michel Fich, and Antony A. Stark. They suggested that it might be part of a complex that includes IC 1470, a slightly brighter nebula 10' to the north-northeast (omitted from the chart on page 282).

The moral of this story is: You, too, could find your name attached to the sky if you're meticulous enough. If you suspect you've found an uncataloged object, enter its coordinates in the SIMBAD search engine (http://simbad.u-strasbg.fr/simbad/sim-fcoo) to see whether any of the objects it displays match your discovery. A group devoted to finding uncataloged deep-sky objects can be found on the web at http://tech.groups.yahoo.com/group/deepskyhunters.

We can catch a few more unsung clusters near 5.4-magnitude 6 Cassiopeiae, which lies 4° northwest of Beta and is the brightest star in the area. First, we'll look 30' north-northeast of the star for **King 21**. The 105mm refractor at 153x reveals a small group of three faint and several extremely faint stars. A slight haze indicates unresolved stars within the cluster. A 10-inch scope at 118x shows me about 20 fairly faint to very faint stars within 3'.

As a Junior Fellow at Harvard College Observatory in the

Five of the clusters the author mentions can be picked out of the rich Milky Way backdrop in this photograph of a 1°–tall field. Photo: POSS-II / Caltech / Palomar

AUTUMN December

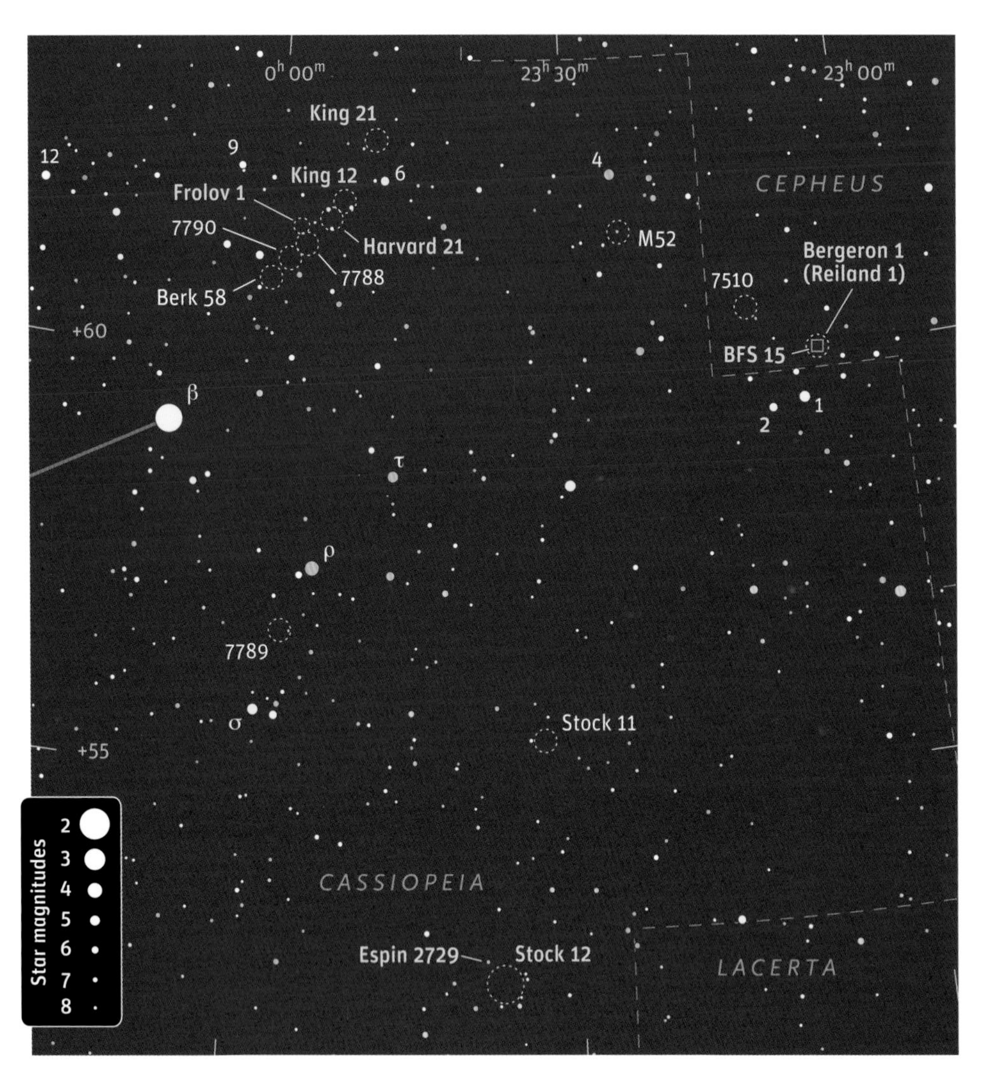

1940s, Ivan R. King had access to plates of the sky surveys taken with the 16-inch Metcalf and 24-inch Bruce refractors. King says that they were fun to look at and that he began to notice clusters that had never been cataloged. His first 21 objects were published in the observatory's bulletin in 1949, although King 3 turned out to be a previously discovered object, NGC 609.

Another King cluster is located 33' east-southeast of 6 Cassiopeiae. Seen through my small refractor at 153x, **King 12** displays a dozen faint to very faint stars in a small gathering elongated northeast to southwest. The two brightest stars are a matched pair, southeast of center, aligned north-northwest to south-southeast. This cluster also seems to show a faint haze of stars just beyond the telescope's grasp.

King 12 can share the field of view with **Harvard 21**, 16' to the southeast. My little scope at 127x simply shows a small U of five faint stars. A larger telescope adds little else. The Harvard clusters were discovered at Harvard College Observatory and were first listed in Harlow Shapley's 1930 monograph, *Star Clusters*. Other astronomers had previously found several of these clusters, but Harvard 21, inconspicuous though it is, appears to be an original discovery. All 21 Harvard clusters are plotted on my Bečvář atlas, which isn't surprising since its source for clusters was Shapley's catalog.

A little knot of stars known as **Frolov 1** sits 23' east-southeast of Harvard 21. In my 105mm scope at 127x, it is a checkmark of five faint stars. The long bar of the check runs north to south. The short bar starts at the north end and slants east-southeast. My 10-inch at 219x shows six 11th- to 12th-magnitude stars in the checkmark plus a scattering of very faint stars covering 2'. The cluster appears sparse but fairly obvious. Russian astronomer Vladimir Frolov discovered this group while studying the proper motions of stars in the nearby clusters NGC 7788, NGC 7790, and Berkeley 58.

Our final target is **Stock 12**, which dangles well south of the clusters we've visited thus far. If you draw a line from Kappa (κ) through Beta Cassiopeiae and continue for 1⅔ times that distance again, you should end up at Stock 12. My small refractor at 47x shows a coarse collection of 60 mixed bright and faint stars, many arranged in eye-catching curves. The group spans about 35' but is very irregular with indefinite boundaries. A golden star perches on the southwest border. Off the northeast edge, the double star **Espin 2729** sports an 8th-magnitude orange primary with a 10th-magnitude companion to its southeast. Stock 12 is improperly positioned on some atlases.

Stock 12 is one of the cluster candidates that Jurgen Stock discovered in 1954 while examining spectral plates taken at Warner and Swasey Observatory in Cleveland, Ohio. Stock said that most of them "do not show a conspicuous concentration of stars, the presence of a cluster being indicated only by the presence of stars of similar spectral types and apparent magnitudes."

We can see why many of our unsung clusters were missed by earlier surveys. They were either too sparse, too loose, too large, or against too rich a background to be easily noticed. When I observe such groups, I often wonder who decided they might be clusters. Now we know the answers for some of them.

First published December 2004

Island Universe

Several fascinating objects surround our neighboring spiral galaxy.

Behold Andromeda's dim aureole of light,
An island universe in yonder sky
Whose rays a million years ago took flight
To reach this very hour my wondering eye.
— George Brewster Gallup, "Andromeda"

Below: Bob and Janice Fera's photograph shows the Andromeda Galaxy and its bright companion M110, at top, and M32, near the bottom. NGC 206 is the topmost and brightest of the starclouds at lower right. Its blue color isn't perceptible through the eyepiece.

This verse was penned early in the 20th century when the controversial "spiral nebulae" became recognized as vast stellar systems external to our Milky Way. Foremost among them was **M31**, the Great Nebula in Andromeda, which we now call the Andromeda Galaxy. Yet distance estimates at the time placed M31 well short of the 2½ million light-years currently favored by modern astronomers. Unfathomably remote though it is, M31 is the nearest large galaxy to our own.

From my semirural home, the Andromeda Galaxy is clearly visible to the unaided eye as a soft, oval glow. It's quite easy to pinpoint the galaxy's place in the sky. An imaginary line drawn from Beta (β) through Mu (μ) Andromedae and continued for the same distance again takes you right to it. In 14x70 binoculars, the galaxy is gorgeous with a small, intense round heart enveloped by a very bright, degree-long oval that fades toward its edges. Faint extensions stretch M31 to nearly 3° x ½°.

The Andromeda Galaxy overspills the field of my 105mm refractor at 47x. A stellar nucleus tacks the center of the galaxy, and the core contains three fairly distinct steps in brightness. The side of the galaxy that faces southeast fades gently into the background sky. The northwestern side abruptly gives way to darkness, beyond which the merest breath of haze films the sky.

These delicate features mark one of the galaxy's dust lanes and the next spiral arm outward. With my 10-inch reflector at low to moderate magnifications, I can detect another dark lane and spiral arm beyond this. These features become less obvious toward the ends of the galaxy.

Next, we'll look for **NGC 206**, a giant starcloud within the Andromeda Galaxy. I always found it somewhat dim in my old 10-inch scope, so I was surprised to see how easy it is to pick out in the little refractor. NGC 206 makes a squat isosceles triangle with the nucleus of the Andromeda Galaxy and the nearby galaxy M32, but the trick is to keep these distractingly bright objects out of the field of view. At 87x, NGC 206 is a faint, north-south oblong about 3' x 1'. This vast stellar association is one of the largest star-formation regions in the Local Group of galaxies. NGC 206 is youthful, astronomically speaking, at a mere 30 million years of age.

Quite a few of the Andromeda Galaxy's family of globular clusters are visible in amateur telescopes. The brightest one, known as **G1** or Mayall II, has been reported in scopes as small as 5 inches in aperture. This globular is located in an unremarkable star field well off M31's southwestern tip, so you'll need to star-hop to it carefully with the help of the chart below. As the accompanying sketch shows, G1 appears markedly nonstellar in my 15-inch reflector at 221x, and it's hugged by two faint stars.

Tales of G1 being seen in small telescopes inspired me to give it a try in my 10-inch. At 118x, a nonstellar object is easily spotted in the correct position — a blend of the globular and its attendant stars. At 171x, two stars sparkle in the blur, but the objects are not individually distinct. G1 was low in the morning sky when I made this observation, so I hope for a better view when Andromeda climbs higher. Are you able to separate the trio?

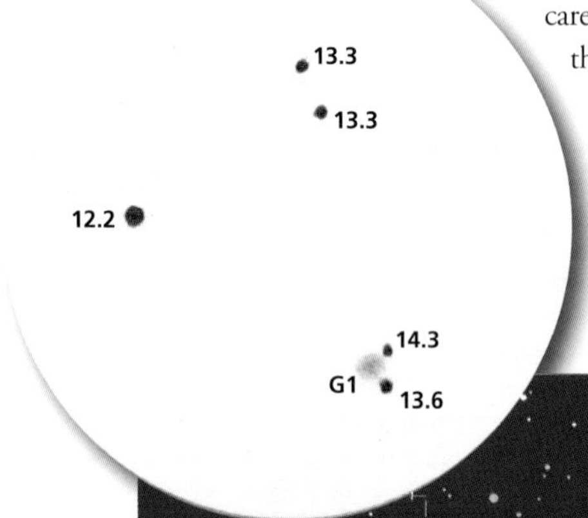

This sketch by the author shows the central 4' of the field when 13.5-magnitude G1 is viewed through her 15-inch scope at 221x. The labels indicate the magnitudes of the field stars.

The Andromeda Galaxy holds several small satellite galaxies in its gravitational sway. Two can be seen in the same lower-power field with their parent galaxy. **M32** has the highest surface brightness of the retinue and is located 24' south of M31's nucleus. In low-power binoculars, you might mistake M32 for an 8th-magnitude star, but in 14x70s, it's a bright little mote that intensifies toward the center. My 105mm refractor at 17x shows M32 pinned to the edge of the Andromeda Galaxy. A magnification of 68x turns M32 into an oval, tilted a bit west of north, with a starlike nucleus. In my 10-inch reflector at 171x, M32 displays brightness steps similar to those of the Andromeda Galaxy. Its faint outer halo measures 3' x 4'. A bright inner halo half as large

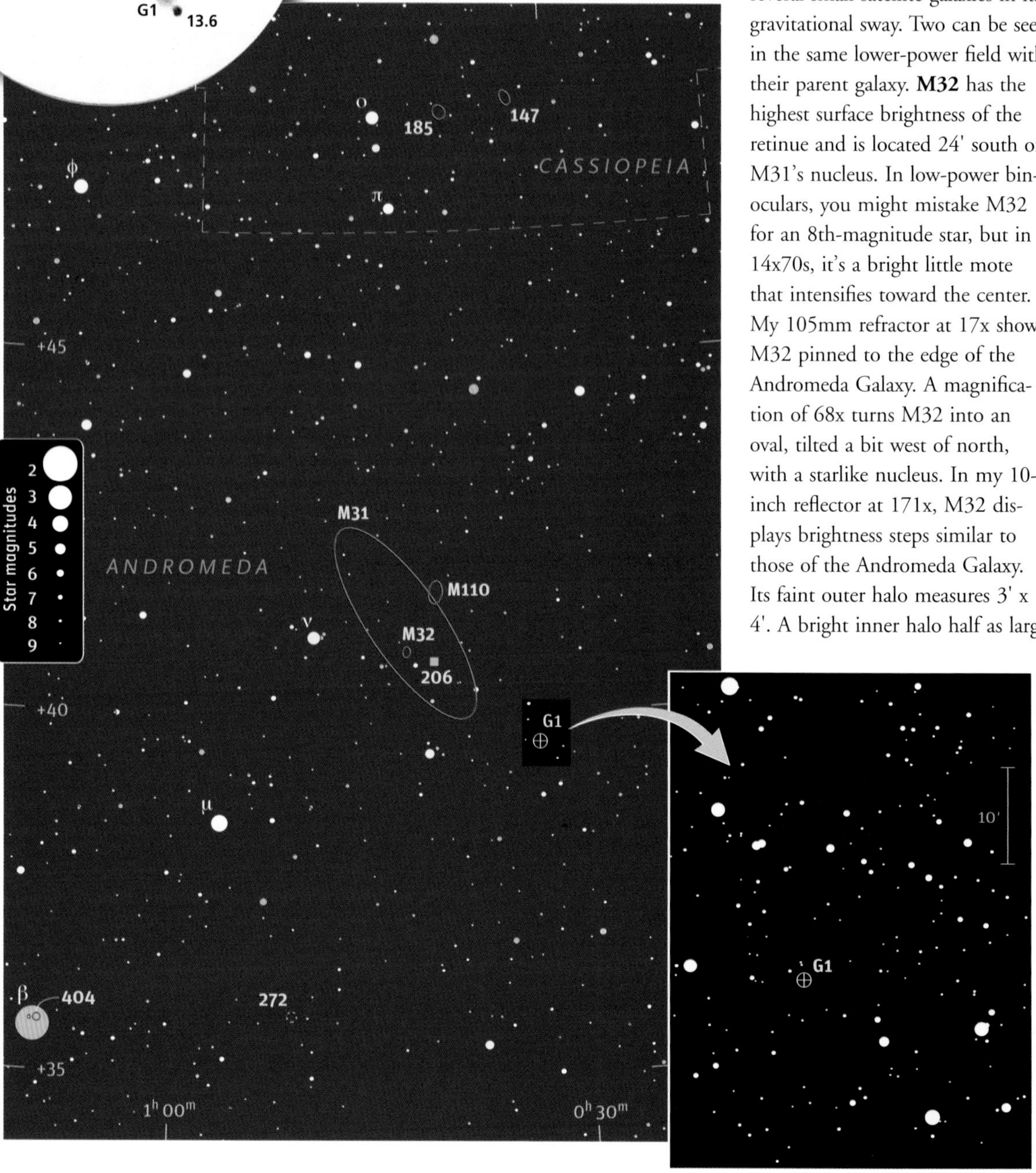

Around the Great Galaxy

Object	Type	Magnitude	Size/Sep.	Right Ascension	Declination	*MSA*	*U2*
M31	Spiral galaxy	3.4	3.2° × 1°	$0^h 42.7^m$	+41° 16'	105	30L
NGC 206	Starcloud in M31	11.9 (blue)	4.0' × 2.5'	$0^h 40.6^m$	+40° 44'	105	30L
G1	Globular cluster in M31	13.5	0.6'	$0^h 32.8^m$	+39° 35'	105	45L
M32	Compact elliptical galaxy	8.1	8.7' × 6.5'	$0^h 42.7^m$	+40° 52'	105	30L
M110	Elliptical galaxy	8.1	22' × 11'	$0^h 40.4^m$	+41° 41'	105	30L
NGC 185	Dwarf elliptical galaxy	9.2	8.0' × 7.0'	$0^h 39.0^m$	+48° 20'	85	30L
NGC 147	Dwarf elliptical galaxy	9.5	13' × 7.8'	$0^h 33.2^m$	+48° 30'	85	30L
NGC 404	Lenticular galaxy	10.3	3.5'	$1^h 09.5^m$	+35° 43'	125	62R
NGC 272	Asterism	8.5	5'	$0^h 51.4^m$	+35° 49'	126	62R

Angular sizes or separations are from recent catalogs. The visual impression of an object's size is often smaller than the cataloged value and varies according to the aperture and magnification of the viewing instrument. The columns headed *MSA* and *U2* give the chart numbers of objects in the *Millennium Star Atlas* and *Uranometria 2000.0*, 2nd edition, respectively.

grows considerably brighter toward a small oval core and a tiny gleaming nucleus. Unlike M31, however, M32 is classed as an elliptical galaxy.

Sitting 36' northwest of the Andromeda Galaxy's nucleus, **M110** is larger than M32 but has a much lower surface brightness. It's visible in 14x70 binoculars as a fairly large, faint oblong. My small refractor at 68x reveals a 15' x 7' oval with the same alignment as M32. It brightens weakly toward a core that looks slightly offset toward the galaxy's northern end. Examining the field with my 10-inch at 70x shows a very faint smear of light that extends from M110 toward the Andromeda Galaxy, like a slightly bent extension of M110's halo. It reaches out for about 8' or 9', narrowing and fading as it approaches M31. This gossamer plume was created by the tidal pull of the Andromeda Galaxy on M110's stars. It has been seen in scopes as small as 4.7 inches in aperture.

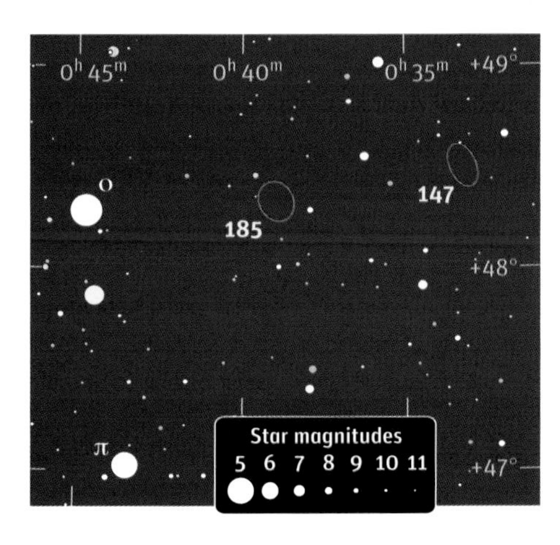

Tim Hunter and James McGaha have made tricolor images of all the Messier and Caldwell objects visible from their observatory in Arizona, including NGC 185, above left, and NGC 147, above right. These are entries 18 and 17, respectively, in the Caldwell catalog. If you're having trouble seeing NGC 185 or NGC 147, make sure you're looking in exactly the right spot.

Two more companions to the Andromeda Galaxy lie farther afield. North of M31 and just across Cassiopeia's border, we find the stars Pi (π) and Omicron (ο) Cassiopeiae. **NGC 185** lies 1° west of Omicron. In my little refractor at 47x, it's a fairly faint northeast-southwest oval with a small brighter core and a 12th-magnitude star just off its northeastern tip. At 87x, the galaxy appears about 4' long and brightens gently toward its 1½' core. My 10-inch scope at 118x stretches the galaxy to 7' x 6' and shows a very faint star in its western edge.

The second companion, **NGC 147**, is a bit farther north

First published December 2005

and 1° west. A 7.5-magnitude star halfway between the two galaxies and a little north of a line connecting them can help you ascertain the galaxy's position. This is crucial when you use a small telescope, because NGC 147 has a very low surface brightness. In my 105mm scope at 47x, its faint oval is visible only with averted vision (that is, by looking a bit off to one side of the object). At 87x, it becomes visible with direct vision but is still rather elusive. Its diaphanous oval slants north-northeast and is about 4' long and one-third as wide. Although still faint in my 10-inch, the galaxy is easily seen at 44x. When I boost the magnification to 118x, it shows a slightly brighter center and sports a few very dim superposed stars. I can trace the galaxy out to around 5' x 2½'.

Contrary to what you might expect from their visual appearance, NGC 185 and NGC 147 are actually closer to us than either M32 or M110.

While hopping from Beta to Mu Andromedae, we bypassed two seldom-visited deep-sky wonders. **NGC 404** shyly hides in the glare of Beta (Mirach), which lies just 7' south-southeast, so it's sometimes called Mirach's Ghost. Nonetheless, the galaxy is visible in my small refractor at 68x as a small, round glow with mild brightening toward the center. The asterism **NGC 272** makes an isosceles triangle with Beta and Mu. My 10-inch scope at 171x shows a 5' group of nine stars, magnitudes 9 through 13. Five of the stars form a tight little arc that includes a close double; the other four lie to the north. The coordinates generally listed for NGC 272 mark the center of the arc, not the center of the asterism.

A Whale of a Tale

Deep-sky observers will have a whale of a time hunting for objects in Cetus.

Oh, the rare old Whale, mid storm and gale,
In his ocean home will be
A Giant in might, where might is right,
And King of the boundless Sea.

— Henry Theodore Cheever,
The Whale and His Captors, 1849

Cetus is portrayed on many old pictorial atlases as an unlikely-looking sea monster, but today he's often thought of as representing a whale. We can see that the latter idea is now firmly entrenched by our use of the term Cetacea for the order of mammals that includes whales. At this time of the year, Cetus breaches the southern horizon, attaining the crest of his leap during the evening hours. The leading light of this denizen of starry deeps is Beta (β) Ceti, or Deneb Kaitos, which marks Cetus's tail and the area of the celestial sea where we'll go whaling.

Our first stop will be the remarkable planetary nebula **NGC 246**, which lies within the triangle formed by Beta, Eta (η), and Iota (ι) Ceti. With a finderscope, you can star-hop from golden Eta through a wavy line made by Phi^4 (ϕ^4), Phi^3 (ϕ^3), Phi^2 (ϕ^2) and Phi^1 (ϕ^1) Ceti. NGC 246 sits south of, and forms an equilateral triangle with, the last two stars in the line. You can also arrive at Phi^1 by sweeping 7.4° north from Beta.

Even at 17x in my 105mm refractor, NGC 246 is easily visible as a little fuzzy spot. At 47x, the planetary looks

A Whale's Bounty

Object	Type	Magnitude	Size/Sep.	Right Ascension	Declination	*MSA*	*PSA*
NGC 246	**Planetary nebula**	**10.9**	**4.1'**	**00^h 47.1^m**	**–11° 52'**	**316**	**7**
NGC 255	**Spiral galaxy**	**11.9**	**3.0' × 2.5'**	**00^h 47.8^m**	**–11° 28'**	**316**	**(7)**
NGC 247	**Sculptor Group galaxy**	**9.1**	**19.2' × 5.5'**	**00^h 47.1^m**	**–20° 46'**	**340**	**7**
NGC 253	**Sculptor Group galaxy**	**7.2**	**29.0' × 6.8'**	**00^h 47.6^m**	**–25° 17'**	**364**	**7**
NGC 288	**Globular cluster**	**8.1**	**13.0'**	**00^h 52.8^m**	**–26° 35'**	**364**	**7**
IC 1613	**Local Group galaxy**	**9.2**	**16.2' × 14.5'**	**01^h 04.8^m**	**+02° 07'**	**267**	**5**
WLM	**Local Group galaxy**	**10.6**	**9.5' × 3.0'**	**00^h 01.9^m**	**–15° 27'**	**318**	**7**

Angular sizes are from recent catalogs. The visual impression of an object's size is often smaller than the cataloged value and varies according to the aperture and magnification of the viewing instrument. The columns headed *MSA* and *PSA* give the appropriate chart numbers in the *Millennium Star Atlas* and *Sky & Telescope's Pocket Sky Atlas*, respectively. Chart numbers in parentheses indicate that the object is not plotted.

A favorite target for observers at low latitudes, NGC 253 is, nevertheless, readily accessible from northern temperate locales, and it's often cited as the most easily observed spiral galaxy after the Andromeda Galaxy. Small telescopes will reveal its mottled appearance. The field is 25' wide with north up. Photo: R. Jay GaBany

nearly round and spans about 3½'. Three stars of about magnitude 11½ in a skinny isosceles triangle are positioned at the nebula's center, west-southwest of center, and at the northwest edge. The view of NGC 246 is enhanced by an oxygen III or a narrowband filter, but the latter keeps the stars visible and gives a nicer overall view. Boosting the magnification to 87x reveals a faint star superposed east-southeast of center. With my 10-inch reflector at 115x, the nebula appears very dim near this star and is weakly annular. The faint galaxy **NGC 255** can be seen 26' to the north-northeast. It has a low surface brightness and appears only half as large as the planetary.

Now return to Beta and then drop 2.9° south-southeast to the more impressive galaxy **NGC 247**. Look for it 1° north of the eastern end of a triangle of 5½-magnitude stars. My refractor at 47x shows a 14' spindle tipped a little west of north. It has a brighter core and a 9½-magnitude star pinned to its southern tip. At 87x, the galaxy appears delicately detailed, especially near the center, and is brighter in its southern half. NGC 247 also has a low surface brightness, so careful study is needed to discern its features.

NGC 247 is about 7 million light-years away and belongs to the Sculptor Group of galaxies, which vies with the Maffei I Group for the title of nearest aggregation to our own Local Group. At 10 million light-years, **NGC 253** is a more distant member of the Sculptor Group. It lies 4½° due south of NGC 247 in the constellation Sculptor, where the artist seems well positioned for sculpting a likeness of our whale.

Despite its greater distance, NGC 253 appears considerably brighter than its cousin. One winter, I enjoyed observing it in the dark and tranquil skies of Little Cayman, where the galaxy climbs 23° higher than it does at my upstate New York home. Even with a 4.5-inch reflector at 35x, NGC 253 looks big and beautiful! Its elongated form covers 24' x 5' and leans northeast. The core is quite mottled and

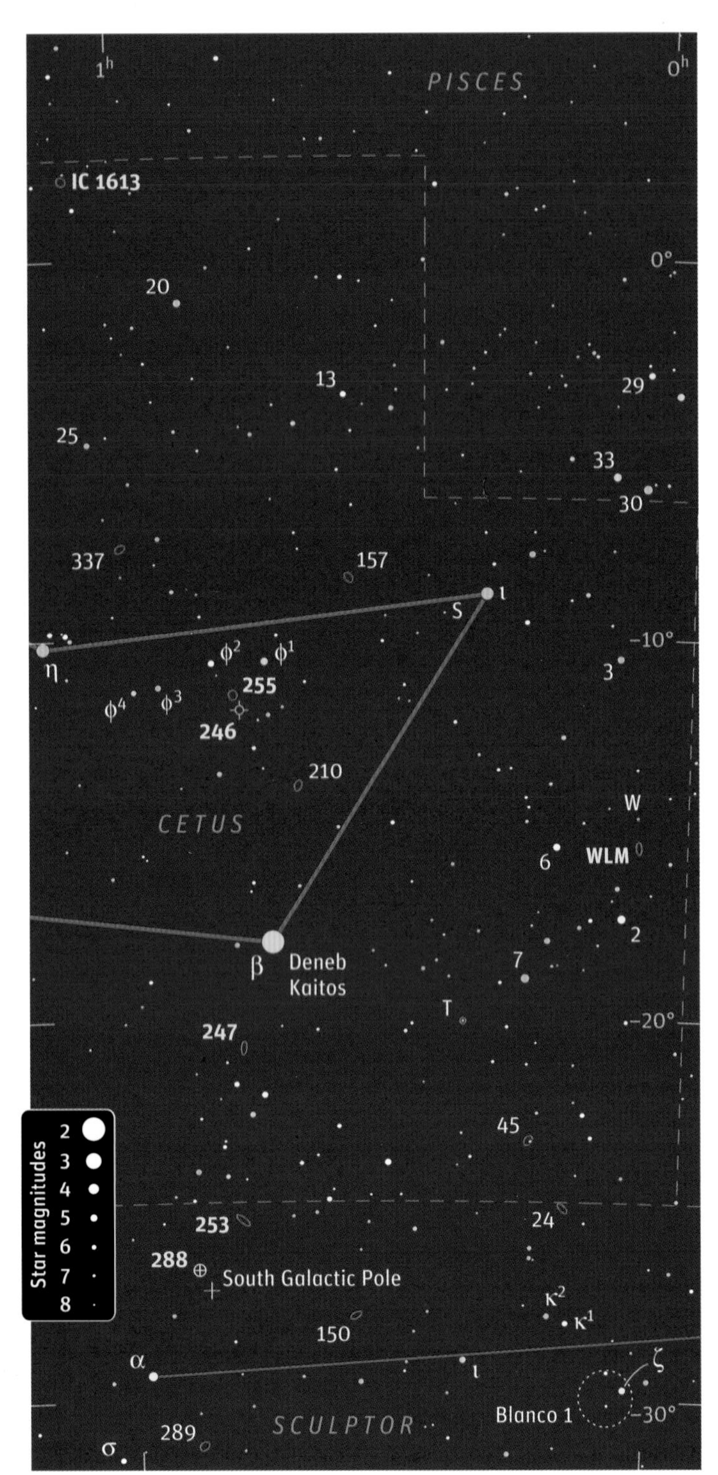

6' long, with a dim star near its southwestern tip. A pair of fairly bright stars lies south of center, the closer one nestled against the galaxy's edge. A faint star decorates the opposite side, west of center.

Now let's return to Cetus, where we'll track down two out-of-the-way challenges, both members of our Local Group of galaxies. The brighter one is **IC 1613**, a dwarf irregular galaxy about 2.4 million light-years away, just a little closer than the Andromeda Galaxy. It's found in northern Cetus near the Pisces border. To locate IC 1613, it's probably easiest to start at Epsilon (ε) Piscium and drop 2½° south-southeast to a ¾° triangle of 6th-magnitude stars. From the southern point of the triangle, sweep southward along a 2°-long, shallow curve of three equally spaced, golden 7th-magnitude stars. The galaxy is just south of the last star in the chain.

Since Arizona astronomer Brian Skiff has viewed this galaxy in his 70mm refractor at low power, I thought it would be fair game for my 105mm. Nonetheless, this elusive galaxy took a long time to spot, and even then, I saw only part of it. Observing was hampered by moisture from my relatively warm face fogging up my cold eyepieces. After sufficiently warming up one eyepiece, I could just barely detect a small bit of haze, but it was not where the center of this fairly large galaxy was plotted. I made a careful sketch and compared it to photos of the galaxy. The little smudge I observed corresponds to a distinctly brighter patch in the northeastern part of the galaxy. I'm eager to give this another try.

With a total visual magnitude of 9.2, how can IC 1613 be such a difficult catch? The answer is found in the galaxy's surface brightness. The light is spread across such a large area that, on average, each square arcminute of its area is only as bright as a 15th-magnitude star. By way of comparison, NGC 247 and NGC 253 have surface brightnesses of 14.0 and 12.8, respectively.

Our other Local Group member is **Wolf-Lundmark-Melotte** (WLM), or MCG-3-1-15, a dwarf irregular galaxy 3.1 million light-years away. This "nebula" was independently discovered on photographic plates by astronomers Max Wolf in 1909 and, in 1926, by Knut Lundmark and Philibert Jacques Melotte.

WLM is located 2.2° due west of the 4.9-magnitude star 6 Ceti. Scanning westward from 6 Ceti with a low-power eyepiece, you'll come to an 8.6-magnitude golden star. The galaxy is only 55' farther west. Look for it halfway along and just west of a north-south line connecting a pair of widely spaced 9th-magnitude stars. This was a very tough target for me with my 10-inch reflector. At 68x, I could faintly see a fairly large, north-south glow. A very small brighter spot sat near the middle and another in the southern end, but I was unable to distinguish whether these were part of the galaxy or foreground stars. Later examination of photos indicated that the object near center is probably a pair of star-forming regions cataloged as HM8 and HM9, but my notes aren't precise enough to help me identify the southern object.

Photographs of WLM show two "stars" just off the western side of the galaxy. The southern one is actually WLM-1, the galaxy's only known globular cluster. It's a challenge for an 18-inch scope.

First published December 2006

Square Dancing

High overhead on December evenings, the Great Square of Pegasus is edged with observing delights.

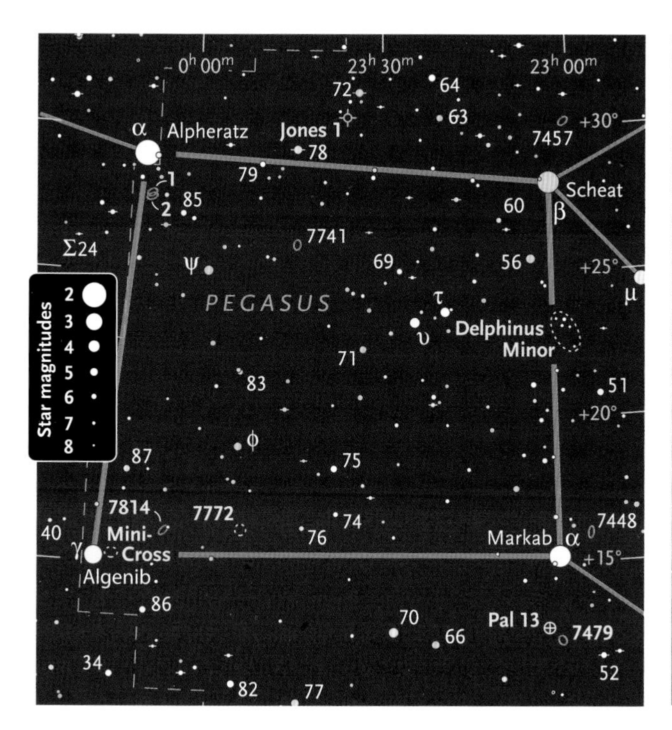

The Great Square of Pegasus is one of the night sky's most prominent asterisms, in part because few bright stars compete for attention in the surrounding sky. There is, however, an array of interesting celestial sights along the Square's perimeter. Here, the author describes some that are easily seen in small binoculars and others that will challenge observers with large telescopes. Photo: *Sky & Telescope* / Richard Tresch Fienberg

In the early evening at this time of the year, the Great Square of Pegasus prances high across the velvet fields of night. As Garrett P. Serviss writes in his classic work *Astronomy with an Opera-Glass*, this distinctive asterism "at once attracts the eye, there being few stars visible within the quadrilateral, and no large ones in the immediate neighborhood to distract attention from it." If we dance our way along the borders of this prominent celestial landmark, we'll encounter a splendid variety of deep-sky wonders.

A smaller asterism is nosed up to the western side of the Great Square two-fifths of the way from Beta (β) to Alpha (α) Pegasi, also known as Markab. It bears a striking resemblance to the neighboring constellation Delphinus, the Dolphin, and even has the same orientation in the sky. Four stars outline the little dolphin's diamond-shaped head while two farther south form its tail.

Observing with 11x80 binoculars, Dana Patchick came across this delightful asterism in 1980. He showed it to fellow Los Angeles Astronomical Society member Steve Kufeld, who dubbed it **Delphinus Minor**. The dolphin's 7th- and 8th-magnitude stars are visible in my 12x36 binoculars with an extra star, fainter than the rest, tucked against the dolphin's cheek. Nose to tail, Delphinus Minor spans 1.1°.

Our next target, the large planetary nebula **Jones 1** (PN G104.2–29.6), hovers over the middle of the Great Square's northern side. You can find it by starting at 72 Pegasi and moving 1° southeast through two 7th-magnitude stars. From the end of this line, hop 12' southwest to an 11th-magnitude star and continue for 8' to reach the center of Jones 1.

Don't be surprised if you can't see the nebula right away. It's 5.3' across with a very low surface brightness. Jones 1 is faintly visible with averted vision at a magnification of 28x through my 105mm refractor using either an oxygen III filter or a narrowband nebula filter. It also appears quite faint through my 10-inch reflector. At 44x with an oxygen III filter, the planetary shows an incomplete ring with a

Dancing With the Horse

Object	Type	Magnitude	Size/Sep.	Right Ascension	Declination	*MSA*	*U2*
Delphinus Minor	Asterism	—	1.1°	23h 01.9m	+22° 53'	1185	64L
Jones 1	Planetary nebula	12.1	5.3'	23h 35.9m	+30° 28'	1162	45R
NGC 1	Galaxy	12.9	1.7' × 1.1'	00h 07.3m	+27° 42'	150	63L
NGC 2	Galaxy	14.2	1.1' × 0.6'	00h 07.3m	+27° 41'	150	63L
Mini-Cross	Asterism	—	16.5'	00h 10.5m	+15° 18'	198	(81L)
NGC 7814	Galaxy	10.6	5.5' × 2.3'	00h 03.2m	+16° 09'	198	81L
NGC 7772	Open cluster	9.6	5'	23h 51.8m	+16° 15'	(1207)	81R
NGC 7479	Galaxy	10.9	4.1' × 3.1'	23h 04.9m	+12° 19'	1233	82L
Pal 13	Globular cluster	13.5	1.5'	23h 06.7m	+12° 46'	1233	82L

Angular sizes are from recent catalogs. Visually, an object's size is often smaller than the cataloged value and varies according to the aperture and magnification of the viewing instrument. The columns headed *MSA* and *U2* give the appropriate chart numbers in the *Millennium Star Atlas* and *Uranometria 2000.0*, 2nd edition, respectively. Chart numbers in parentheses indicate that the object is not plotted. All the objects this month are in the area of sky covered by Chart 74 of *Sky & Telescope's Pocket Sky Atlas*.

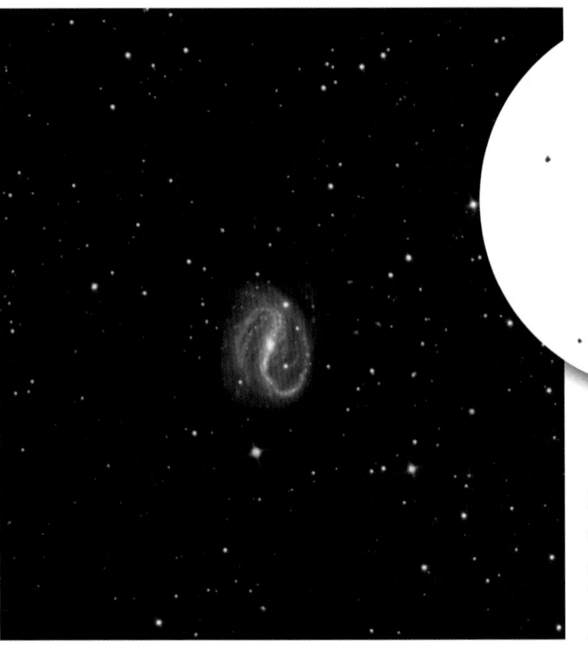

LOOK FOR THE HOOKS The barred spiral galaxy NGC 7479 is bright enough to be seen in small telescopes, but you'll need a fairly large aperture if you want to see the spiral arms at the tips of the galaxy's central bar. Arizona amateur Bill Ferris caught sight of them in his 10-inch reflector at 191x and made the accompanying sketch. The field of the photograph is ⅛° square with north up. Photo: Don Goldman

large, dimmer center. The annulus displays brighter arcs northwest and south-southeast but fades away in the east. Rebecca B. Jones first noted this planetary in 1941 on a photograph made with the 16-inch Metcalf camera at Oak Ridge Observatory in Harvard, Massachusetts.

Waltzing over to the eastern side of the square, we find the first two objects in the *New General Catalogue of Nebulae and Clusters of Stars*, **NGC 1** and **NGC 2**. They're located 1.4° south of Alpheratz (α Andromedae), which forms one corner of the Great Square but "officially" belongs to the constellation Andromeda. Follow a line of three 6½-magnitude stars that trends south-southwest from Alpheratz for 1.6°. The galaxy duo lies 30' due east of the final star.

In my 10-inch scope at 68x, NGC 1 is dimly seen halfway between an 11.6-magnitude star north-northeast and a 13.2-magnitude star on its opposite side. It's much easier at 115x, and at that magnification, NGC 2, south of NGC 1 and east of the fainter star, becomes visible with averted vision. At 166x, NGC 2 appears very small, faint, and uniformly bright with direct vision. Its elongated form tips east-southeast. Slightly oval NGC 1 has the same orientation and sports a tiny, faint nucleus. A relatively large and slightly brighter core is noticeable at 213x.

NGC 1 earned its first place by having the "lowest" right ascension (RA) in the original *New General Catalogue*: 00h 00m 04s. That, however, was for equinox 1860.0 coordinates, and precession has now carried the zero-line of right ascension farther from NGC 1. In equinox 2000.0 coordinates, a few dozen NGC galaxies now bear lower RA values than NGC 1.

Now dip down to Gamma (γ) Pegasi at the southeastern corner of the Great Square. Another constellation mimic lies 40' west and a little north of Gamma. This asterism, readily spotted in my 4.5-inch reflector at low power, reminds me of the Northern Cross, which dominates Cygnus, the Swan.

Split down the middle, a thin band of obscuring dust, challenging to glimpse visually, neatly cleaves our view of the edge-on spiral galaxy NGC 7814. The field is 10' wide with north upper right. Photo: Adam Block / NOAO / AURA / NSF

The 16.5' **Mini-Cross** comprises five stars, magnitudes 8 to 10½. Its short bar runs approximately north-south and its long bar east-west. Unlike the Northern Cross, whose brightest star crowns its top, the Mini-Cross is capped by its faintest star.

Sweeping 2° west-northwest brings us to the galaxy **NGC 7814**. It rests 12' southeast of a 7th-magnitude star, and its oval profile points toward the star. NGC 7814 is fairly bright in my 105mm refractor. At 87x, it appears about 4' long, one-third as wide, and slightly mottled. It harbors an oval core and a tiny bright nucleus. While some observers claim that the wafer-thin dust lane girding this galaxy can be glimpsed in an 8-inch scope at high magnification, it can be a challenge even with twice that aperture.

The small open cluster **NGC 7772** sits 2.8° farther west. With my little refractor at 17x, I see a fuzzy spot just northeast of a 10th-magnitude star. At 153x, it becomes a broad, squat W of five stars plus one additional star to its southeast. The stars range from magnitudes 11.3 to 13.5. My 10-inch scope at 166x adds a seventh star that turns the group into a catchy north-south zigzag, which is 2' tall and grows wider but more compressed toward the south. NGC 7772 may be the aged remnant of an open cluster that lost its gravitational grip on many of its former members.

Let's complete our square by returning to Alpha Pegasi. From there, drop 2.9° south to the barred spiral galaxy **NGC 7479**. It's perched at the northeastern end of a ½°-long line of five stars, magnitudes 8 to 11. My little refractor at 28x shows an oval glow about 2' long and one-third as wide, running nearly north-south. A faint star adorns the northern tip at 127x. With my 10-inch reflector at 115x, the galaxy spans 2½' and is half as wide. The bright, mottled bar runs lengthwise through most of the galaxy. With his 10-inch scope, Arizona amateur Bill Ferris has detected the hook of a spiral arm at each end of the bar.

Climbing 38' northeastward takes us to the ghostly globular cluster **Palomar 13** (Pal 13). Look for five 7½- to 10½-magnitude stars that make a skinny, east-west kite about 13' long. North of the kite, a gentle curve of 11th- and 12th-magnitude stars points to Palomar 13, which sits 1.7' west of the brightest star. With averted vision through my 10-inch scope at 213x, I see a small hazy spot with an extremely faint star intermittently visible at its northern edge. Amazingly, Virginia amateur Kent Blackwell has managed to spot Palomar 13 with a 4-inch refractor. He says the scope's clock-driven mount gave him the extra edge needed to capture this elusive globular.

Palomar 13 travels a rosettelike orbit that lofts it high into the galactic halo and then plunges it deep into the core. The cluster is already so highly disrupted that it may not survive another pass through the heart of our galaxy.

As you trip the light fantastic around the Great Square, you might wonder what music goes with our celestial dance. If so, you can hear a clip from Jasper Wood's recording of "The Great Square of Pegasus" at the website www.musiccentre.ca.

The jiggle factor

Because of the nature of human vision, it's often easier to see a faint, diffuse object like a nebula or galaxy when it's moving in the field of your telescope's eyepiece. So experienced observers sometimes tap their telescopes to make the field jiggle when searching for the faintest nebulae.

First published December 2007

The Darkling Fish

Dim Pisces teems with fine galaxies and double stars.

This way a goat leaps, with wild blank of beard;
And here fantastic fishes duskly float.
— Elizabeth Barrett [Browning], *A Drama of Exile*

Although Pisces, the Fishes, floats ever so duskily upon the sky, rural stargazers can swiftly espy the constellation's western fish. It swims beneath the Great Square of Pegasus and includes the asterism known as the Circlet of Pisces. But the eastern fish presents no distinctive pattern and glimmers more feebly as it nudges Andromeda's waist. Though lacking bright guide stars, its celestial waters serve up many deep-sky wonders.

Let's work our way toward the eastern fish by starting in the ribbon of stars that joins the Fishes together. Our springboard will be the pretty double star **Zeta (ζ) Piscium**. My 105mm refractor at 17x reveals a white, 5th-magnitude primary cozied up to a yellow, 6th-magnitude companion east-northeast.

The galaxy **NGC 524** is 3.4° northeast of Zeta. Trying to hide in a patch of 9th- to 12th-magnitude stars, it's still easy to spot through my little refractor at 47x. There's a filled-in wedge of stars east of the galaxy and a more haphazard scattering near it and to the west. A wide, yellow-white and orange star pair sits off the southwestern edge of the group. Small and round, NGC 524 intensifies toward the center and has an 11th-magnitude star near its southern edge. At 87x, two more stars are visible north-northeast and southeast of the galaxy. They join the first to make an isosceles triangle, with NGC 524 occupying much of its western side. The galaxy grows sharply brighter toward a tiny nucleus.

In my 10-inch reflector at 115x, NGC 524 spans about 2.5' and holds a faint star at its east-southeastern edge. It sits in a westward-pointing V of five dimmer galaxies that would share the field of view, but I see only three little smudges at this magnification. NGC 525 sits 10' north, NGC 516 the same distance west, and NGC 518 is 15' south-southwest. Boosting the power to 213x lets me spot NGC 509 at the point of the V and NGC 532 at the end of its southern branch, but the field is now too small to encompass all six galaxies. The NGC 524 Group is about 90 million light-years away and incorporates several additional galaxies in the immediate area.

On the opposite side of the Fishes' ribbon, 1.5° west of 95 Piscium, we find **NGC 488**, a beautiful galaxy that displays tightly wound spiral arms in astrophotos. In my little refractor at 47x, it's simply a softly glowing north-south oval

Below, left: M74, the most photogenic galaxy in Pisces, displays its stunning spiral arms in this half-hour exposure from the 8.1-meter Gemini North Telescope in Hawaii, but the galaxies can be hard to discern through the eyepiece of a backyard telescope. Photo: Gemini Observatory / GMOS Team. Below, right: The *New General Catalogue* includes five galaxies within 25' of NGC 524. This image, synthesized from red, blue, and green plates of the Second Palomar Observatory Sky Survey, shows several that were too faint to be visible when the catalog was compiled. Photo: POSS-II / Caltech / Palomar

Above: Traditionally, the constellation Pisces was visualized as a pair of fish, each dangling on its own fishing line or ribbon. Right: Technically classified as a ringed spiral galaxy, NGC 488 has extraordinarily subtle and tightly wound spiral arms Photo: Adam Block

that becomes considerably more luminous toward the center. At 87x, I catch glimpses of an elusive starlike nucleus. Four stars, magnitudes 10 to 13, form a tangent to the galaxy that slopes east-northeast. The line's second-brightest star touches the galaxy's south-southeastern edge. Regrettably, I see no trace of the lovely spiral arms even in my 10-inch scope. Can you detect them?

Sweeping 1.9° east-northeast of Nu (ν) Piscium brings us to **NGC 676**, which makes a nice isosceles triangle with golden 8th-magnitude stars 30' east-northeast and 19' south-southeast. The galaxy is very easy to overlook because there's a 10.5-magnitude star smack-dab in the middle of it. The magnitude listed in the table below excludes this star. In my 105mm scope at 87x, the galaxy is a faint, 2'-long streak tipped a little west of north. There's an 11th-magnitude star 5' east-southeast and a fainter star a bit closer and a shade east of north. In my 10-inch scope at 216x, NGC 676 brightens near the core, and a dim star is intermittently visible just off the galaxy's side, approximately east of center.

Alpha (α) Piscium marks the bend in the ribbon connecting the Fishes. It consists of a pair of white stars, magnitudes 4.1 and 5.2, only 1.8" apart. My little refractor at 122x splits this pair by a hair's breadth, showing the companion west of its primary. An 8.1-magnitude orange star lies 14' west, and I see slightly fainter, yellow stars 6.7' east-northeast and 7.2' north-northwest. The three form a straight line, and the last two are listed as components C and D in the *Washington Double Star Catalog*, though they are probably not physical members of the star system.

The only Messier object in Pisces is the galaxy **M74**. Because it lies just 1.3° east-northeast of Eta (η) Piscium and has a total magnitude of 9.4, novices often expect that

Pleasures of Pisces

Object	Type	Magnitude	Size/Sep.	Right Ascension	Declination
ζ Piscium	Multiple star	5.2, 6.2	23"	$1^h\ 13.7^m$	+7° 35'
NGC 524	Galaxy	10.3	2.8'	$1^h\ 24.8^m$	+9° 32'
NGC 488	Galaxy	10.3	5.2' × 3.9'	$1^h\ 21.8^m$	+5° 15'
NGC 676	Galaxy	11.9	4.0' × 1.0'	$1^h\ 49.0^m$	+5° 54'
α Piscium	Multiple star	4.1, 5.2, 8.3, 8.6	1.8", 6.7', 7.2'	$2^h\ 02.0^m$	+2° 46'
M74	Galaxy	9.4	10.5' × 9.5'	$1^h\ 36.7^m$	+15° 47'
ψ^1 Piscium	Multiple star	5.3, 5.5, 11.2	30", 91"	$1^h\ 05.7^m$	+21° 28'
φ Piscium	Multiple star	4.7, 9.1	7.8"	$1^h\ 13.7^m$	+24° 35'
HD 4798 Group	Asterism	—	5.6'	$0^h\ 50.1^m$	+28° 22'
Renou 18	Asterism	7.0	18'	$1^h\ 14.5^m$	+30° 00'

Angular sizes and separations are from recent catalogs. Visually, an object's size is often smaller than the cataloged value and varies according to the aperture and magnification of the viewing instrument.

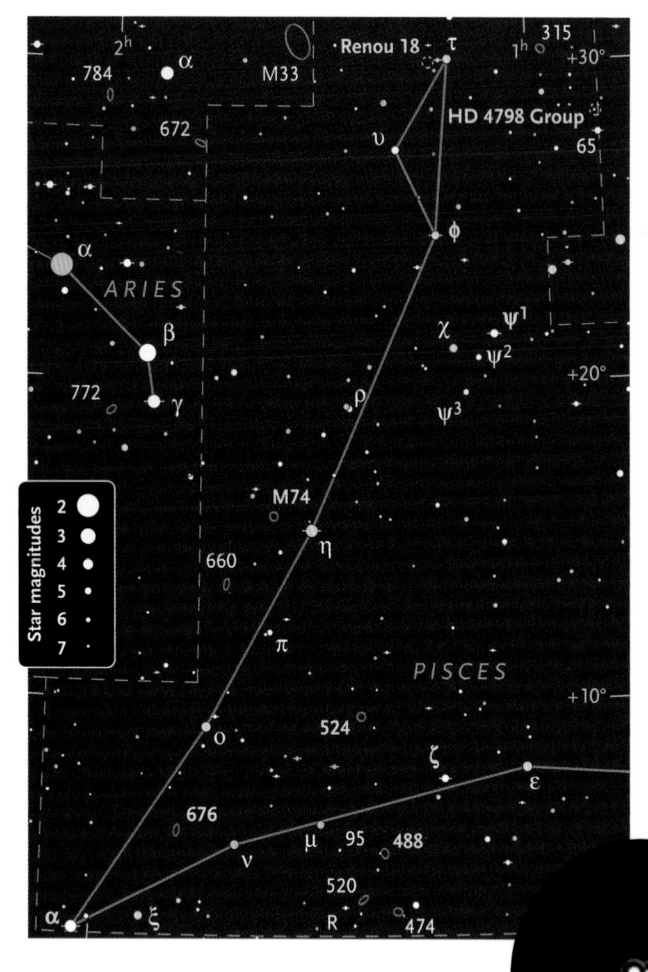

Alpha Piscium is a tight split at 240x in Arizona stargazer Jeremy Perez's 6-inch Newtonian reflector. Note the diffraction rings in this close-up view.

M74 will be easy to spot. Not so! The galaxy's light is spread over a comparatively large area of the sky, so its surface brightness is low. Keep your eyes peeled for a sizable ghostly glow. Having seen M74 many times, I no longer find its phantom fuzz difficult to nab from my semirural home. My 105mm refractor at 17x reveals a faint halo enfolding a relatively large, brighter core. At 47x, I see a faint star at M74's eastern edge, which puts the galaxy's apparent size at 8'. A magnification of 87x shows a second star 1½' south of the first and a superposed star 3' southwest of the galaxy's center. M74 is subtly patchy, and with careful study, I can trace some spiral structure — especially an arm that seems to start north and then wrap outward to the east.

With my 10-inch scope at 213x, I can make out a second spiral arm opposing the first. M74's halo is slightly oval north-south and grows progressively brighter toward the center. The heart of the core is a very short bar with the same orientation, but the arms spiraling into the core make it look elongated east-west farther out. Two nearly parallel lines of three stars each cut through the outer reaches of M74 west and northeast. A two-star line closer to the galaxy's center parallels the latter, its northern star placed squarely on a spiral arm.

According to some pictorial atlases, we reach the tail of the eastern fish at Eta Piscium, but others show it beginning at Rho (ρ) or Chi (χ). In any case, the multiple stars **Psi¹** (ψ¹) and **Phi** (φ) are well within the fish's domain. Through my little refractor at 17x, Psi¹ shows as a pair of white, 5th-magnitude stars with the companion wide to the south-southeast of its primary. The 11th-magnitude third component is three times as distant to the east-southeast. At 68x, Phi displays a 5th-magnitude deep yellow star with a 9th-magnitude companion, perhaps reddish, nestled closely against it to the southwest.

California amateur Robert Douglas told me about a charming asterism in the far western side of the fish. It's named the **HD 4798 Group** after its brightest star, but Douglas likes to call it the Flying Wing. Look for it 40' north of 65 Piscium. Through my 105mm scope at 47x, I see seven stars outlining a southward-pointing triangle. The "wing" spans 5.6' tip to tip and is composed of stars ranging from magnitudes 7.2 to 12.8, with the namesake star boasting a deep yellow hue. In light-polluted skies, you may need a larger scope to see the dimmest star, which marks the western wing tip.

Our final target is the asterism **Renou 18**, named after Alexandre Renou, who writes for the French journal *Astronomie Magazine*. It lies in the eastern fish 37' east of Tau (τ) Piscium and marks the western point of a 20' triangle formed with two yellow stars, magnitudes 6.2 and 6.7. I can see about 25 stars scattered across 18' in my little refractor, but it's the view in my 10-inch scope that catches my imagination. Half the stars are gathered into a shape that strongly reminds me of the S shape in the iconic emblem of the fictional character Superman. The S covers 10' x 8½', and the top (wider half) is to the east. Maybe the planet Krypton circled one of these stars.

Avert your gaze

Faint, diffuse galaxies often look boldest at low power — but boldest isn't necessarily best. Higher magnifications spread out the light from these objects and make them look even more ethereal. But if you scan them carefully with averted vision, looking a little away from the subjects of interest, you will usually see the most details and the faintest features at surprisingly high powers — sometimes as much as 25x per inch of aperture.

First published December 2008

Zeroing in on Cassiopeia

Many fascinating but rarely visited treasures lurk among this constellation's dazzling star clusters.

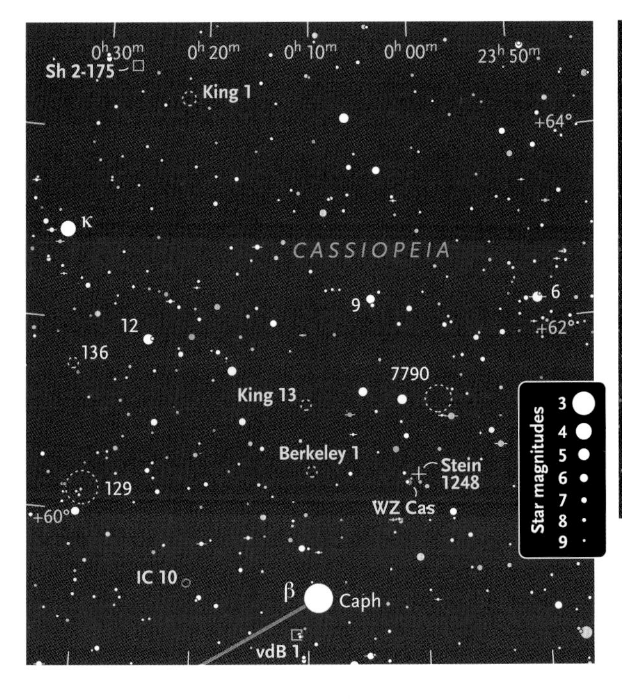

Reflection nebulae such as van den Bergh 1 (pictured above) often appear strikingly blue in photographs, but their color is not visible to the human eye. Note also the two much fainter nebulae northeast of van den Bergh 1. Photo: Thomas V. Davis

The starry realms of Cassiopeia are teeming with hundreds of deep-sky wonders. With such a surfeit to feast upon, it's easy to neglect some of the lesser-known morsels that enrich this opulent constellation. Let's literally zero in on a small region of Cassiopeia by nibbling on treats, newly served here, that are laid out at a right ascension of zero hours.

We'll start at the beginning of the zero-hour zone near Beta (β) Cassiopeiae, also known as Caph. Just 26' south-southeast, you'll find a skinny little triangle of 8th- and 9th-magnitude stars enshrined in the reflection nebula **van den Bergh 1** (vdB 1). Most reflection nebulae are quite faint in a small telescope, but vdB 1 is relatively bright through my 105mm refractor. I prefer a magnification that places dazzling Caph out of the field of view while keeping vdB 1 near the center. The nebula covers about 3' and is most intense near the embedded stars that illuminate it.

Star-gilded vdB 1 is the first entry in a catalog of 158 reflection nebulae compiled in 1966 by the Dutch-born Canadian astronomer Sidney van den Bergh.

Deep images of vdB 1 expose two small reflection nebulae northeast. The larger one has an unusual, oval-loop shape. The star couched in its northwestern end is the primary source of illumination, but the nebula itself was probably sculpted by outflow from another nearby star. A fainter ring, offset southeast, may indicate a previous outflow episode. The illuminating star is designated LkHα 198 (a Lick Observatory hydrogen-alpha emission star), and its nebula shares the name. I don't know of any visual observations of this nebula, but it might be possible for observers pushing the envelope with some of today's impressively large amateur telescopes.

Now we'll turn to the Local Group galaxy **IC 10**, located 1.4° east of Caph. My 105mm refractor shows a very faint star at the western end of a little patch of fuzz about 1' across. It grows twice as large with averted vision. My 10-inch reflector reveals that this is merely the brightest, southeastern region of the galaxy. At low power, IC 10 is an easily visible, uneven glow at the northern end of a 12'-long chain of 7th- to 12th-magnitude field stars. At 118x, the galaxy is 3½' long and leans northwest. Two faint stars are superposed on the northern reaches of the galaxy, one on each flank. Little spots that look nonstellar lie southeast of the eastern flank star and east of the star in the bright region. Images unmask them as foreground stars juxtaposed with star-forming regions in the galaxy. IC 10 is approximately 2.6 million light-years distant and belongs to the small group of galaxies that are dominated by the Milky Way and the Andromeda Galaxy. It's a dwarf irregular galaxy and is heavily obscured because we see it through the dusty plane of our galaxy.

Because it lies near the plane of the Milky Way, IC 10 is heavily obscured and reddened by interstellar dust. Photo: Adam Block / NOAO / AURA / NSF

Sweeping 1.6° northwest of Caph takes us to the semi-regular variable star **WZ Cassiopeiae** (WZ Cas). The star is thought to have superposed periods of variability. The two main cycles, roughly one year and a half year, are due to radial pulsations of the star. WZ Cas is a deeply red-orange carbon star, which makes estimates of its magnitude so challenging that some observers call it the Red Beast. Most of the time, the star ranges between visual magnitudes 7 and 8½.

As a welcome bonus, WZ Cas is a double star with a blue-white, 8.3-magnitude companion a spacious 58" east of the ruddy primary. The designation of the pair is ΟΣΣ 254, the Greek letters signifying its place in Otto Struve's Supplement to the Pulkovo catalog. Although only a chance alignment of unrelated stars, it's a lovely duo when viewed through a telescope.

The nicely contrasting colors of ΟΣΣ 254 show well through my little refractor at 28x. When observing the pair through my 10-inch scope at 44x, I also noticed the nearly equal double **Stein 1248** about 7.6' northwest. The 10th-magnitude primary appears yellow, and its companion 12" northeast shines with a yellow-orange hue.

Residing 1° east of WZ Cas, the open cluster **Berkeley 1** is a little sprinkling of extremely faint stars through my 105mm refractor at 127x. Two brighter stars, 11th magnitude, lie in the group's western side. Through my 10-inch reflector at 115x, this pair is part of a U of stars that dominates the cluster's eastern side and opens northward. The U overlays a 4' mist glittering with faint sparks of light. A bowl-shaped group of stars, 10th to 12th magnitude, is balanced atop the cluster.

Berkeley 1 is the first object in a catalog published in 1960 by Arthur Setteducati and Harold Weaver, astronomers at the University of California, Berkeley. Of the 104 objects on the list, 85 were original discoveries.

Climbing 42' north brings us to the open cluster **King 13**. My little refractor at 127x displays a 5' foggy patch holding several faint to extremely faint stars. The fog is enshrined in a parallelogram formed by a row of three stars

Lesser-Known Treasures of Cassiopeia

Object	Type	Magnitude	Size/Sep.	Right Ascension	Declination
vdB 1	**Reflection nebula**	—	**5'**	**0h 10.7m**	**+58° 46'**
IC 10	**Local Group galaxy**	**10.4**	**6.3' × 5.1'**	**0h 20.3m**	**+59° 18'**
WZ Cas	**Carbon/Double star**	**6.9–8.5, 8.3**	**58"**	**0h 01.3m**	**+60° 21'**
Stein 1248	**Double star**	**10.4, 10.8**	**12"**	**0h 00.4m**	**+60° 26'**
Berkeley 1	**Open cluster**	—	**5'**	**0h 09.7m**	**+60° 29'**
King 13	**Open cluster**	—	**5'**	**0h 10.2m**	**+61° 11'**
King 1	**Open cluster**	—	**9'**	**0h 21.9m**	**+64° 23'**
Sh 2-175	**Emission nebula**	—	**2'**	**0h 27.3m**	**+64° 42'**

Angular sizes and separations are from recent catalogs. Visually, an object's size is often smaller than the cataloged value and varies according to the aperture and magnification of the viewing instrument.

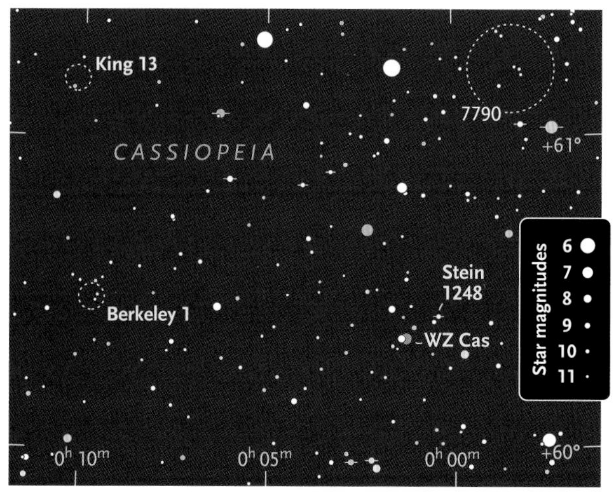

Above: The saturated red color of carbon stars such as WZ Cassiopeiae is strikingly different from the relatively subtle tint of normal red-giant stars. It's due to preferential absorption in the carbon star's atmosphere. Photo: POSS-II / Caltech / Palomar. Left: Faint little Sharpless 2-175, the emission nebula at upper left, forms an unlikely pair with the delicate open star cluster King 1 (lower right) in a low- to medium-power telescopic field.

that clips its southwestern edge and a parallel line of two stars on the opposite side, all 10th or 11th magnitude.

Since we've visited the first of the van den Bergh nebulae and the first of the Berkeley clusters, let's rocket 3.5° north-northeast to **King 1**. In my 105mm scope at 76x, this open cluster is a very faint, slightly mottled haze 5½' across. A 10th-magnitude star guards the northeastern edge, and another is nestled in the opposite side. When viewed at 122x, the latter is revealed as a double star (Stein 40) with a 12th-magnitude companion 13" southeast of its primary. At least a half dozen pinprick stars are entangled in the haze.

King 1 is delicately pretty through my 10-inch scope at 115x. Many minute flecks are irregularly scattered over 7½' of sky. At 187x, I count 45 stars, 13th-magnitude and fainter.

King 1 and King 13 were discovered by Ivan R. King on sky survey plates taken with the 16-inch Metcalf and 24-inch Bruce refractors at Harvard College Observatory. King enjoyed looking at the plates and began to notice uncataloged clusters. In 1949, he published a list of 21 clusters either found by him or marked on the plates by previous workers. Two subsequent papers added six more clusters to the roster. Of the 27 King clusters, 22 were original discoveries.

Our final target is the little emission nebula **Sharpless 2-175** (Sh 2-175), which rests 40' east-northeast of King 1. It's framed by an 8th-magnitude golden star 14' west-southwest and a 7th- and 8th-magnitude pair 20' east-northeast. With my 105mm scope at 47x, I see a small nimbus of light haloing an 11th-magnitude star. Sh 2-175 is also noticeable when I use a narrowband nebula filter, but the star is dim and the nebula little improved. Boosting the magnification to 76x, I estimate the nebula's size as 1½'. Seen through my 10-inch scope at 187x, this subtle glow is slightly elongated northwest-southeast and brighter on its northeastern side.

The second Sharpless (Sh 2) catalog was published in 1959 by Stewart Sharpless of the U.S. Naval Observatory Flagstaff Station. It listed 313 objects thought to be H II regions (clouds of ionized hydrogen). If seemingly detached portions of nebulosity were thought to be ionized by the same stars, they were deemed to be a single H II region.

First published December 2009

Star Maps

How to Use the All-Sky Star Maps

Even if you're a novice stargazer, you can usually find the Big Dipper or Orion, the Hunter, without too much trouble. But what about the lesser-known star patterns, such as Delphinus, Sagitta, or Monoceros?

If you're not familiar with the night sky, the 12 all-sky star charts (one for each month) beginning on page 302 will help you locate the bright stars and major constellations visible this evening or any time during the year. The charts may look daunting, but they're nothing more than a representation of the starry dome, flattened and shrunk onto a page.

GETTING STARTED

Leaf through the following pages and select the all-sky chart that's good for the date and time you want to observe. Make sure you use the star map within an hour of so of the listed times. (Note that standard time is used throughout. If you're on daylight saving time, don't forget to add one hour to the listed times.) You may have to flip back and forth between months if your observing time is late in the evening (or after midnight). For example, to see how the sky appears in early August at 1 a.m. daylight saving time (midnight standard time), you'll want to use the October star chart.

When you venture outside to stargaze, you'll need to know in which direction you're looking. (If you aren't certain, face west, the direction where the Sun sets; north is to your right.) Hold the map out in front of you, and turn it so that the label along the curved edge which matches the direction you're facing is right side up. That curved edge represents the horizon, and the stars above it are now oriented to match the sky. The map's center is the zenith, the point in the sky directly overhead.

At the lower left on each chart is a scale depicting the *magnitude* (brightness) of the stars. The 12 all-sky maps show stars to magnitude 4.5, about as faint as you can see from a suburban environment. At lower right is a key to the different types of stars and deep-sky objects plotted on the charts. These sights and many others are described in the chapters throughout this book, where the inclusion of more detailed finder charts with each essay makes it easy to locate the various celestial wonders.

FINDING THE BIG DIPPER

As an example of how to use the all-sky charts, turn to the September star map on page 310 and locate the Big Dipper. Rotate the page and hold it so that the "Facing NW" label is right side up. The Big Dipper is about one-quarter of the way between the horizon and the zenith. (It looks like a giant spoon with three stars in its handle emerging from the left of a bowl-shaped group of four stars.)

If you go outside around one of the dates and times listed on page 310, face northwest, and look one-quarter of the way up from the horizon, you'll see the seven stars of the Big Dipper — assuming that no trees, houses, or clouds block your view.

TIPS FOR SUCCESS

When you first step outside, look for the brightest stars on the map — those shown with the biggest dots. Ignore the fainter ones initially, particularly if you live in a city or suburb (or if there's a bright Moon in the sky), because they'll be invisible through all the light pollution. Also remember that there's a much bigger difference between the bright and faint stars in the real sky than is suggested by the charts.

Something else to keep in mind is that the star patterns look much larger in the sky than they do on paper. Try this experiment to see how much larger. Find the Big Dipper in the April all-sky map on page 305, and then hold the chart at arm's length in front of you. Fully extend your other arm with your fingers splayed as wide open as possible. From the tip of your thumb to the tip of your little finger covers about 20° of sky — a little less than the width of the Big Dipper in the sky. Now look at the size of the Dipper on the chart... rather tiny by comparison.

These all-sky charts are drawn for skygazers who live anywhere in the world between 35° and 45° north latitude. But even if you don't live within this region, the charts are still useful. If you're south of 35° north latitude, stars in the southern part of the sky appear higher than the maps show and stars in the north are lower. If you live north of 45° north latitude, the reverse is true.

Although many of the brighter Messier deep-sky objects are plotted on the charts, you won't find markers for the much brighter planets. That's because the planets always change their positions. Instead, note the green line cutting through each chart. It's called the *ecliptic*, the path along which the Sun, Moon, and planets travel. If you notice a bright "star" near the ecliptic that's not on the chart, you've spied a planet.

Greek letters on star maps

On both the large all-sky maps and the small star charts throughout the rest of this book, many of the stars in each constellation are identified with Greek letters. A constellation's most brilliant star is usually called Alpha, the first letter in the Greek alphabet; the second-brightest is Beta, and so on.

α Alpha	**ι Iota**	**ρ Rho**
β Beta	**κ Kappa**	**σ Sigma**
γ Gamma	**λ Lambda**	**τ Tau**
δ Delta	**μ Mu**	**υ Upsilon**
ε Epsilon	**ν Nu**	**φ Phi**
ζ Zeta	**ξ Xi**	**χ Chi**
η Eta	**ο Omicron**	**ψ Psi**
θ Theta	**π Pi**	**ω Omega**

Numerous bright stars have Arabic names that have remained in common usage. For instance, the brightest star in Cygnus is Alpha (α) Cygni, which is better known as Deneb.

January

WHEN TO USE THIS MAP

JANUARY EVENINGS

Late January	**Dusk**
Early January	**8 p.m.**

These are standard times. If daylight saving time is in effect, add one hour.

Other Times

Late December	9 p.m.
Early December	10 p.m.
Late November	11 p.m.
Early November	Midnight
Late October	1 a.m.
Early October	2 a.m.
Late September	3 a.m.
Early September	4 a.m.

WHEN TO USE THIS MAP

FEBRUARY EVENINGS

Late February	**Dusk**
Early February	**8 p.m.**

These are standard times. If daylight saving time is in effect, add one hour.

Other Times

Late January	9 p.m.
Early January	10 p.m.
Late December	11 p.m.
Early December	Midnight
Late November	1 a.m.
Early November	2 a.m.
Late October	3 a.m.
Early October	4 a.m.

March

WHEN TO USE THIS MAP

MARCH EVENINGS

Late March	**Dusk**
Early March	**9 p.m.**

These are standard times. If daylight saving time is in effect, add one hour.

Other Times

Late February	**10 p.m.**
Early February	**11 p.m.**
Late January	**Midnight**
Early January	**1 a.m.**
Late December	**2 a.m.**
Early December	**3 a.m.**
Late November	**4 a.m.**
Early November	**5 a.m.**

WHEN TO USE THIS MAP

APRIL EVENINGS

Late April	**Dusk**
Early April	**9 p.m.**

These are standard times. If daylight saving time is in effect, add one hour.

Other Times

Late March	10 p.m.
Early March	11 p.m.
Late February	Midnight
Early February	1 a.m.
Late January	2 a.m.
Early January	3 a.m.
Late December	4 a.m.
Early December	5 a.m.

May

WHEN TO USE THIS MAP

MAY EVENINGS

Late May	**Dusk**
Early May	**10 p.m.**

These are standard times. If daylight saving time is in effect, add one hour.

Other Times

Late April	**11 p.m.**
Early April	**Midnight**
Late March	**1 a.m.**
Early March	**2 a.m.**
Late February	**3 a.m.**
Early February	**4 a.m.**
Late January	**5 a.m.**

WHEN TO USE THIS MAP

JUNE EVENINGS

Late June	**Dusk**
Early June	**10 p.m.**

These are standard times. If daylight saving time is in effect, add one hour.

Other Times

Late May	11 p.m.
Early May	Midnight
Late April	1 a.m.
Early April	2 a.m.
Late March	3 a.m.
Early March	4 a.m.
Late February	5 a.m.

WHEN TO USE THIS MAP

JULY EVENINGS

Late July	**Dusk**
Early July	**10 p.m.**

Other Times

Late June	11 p.m.
Early June	Midnight
Late May	1 a.m.
Early May	2 a.m.
Late April	3 a.m.
Early April	Dawn

These are standard times. If daylight saving time is in effect, add one hour.

August

SUMMER

WHEN TO USE THIS MAP

AUGUST EVENINGS

Late August	**Dusk**
Early August	**9 p.m.**

These are standard times. If daylight saving time is in effect, add one hour.

Other Times

Late July	10 p.m.
Early July	11 p.m.
Late June	Midnight
Early June	1 a.m.
Late May	2 a.m.
Early May	3 a.m.

September

WHEN TO USE THIS MAP

SEPTEMBER EVENINGS

Late September Dusk

Early September 8 p.m.

These are standard times. If daylight saving time is in effect, add one hour.

Other Times

Late August	9 p.m.
Early August	10 p.m.
Late July	11 p.m.
Early July	Midnight
Late June	1 a.m.
Early June	2 a.m.
Late May	Dawn

WHEN TO USE THIS MAP

OCTOBER EVENINGS

Late October	**Dusk**
Early October	**8 p.m.**

These are standard times. If daylight saving time is in effect, add one hour.

Other Times

Late September	9 p.m.
Early September	10 p.m.
Late August	11 p.m.
Early August	Midnight
Late July	1 a.m.
Early July	2 a.m.
Late June	Dawn

AUTUMN

November

WHEN TO USE THIS MAP

NOVEMBER EVENINGS

Late November 7 p.m.

Early November 8 p.m.

These are standard times. If daylight saving time is in effect, add one hour.

Other Times

Late October	9 p.m.
Early October	10 p.m.
Late September	11 p.m.
Early September	Midnight
Late August	1 a.m.
Early August	2 a.m.
Late July	3 a.m.

December

AUTUMN

WHEN TO USE THIS MAP

DECEMBER EVENINGS

Late December 7 p.m.

Early December 8 p.m.

These are standard times. If daylight saving time is in effect, add one hour.

Other Times

Late November	**9 p.m.**
Early November	**10 p.m.**
Late October	**11 p.m.**
Early October	**Midnight**
Late September	**1 a.m.**
Early September	**2 a.m.**
Late August	**3 a.m.**

Resources

Atlases

The Cambridge Star Atlas, 3rd Edition
Wil Tirion (Cambridge University Press)
This basic atlas is a good choice for skygazers who are still learning their way around the constellations. It contains a Moon map and monthly sky maps and shows stars to magnitude 6.5 (plus nearly 900 deep-sky objects). A table of interesting telescopic targets accompanies each chart.

MegaStar 5 (Willmann-Bell)
This software atlas has more than 208,000 deep-sky objects, many of which can be overlaid with photographic images. Stars can be plotted from any of three included star atlases, which contain more than 15 million stars. The easy-to-use filters help you tailor the charts to your needs. (Requires Windows 95 or higher, 32 megabytes RAM, 40 megabytes hard-disk space, CD-ROM drive.)

Norton's Star Atlas and Reference Handbook, 20th Edition
Ian Ridpath, editor (Dutton)
This reference is similar in scope to the *Cambridge Star Atlas*, though it lacks the monthly sky maps. It does include a lengthy astronomy handbook.

Sky Atlas 2000.0, 2nd Edition
Wil Tirion and Roger W. Sinnott (Sky Publishing)
This is a good atlas to use in conjunction with *Deep-Sky Wonders*. It plots stars to magnitude 8.5 as well as 2,700 deep-sky objects. Included are handy close-up charts of some of the crowded regions of the sky.

Sky Atlas 2000.0 Companion, 2nd Edition
Robert A. Strong and Roger W. Sinnott (Sky Publishing)
The text lists and briefly describes every object plotted on *Sky Atlas 2000.0*.

Sky & Telescope's Pocket Sky Atlas
Roger W. Sinnott (Sky Publishing)
This star atlas is great to use at the telescope because it's compact (6 by 9 inches) and spiral-bound. It has 80 charts, more than 30,000 stars to magnitude 7.6, and some 1,500 deep-sky objects. Also included are close-up charts of the Orion Nebula region, the Pleiades, the Virgo Galaxy Cluster, and the Large Magellanic Cloud.

Sky & Telescope's Star Wheel (Sky Publishing)
This planisphere is a simplified "atlas" that will show you which constellations are visible on any date at any time. It's available for different latitudes.

Uranometria 2000.0 Deep Sky Atlas, 2nd edition
Wil Tirion, Barry Rappaport, and Will Remaklus (Willmann-Bell)
This is an advanced star atlas with stars down to magnitude 9.75 and more than 30,000 nonstellar objects. It's great for pinning down those faint objects. It includes both 5th- and 6th-magnitude star charts to help you navigate the main atlas and has 26 close-up charts of crowded regions of the sky, and the index gives the chart numbers for all Messier, NGC, and IC objects (as well as those for bright stars and objects with common names). *Uranometria 2000.0* comes in two overlapping volumes: the northern sky and the southern sky.

Uranometria 2000.0 Deep Sky Field Guide
Murray Cragin and Emil Bonnano (Willmann-Bell)
This text contains useful data and a chart index for all the deep-sky objects included in *Uranometria 2000.0*.

Observing Guides

Deep-Sky Companions: The Messier Objects
Stephen James O'Meara (Sky Publishing)
With the help of this detailed guide, you can eke out all manner of detail from the Messier objects.

Deep-Sky Companions: The Caldwell Objects
Stephen James O'Meara (Sky Publishing)
Reach beyond the Messier objects with this in-depth look at some of the sky's most interesting deep-sky delights.

Deep-Sky Wonders
Walter Scott Houston, edited by Stephen James O'Meara (Sky Publishing)
Here are selections from 48 years of the *Sky & Telescope* column of the same name penned by Walter Scott Houston, who charmed us all with his journeys into the night sky.

Observing Handbook and Catalogue of Deep-Sky Objects
Christian B. Luginbuhl and Brian A. Skiff (Cambridge University Press)
This excellent catalog describes more than 2,000 deep-sky objects (in constellations north of declination –50°) as seen through 60mm to 12-inch telescopes.

The Night Sky Observer's Guide
George Robert Kepple and Glen W. Sanner (Willmann-Bell)
This two-volume guide to more than 5,500 double stars, variable stars, and nonstellar objects in 64 constellations visible from the Northern Hemisphere includes descriptions and sketches of objects as seen through 50mm to 22-inch telescopes. However, note that many sights listed are not visible through a small telescope.

Touring the Universe through Binoculars
Philip S. Harrington (Wiley)
There are more than 1,100 deep-sky objects listed in this guide, and more than 400 are described in some detail. Most of these sights are great for small scopes too.

Turn Left at Orion
Guy Consolmagno and Dan M. Davis (Cambridge University Press)
A fine observing guide to use with your small telescope. It lists 100 deep-sky sights and includes sketches of how they appear in the eyepiece.

General Reference

NightWatch: A Practical Guide to Viewing the Universe, 4th Edition
Terence Dickinson (Firefly Books)
This is one of the best introductions to observing the night sky ever written, delightfully touching on a wide range of astronomical pursuits.

The Backyard Astronomer's Guide, 3rd Edition
Terence Dickinson and Alan Dyer (Firefly Books)
Here's an in-depth look at the world of amateur astronomy, including the ins and outs of equipment.

Atlas of the Messier Objects: Highlights of the Deep Sky
Ronald Stoyan (Cambridge University Press)
A beautiful book on the history, science, and observational aspects of the night's sky's most well-known objects.

Websites of Interest

American Association of Variable Star Observers (AAVSO)
www.aavso.org
This site contains a wealth of information on variable stars, including comparison charts and light curves.

Cartes du Ciel (Sky Charts)
www.ap-i.net/skychart
This is a software atlas that can be downloaded for free.

Internet Amateur Astronomers Catalog (IAAC)
www.visualdeepsky.org
Read observations by other amateur astronomers with a wide range of observing equipment and different levels of experience; post your own observations here too!

Messier45.com
http://messier45.com
The site contains information for about 500,000 deep-sky objects and more than 2 million stars. It includes maps and images for every object and has powerful search capabilities.

SEDS (Students for the Exploration and Development of Space)
www.seds.org/messier
http://spider.seds.org/ngc/ngc.html
SEDS hosts fascinating information on deep-sky objects in the Messier and NGC catalogs.

Sky & Telescope Publishing
www.skyandtelescope.com
This site supports *Sky & Telescope* magazine and contains a wealth of information on many topics of interest to amateur astronomers.

The Washington Double Star Catalog
http://ad.usno.navy.mil/wds/wds.html
This catalog contains data for over 100,000 star systems and is the world's principal database of double- and multiple-star information.

Index

ADDITIONAL PHOTO CREDITS:

Page 4-5, from left to right: POSS-II / Caltech / Palomar; Ross and Julia Meyers / Adam Block / NOAO / AURA / NSF; Johannes Schedler; NASA / ESA / Hubble SM4 ERO Team; R. Jay GaBany; Robert Gendler

Page 8-9: Top, from left to right: Ross and Julia Meyers / Adam Block / NOAO / AURA / NSF; Daniel Verschatse / Observatorio Antilhue / Chile; R. Jay GaBany; Robert Gendler. **Middle, from left to right:** Cord Scholz; Doug Matthews / Adam Block / NOAO / AURA / NSF; Robert Gendler. **Bottom, from left to right:** Robert Gendler; Sean Walker; Robert Gendler; Robert Gendler; Robert Gendler

Page 78-79: Top, from left to right: Johannes Schedler; Robert Gendler; Adam Block / NOAO / AURA / NSF; Martin Pugh. **Middle, from left to right:** Sloan Digital Sky Survey; Gary White / Verlenne Monroe / Adam Block / NOAO / AURA / NSF; Johannes Schedler. **Bottom, from left to right:** Daniel Verschatse / Observatorio Antilhue, Chile; NASA / ESA / Hubble Heritage Team / STScI / AURA; Bernhard Hubl

Page 152-153: Top, from left to right: Johannes Schedler; Brian Lula; NASA / ESA / STScI / Jeff Hester / Paul Scowen; Bernhard Hubl. **Middle, from left to right:** Akira Fujii; Daniel Verschatse / Observatorio Antilhue / Chile; William McLaughlin; NASA / ESA / Hubble SM4 ERO Team. **Bottom, from left to right:** Akira Fujii; Bernd Flach-Wilken / Volker Wendel; B. Balick / V. Icke / G. Mellema / NASA; NASA / STScI / WikiSky; Johannes Schedler

Page 314-315: Top, from left to right: European Southern Observatory; Sean Walker and Sheldon Faworski; Davide de Martin / POSS-II / Caltech / Palomar; *Sky & Telescope* / Dennis di Cicco / Sean Walker. **Middle, from left to right:** Robert Gendler; Adriano Defreitas; POSS-II / Caltech / Palomar; R. Jay GaBany. **Bottom, from left to right:** Russell Croman; Robert Gendler; Anthony Ayiomamitis; Robert Gendler

Page 298-299: Top, from left to right: NASA / ESA / Hubble Heritage Team / STScI / AURA; Martin Pugh; Robert Gendler; NASA / ESA / Hubble SM4 ERO Team. **Middle, from left to right:** Cord Scholz; Johannes Schedler; Ross and Julia Meyers / Adam Block / NOAO / AURA / NSF; Daniel Verschatse / Observatorio Antilhue / Chile. **Bottom, from left to right:** European Southern Observatory; R. Jay GaBany; Robert Gendler; Bernd Flach-Wilken / Volker Wendel

Page 300-313: *Sky & Telescope* Publishing